Schrödinger's Killer App

Race to Build the World's First Quantum Computer

Jonathan P. Dowling

CRC Press
Taylor & Francis Group
Boca Raton   London   New York

CRC Press is an imprint of the
Taylor & Francis Group, an **informa** business

Taylor & Francis
Taylor & Francis Group
6000 Broken Sound Parkway NW, Suite 300
Boca Raton, FL 33487-2742

© 2013 by Taylor & Francis Group, LLC
Taylor & Francis is an Informa business

No claim to original U.S. Government works

Printed on acid-free paper
Version Date: 20130226

International Standard Book Number-13: 978-1-4398-9673-0 (Paperback)

Library of Congress Cataloging-in-Publication Data

Dowling, Jonathan P.
  Schrödinger's killer app : race to build the world's first quantum computer /
Jonathan P. Dowling.
  pages cm
  Includes bibliographical references and index.
  ISBN 978-1-4398-9673-0 (pbk. : alk. paper)
  1. Quantum computers--Research--United States. 2. Dowling, Jonathan P.--Career
in information science. 3. Information science. 4. Quantum theory. 5. Schrödinger
equation. I. Title.

QA76.889.D69 2013
004.1--dc23   2013000953

Visit the Taylor & Francis Web site at
http://www.taylorandfrancis.com

and the CRC Press Web site at
http://www.crcpress.com

Schrödinger's Killer App

Race to Build the World's First Quantum Computer

Dedicated to My Parents

Patricia Mary Dowling
and
John Philip Dowling

Contents

PREFACE

Since 1994, I have been closely involved in the development of quantum information science in the United States. I was co-organizer of the first Department of Defense (DoD) workshop on the topic (held in Tucson in 1995), and from 1994 through 1998, I served as a DoD reviewer of the program. From 1998 through 2004, I continued in this role as a scientist at the NASA Jet Propulsion Laboratory. From 2004 to the present, I have served as a scientific academic reviewer of quantum technologies for the government in my current post at Louisiana State University. (Geaux Tigers!) This book consists of my personal history of the development of the field of quantum computing in the United States, constructed primarily from memory and detailed notes, from my point of view as a quantum physicist and a member of the DoD advisory programs.

The "killer app" (or "killer application") in the title refers to Shor's quantum factoring algorithm, which—if only a quantum computer could be built to run it—would unveil the encrypted communications of the entire Internet for all to see. (Your credit card number would never be safe again.) I call this algorithm Schrödinger's Killer App, instead of Shor's Killer App, because Schrödinger's notion of quantum entanglement is at the heart of it all (and as an alliterative allusion to Schrödinger's infamous half-alive and half-dead cat).

From international multi-billion-dollar financial transactions to top-secret government communications, all would be vulnerable to such a quantum code breaker. Thus, the race is on to be the first to construct one, as the winner will hold the key to the entire Internet. My particular history leads me to focus on the US government's role in all this, specifically the DoD and other intelligence agencies that are vitally interested in the secret-code-breaking ability of the quantum computer.

To keep myself out of hot water (and keep my friends and colleagues at such agencies from despair), when referring to real persons in this book, particularly those in the intelligence agencies, I resort to a first-name-only reference (or first name with last initial in case of confusion). I take full responsibility for any errors in fact or fiction contained herein. (I am Irish, and under Irish rules, the truth or falsehood of a story always plays second fiddle to the entertainment value that it has and the moral that it provides. And yes—I have kissed the Blarney stone.)

In these pages, I also discuss the remedy to the potential threat posed by the quantum code breaker—quantum cryptography, which is unbreakable even by the quantum computer. While code breaking is the focus of my work here, I also discuss applications to important yet more mundane tasks such as the development of quantum physics simulators, synchronized clocks, quantum search engines, and quantum sensors and imaging devices. (I also tell embarrassing stories about myself as well as embarrassing stories about others.)

This book is meant to appeal to the layperson who is interested in quantum physics and quantum computing but who has no formal training in either physics or advanced mathematics. I have therefore attempted to explain somewhat difficult notions—such as quantum entanglement, Schrödinger's cat, Bell's inequality, and quantum computational complexity—by simple analogies, which require no physics schooling and little or no mathematics to grasp. Since mathematics is what gives accuracy to such physical notions, I realize that my analogies will not be completely technically accurate. (For this I am sure to get complaints from my physics colleagues, and I apologize to them in advance.) But I assure you my analogies are correct in spirit, and it is in this spirit that I intend to engage the less technically inclined reader. (Quantum physicists should only read this for the jokes.) I know I will not make both camps absolutely happy. Making all parties *maximally happy* does not optimize human interaction. Making all parties *minimally unhappy* always solves the human equation. This is my goal.

This book is not intended to be a complete scientific review of the field, and so detailed information and clarification on scientific topics will be relegated to the endnotes. I will try to reference comprehensive reviews, written by others, for those readers wishing to pursue these concepts further. I apologize in advance to my colleagues whom I have cited herein—much to their embarrassment—or to those I have neglected to cite—much to their chagrin. To paraphrase Bilbo Baggins' speech at his eleventy-first birthday party: I don't cite half of you half as well as I should have liked; and I cite less than half of you half as well as you deserve.[1] For those I have left out of my anecdotes, and who are thereby dismayed (instead of relieved), take solace in the immemorial words of Groucho Marx: "I don't want to belong to any club that will accept people like me as a member."[2] In the interest of instantaneous intellectual gratification, I will provide Internet links as much as possible to supplemental material in the endnotes, material that may only be a mouse click away in an e-book. I will try to use Google Books, when searching the book contents online is useful and available, but otherwise I'll stick to the noncommercial WebCat, which allows you to find the book in a local library. Even *Wikipedia* has its place, but there I will try to only cite pages that seem well written, stable, unbiased, and sane.

Many thanks to my sister, Ellen Dowling, PhD, President of Dowling & Associates, for her boundless sense of humor and her many fine suggestions for editorial and stylistic improvements to this work. To quote her (when she was midway through reading Chapter 1), "Well, you of all people should know that in order to be funny you have to take some risks."

NOTES

1. The original quote is: "I don't know half of you as well as I should like; and I like half less than half of you half as well as you deserve." From Bilbo's farewell speech in *The Fellowship of the Ring* by J.R.R. Tolkien (Harper Collins, 2009), from Chapter 1, "A Long Expected Party," http://books.google.com/books?id=pK43Jn0RmTcC.
2. *Groucho and Me* by Groucho Marx (Da Capo Press, New York, 1995), page 321, http://books.google.com/books?id=iRmxmjZO1wAC.

The Early Years—When Einstein Attacks!

This is the story of the quantum computer and the international race to build such a device (as it does not actually exist yet and may not for many years). This story is also my own personal perspective of the development of the quantum computer. I have been involved in the development of this machine since 1994, first primarily as a government scientist involved in the US race to construct one, and then more recently as a university scientist who continues to conduct research in the field. This background gives me a unique vantage point and many amusing anecdotes.

A common theme of this book is that quantum computers are much different than classical computers (such as your personal computer), because they require quantum mechanics for their working (and classical computers do not). To understand just how different a quantum computer is from a classical

---

* Photo: Albert Einstein statue, at the Griffith Observatory, Los Angeles, California (2006). Photo taken by Elliot Schwartz for StudioEIS.

computer, it is necessary to understand just how different quantum mechanics is from classical mechanics. I could just tell you they are different and get right to work on the story of the quantum computer, but then you would just have to take my word for this difference. However, I don't want you to just take my word for it. I don't want you to just *believe* quantum mechanics is different than classical mechanics; I want you to *know* it. So my goal in this chapter is to show you that the quantum theory of Bohr and Heisenberg and Schrödinger is really different (and much stranger) from the classical theory of Kepler and Newton and Maxwell.

In this first chapter, I will take you through a brief tour of the history of quantum theory and particularly the foundations of quantum theory, focusing on the debates and experiments carried out over the past 80 years or so that prove that quantum theory is not only stranger than we think, it is stranger than we *can* think (and a bit like magic). However, unlike magic, quantum theory has had its strangeness proofed in the crucible of scientific experiment, and by that proof we know it to be true. Mathematics is the tool that gives the statement of physical theory its accuracy, precision, and conciseness. Because I promised that you would not have to be a mathematician to follow my argument, I will resort to analogies, which may at times seem long-winded or even tedious. However, the rewards will come in Chapter 2 and beyond, where I will not have to spend so much time arguing and rearguing the strangeness of quantum theory each time I introduce a new quantum technologies concept—I will have laid the foundation for your understanding here in Chapter 1.

The modern field of quantum information science, on which the quantum computer is based, originates from a much earlier field of study about the foundations of quantum mechanics. Beginning with the work of German physicist Max Planck in 1900, scientists developed the theory of quantum mechanics through 1930, when the theory evolved into what is now considered its modern form as encapsulated by the English quantum physicist Paul Dirac. While the mathematics of the theory and the mechanism for using it were in place by 1930, debates continue to this day as to what the theory means, how it should be interpreted, and why it is so strange. This interpretation and philosophy of quantum mechanics is covered in numerous other books, and so I will just hit on some relevant highlights, citing other books as needed.[1]

The famous Swiss physicist Albert Einstein made early contributions to quantum mechanics, particularly in the understanding of light and matter, which won him the 1921 Nobel Prize "for his services to Theoretical Physics, and especially for his discovery of the law of the photoelectric effect."[2] The photoelectric effect (the process of conversion of light into electricity, through the mediation of a solid metal or semiconductor) is the theory behind how Xerox machines, digital cameras, and solar panels work: Incoming light is converted to an outgoing electrical current that can then be used to make a copy, take a photograph, or power your house.

The German physicist Heinrich Hertz, the grandfather of radio waves, carried out the first photoelectric effect experiments in 1887, in which he measured the energy of electrons ejected from metal surfaces after being exposed to light. The results were considered mysterious at the time, as the energy of the ejected electrons seemed to depend on the color of the light and not, as expected from classical theory, on the brightness of the light.

Einstein cleared things up in one of three papers he published in 1905 in a single volume of the German journal *Annalen der Physik* (*Annals of Physics*). (The hard copy version of this single Volume 17 currently sells for approximately $15,000 at auction and was for a time the *most stolen* physics journal volume in libraries worldwide until the librarians wised up and put it under lock and key.) Einstein's three papers led to three breakthroughs in physics: in quantum mechanics (the photoelectric effect), statistical mechanics (Brownian motion), and classical mechanics (the special theory of relativity). Curiously, it took the Nobel committee 16 years to recognize any of these, and even then they felt only the photoelectric effect deserved top billing.

Einstein's contribution to the photoelectric effect was to postulate that the energy of the light field was quantized into small packets that carry both energy and momentum. The key idea was that the energy was proportional to the color of the light and that the proportionality constant was "Planck's constant h" from the new quantum theory of that German scientist. Einstein postulated that the incoming light was not a continuous flow, like water from a hose, but was quantized into little energetic balls of light, like bullets from a machine gun. (These little balls of light are now called photons.) Einstein predicted that the incoming photons kick out the electrons in much the same way that an overzealous pool player scratches on a break shot so that the white cue ball (photon) knocks one of the colored balls (electrons) completely off the pool table (metal surface). The key point is that the energy and momentum are transferred from the photon to the electron in a chunk or packet (or "quantum"), not in a slow continuous way as the classical wave theory would suggest, but in one swift kick.

The classical prediction would be more like filling a bathtub to the brim with ping-pong balls (electrons) and then slowly adding tap water (light) until the ping-pong balls begin to overflow the sides at a rate equal to the speed of the water filling the bath. Here, the light is flowing in continuously and not in chunks.

Physicists since the time of Isaac Newton, the primary inventor of classical mechanics, had argued endlessly about whether light was a particle (Newton) or a wave (Dutch scientist Christiaan Huygens).[3] Just when (by the late 1800s) the consensus seemed completely settled on "wave," Einstein came along in

1905 and told us once again that light is sometimes a particle. (The current consensus is that it is *both* a wave and a particle. This wave–particle duality is the backbone for the primary interpretation of quantum theory—the Copenhagen interpretation.)

The photoelectric effect was Einstein's first notable contribution to quantum theory. His last notable contribution was his 1925 prediction of what we now call the quantum mechanical Bose–Einstein condensation of supercold atoms in a gas cloud in his paper published in *Sitzungsberichte der Preussischen Akademie der Wissenschaften*.[4] In Bose–Einstein condensation, all the atoms, when supercooled, condense into the same spot, like a bunch of cows bunched together against the cold in the center of a field in Wisconsin in January.[5]

By 1925, Einstein was pretty fed up with quantum theory, and after the Bose–Einstein paper, he made no further significant contributions toward improving it. But the 1905 photoelectric effect paper puts him down on record as one of the founders of the theory. What went wrong? Why did Einstein abandon it? Even more interesting, why did he end up attacking it? In the end, Einstein did not like the philosophical interpretations of the quantum theory, which were promoted by a majority of the other physicists working in the field. By 1925, a somewhat dogmatic interpretation of quantum theory, attributed to the Danish physicist Niels Bohr and known as the Copenhagen interpretation, held sway, and Einstein would have none of it. After 1925, he continued to publish on the quantum theory he helped erect, but only now his goal was to attack the very foundations of his own construction.

Einstein did not like a number of features of quantum mechanics, but particularly he did not like the probabilistic interpretation of the theory. In classical Newtonian mechanics, a teaspoon can either sit here in the coffee cup on my desk or next to the sink across the hall in the bathroom where I wash out my coffee cup. It cannot be in both places at the same time. In quantum mechanics, a small particle, say a photon or atom or caffeine molecule, can indeed be in two places at once—in principle even the teaspoon can simultaneously be in my coffee cup in my office and across the hall on the sink in the bathroom. This business of being in two places at once is called a quantum superposition. Even stranger, the theory says that for a suitably prepared superposition, 50% of the time when I look into my coffee cup on my desk, I will find the teaspoon there, and 50% of the time when I look, the teaspoon will materialize next to the sink across the hall in the bathroom (giving Prof. Hwang Lee a mild heart attack). This may seem crazy, but this exact effect has been seen in the laboratory of the Austrian physicist Anton Zeilinger, not with teaspoons, but with biomolecules much smaller than teaspoons but still much larger than a caffeine molecule, separated by a distance of a few millimeters![6] In quantum mechanics, things can be in two places at the same time, and most of us quantum physicists have just come to accept this. Einstein never did.

CERTAIN UNCERTAINTIES

Einstein's first most notable public attack against quantum theory occurred during the October 1927 *Solvay International Conference on Electrons and Photons* held in Brussels, Belgium. (These conferences have run about every 3 years continuously since 1911.) At the 1927 conference, all the quantum intelligentsia were gathered, among them Danish scientist Niels Bohr, Einstein, German scientist Werner Heisenberg, and the Austrian scientist Erwin Schrödinger—the most notable founders of the new quantum theory. It was at this meeting that Einstein, in a series of heated public debates with Bohr, tried to show that not only did quantum mechanics violate common sense—but also that it was just plain *wrong!*[7]

Einstein focused on what is known as the *Heisenberg Uncertainty Principle*, which puts a limit on what is knowable in quantum mechanics. For example, this principle says that you cannot simultaneously know the speed and location of your car with infinite precision. Tell that to the global positioning system (GPS) receiver in my Saturn VUE automobile![8]

My GPS receiver routinely tells me my speed and location, but it is only accurate to within maybe plus or minus a few kilometers per hour (miles per hour) and plus or minus a few meters (yards). (This uncertainty is related to the uncertainty in the velocity of the propagation of the radio waves through the Earth's turbulent atmosphere.) The point is that the quantum uncertainty is very, very, very, VERY much smaller than that; it is immeasurably small compared to an object with the mass and speed of my car, much less than the accuracy of my GPS reading. However, for an object with the mass of a caffeine molecule, this error becomes substantial and can be measured.

> The quantum mechanical car,
> Is really the most bizarre.
> For as soon as you're knowing,
> How fast you are going,
> *You can't know where you are!*[9]

Heisenberg explained his uncertainty principle by means of a thought experiment with a microscope. Suppose I try to measure the speed of a caffeine molecule in Anton Zeilinger's molecular interferometer laboratory by bouncing a photon off the molecule and measuring the photon's Doppler shift, just like a police radar detector. In order to get a good reading on the molecule's speed, I have to use a high-energy (short-wavelength) photon for accuracy, and then (just like the photoelectric effect with the cue ball) I knock the hell out of this caffeine molecule so I have no idea where it is located anymore. Similarly, if I try to use the photon to measure its position, I also need to use a short-wavelength photon for good resolution (which is again a high-energy photon), which then kicks the molecule so its speed is now also uncertain. I can try to balance these two effects, but then I'm left with a little uncertainty in both velocity and in position, an uncertainty that is related to Planck's constant h, which is a very, very, very small number when expressed in ordinary human-sized units like meters (yards), kilograms (pounds), and seconds.[10] This is why quantum effects were not observed until around the year 1900—up until then, the technology was just not available to see such small effects.

Avoiding any equations, I can give you the answer about the uncertainty of the caffeine molecule in Anton Zeilinger's laboratory. The caffeine molecule can only be localized in position to within about plus or minus one one-hundredth of a centimeter (one four-thousandth of an inch), at the price of not knowing its velocity to within plus or minus two ten-thousandths of a kilometer per hour (one one-thousandth of a mile per hour). This is some pretty small change for a caffeine molecule. However, according to quantum theory, this is an absolute lower bound; I cannot do any better than this accuracy. These quantum uncertainties are inversely proportional to the mass of the object. My car, because it is much heavier than a single caffeine molecule, has even smaller (and totally unmeasurable) uncertainties in its speed and position. A single electron, because it is much lighter than a caffeine molecule (a caffeine molecule weighs approximately as much as 323,980 electrons), has much larger uncertainties in its velocity and position as it orbits the hydrogen nucleus. The bigger the mass is, the smaller the uncertainties in position and velocity are. The smaller the mass is, the bigger the uncertainties are. This is what the Heisenberg Uncertainty Principle tells us.

Einstein could not agree that there were any uncertainties in nature at all! Electron or molecule or teaspoon or Saturn VUE—the "Father of Classical Mechanics," Isaac Newton, implores us (from his tomb in Westminster Abbey) that position and velocity can always be measured simultaneously with infinite precision. No uncertainties! Or, as Einstein intoned in his oft-repeated injunction, "God does not play dice with the Universe." (Bohr responded, "Einstein—stop telling God what to do!") In the Solvay debates, Einstein attempted to show, in a series of "thought experiments,"[11] that the Heisenberg Uncertainty Principle (and hence quantum mechanics itself) was ... just ... plain ... *wrong!*

Each day of the conference, Einstein would cook up a little toy model of the Heisenberg microscope—a thought experiment—and set it up and run it out on the chalkboard and attempt to show that Heisenberg's principle was wrong.

No uncertainties! Each sleepless night, Bohr would worry and fume and ruminate about Einstein's attack, and then he would respond the next day with a keen rebuttal, showing where Einstein had missed something, and salvage Heisenberg's principle. This debate went on for days at that Solvay conference and continued on 3 years later at the next conference.

The coup de grâce occurred in the very last round, at the Solvay Conference in 1930, where Einstein attacked the Heisenberg Energy–Time Uncertainty Relation, which states that a particle's energy cannot be measured with infinite precision in any finite time, also via an inequality involving Planck's constant. (Here, energy and time loosely replace position and velocity in the old version.) Bohr then exploited *Einstein's own theory of relativity* in his counterattack. Using Einstein's famous equation[12] against him, Bohr successfully derailed Einstein's final salvo against the Heisenberg Uncertainty Principle (at least at the Solvay Conferences). Oh poor Albert Einstein! His own famous formula used against him! How embarrassing!

Even Einstein (*Time* magazine's "Person of the Century"[13]) could not prove that quantum mechanics was wrong. So he slowly stewed over his discontent with the theory for another 8 years, collected his thoughts, and then launched his final and most famous attack against quantum mechanics. Because he could not prove that the theory was wrong, he decided instead to move the goalpost and prove instead that it was "incomplete"—that quantum mechanics was not the whole story and that we were missing something hidden! Einstein thought he had a surefire way to stop God from playing with those dice.

ALL I'M EVER GOING TO BE IS INCOMPLETE

Einstein's final word on the matter appeared as a scientific article published in the prestigious American physics journal *Physical Review* on May 15, 1935. The article was titled, in a very philosophical fashion for a physics journal, "Can Quantum-Mechanical Description of Physical Reality Be Considered Complete?"[14] The resounding conclusion of the paper was—"No!" The paper was authored by Einstein and his collaborators at the Princeton Institute for Advanced Study, Boris Podolsky and Nathan Rosen, and is now universally called the Einstein–Podolsky–Rosen (or "EPR") paper.[15]

It is typical, when announcing an important new result, for scientists to issue a press release to various news sources in advance of the official publication of the science article. But the press release is always supposed to be held back until the date that the scientific journal article actually appears in print, so that the press announcement and the journal article appear simultaneously. In this case, a snafu occurred when Boris Podolsky leaked a version of the EPR paper, as well as some of his own comments, quite a few days *early* to the *New York Times*—and without Einstein's permission! Hence, there appeared 11 days before the publication of the *Physical Review* paper an article in the *New York Times* with the headline, "EINSTEIN ATTACKS QUANTUM THEORY—Scientist and Two Colleagues Find It Is Not 'Complete' Even Though 'Correct'."[16] Einstein was furious with Podolsky about this leak to the *New York Times* and consequently never spoke to him again.[17] Podolsky was forced to flee Princeton (and Einstein's wrath) on the wings of a quantum mechanical storm. Oh, how embarrassing! Poor Boris Podolsky.

What was all the fuss about? We see from the headline that Einstein had given up on demonstrating that quantum mechanics was incorrect. By 1935, experimental tests of quantum mechanics were numerous and there were no incorrect predictions of quantum theory that disagreed with anything any experimental physicist had yet measured in the laboratory. Going after "incorrect" was hopeless. So Einstein, Podolsky, and Rosen took a different tact with this business of "incomplete." The EPR paper laid out a very philosophical and nuanced argument, an argument whose guts also lie at the heart of our quantum computer, and which I will try now to explain using a few simple thought experiments employing timekeeping devices, in honor of Albert Einstein (whose papers on relativity were filled with thought experiments with clocks and rulers and trains and elevators).

Let's suppose that Alice and Bob are fraternal twins who were born and raised in Hyde Park, Chicago. After each graduates with a PhD in quantum physics from the University of Chicago, Alice takes a job at Fermilab, in Batavia, Illinois, and rents an apartment in nearby Aurora, Illinois, and Bob takes a job near London, at the Royal Observatory in Greenwich in England, and rents a flat in the nearby town of Brockley. Being very close siblings, just before Bob leaves for the United Kingdom, he and Alice carefully synchronize their watches, but with a time offset of 6 hours, appropriate for the conversion from central standard time in Aurora to Greenwich mean time in Brockley. They both have accurate *E. Howard & Company Atomic Analog Watches*, which are actually digital watches with old fashioned–looking analog hour and minute hands. (I am wearing one of these now.)

These watches have little antennas in them that allow them to synchronize with the atomic clock at the National Institute of Standards and Technologies (NIST), in Boulder, Colorado, via their radio control signal broadcast from radio station WWVB in Fort Collins, Colorado.[18] (Alice in Aurora should get good synchronization every few hours, whereas Bob in Brockley may have to wait every few days for a clear night.[19])

So now Alice and Bob know, when they arrange their weekly Voice over Internet Protocol call on Skype, just what time it is on each end. If Alice's hour hand points to 3:00 p.m., then she knows Bob's hour hand points to 9:00 p.m.

and vice versa. This goes for any time by the hour hand: 12:00 noon for Alice is always 6:00 p.m. for Bob and 2:00 p.m. for Bob is always 7:00 a.m. for Alice. No matter where Alice's hour hand is pointing, she is always sure Bob's hour hand is pointing 180° away on the opposite side of the dial. This is what the laws of classical physics tells us will happen and nobody is surprised if Alice and Bob look at their watches at random times during the day or night, then call each other on Skype, and always report that their hour hands are pointing in opposite directions.

Everybody, including Einstein, would be very happy with this scenario. The EPR paper pointed out that there are three important points or conditions to notice about this thought experiment, which are obvious and make common sense:

1. *The reality condition:* Each watch keeps its own time independent of the other watch. Even though the watches are synchronized, Alice only needs to consult her own watch to find out what time it is in Aurora. Alice's hour-hand position has its own reality associated with her watch *only*, and Bob's hour-hand position has a reality associated with his watch *only*.

2. *The locality condition:* Any action performed on one watch cannot affect the time on the other watch. If Bob accidentally drops his watch out of the dome of the Greenwich Observatory onto the pavement below, where it smashes to bits, Alice will not even notice. She won't even know that Bob broke his watch until he calls to tell her about it.[20]

3. *The certainty condition:* Alice and Bob can in principle measure the time on their clocks with arbitrary accuracy and the NIST synchronization signal can be in principle made to synchronize them both to within arbitrary accuracy. (Precise measurements produce precise outcomes.)

The certainty condition is Newton's classical mechanical way of telling us that we can always remove technical sources of noise and improve our measurements to arbitrary accuracy, and that the measurement—if made carefully enough—does not disturb the thing being measured. You would not buy a wristwatch if every time you raised your arm to look at it, the motion of your wrist caused the hands to unpredictably swing around the dial and point at random times. That would be *uncertainty*—precision measurements that produce unpredictable and random outcomes.

Consider flipping a coin onto the back of your left hand and covering it up with your right in a game of "heads or tails." Classically, the certainty principle implies that before you look it is certainly *either* heads or tails, but you don't know which until you lift your hand to peek. Before looking, you can only say it has a 50–50 chance of being heads or tails, because you did not take all the details of the coin-flipping process into account to know which would be which. According to Newton and classical mechanics, you could in principle have predicted the outcome, heads or tails, with absolute certainty, if only you had carefully calculated its trajectory in the flip and had measured all the forces on it from your hand, gravity, air friction, and so forth. If you carefully measured and calculated everything, the game would not be random at all!

In contrast, the quantum theory states that a quantum coin is *both* heads *and* tails until you look at it, and then the mere act of peeking at it "collapses" the coin randomly into heads *or* tails with a 50–50 probability. In quantum theory, there is no way to predict the exact outcome with certainty, no matter how carefully you measure or model the coin-flipping process. The uncertainty is not just a lack of your knowledge about how the coin was flipped, but the uncertainty is built into the very guts of Mother Nature herself. This is what so disturbed Einstein and led to his many proclamations about God not playing dice. (God should also not play "heads or tails.")

Another way to express the *certainty condition* is to say that if Alice and Bob prepare the wristwatches exactly the same way, each time they meet and synchronize them, then they can predict the outcome of the time when they look at the watch with absolute certainty; under identical circumstances, the watch reading will be identical to any previous watch reading prepared the same way. No random and unpredictable readings from absolutely identical setups. This is the certainty provided by classical Newtonian mechanics—if you know *everything* about the coin flip, you can predict which side will land up. If you know everything about how the watches are prepared, you can predict with certainty what time it is and consequently the position of Alice and Bob's hour hands.

Okay, why did I take you all through this long-winded thought experiment about synchronized wristwatches, which led in the end to what seems like a bunch of common sense? Because, following Einstein, Podolsky, and Rosen, I'll now take you through the quantum version of this thought experiment, which E., P., and R. claimed defied common sense (by which they meant the predictable, local, realistic, and certain laws of classical physics). They attempted to encode common sense in these three apparently unassailable conditions: *reality*, *locality*, and *certainty*. To see clearly *what is strange* about quantum mechanics, we have to clearly state *what is not strange* about classical mechanics. But first we have to decide—what is common sense?

Common sense is our ability to predict future events on the basis of our past experiences. When I look out my window on a clear day, I know that the sky will be blue because it always has been blue whenever I have looked out the window on a clear day. That's just common sense! I don't ever expect it to be pink. That would defy *my* common sense. But the common-sense notion of a blue sky may not be common sense at all to a Martian, where on a clear day on Mars the sky is butterscotch. Common sense depends on the personal previous experiences of the observer.

Einstein's formal education ended in 1905, just 4 years after the first paper on quantum theory was published by Max Planck. Most of Einstein's past experiences up until 1905 were with classical mechanics (the mechanics of Isaac Newton) and statistical mechanics (the mechanics of the Austrian scientist Ludwig Boltzmann). Einstein never developed a common sense for quantum mechanics, despite having invented bits of it early on, because he had so few prior experiences in dealing with it in his early career and academic training.

Now, over 100 years later, those of us who are trained in the field of quantum mechanics have developed a common sense that allows us to make predictions with quantum theory, on the basis of our previous experiences with it, even though these predictions often seem at great odds from what Newton's classical theory may have led us to believe. Still, many present-day quantum physicists are very unhappy with quantum theory and continue to think that it defies common sense. As for me, I just adapt my own notions of what is common sense, just as I would have to do if I were to emigrate to Mars.

A slightly relevant side note: In August of 1999, I attended a *Workshop on Fundamental Problems in Quantum Mechanics*, held at the University of Maryland, in Baltimore County. This was a very raucous workshop, attended by the world's experts on the foundations of quantum mechanics, with different quantum factions and sects and cults, whose individual proponents were holding forth on different interpretations of quantum theory. They were all shouting at each other in the lectures—proof by intimidation!

The conference organizers asked me to chair a session that everyone expected to be particularly lively. The Organizer-in-Chief, Chinese-American physicist Yanhua Shih, implored me, "Help us, Jonathan Dowling, you're our only hope. You have to avoid total chaos!" (I am now the "bouncer" at quantum physics workshops.)

My session was particularly chaotic in that a bunch of the quantum theorists had ganged up on the lone troublemaking philosopher, David Albert, from Columbia University. Albert stood at the podium (with no audiovisual aids whatsoever) and held up his paper and read it aloud—every single word—even the equations! *"Blah, blah, blah,"* intoned Albert. *"Blah, blah,* equals *blah, blah, blah*—nabla!—*blah, blah, blah,* open curly bracket, open square bracket, open parenthesis, *blah, blah, blah, blah,* close parenthesis, close square bracket, close curly bracket... *quod erat demonstrandum."* After 30 minutes of this, I was brain dead.

It is hard to imagine anybody wanting to attack such a soul-stealing lecture, but the quantum theorists were after payback for Albert's acerbic comments during their own talks. During the question-and-answer period at the end, I had to wrestle a brilliant 2-meter-tall (6-foot-3-inch-tall) and 90-kilogram

(200-pound) army scientist, my friend and colleague Howard Brandt (who looks like a line-backing Santa Claus in a black suit and white tennis shoes). I gave each questioner 1 minute. After Howard's time was up, I announced, "Howard, your minute is up, give me back the microphone." Towering over me, Howard shook the microphone in my face and ranted, "I have *not* finished my *point!*" (He made the mistake of trying to "out loud" me—nobody out louds Dowling!) I grabbed the microphone out of his hand and bellowed, "*HOWARD BRANDT! Your time is UP and your point is MOOT! Now SIT down!*" Howard instantly deflated himself and collapsed back into his chair in astonishment. (Who knew quantum physicists could be such an emotional bunch?)

During this cacophonous conference, we all took a vote on what each participant believed to be the one true interpretation of quantum mechanics. No interpretation—not even Niels Bohr's traditional Copenhagen interpretation— got a simple majority! (I introduced my own *Many-Beer* interpretation of quantum mechanics: With zero beers, quantum mechanics makes no sense; with one, you get an inkling; with two, things seem clear; with three, all mysteries are revealed; with four, things get foggy; with five, quantum mechanics makes no sense once again.[21] It got three votes.) What is the world coming to when profound questions in quantum physics are decided by an international collection of experts in a rigged election? It makes me think of religious schisms, where one religion splits in two, and then they both split in two, until you have all sorts of competing sects. In the end, as Carl Sagan said about religion, "I do not want to *believe*; I want to *know*." That is how I feel about quantum theory. I do not want to believe—I want to know.

Okay, enough of that. We return to our analogy of Alice and Bob's watches. In the quantum version of our wristwatch thought experiment, NIST now gives Alice and Bob shiny, plutonium-encased,[22] quantum-atomic pocket watches, the kind with the snap-over cover on the case that does not let you look inside to see the time until you pop it open.[23] For the purpose of this argument (and to simplify things), we'll say these pocket watches have hour hands, but no minute or second hands. (Remember, these pocket watches are not for telling time but for illustrating some points about quantum theory.) A quantum physicist named Dave, who runs the National Institute of Quantum Information Standards and Technology (NIQuIST), a subdivision of NIST that is also located in Boulder, gives the quantum pocket watches to Alice and Bob.[24]

Inside the clockwork of these quantum pocket watches are single calcium ions.[25] The calcium ions are held in the clockwork on an ion chip, a computer chip–like gizmo that pins the ion down and traps it in place with electric and magnetic fields and laser beams. The ion itself has a magnetic field and can be visualized as a tiny spinning ball that, just like the Earth, has its own north and south pole. There is for now one calcium ion trapped in each quantum pocket watch. The little itty-bitty laser beams that help cool the ion down and hold it in place can also be used to read in and out information about the ion and to calibrate, set, and synchronize the quantum pocket watch. Finally, the pocket watch is set up so that when you try to open the cover, the latch momentarily jams until the lasers inside the watch measure which way the north pole of the spinning ion is pointing in the clockwork chip. The lasers then relay this information to the clockwork, which then rapidly moves the hour hand to point in that same direction as the ion points. This all happens *before* the cover will fully release to allow you to look at the hour hand.[26]

The two quantum pocket watches are quantum synchronized in a *very particular way* before leaving the NIQuIST shop and shipped to Alice and Bob. The good folks at NIQuIST prepare the two ions in a peculiar—highly correlated—quantum mechanical state of affairs and then carefully (using the laser beams) place one ion in Alice's pocket watch and the other ion in Bob's pocket watch, which they then ship off to Aurora and Brockley—without in any way disturbing the contents of the watches. This *peculiar quantum mechanical state of affairs* is the one that Einstein, Podolsky, and Rosen had proposed to set up in a thought experiment in their paper,[27] and it puts a very strong correlation on the outcomes of any readings of the quantum pocket watches. It also defies common sense. As we shall see, no classical theory can mimic quantum theory's predictions of this quantum pocket watch synchronization thought experiment.

The correlations of the watch hands, on *face* value, look just like the correlations of the classical wristwatches. If Alice opens her quantum pocket watch and finds the hour hand is pointing at 12:00 noon, then when Bob opens his watch, he will find that his hour hand points at 6:00 p.m. If Bob opens his watch and sees 9:00 p.m., then Alice will see 3:00 p.m. when she opens hers. For sure, whenever Alice opens her watch and looks, her hour hand will point to the opposite side of the watch dial as Bob's does, which she can confirm via Skype or a text message to Bob. This all sounds just like the commonsensical classical wristwatch thought experiment all over again. What is different is that *in the quantum mechanical description of what is happening*, which was clearly spelled out in the EPR paper, and which is clearly predicted by the theory of quantum mechanics, is now the correlations are of a new and very strange type—correlations that are never seen in classical experiments. It all boils down to that *peculiar quantum mechanical state of affairs* that correlates the ions in the pocket watches. Here's what quantum theory predicts:

1. *Unreality:* The north pole of *each* ion in each quantum pocket watch, separated by thousands of kilometers, *is pointing in all directions simultaneously*, until either Alice or Bob opens the cover. The act of opening the watch causes the ion to jump into a particular alignment, say indicating 9:00, and then stay there. Before the cover

is opened, *the ion points in all 12 directions*, like the spoon that is simultaneously on my desk and in the bathroom. (In EPR lingo, the notion of "direction" is not an "element of physical reality" associated with the ion.)

2. *Nonlocality:* The readings of the remote quantum pocket watches are not independent of each other—indeed, they are highly correlated. If Alice opens her watch first in Aurora and gets 3:00, this act of observation on Alice's part *instantaneously* causes Bob's ion to collapse and his hour hand to swing to 9:00, even if he does not open the case to look at it. Similarly, if Bob opens his watch first and reads 6:00, then *instantaneously* the ion in Alice's watch will collapse and her hour hand will swing to point in the opposite direction corresponding to 12:00, even if she does not open her watch. What Alice does *locally* to her watch affects *nonlocally* the outcome of Bob's and vice versa.[28]

3. *Uncertainty:* For each run of this experiment, the directions the hour hands point to will be completely random, possibly any of the 12 hours, all over the watch dial, no matter how carefully the observation is made. The only constraint is that the hands always point in opposite directions; *which* opposite direction they point to on any given run of the experiment is completely unpredictable. (These would not make very good synchronized watches for timekeeping.)

There is no longer any correlation between the hour hands and the actual time, and this thought experiment is no longer about timekeeping but about illustrating the weirdness of quantum theory. By "each time the experiment is run," I mean that after Alice and Bob open their watches once and compare their hour hands, they have to either send them back to NIQuIST for resynchronization or order a new pair of pocket watches. They could also order the advanced Mark II quantum pocket watch model, which stores a whole bunch of ions at a time, but the point is that the quantum-synchronized ions, once measured, have irreversibly lost the synchronization—that peculiar quantum mechanical state of affairs—put into them by NIQuIST and must now be sent back for quantum resynchronization. Alice and Bob then have to start over with a new pair of watches or a new pair of ions.

Such a thought experiment with pocket watches has not been done (yet), but similar experiments with pairs of ions in small traps separated by a few centimeters (a few inches) have been conducted, and they bear out the predictions of quantum theory.[29] Such experiments have been also done with light—with pairs of quantum-synchronized photons over distances of around 100 kilometers (60 miles) flying along optical communication fibers under Lake Geneva in Switzerland.[30] There is nothing in principle to prevent one from conducting an experiment with ions in quantum pocket watches, separated by thousands of kilometers (miles), but for now, that is just a thought experiment on the basis of what quantum theory would predict.

Quantum theory, which agrees with all experiments done so far, predicts the outcome of the thought experiment with the three properties *unreality*, *nonlocality*, and *uncertainty*. The point Einstein, Podolsky, and Rosen made in their *New York Times*–heralded paper is that in addition to *uncertainty* (which Einstein never liked anyway), quantum theory has these other two new features they didn't like either: *unreality* and *nonlocality*. So the EPR paper was a succinct, concentrated, boiled down, and saucy reduction of the three main features that Einstein did not like about quantum theory—all of which he thought defied common sense.

It is only common sense that the ions have some element of physical reality and have some definite direction, even when you are not peeping in through the cracked open cover—*reality!*

It is only common sense that Alice, looking at her watch, can have no effect on Bob's watch thousands of kilometers (miles) away—*locality!*

And finally, it is only common sense that if the watches are prepared exactly the same way, each time you open the cover and look, you get the same result each time with no randomness—*certainty!*

Bohr and Einstein used to take long walks together, discussing the quantum theory, and as Bohr tells it, "I recall that during one walk Einstein suddenly stopped, turned to me, and asked whether I really believed that the Moon exists only when I look at it."[31] So now here is a subtle philosophical point (and indeed the entire EPR paper is filled with subtle philosophy points). These crazy elements—*unreality, nonlocality*, and *uncertainty*—are unacceptable properties of the quantum theory; Einstein decreed that they are misplaced philosophical baggage layered on the theory's mathematics. Can't we save the mathematical apparatus but jettison this unacceptable philosophical baggage? No, we can't.

In order to agree with real experiments, any attempt to remove this one set of unacceptable quantum mechanical baggage must be replaced with an equally unacceptable quantum luggage set. You don't like your pink polka-dotted portmanteau? You can only trade it in for a frighteningly fuchsia fortnighter. Or perhaps you would like this horribly hemp-hued haversack? Quantum theory has built into it these non-commonsensical features of *unreality, nonlocality*, and *uncertainty*. Any theory that replaces quantum theory must also have these same gosh-awful features in order to agree with every real experiment! To do so, any theory that makes all the same predictions as the ordinary Copenhagen quantum theory will need to contain an equivalent lump of quantum weirdness that will be just as hard to swallow. Let's see just why.

Suppose Alice and Bob perform their experiment 10 times and jot down in their laboratory notebooks the results of their hour-hand observations for each of the 10 runs, using 10 pairs of pocket watches, carried out in sequence, and then compare the lists, as shown in Table 1.1. (Because there is no correlation to the actual time anymore, the use of a.m. and p.m. becomes irrelevant, and we drop it.)

Recall that because of the *uncertainty condition*, although these pocket watch observations were carried out in a sequence of 10 measurements, they no longer bear any relationship to the actual time at measurement. The hour hand of the watch just becomes a convenient way to measure the correlations in our thought experiment.[32] This list agrees with the predictions of the quantum theory. The hands always point in opposite directions but the opposite directions appear at random.

So while the quantum theory that goes into predicting this correlated list of numbers has some very peculiar features (*unreality, nonlocality*, and *uncertainty*), the list itself does not look all that peculiar. It is a simple list of random but highly and oppositely correlated numbers. Could there be (as posited by Einstein, Podolsky, and Rosen) another simpler way to explain this list that is not so "spooky" as quantum theory?

The EPR paper, remember, claimed that any sensible physical theory should have the common-sense properties of satisfying the *reality, locality*, and *certainty* conditions—which had been found in all sensible physical theories prior to the introduction of quantum mechanics. The EPR paper then goes on to claim

TABLE 1.1 ALICE IN AURORA AND BOB IN BROCKLEY LIST THE RANDOM BUT HIGHLY CORRELATED HOUR-HAND POSITIONS OF THEIR QUANTUM POCKET WATCHES IN A SERIES OF 10 EXPERIMENTS CARRIED OUT WITH 10 PAIRS OF WATCHES

Run Number	Alice	Bob
1	6:00	12:00
2	10:00	5:00
3	3:00	9:00
4	9:00	3:00
5	8:00	2:00
6	9:00	3:00
7	5:00	10:00
8	6:00	12:00
9	4:00	10:00
10	9:00	3:00

Note: The random times were generated using the random integer number generator in my Mathematica software.

that any theory satisfying *reality, locality,* and *certainty* conditions should be considered to be "*complete*"—and the last word on the subject. Contrariwise, the EPR paper then states that any theory that has features such as *unreality, nonlocality,* or *uncertainty*—quantum theory—must be considered "*incom-*plete." *That* is the head-spinning logic that made the *New York Times.*

Einstein continued to hold out hope that quantum theory could be replaced with a more sensible and complete theory. It took more than 30 years until his last hope was finally and absolutely dashed.

Let's take a concrete example by modifying our thought experiment with Alice and Bob's pocket watches. Suppose a NIQuIST engineer named Eve, in a plutonium-induced rage, decides to cut corners in the quantum pocket watch racket, and instead provides them with perfectly classical pocket watches, which she labels "authentic NIQuIST quantum pocket watches." Instead of the careful choreographed, double-spinning, ion-synchronized quantum pocket watches, Eve instead gives Alice and Bob ordinary one-handed pocket watches. Then, Eve just randomly (using a random number generator like I did to make the list in Table 1.1) sets Alice's pocket watch to the 1-hour hand setting and then Bob's to that same setting plus 6 hours on the opposite side of his dial. No ions, no clockwork, no quantum, no nothing.

Eve just sets the watches up when nobody is looking and then ships them to Alice and Bob and tells them they are "quantum synchronized." Think of how much money Eve can shave off of her NIQuIST operating budget without the need to provide all the quantum gadgetry! Alice and Bob order from NIQuIST 10 times in 10 weeks 10 newly synchronized watches, and each time, when they open them in Aurora and Brockley, they will get correlated and seemingly random lists of opposite clockface numbers, just as in Table 1.1. The explanation now is that the weirdness of quantum mechanics has been replaced by the devious Eve's perfectly classical and sensible—but unknown (at least to Alice and Bob)—random and opposite hand settings. Eve's skullduggery meets *all three* of the EPR requirements for a complete theory:

1. The Eve-rigged pocket watches satisfy the reality condition: The hands of each watch are fixed in advance on a known position on the dial (known to Eve) where they remain as they are shipped to Alice and Bob.
2. The Eve-rigged pocket watches satisfy the locality condition: If Alice pops open the case and peeps at her pocket watch, it has no effect on Bob's setting or vice versa. Even if Bob throws his pocket watch off the roof of the Greenwich Observatory again, Alice's hour hand stays fixed on her clockface where it always was.
3. The Eve-rigged pocket watches satisfy the certainty condition: Eve (if she keeps good laboratory notebooks) will know for certain what "random" hour-hand positions Alice and Bob will see on each run, even if

Alice and Bob don't know this. The best Alice and Bob can do, without help from Eve, is just average over all the runs and conclude that the hands seem to point randomly and oppositely every time they spring open the lids.

Pitched like this, the theory begins to sound more like the theory of statistical mechanics (a favorite of Einstein's), where Alice and Bob could, *in principle*, know on each run where the hands *would* point (if they interrogated Eve). But without this information from Eve, Alice and Bob are forced to just make statistical averages over many events that seem to them to be random. (This setup is akin to my previous example of classical "heads or tails" where you could know in advance which way the coin would land if you just tracked everything carefully.) The random hand settings that Eve makes on each run are "hidden" from Alice and Bob, and when those hidden settings are included in the analysis of the data in Table 1.1, that particular experiment can be explained without using quantum theory and in a way that satisfies the *reality*, *locality*, and *certainty* conditions. Einstein would be so proud.

Einstein, Podolsky, and Rosen actually suggest in their 1935 paper that such an internal and random but hidden feature—what we now call a *hidden variable theory*—would be a way out of the quantum weirdness of the *unreality*, *nonlocality*, and *uncertainty* conditions. *Hidden variable theory!* Mother Nature obeys classical common sense, but she is just hiding some of the data from us. Einstein would thus consider the quantum mechanical explanation of the quantum pocket watch experimental data, in Table 1.1, to be *incomplete* because it contains elements of *unreality*, *nonlocality*, and *uncertainty* that no sensible complete theory should contain. This point is the origin of the (questionable) title of the EPR paper, "Can Quantum-Mechanical Description of Physical Reality be Considered *Complete*?" (Italics mine.) But if quantum mechanics is incomplete, what do we replace it with? Einstein proposed to replace quantum theory with a setup similar to Eve and her secret bag of fraudulent clockwork—a local, real, certain, and complete *hidden variable theory*. Hidden?—Yes. Spooky?—No!

The EPR paper ends with this very Einstein-like call to arms, "While we have thus shown that [quantum theory] does not provide a complete description of the physical reality, we left open the question of whether or not such a description exists. We believe, however, that such a theory is possible." Einstein suggests that it should be possible to replace quantum theory, and all its spooky features, with a mundane Eve (hiding under Mother Nature's petticoat), whose hidden (but very classical) handiwork is lurking in our supposedly quantum mechanical clockwork. If only Alice and Bob knew the details of Eve's tampering with the hour hands, they would not need to invoke quantum theory at all to explain their 10 experimental data runs in Table 1.1.

Such a classical theory is called a *hidden variable theory*, because in this example, the hidden variables are the random numbers Eve uses to set the hour hands.

We can summarize the EPR call for the search for such a *complete* local hidden variable theory thusly: It should be possible, said the gang of three, to replace incomplete quantum theory, with its horrible and spooky *unreality*, *nonlocality*, and *uncertainty* conditions, with a lovely, fresh, and complete classical hidden variable theory, which obeys the commonsensical *reality*, *locality*, and *certainty* conditions. We have just seen that it is possible to do this in our particular thought experiment that produces the data in Table 1.1. But this experiment in Table 1.1 is a single simple example, not a general proof. What we would need to demonstrate, to assuage Einstein, is that there exists a commonsensical hidden variable theory that can reproduce *all* the predictions of quantum theory in *any* experiment. That is a very much taller order, and it turns out that is an order that can never be filled.

FAIRIES, GREMLINS, AND MAGIC DICE

To illustrate just how difficult a hidden variable theory replacement of quantum theory would be, I hereby subject you to yet another thought experiment with these infernal quantum pocket watches. Let's say that Dave, the Director of NIQuIST, notices one day in the NIST parking lot that Eve is driving a brand new, shiny, plutonium-plated Corvette Z06.[33] Dave is suspicious that Eve can afford such a car on her paltry federal government salary and launches an investigation into the quantum pocket watch facility, whereupon he uncovers Eve's scheme of replacing the expensive quantum pocket watches with the much cheaper fraudulent classical ones. (Eve had been mocking up the clockwork and cooking the books!) Dave tries to fire Eve, but as we all know, it is impossible to fire a federal employee. So he instead encourages her to take early retirement with a plutonium parachute. Dave then hires Charlie, an IBM quantum information theorist, as a consultant to help design a better quantum pocket watch that Alice and Bob can test themselves for true quantum behavior and which cannot be tampered with by any future nefarious Eves.[34]

Charlie points out that the quantum pocket watches can be easily fixed in a way so that Alice and Bob can always be sure that they have paid for the expensive, truly quantum watches and not some cheap Eve-tampered classical knockoffs. Charlie proposes just modifying the snap-on cover in such a way that Alice and Bob cannot view the entire watch face at one time. Instead, Charlie designs a new cover with a ratchet mechanism and single radial viewing slit in the cover (see Figure 1.1). The new cover can be rotated so that the viewing slit aligns to any of the 12 hour watch dial positions, and then once in place over an

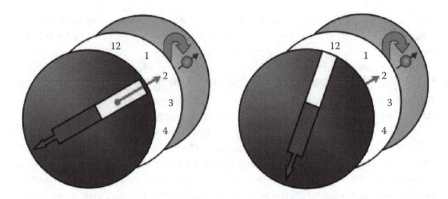

Figure 1.1   Charlie's improved quantum atomic pocket watch. Alice rotates the plutonium slit cover (rectangle with arrow) randomly to 1 of the 12 clock numbers and attempts to then slide it open to view the dial. This movement of the slit cover triggers the laser beam (curved arrow) to measure the direction the ion is pointing (ball with arrow through it). The measurement causes the ion to collapse from a quantum superposition of all 12 hour-hand directions into just one of them at random. The laser records the result of the ion's collapse and activates the clockwork to move the hour hand (arrow pointing towards 2) to point in the same direction as the collapsed ion. Only then is the locking mechanism fully released on the slit cover, and only then can Alice slide it all the way open to look at the watch dial. There is a probability of 1 in 12 that Alice's slit will actually be where the hour hand is located, and then Alice will see the hour hand through the slit (left). However, most of the time, with a probability of 11 in 12, Alice's slit opening will not be aligned over the hour hand and she will see nothing at all (right).

hour position, the cover of the slit can be slid open to observe the hour number and to also see if the hour hand is pointing to that hour number or not. If not, the hour hand will not be visible, and it will be pointing somewhere else on the dial, hidden by the cover.

For example, Alice can rotate the slit to 12:00 and open it to see if the hour hand is pointing to 12:00 or not. Sometimes, the hour hand will be there, visible through the slit, but much of the time when Alice opens the slit, she will not see the hour hand. (Remember—in this new thought experiment, there is no correlation with the real time anymore, and there is only some random chance that the hour hand will point to any given marking on the watch dial face.) Charlie also suggests a cost savings by loading up the quantum pocket watches with 10 ions at once, producing the Mark II Quantum Pocket Watch, so that NIQuIST can save on shipping charges in any test run of 10 measurements. The idea is that each of the quantum-synchronized ion pairs is used up and discarded in each step of a 10-step data run and the lasers move on to the next ion pair for the next step.

Charlie's protocol for testing the quantum pocket watches goes as follows: Once each hour, at roughly the same time, Alice and Bob should independently rotate their watch cover slit to a random hour position on the watch dial and then open the slit on their respective watches to see if the hour hand is actually there or not. "It is critical," admonishes Charlie to Dave, wagging his finger, "that Alice and Bob each perform a *delayed* choice—that they choose their own random hour positions long after the watches have shipped and without communicating with each other about their respective choices."

After all 10 observations have been made, Alice will call up Bob on Skype and they will exchange the results of the data run and perform a statistical analysis on the combined results. Charlie will then provide a computer code that Alice and Bob are to use for the analysis. Charlie claims that after the data analysis is run, the computer code will output the result "true" if it is a *true* pair of expensive quantum pocket watches and it will spit out the word "false" if they are a *false* pair of evil Eve's classical knockoffs.

Dave dutifully produces the new quantum pocket watches, carefully following Charlie's instructions, synchronizes the 10 pairs of ions in each watch, and then ships the two watches off to Alice and Bob with little pamphlets describing Charlie's testing procedure, including a copy of the statistical analysis computer code in C++. Alice and Bob, who are very upset with NIQuIST for charging them for expensive quantum watches and then providing them with classical frauds, are eager to try the new test. Starting at 12:00 noon in Aurora (6:00 p.m. in Brockley), they open their watches on the hour 10 times in a row, choosing their own individual random settings of the 12 possible watch dial positions of the observing slits. (The actual time is just used now to synchronize their measurements.) Alice uses her own true random number generator, purchased from MagiQ in the United States, to choose her *delayed* random choice of the slit settings. Bob uses his own true random number generator, purchased from ID Quantique in Switzerland, to choose his *delayed* random choice of the slit settings.

Once each dial position is chosen, they each slide open the slit cover and look to see if the hour hand is there or not. For each observation, they record the clockface dial setting of the slit (12:00, 1:00, 11:00, etc.) and also whether the hour hand is visible ("yes") or not visible ("no"). The data now look something like Table 1.2. (Because a.m. and p.m. become irrelevant, as the "times" do not correspond to real time, we again drop them.)

To get good statistics, Alice and Bob must actually do this test hundreds or thousands of times, but Table 1.2 illustrates a number of key points. The primary point is that if Alice sees "yes" in her slit, and by chance Bob has chosen the 180° slit, Bob must also see "yes"—as in test number three where Alice sees 8:00 and Bob sees 2:00. If the hands are there, visible in both slits by random chance, then they must be anti-correlated—that is, they must point in opposite directions as before.

TABLE 1.2   ALICE IN AURORA AND BOB IN BROCKLEY CARRY OUT CHARLIE'S TEST EXPERIMENT WITH 10 PAIRS OF NIQUIST-SYNCHRONIZED IONS IN THE QUANTUM POCKET WATCHES

Test Number	Alice's Slit Setting	Alice's Hour-Hand Observation	Bob's Slit Setting	Bob's Hour-Hand Observation
1	2:00	No	12:00	Yes
2	6:00	No	9:00	No
3	**8:00**	**Yes**	**2:00**	**Yes**
4	6:00	No	9:00	No
5	5:00	No	12:00	No
6	**3:00**	**No**	**9:00**	**No**
7	7:00	Yes	10:00	Yes
8	11:00	Yes	10:00	Yes
9	1:00	No	2:00	Yes
10	2:00	No	5:00	Yes

Note: The random times were generated using the random integer number generator in Mathematica, as were the "yes" and "no" results. The actual probability of Alice or Bob getting a "yes"—seeing the hand when they open the slit—is only 1/12 or less than 10%—but I have made it 50% for the sake of argument. In a run of 10 tests, Alice and Bob should see mostly "no." Data in boldface indicate results when they (at random) open the slits 180° from each other.

The secondary point in Charlie's protocol is that there are now many other things that can happen. For example, the hand may not be there at all. Or even if it is, Alice and Bob may not have chosen the slits that are 180° opposite to each other, because they each make this choice at random. Hence, Alice may see a hand and Bob nothing, or both Alice and Bob may see nothing. It is even possible that by random chance, Alice and Bob will both choose slits that are not 180° apart and both see the hour hand, but these events will happen at random with no apparent anti-correlation. To be clear, if they randomly choose slits 180° apart and Alice sees the hour hand, then Bob *must* see the hour hand. However, if they randomly choose slits not 180° apart, say the 3:00 and the 6:00 slit settings, if Alice sees the hour hand at 3:00, then Bob may or may not see the hour hand in an apparently random fashion. (Such types of events are seen in runs 7 and 8 in Table 1.2.) All event outcomes from Table 1.2 are fed into Charlie's code.

My point is that there are many more possible outcomes in Charlie's test, and only quantum theory—or a similar *unreal, nonlocal, uncertain,* and *incomplete* theory—can keep up with the test and produce a "true" result. Any *real, local*, *certain*, and *complete* hidden variable theory cannot keep up with the statistics in the long run. Any counterfeit pocket watches that embody a local hidden variable theory, such as Eve's list of random numbers (the hidden variable) is doomed to fail. For Eve's crooked watch factory, Charlie's test will always produce the output "false." So now, Alice and Bob call each other up and only then share each other's data, including the slit setting and the observation result for each test. They then feed all these results into Charlie's C++ computer code and then Charlie's code spits out a single result: "true" if they have true quantum pocket watches, or "false" if they have the false classical knockoffs. What on earth is Charlie's code doing and how can it possibly tell?

To answer this, we have to go back to Einstein, Podolsky, and Rosen, who (you will remember) conjectured that it could be possible to cook up a commonsensical classical-like *hidden variable theory*, satisfying the *reality*, *locality*, and *certainty* conditions and reproducing *all* the predictions of quantum theory. With such a hidden variable theory, Einstein would be able to rest easy for all eternity. Once found, we would simply replace the old non-commonsensical quantum theory with the new commonsensical hidden variable theory and be done with the spookiness once and for all!

It took nearly 30 years to prove that Einstein was wrong. In 1964, John Bell, a Scottish physicist working at the giant CERN atom smasher on the border of Switzerland and France,[35] proved that Einstein's hoped-for hidden variable theory—the replacement for quantum theory—couldn't exist. Because Einstein had died 11 years earlier, on April 18, 1955, he never lived to see this result. (Einstein now sleeps restlessly for all eternity, his cremated ashes scattered to the wind over the New Jersey shore of the Delaware River, and his brain barreling about the country in a formaldehyde-filled Tupperware bowl in the trunk of a Buick Skylark.[36])

What Bell proved is that the correlations predicted by the incomplete quantum theory are *much stronger* than those predicted by any complete classical local hidden variable theory. He did this by performing a statistical analysis of the predictions of a generic (but general) class of classical local hidden variable theories, all of which obeyed all of Einstein's *reality*, *locality*, and *certainty* conditions. He then compared the results with those predicted by quantum theory, for a similar general setup, and found that the predictions of any of these local hidden variable theories disagreed with each of the predictions of quantum theory! This result is called "Bell's Theorem," and in his paper he provided a formula that quantified the level of disagreement, which is called "Bell's inequality."

Bell's single formula demonstrated a clear demarcation between *all* local hidden variable theories and *the single* quantum theory. In a particular experimental setup, you could plug your data into this single formula, and if the formula spits out a number *less than or equal to two* (an inequality), then the experiment could be explained with a *local hidden variable theory*. However, if the formula spits out *a number greater than two*, then *only* quantum theory could explain the result. This formula is at the heart of our IBM-Charlie's test of the NIQuIST quantum atomic pocket watches! Bell's conclusion was that any theory that could match quantum theory in predictive power must have built in itself the very conditions of *unreality*, *nonlocality*, or *uncertainty* that Einstein so loathed. In order to agree with his experiment, you could throw out quantum theory only by replacing it with something just as bizarre, such as Bohm theory, Many-Worlds theory, or Magic-Dice-Throwing-Telepathic-Gremlin theory (see below).

There are several important points to make about Bell's result. Either classical hidden variable theory is right or quantum theory is right. Both cannot be right and *the decision about which is right can now be tested in an experiment.* This takes the discussion out of the realm of philosophy, where it had languished for 30 years, and puts it smack dab in the realm of a scientific physical hypothesis that can be tested in an experiment. A test of your theory in an experiment is the gold standard of the scientific method.[37]

The grandfather of the scientific method was Galileo Galilei who, while avoiding being burned at the stake by the Vatican for his heretical idea that the Earth moved around the Sun, managed to (while under Church-imposed house arrest) codify the need for experiments to test hypotheses. That is the cornerstone of the scientific method. Only after a baptism of fire, in the crucible of a laboratory experiment, is a hypothesis christened a scientific theory. The Greek philosopher Aristotle, on the other hand, was not so constrained and hypothesized away to his heart's content, often appealing to common sense, without ever bothering to check any of it in the laboratory or the real world. Aristotle's commonsensical hypotheses often disagreed with experiment and were eventually ruled out as just plain wrong. (Bowling balls fall faster than softballs when dropped off the Leaning Tower of Pisa—wrong![38])

Quantum theory may disagree with common sense, but if it agrees with experiment when put head to head with a local hidden variable theory, then quantum theory must be accepted as right and hidden variable theory ruled as wrong. If quantum theory is right, then nature is not only weirder than we think; it is weirder than we *can* think. The spookiness of quantum theory is then here to stay and there is nothing Einstein or anybody else can do about it. If we insist on constructing an alternative theory for quantum mechanics, it will not obey common sense either, and will be just as strange as quantum theory (if it is to have any hope of giving correct experimental predictions). Let's now do just that—construct an alternative to quantum theory that agrees with experiment and face the weirdness head on.

After Eve's departure and IBM-Charlie's consult, Dave of NIQuIST hires Evan—a replacement for the now retired Eve—to build and ship Charlie's newly redesigned quantum pocket watches. However, the federal background check fails to reveal that Evan is in fact Eve's evil twin brother! After hearing Eve's stories about the plutonium-plated Corvette, Evan decides he wants a piece of the action and plans to doctor the pocket watches and skim the profits off the top. However, when he arrives at work, Evan finds the new watches with IBM-Charlie's improvements. The fix is that Charlie used John Bell's theorem to design his test for the new tamperproof quantum pocket watches. If Bell's formula spews out a number less than or equal to two, then Charlie's code will output "false," thus revealing the local hidden variable clockwork.

Looking at the list of numbers and results from Table 1.2, Evan realizes he can't mock up the pocket watches just by having the watch hands always point in opposite directions, as Eve had done. Without going into the some-what mind-numbing detail of the Bell analysis, Bell's inequality, and thence Charlie's code, would catch that up in the statistics. You have to trust me on this—the average number of times the watch hands would show up in a slit, particularly in the cases where Alice and Bob's slits are not 180° apart, would be off. Evan could try to cook up all sorts of fancy (but classical) local hidden variable tricks to hide in the clockwork, but if all these tricks obeyed Einstein's *reality*, *locality*, and *certainty* conditions, then they would never reproduce the actual predictions of quantum theory. This is Bell's result. And Alice and Bob, running Charlie's test, would always get a "false" on any test runs of Evan's shipments and know for sure they had been defrauded with non-quantum pocket watches.

A sophisticated local hidden variable theory that Evan could deploy, which would be much better than just randomly anti-aligning the ions and the watch hands as Eve did, would be to hide two invisible fairies, named Alfreeda and Breena, inside each of the pocket watches. Let's say Alfreeda is in Alice's pocket watch and Breena is in Bob's. Evan gives the fairies tiny ham radios and 12-sided dice.[39] Whenever Alice attempts to dial up a clockface position and pulls back her slit cover to look for the hour hand, Alfreeda throws her 12-sided die. Say Alfreeda's die yields 9:00. Then Alfreeda rapidly moves Alice's hour hand to 9:00, and then she calls up Breena on the ham radio and tells her to move Bob's hour hand to 3:00. Alice and Bob would start getting results that *looked* like those in Table 1.2.

The fraudulent fairy clockwork is *local*—the communications between Alfreeda and Breena take place via ham radio transmissions that move at or *less than the speed of light* so no instantaneous action at a distance is allowed.

The fraudulent fairy clockwork is *real*—the hour hands have a *real* position on the clockface immediately before Alice and Bob slide open the slits, and that position does not change when Alice or Bob look through them.

The fraudulent fairy clockwork is *certain*—we could carefully track the motion of the 12-sided die on a supercomputer and predict with absolute certainty any outcome, such as 9:00.

By his own hand, Einstein would have to accept Evan's local-hidden-variable-fairy theory as a legitimate possible alternative to quantum mechanics to be tested. Of course, nobody knows what such a hidden variable theory would look like. Physicists usually just put general constraints on such theories, like whether they are local (influences move at or less than the speed of light) or nonlocal (influences move faster than the speed of light). Because nobody actually knows how Mother Nature would implement such a local hidden variable theory, why not just use fairies?

However, in a detailed statistical analysis, IBM-Charlie's test would reveal even this fraudulent fairy clockwork to the suspicious Alice and Bob. Bell showed that the statistical correlations predicted by quantum theory for two highly correlated quantum systems (the two ions in the pocket watches) are much stronger than those of any classical hidden variable theory of the same two systems (the fraudulent fairy clockwork). By analyzing those correlations in the test run experiment, IBM-Charlie's black box can tell which is which. The *local*, *real*, and *certain* fraudulent fairy clockwork can simply not keep up with all the different possible independent settings of the two independent slits and the statistics of the outcomes that actual quantum theory predicts. Anything Evan tries to do to mimic the quantum theory with a *realistic, local, certain,* and complete classical fairy clockwork mockup, is doomed to fail. (For the rest of this book, we'll use fairies as a stand-in for the local hidden variable theories.) What would Evan have to do to successfully fake quantum theory?

There is a hidden variable theory, proposed by the British quantum physicist David Bohm, which does produce all the predictions of quantum theory, but at the price of violating Einstein's *locality* condition.[40] One way that Evan could mimic this *nonlocality* condition of quantum theory, and hence reproduce the predictions of quantum theory, would be to hide—instead of the two fairies— two invisible, identical twin telepathic gremlins, named Aagar and Brashnak (who are undetectable by any means), in the jury-rigged pocket watches.[41] Let's say Aagar sits in Alice's pocket watch and Brashnak is in Bob's.

The gremlins would be given a set of complex instructions allowing them to strongly correlate Alice and Bob's observations. For example, if Alice randomly sets her slit to 3:00 and observes the hour hand, Aagar would throw a *magic* 12-sided die and *instantaneously* and telepathically alert Brashnak to throw his magic 12-sided die. The dice will always (magically) give random but opposite answers, if Alice and Bob choose random and opposite slit positions, so if Aagar sees "3:00" and rapidly moves his hour hand to 3:00, Brashnak is guaranteed to always get the 180° number and he would thus set Bob's hour hand to 9:00. Also, again because the dice are magic, the random outcome is unpredictable by any means. So if Bob looked even a split billionth of a second after Alice saw the hour hand at 3:00, then Bob would see it at 9:00 (if he had randomly set the slit to 9:00).

That would take care of the correlations in Table 1.1 but would not be enough to fool Charlie's test, which involves also including the events when Alice and Bob don't randomly choose 180° opposite slits. There are many events where

Alice sees nothing and Bob sees an hour hand or Bob sees nothing and Alice sees an hour hand, or both see nothing, or both see an hour hand but they are not 180° apart. The statistics of how often such events occur need to be tabulated and fed into Charlie's code. To fool Charlie's code, on the basis of Bell's theorem, there would need to be a much more elaborate set of instructions for what Aagar would alert Brashnak to do with even more elaborately correlated magic dice, in order to get the same statistics as quantum theory would predict when Alice and Bob do not randomly choose opposite slits. In the limit of sufficiently magic dice and sufficiently telepathic twin gremlins, Evan's magic-dice-throwing-telepathic gremlin clockwork could be jury rigged to reproduce all the predictions of quantum theory.

The magic-dice-throwing-telepathic-gremlin clockwork is *nonlocal*— the telepathic communications between Aagar and Brashnak take place instantaneously.

The magic-dice-throwing-telepathic-gremlin clockwork is *unreal*—the magic 12-sided dice have no correlation until Aagar and Brashnak throw them.

The magic-dice-throwing-telepathic-gremlin theory is *uncertain*—the magic 12-sided dice produce absolutely random (but always anti-correlated) results that are *unpredictable by any means.*

Critical to Evan's invisible telepathic gremlin fraud is that the gremlins violate the *locality condition*: Aagar can alert Brashnak to throw his magic die to instantaneously affect the outcome of Bob's measurement made only a billionth of a second after Aagar learns of Alice's measurement. This would seem to violate Einstein's own theory of relativity in that Aagar influences the behavior of Brashnak with some sort of signal that moves at infinite speed—faster than the speed of light. A signal traveling at the speed of light moving over the surface of the earth through a fiber optic cable would take a fraction of a second to get from Aurora to Brockley. The gremlins are influencing each other instantaneously, so whatever this influence mechanism is, it is faster than the speed of light (which would of course make Einstein cringe).

To agree with quantum theory, this instantaneous influence would have to be independent of distance, even if Alice was in the star system Alpha Centauri and Bob was in Beta Pictoris—tens of light-years apart. But instantaneous gremlin influences over distances of tens of light-years violate Einstein's relativity theory—no signal can travel faster than light. The gremlins' magic dice also violate the *reality condition*. The dice somehow always give correctly anti-correlated (opposite) results for all possible settings of the watches when opposite slits are chosen. Lastly, the magic dice are *uncertain*—their outcome cannot be predicted by any physical means.

So it is then true that we could, if we wanted to, replace quantum theory with a *nonlocal, unreal, uncertain* theory based on telepathic gremlins throwing magic dice. That gremlin theory would be a perfectly respectable *unreal, nonlocal*, and *uncertain* hidden variable theory and would reproduce all the predictions of quantum theory. (The gremlins, their dice, and their instruction booklet would be the hidden variables.) Hence, it would be indistinguishable from quantum theory in any experimental test. (For the rest of this book, we'll use gremlin theory as a stand-in for a nonlocal, unreal, uncertain theory that reproduces the same results as quantum theory.) But are we any better off rejecting quantum theory and adopting gremlin theory? No.

Both gremlin and quantum theories are incomplete—they both violate the *locality*, *reality*, and *certainty* conditions. If Einstein was unhappy with quantum theory, then by his own standards, he would be equally unhappy with the magic-dice-throwing-telepathic-gremlin theory. As Arthur C. Clarke tells us in his *Third law*, "Any sufficiently advanced technology is indistinguishable from magic."[42] Indeed, we see that quantum theory has features that—if explained in other ways such as by invoking gremlins—would require something akin to magic.

THE INVERTED EARTH SOCIETY

I would like to insert here a philosophical aside about the scientific method and just how physicists decide if a theory is true or not. How do we know that the quantum theory is true and the magic-dice-throwing-telepathic gremlin theory is not true? Well, we don't. The scientific method is supposed to help us distinguish between two theories, but only if the theories make different predictions in the laboratory. The scientific method then orders us to go to the laboratory and take measurements and then decide which theory agrees with the data. We then consider the theory that agrees with the data to be true, as least for the time being, and toss out the theory that disagrees with the experiment. However, in quantum theory, we have a whole slew of different "interpretations" that are layered onto the solitary mathematical framework. These interpretations plus the mathematics give rise to a whole slew of different quantum theories—each and every one manipulated to agree with the usual and popular Copenhagen theory.

In this book, I will call the mathematical framework, together with a particular interpretation, a quantum theory. While most everybody agrees on the quantum math, most everybody disagrees on the quantum interpretation, giving rise to many different quantum theories. In almost all cases, these theories are—by design—cooked up to always give the same predictions as the original theory—quantum math and the Copenhagen interpretation of Niels Bohr's camp. Critical for my discussion here is that, in physics, a theory is not just a mathematical cookbook for generating numbers that can be compared to an experiment; it also has a certain amount of philosophical baggage associated with what the theory "really means"—the interpretation.

Some physicists, Heisenberg in particular, took a bare bones approach. You should not ask, implored Heisenberg, what the theory really means—no interpretations! You should just use the mathematical apparatus to calculate things you might see in the laboratory and then check if you really see those things, and if you do, then you say the theory is—for now—correct. This is an interpretationless theory that mostly only requires math to work with little or no philosophical interpretation. (I say "for now" to remind us that just because the theory agrees with the experiment in a particular case, it does not mean that there will not be some future experiment that will disagree with the theory.)

Newton's theory of gravity agreed with every astronomical observation ever made for about 150 years, until in 1859 the French astronomer, Urbain Le Verrier, pointed out that Newton's theory did not seem to predict the correct orbit for the planet Mercury.[43] We now know that Newton's theory was wrong, that it disagreed with the experiment in this prediction about Mercury, owing to effects that could only be explained by Einstein's 1915 general theory of relativity (which came to supplant Newton's theory of gravity). The scientific method kicked in: Einstein's theory of gravity could be tested against Newton's, and the data showed Einstein's theory was right and Newton's was wrong. That was it.

Le Verrier, 50 years before Einstein's theory of relativity had been put forth, instead tried to fix this problem by suggesting that there was a small unseen planet he called Vulcan, which orbited the sun inside Mercury's orbit. The gravitational pull of Vulcan was supposedly messing with Mercury and would explain that orbit discrepancy and save Newton's theory. Le Verrier looked and looked and looked for this planet Vulcan in vain until he died in 1877. He was actually convinced by some unrepeatable astronomical observations that Vulcan did exist, and he died believing in it. If you go look at the statue of Le Verrier in the gardens of the Royal Observatory in Paris, on the base of the statue is an engraving of the nine planets (counting Pluto). If you look very carefully, you can see where a 10th planet, Vulcan, has been carefully plastered over—now relegated into the historical dumpster of wrong theories, only to survive as the fictional home planet of Mr. Spock in the television series *Star Trek*. (Live long and prosper.)

The correct fix to Mercury's mercurial orbit was not to introduce a new planet, but a new theory of gravity—Einstein's. We expect that someday, in situations where measurements are made at the very smallest distance scales, Einstein's theory too will fail and will disagree with some future experiment and have to be replaced with some new quantum theory of gravity, such as loop-quantum gravity or—heaven forbid!—superstrings.

In Newton's gravity theory, there are math equations and also philosophical baggage: Time and space are *absolute* and gravity *acts instantly at a distance*.

In Einstein's gravity theory, there are math equations and also philosophical baggage: Time and space are *relative* and gravity acts *at the speed of light* through the curvature of space.

These associated gravity theory baggage are almost universally accepted and used by all physicists studying gravity, and so there is not a lot of discussion about them. It is an anomaly then, in the physics community, that the quantum physicists do not all agree on the same philosophical baggage for the same quantum mathematical equations.

While many are still in the camp of Niels Bohr and the Copenhagen interpretation, with its use of notions of complementarity and wave–particle duality (the electron is both a wave *and* a particle; the teaspoon is both in my coffee cup on my desk *and* in the bathroom sink down the hall), there are many, many newer interpretations with entirely different sets of metaphysical baggage. Take, for instance, the Bohmians (not to be confused with the Bohemians), who are slavishly devoted to David Bohm's theory that electrons are classical particles, like ping-pong balls, which are then guided through space by invisible (and nonlocal) guiding fields, undetectable by any means, which fill all empty space and which instantaneously rearrange themselves out at distances of light-years the moment you adjust a gadget in your laboratory here on Earth.

To my mind, the Bohm theory is not any more satisfying than my magic-dice-throwing-telepathic-gremlin theory. Both the Bohm and the gremlin theories violate the *locality* condition, a violation that we know is needed to get quantum theory to agree with experiment. I can further narrow down my gremlin theory so that it matches the mathematics of the quantum theory at every turn, and then I can also endow the gremlins with the properties that they are invisible and undetectable by any means.[44] This ensures that no experiment can reveal their existence. The gremlin theory, the Bohm theory, and the Copenhagen theory would all agree with all experiments—by design—but the philosophical baggage associated with each theory is radically different.

Such quantum theories, with identical math but different philosophical baggage, are called "interpretations of the quantum theory."[45] The math remains the same, but the number of interpretations continues to grow as each sect and subsect and cult finds some undesirable feature with a previous belief system and goes off on a new tangent. This explains why the Copenhagen no longer gets the majority vote. Which interpretation is really real? This is impossible to tell.

If all the interpretations are hardwired to give the same exact predictions in every experiment, then there is no way to tell them apart at all. The only cleaver we have to split false theories from true theories is the scientific method—to test each against experiment. But if all theories with their different interpretations agree with each other in their predictions in all experiments, then there is no way to distinguish them in an objective way using the scientific method.

This brings us to *The Inverted Earth Society*. A spin-off of *The Flat Earth Society*, the inverted Earthers believe that the Earth is stationary and is the center of the universe. So far so good. But even better, they believe that the Earth is inverted—turned completely inside out around its center—something like a single hollow spherical bubble of air and space inside an infinite block of Swiss cheese. We all live on the inside of the bubble, where the air meets the cheese. The cheese is the dirt of the earth stretching out to infinity with the earth's core spread out like some hot dense hard Swiss cheese coating at the infinity point. Inside the bubble is the atmosphere, and then inside that is outer space (now inner space) and all the planets and the stars and the sun, which move on crazy loopty-loop orbits, whizzing like angry bees around and around inside a hollow spherical beehive. At the very center of the bubble are the farthest galaxies and the end of the observable outer space (see Figure 1.2). You would think it would be easy to disprove this crazy theory but it's not.

The sophisticated inverted Earthers are very smart. They take Newton's mathematics for gravity (and every other theory of physics) and perform a mathematical inside-out transformation around the center of the moving, solid, ball-of-dirt earth we love and know and come up with a crazy inverted

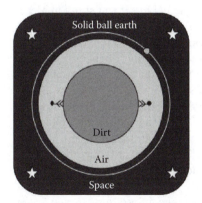

Figure 1.2 The usual solid-ball-earth model (left) is compared to the unusual inverted-hollow-earth model (right). For the solid-ball earth, the infinite outer space with stars and an orbiting satellite is on the outside. Alice in Mururoa and Bob in Mecca stand on the outside of the dirt on the ball on opposite sides of the earth, their heads pointing away from each other, and the dirt makes up the inside. For the inverted-hollow earth, everything is inside out, with infinite dirt on the outside of the hollow-bubble earth. Alice in Mururoa and Bob in Mecca now stand on the inside wall of the hollow earth on the dirt with their heads pointing toward each other. The satellite orbits inside the hole and the distant stars are at now at the center. By construction, no experiment can distinguish which of these models is really real, and so both should be considered viable physical theories as far as the scientific method is concerned.

bubble earth. But they do this in such a way that all the predictions of hollow inverted bubble earth theory *are identical* to those of the regular solid-ball-of-dirt earth theory. The inverted-hollow-bubble earth's Newtonian gravity theory is set up to give exactly the same predictions as the regular solid-ball-of-dirt-earth Newtonian gravity theory we are so fond of.[46] There is no way, by any experiment, to tell the two theories apart!

The claim then of the inverted Earthers is that space and time are curved in such a way—owing to their inside-out math—that when we launch a satellite it only *looks* like it is orbiting around the outside of a solid ball: It really is orbiting around inside a hollow bubble, like a bee buzzing around the inside wall of its hollow hive. And if you start in the holy city of Mecca in Saudi Arabia and drill straight down through the Earth, you still end up coming out in the Pacific on the antipodal point of the unholy island of Mururoa in Polynesia (where the French used to do aboveground nuclear weapons testing). Only now, in the inverted Earth theory, you only *think* you are drilling on a straight line from one side to the opposite side of a solid ball.

What you are really doing is drilling up away from the bubble into the inverted Earth's Swiss cheese on an infinite loop that bends back around at infinity to take you back to the other side of the bubble. If you stand in Mecca and look straight up with a telescope, you would see your radioactive compatriot standing upside down on Mururoa, if only light moved in straight lines at constant speed. But in the inverted Earth theory, your line of sight now stretches off to the edge of the known universe at the center of the bubble and stops. If you shot a laser straight up, the light would move slower and slower and slower until it reached the edge of the universe, at the center of the bubble, where it would come to full stop, and never get to Mururoa on the other side of the bubble. (For that, you have to beam the light straight down into your tunnel, where it curves around the bubble in the dirt at infinity and comes back to Mururoa through the cheese to tickle your radioactive colleague's glowing feet.)

The mathematical bubble earth inversion applies to your compass, inertial guidance systems, light rays, and so forth to fool you into *thinking* you are on the outside of the surface of a solid ball, when in reality you are on the inside of a hollow bubble. The theory can even be improved to include Einstein's relativity theory and get the loopty-loop orbit of Mercury just right as it spirals about the sun inside the bubble like an unstable drunk staggering around and around a moving lamppost.

There is no way to tell, *based on any experiment*, whether the regular solid-ball-earth theory we know and love is right or if the crazy inverted-bubble-earth theory is right!

There is no way to tell, *based on any experiment*, whether the Copenhagen quantum theory or the Bohm quantum theory or the magic-dice-throwing-telepathic-gremlin theory is right!

It appears that Galileo's cleaving sword of truth and falsehood—the scientific method—has failed us. At this juncture, we physicists tend to launch into heuristic and subjective diatribes, such as invoking Occam's razor—the dictum that the simplest theory that explains all the data must be the true theory. But who decides which theory is the simplest? What is "simple" may be very personal and subjective. While the inverted-bubble-earth theory seems very complicated, the inverted Earthers love it and they are quite happy with what they perceive as its beautiful simplicity—dirt to infinity (and beyond!). We may take some solace in the fact that the inverted Earthers are a tiny minority of a dying breed; but that in and of itself does not make them wrong.

The Bohmians are perfectly happy with their classical ping-pong ball–like electrons guided hither and yon by a mysterious and invisible nonlocal guiding field that instantaneously readjusts itself for light-years in all directions at the slightest disturbance. The guiding field reacts like a boundless beached flounder that flails about on the shore of an endless sea ... whenever you drop a screwdriver in your laboratory. Who is to say that the magic-dice-throwing-telepathic-gremlin theory is not simpler than the flailing-flounder theory?

Then there is the Many-Worlds interpretation of quantum theory, which has many adherents, and which proposes that each time Alice looks at her pocket watch, the entire universe separates into 12 nearly identical copies, like a slime mold that undergoes fission into 12 gooey bits. Many universes then arise, each containing different copies of Alice and Bob, all getting different results on their different pocket watches, with statistics just as the Copenhagen or gremlin version of quantum theory would predict. The Many-Worlds interpretation forms a terrific plot for many a *Star Trek* episode,[47] and even a love story,[48] but is it the simplest theory? Is a splitting-slime-mold-like multi-universe theory any *simpler* than a magic-dice-throwing-telepathic-gremlin theory? (It may be easier to slit your own throat with Occam's razor....)

It is indeed a strange business that, in the history of science, not everybody agrees on the one true collection of philosophical baggage that should be bestowed on the mathematical apparatus of the quantum theory (unlike the near-unanimous agreement about gravity theory). This disagreement, at its heart, has to do with the strange properties of quantum theory that violate our common sense and so give us the freedom to cook up interpretations that defy somebody else's common sense. But *somebody's* common sense *always* has to be violated so as to agree with experiment. And because everybody's common sense is somewhat personal, everyone cannot agree on which is the *simplest* quantum theory that defies common sense, and thus the schisms continue unabated.

At conferences on the foundations of quantum theory, I am often asked (quietly in the hallway during the coffee break) which interpretation do I *believe* in? If I say that I'm a born-again Copenhagian, then the Bohmians and Many-Worldsians will howl and flee—spilling the coffee out of their little white Styrofoam cups in disgust. This must be what it feels like to be at an international conference on comparative theology. Should I be an atheist, like Heisenberg, and believe in no interpretation? Such heathens, called operationalists, are treated poorly. It is also a very unsatisfactory state of mind to just compute with the math and then run into the laboratory and check it out—without caring at all about what it all really means. That is a barren state of mind. My own quantum theoretical common sense, my ability to make intuitive predictions on the basis of previous experiences, hinges mightily on having an interpretational framework onto which I can sew my own quantum theoretical clothing. My ability to make new predictions and think of new ideas, then try them out in the laboratory, requires an interpretation to frame the questions and pose the answers. I need an interpretation to do my job—but which one? Pascal's wager comes to mind.

Blaise Pascal, the French philosopher, mathematician, and theologian (of Pascal's triangle fame), once posed a wager stating that it is better to believe in God than not. If you don't believe in God and there is one after all, then when you die you'll be in real trouble for not going to confession and saying your prayers for all those years you were alive and you'll end up going to hell. However, if you do believe in God, and do all that is required to get into heaven, nothing much is to be lost if there is not a God, as when you die you'll not be going anywhere anyway. Nothing to lose!

Here is my problem with Pascal's wager—*which* God should I choose to believe in? There are many religions. Most are mutually exclusive. If you spend your life as a devout Buddhist, then you're not getting into heaven. If you spend your life as a devout Christian, then you will never reach nirvana. There are, sadly, many places in this world where the public proclamation of a belief in the "wrong" God is much more dangerous to one's personal health than quietly admitting a belief in no God at all.

When it comes to interpretations of quantum theory, I am a pantheist. The pantheist approach was the one taken by the character Beni Gabor (played by Kevin J. O'Connor) in the 1999 remake of the movie, *The Mummy*. When Beni first encounters the reanimated mummy of the High Priest Imhotep, Beni pulls out from under his shirt a plethora of grimy religious talismans and begins praying aloud with each of them proffered up, one after the other, to ward off the mummy. Beni finally hits upon the Star of David with a Hebrew prayer, which the mummy recognizes, and instead of sucking out his life force, the mummy turns him into a slave. (Spoiler alert! Beni *is* eventually eaten alive by a swarm of giant carnivorous scarab beetles, but that doesn't happen until the end of the movie.)

Pantheism is not without its pitfalls, but unlike Pascal's original wager, you are hedging more of your bets. I'm happy to revel in the Copenhagen interpretation on Monday, Wednesday, and Friday; exploit the Many-Worlds interpretation on Tuesday, Thursday, and Saturday; and on Sunday turn in desperation to Bohm theory. Unlike Pascal's wager, however, I have nothing to lose by this strategy and everything to gain. Often a perplexing quantum problem becomes completely clear to me when I switch from Copenhagen to Many Worlds. (Nothing is ever cleared up when I resort to Bohm theory, but then I usually sleep in on Sundays.) I don't believe any of them!—or perhaps I believe all of them. The physicist Edwin T. Jaynes once said, "So long as I can use and teach a physical theory, I don't have to believe it." Well played!

THE CAT IN THE APP

After the appearance of the EPR paper, also in 1935, the Austrian quantum physicist (and notorious lothario) Erwin Schrödinger wrote his own follow-up paper, which was supposed to be an extension and clarification of the points made by Einstein, Podolsky, and Rosen. In reality, Schrödinger was also greatly discomfited by quantum theory—a theory that he also helped to invent (Schrödinger's equation)—and this follow-on paper to EPR was more of a litany of his own personal complaints about the situation. Translated from German into English, the paper's title is, "The Present Situation in Quantum Mechanics."[49] In this paper, Schrödinger introduced two concepts that are now mainstays of modern quantum physics perplexities: the notion of *quantum entanglement* and paradox of *Schrödinger's cat* (see Figure 1.3).

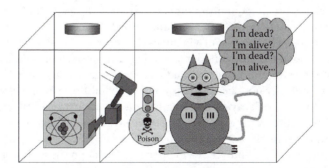

Figure 1.3 Schrödinger's cat is on the right of the box. On the left is an atom that has a 50–50 probability of decaying and emitting an alpha particle that triggers an electric relay (lightning bolt) that activates the gearbox that causes the hammer to smash the flask filled with hydrogen cyanide thus killing the cat. After 1 minute, the atom is in a 50–50 superposition of decayed and not decayed and thus the cat is in a 50–50 superposition of dead and alive. Observers may open the portholes on the top of the box and observe either the state of the atom (left) or the state of the cat (right).

The cat thought experiment goes like this: Suppose you have a radioactive atom, a flask of cyanide, a hammer, and a live cat, all locked up together inside an airtight steel box. The radioactive atom has a half-life so that, after an hour, the atom has had a 50–50 probability of undergoing nuclear decay and emitting a charged alpha particle. A steel door, through which the hammer acts via an electronic relay, separates the decaying atom from the cat with the flask of cyanide. There are also two small portholes, with opaque snap-on covers, in the top of the box. The left porthole allows you to look at the atom (with an electron microscope) and see if it has decayed or not. The right porthole allows you to look at the cat and see if it has died or not.

If the atom *does* emit a charged alpha particle in that hour, then a Geiger counter will detect the alpha particle and activate a relay that flips a switch that releases the latch on a Rube-Goldberg-like contraption that causes a hammer to smash the flask so that the cyanide kills the cat. If the atom does not emit that charged alpha particle in that hour, then nothing happens, and the cat is still alive. At the end of the hour, Erwin randomly peeks into either one of the portholes. According to quantum theory, *if the box is isolated from the entire universe from the beginning until the end of the hour*, then we must conclude four things, according to the standard Copenhagen interpretation of quantum theory:

1. At the end of the hour, before anybody looks in the portholes, the atom is in a quantum bipolar state of 50% decayed and 50% not decayed. Hence, its state is *uncertain* and *unreal.*

2. At the end of the hour, *before anybody looks in the portholes*, the cat is consequently in a quantum bipolar state of 50% dead and 50% alive. Hence, its state is also *uncertain* and *unreal.*

3. At the end of the hour, if Erwin opens the cover on the *left* porthole and looks *only* at the atom, the act of looking at the atom will instantly collapse the atom into a *certain* and *real* condition of decayed (with 50% probability) or not decayed (with 50% probability). If Erwin observes the atom to be decayed, that measurement will instantly kill the cat. If Erwin observes the atom to be not decayed, that measurement will instantly yank the cat out of limbo and back to life.

4. At the end of the hour, if Erwin opens the cover on the *right* porthole and looks only at the cat, the act of looking at the cat will instantly collapse the cat into a *certain* and *real* condition of dead (with 50% probability) or alive (with 50% probability). If Erwin observes the cat to be dead, this measurement will instantly collapse the state of the atom into "decayed." If Erwin observes the cat to be alive, this measurement will instantly collapse the state of atom into not decayed.

This is the atom,
That triggered the counter,
That smashed the flask,
That killed the cat,
That sat in the box that Erwin built.[50]

Erwin Schrödinger had a number of issues with his diabolical machine. All three of Einstein's issues from EPR are fully in play here. The quantum theoretical *uncertainty* condition states that it is impossible to predict with certainty in any run of this experiment whether the cat is dead or alive at the end of the hour. Kitty is both dead *and* alive until you look at it and then it collapses uncontrollably into either dead *or* alive with 50–50 odds. Again, compare this to a classical hidden variable theory, which states that the cat is definitely alive or definitely dead but we just don't know which until we look. In quantum theory, the act of looking at the cat instead changes its quantum mechanical state suddenly—its state collapses. In the classical hidden variable theory, the act of looking at the cat changes nothing but our own knowledge of the cat's state (like our classical game of heads or tails).

The *unreality* condition is here too—the cat is both dead *and* alive until you look. *Nonlocality* is less obvious, but in principle you could, at the beginning of the hour, carefully cleave the box in two—without in any way disturbing the innards—and put the right-hand side of the box with the decayed and undecayed atom on a plane headed west and the left-hand side containing the half-dead half-alive cat on a plane headed east. After the hour, the two planes would be approximately 1600 kilometers (1000 miles) apart and still a flight attendant in the west, peeking in at the atom instantly and *nonlocally*, murders (or resurrects) the cat in the east.

Schrödinger also brought up his own new complaint. Why are quantum theorists perfectly happy about tiny things like atoms being in a superposition of decayed *and* undecayed or itty-bitty ions in a superposition of pointing simultaneously up *and* down, but the same theorists recoil when this language is applied to a large object like a cat? There is a subtle new attack on quantum theory in this atom versus cat business. In physics, whenever a new theory replaces an old theory, it is typical for the new theory to explain everything in the old theory and a few new things the old theory did not. For example, when Einstein's theory of gravity replaced Newton's, it could be shown that Einstein's new theory predicted everything Newton's old theory did, but also a few more things that Newton's theory did not—such as the whacky orbit of Mercury that so plagued the planet Vulcan–obsessed Le Verrier.

Compare that gravity theory situation to quantum theory, which now explains the behavior of atoms correctly in a way that the old Newtonian theory could not. Good. But then quantum theory does not apparently explain the behavior of such things as cats, which the old Newton theory handled just fine. This is not a satisfactory state of affairs, opined Schrödinger, if the new theory explains new things but cannot handle some of the old things.

A number of modern explanations that resolve this Schrödinger's cat paradox do exist; the most popular invokes practical issues such as the environment surrounding the box. In order to carry out Schrödinger's thought experiment, it is critical that the cat and the box be *totally isolated from the environment.* No heat from the cat's body can escape, no molecules from the air can strike the box, and no cosmic rays from outer space can pierce through it. Arranging such a protective shield from the environment is technically fairly easy to do for an atom but extraordinarily and technologically difficult to do with a cat. The least interaction with the outside world (say, a cosmic ray shooting through the box) acts like an inanimate "observer" that carries away information about the state of the cat. Each piercing cosmic ray, each thermal photon the cat emits, carries a little information about whether the cat is alive or dead into the environment. The environment acts like an inhuman observer that takes in all this information then impersonally and very, very, very rapidly collapses the state of the cat into either dead *or* alive—long before we have a chance to look through the porthole—and so fast that we cannot see the effect of the cat being dead *and* alive on any human time scale.

Roughly, this environmentally induced collapse takes place at a rate that is exponentially dependent on the size of the object—the bigger the cat, the faster it collapses on its own into dead *or* alive. For an atom, such environmental degradation (with good isolation) can be forestalled for seconds or minutes—enough to record the behavior of an atom existing as both decayed and not decayed simultaneously. For a large object like a cat, the speed of the collapse in the box, unprotected from the environment, is less than a septillionth of a second—too small for any human measuring device to record. All we see is the end result—the cat is either definitely dead *or* definitely alive. Never do we see signs of it being both dead *and* alive, as the environment has removed this information long, long, long before we have had a chance to look at it or do anything else. (Schrödinger's cat experiments have been done with medium-sized objects, such as laser beams and large molecules, where the laser beam and the molecule is a stand-in for the cat. Such medium-sized "kittens" do indeed decay very quickly upon contact with the environment, but they are not so big, and we can see them for a short period being dead *and* alive and watch them roll over quickly into dead *or* alive.)

In the end, Schrödinger pointed out that the essence of the strangeness of quantum mechanics was tied up with this strange quantum mechanical state of affairs, where the state of the radioactive atom was intimately and strongly linked with the state of the cat. Writing in German, Schrödinger gave this strong quantum mechanical correlation between the atom and the cat a name: *Verschränkung!* This is a German word that is colloquially used to describe tightly folded or crossed arms. Reasonably good English translations are "interleaved" or "interconnected" or even "entwined," like the orderly weaving of strands of hemp to make a sturdy rope. However, later that same year, Schrödinger himself translated *Verschränkung* into English as "entanglement."[51] This might not have been the best translation, as it suggests a much less orderly state of affairs. (I typically use "entanglement" in the context of a ball of fishing line in an untidy tackle box or the jumble of Christmas lights snarled inside a crate in my attic. What is going on between the atom and the cat, or Alice and Bob's two pocket watches, although very strange, is much more orderly and simple and well described mathematically than the Christmas lights in my attic.)

Interestingly, when you translate "entanglement" back into German, the first definition is not Schrödinger's original *Verschränkung*, but instead *Verfangen* ("inter-gripped"), and the second is *Durcheinander* ("in through each other")—like crossed arms. Schrödinger's *Verschränkung* is the *last* definition in the English–German dictionary, with a warning that it is "physics jargon." So *Verschränkung* was mistranslated into English as "entanglement," which then took on a specific quantum physics meaning, and then was translated back into the German as *Verschränkung*, but now only in the case when it has a particular physics definition.[52]

Nevertheless, the English word stuck, and *quantum entanglement* is now used to describe this strange type of two-body correlation—an orderly but much stronger correlation than allowed by classical theory.[53] The quantum mechanical states considered by Einstein, Rosen, and Podolsky in 1935 were just such entangled states in the position and momentum degrees of freedom of an imaginary particle in their thought experiment. In Schrödinger's 1935 follow-up paper, the entanglement was between the atom and the cat. Schrödinger wrote, "Quantum entanglement is the characteristic trait of quantum mechanics, the one that enforces its entire departure from classical lines of thought."[49] In our story above about the quantum pocket watches, the two ions prepared by Dave at NIQuIST in a *particular quantum mechanical state of affairs* in Alice and Bob's quantum pocket watches were in what the cognoscenti would call a two-particle *spin-entangled* state, which we can write in shorthand notation like this: $|\uparrow\rangle_A |\downarrow\rangle_B + |\downarrow\rangle_A |\uparrow\rangle_B$.

(We'll denote classical states or outcomes with parentheses (↑) and quantum states with the funny angular brackets $|\uparrow\rangle$. The angular bracket is actually a quantum notation called a "ket" introduced by the British quantum physicist Paul Adrien Maurice "P.A.M." Dirac and has a special meaning in quantum theory, but here we just use it as a bookkeeping device to track quantum versus classical states. The state $\langle\uparrow|$ is called a "bra" and together they make a $\langle\uparrow|\uparrow\rangle$ or "bra-ket" or "bracket.") We can visualize the ions as little spinning magnets, like miniature earths, with the arrows pointing in the direction of their north poles—up or down. The letters A and B stand for Alice and Bob's pocket watch, respectively. This notation suggests the quantum *unreality* principle that we cannot know if Alice's ion is definitely pointing up and Bob's is definitely pointing down, $|\uparrow\rangle_A |\downarrow\rangle_B$, or the reverse, $|\downarrow\rangle_A |\uparrow\rangle_B$. We only know that if one is up the other is down and vice versa—they are anti-correlated.[54] The notation also encapsulates the quantum *uncertainty* principle: If Alice and Bob both choose (perhaps independently) to measure the ions in the vertical direction ($\updownarrow$), Alice will randomly (with a 50–50 probability) get up ($\uparrow$) and Bob will get down ($\downarrow$). With the same probability, Alice will get down and Bob will get up. According to quantum theory, there is no way in any single experiment to predict with certainty *who* will get up and *who* will get down—quantum theory only gives you the 50–50 odds. God *is* playing dice.

The quantum *nonlocality* principle is here as well. Alice could be on Alpha Centauri and Bob on Beta Pictoris, and still the measurement by Alice that randomly collapses her ion to down instantaneously collapses Bob's ion to up— even though Bob is tens of light-years distant from her.

Mathematically, the state $|\uparrow\rangle_A |\downarrow\rangle_B$ is *un*-entangled because it can be cleanly separated into Alice's ion state $|\uparrow\rangle_A$ alone, which is definitely and really pointing up at all times, and into Bob's ion state $|\downarrow\rangle_B$ alone, which is definitely and really pointing down at all times. The state $|\uparrow\rangle_A |\downarrow\rangle_B + |\downarrow\rangle_A |\uparrow\rangle_B$ is *entangled* because it cannot be split cleanly in two—Alice's bit cannot be separated from Bob's bit— that very important plus sign in the middle (that you should read as "and") gets in the way. This signifies that Alice's ion is unrealistically pointing *both* up *and* down and Bob's is unrealistically pointing *both* down *and* up.

NOTES

1. A modern and reasonably accessible book is *Foundations of Quantum Mechanics: From Photons to Quantum Computers* by Reinhold Blümel (Jones and Bartlett, 2010), http://www.worldcat.org/oclc/319498438. See also *Schrödinger's Machines: The Quantum Technology Shaping Everyday Life* by Gerard J. Milburn (W.H. Freeman, 1997), http://www.worldcat.org/oclc/36083605. For a somewhat out-of-date introduction to the quantum computer in particular, see *The Feynman Processor: Quantum Entanglement and the Computing Revolution*, also by Gerard J. Milburn (Basic Books, 1998), http://www.worldcat.org/oclc/40002991.

2. Einstein was born a German citizen in Ulm, Germany. The house of his birth was destroyed in World War II and replaced with a McDonald's hamburger restaurant but now a plaque on the wall commemorates his birthplace. However, Einstein (a lifelong pacifist) renounced his German citizenship in 1896 at the age of 17 in protest of the militaristic policies of the Prussian-dominated German government and perhaps also to avoid German military service. Effectively stateless for 5 years, he was eventually granted Swiss citizenship in 1901. He remained a Swiss citizen until

the end of his life but became a dual Swiss–American citizen in 1940 after his immigration to the United States. Since Einstein went to so much trouble to identify himself as Swiss (and not German), I follow his wishes and herein also identify him thusly. Einstein, who was Jewish, was poorly treated by the Nazis and fled Germany in 1933 when Adolf Hitler became Chancellor. In April of 1933, the Nazi government passed a law prohibiting Jews from holding government jobs, including university professorships. Einstein, who was on travel in the United States at that time, was then a professor at the Humboldt University of Berlin. He never again returned to Germany and eventually settled at the Institute for Advanced Study in Princeton, New Jersey, where he remained for the rest of his life. In my opinion, the best biography of him is *Einstein: His Life and Times* by Walter Isaacson (Simon and Schuster, New York, 2007), http://www.worldcat.org/oclc/76961150. Most of Einstein's personal letters and correspondence were sealed for 50 years after his death, on April 18, 1955, and so this biography by Isaacson is one of the first to have access to those papers once they became available in 2005.

3. Never expect a Dutchman to pronounce "Christiaan Huygens" without expectorating.
4. I would not ask a Dutchman to pronounce this either, unless you provided him with the English version: *Proceedings of the Prussian Academy of Sciences*.
5. See the nice overview of Bose–Einstein condensation at the University of Colorado website: http://www.colorado.edu/physics/2000/bec/.
6. See "Probing the Limits of the Quantum World" by Markus Arndt, Klaus Hornberger, and Anton Zeilinger in *Physics World*, Volume 18 (2005) pages 35–40, http://www.tiptop.iop.org/full/pwa-pdf/18/3/phwv18i3a28.pdf.
7. For more details on the Einstein–Bohr debates, see *Einstein, Bohr and the Quantum Dilemma: From Quantum Theory to Quantum Information* by Andrew Whitaker (Cambridge University Press, 2006), http://www.worldcat.org/oclc/63186198. A slightly more tractable introduction, *Spooky Physics: Einstein vs. Bohr* by Andrea Diem-Lane (MSAC Philosophy Group, 2008), http://books.google.com/books?id=YTZ3lG0MFqoC, is available in paperback or as an electronic book.
8. "Turn *right!* Turn *right!* Turn *right!* Recalculating...."
9. See *Schrödinger's Killer App: Race to Build the World's First Quantum Computer* by Jonathan P. Dowling (Taylor and Francis Press, 2013), page 5.
10. Planck's constant h is equal to 0.000000000000000000000000000000000662606896 in metric units of mass (kilograms) multiplied by velocity (meters per second) multiplied by distance (meters). In any case, it is very, very, very small in these human-sized units. It is because h is so small that quantum mechanical effects in ordinary life are so hard to observe, and you have to go down to molecules and atoms and electrons to notice quantum effects. There is a story by the Russian physicist, George Gamow, about a fellow named Mr. Tompkins, who goes on a safari to a jungle in Africa where Planck's constant is around 1 (instead of that small number) and quantum effects become visible to the naked eye. This story is titled, "The Quantum Safari," which can be found on page 113 in the collection *The New World of Mr. Tompkins* written by George Gamow and edited by Russell Stannard (Cambridge University Press, 1999), http://books.google.com/books?id=rX6K6MvOGzMC. In one scene, where Mr. Tompkins is about to mount a quantum elephant in order to go hunt a quantum lion, Gamow tells us, "In the quantum jungle, on the other hand,

Planck's constant is large. But even there it is not large enough to produce striking effects in the behaviour of such a heavy animal as an elephant. The uncertainty of the position of the quantum elephant can be noticed only by close inspection."

11. A "thought experiment" or a *Gedanken* experiment (*Gedanken* is the German word for thought) is a simplified set of mental gymnastics that is used to illustrate, explain, or—in this case—discredit a scientific hypothesis or theory. Typically, thought experiments are not something that can be easily done in the laboratory. Curiously, there is an Israeli physical chemist from Bar-Ilan University named Aharon Gedanken, who is indeed an experimentalist, http://ch.biu.ac.il/gedanken. So whenever he does an experiment, it is in fact a *Gedanken* experiment. To be perfectly clear, *Gedanken* experiments are not named after Prof. Gedanken, and Prof. Gedanken is not named after *Gedanken* experiments. It is all just a happy coincidence.

12. $E = mc^2$. In 1980, my first year in graduate school, the English physicist, Paul Dirac, Nobel Prize 1933 awardee, came to the University of Colorado to give a popular talk at The Gamow Memorial Lecture. As a big fan of Dirac, I dragged all my non-physicist friends to the "popular" lecture early to get good seats in the middle and second row from the front. The place was packed with the mayor, the chancellor, the provost, the deans, all the physics professors, a blonde woman from the Sufi community dressed in a turban and a white cloak sporting ceremonial dagger in her waistband, and so forth. (This is Boulder, Colorado, after all.) Dirac gave what I thought was a very interesting talk on the history of quantum theory, but with no slides, no notes, no audiovisual aides, and no nothing. He just stood at the podium and talked for an hour. He was 78 years old at the time and he spoke in a very soft high-pitched, English-accented, mouse-like voice. So soft it was that you could barely hear him at all and the technicians kept cranking up the amplifiers until it screeched periodically from the feedback. The talk put all the non-physicists in the audience immediately to sleep. Then Dirac got to the part where he discovered the Dirac equation predicting the existence of antimatter. He clearly gets a bit excited and impossibly goes up an octave, whereupon the feedback kicks in waking everybody up, and Dirac says, "I was led to the idea of the discovery of antimatter by considering Einstein's most famous equation, $E = ...$" All my buddies from the English department began to nudge me and the crowd visibly perked up. They had not understood a goddamn thing but for sure even the English majors knew what "...Einstein's most famous equation, $E = ...,$" was going to be. Dirac continues triumphantly onward to the hushed auditorium, $E = ... \pm \sqrt{p^2 c^2 + m^2 c^4}$! (Einstein's *least* famous equation?) The audience visibly collapsed upon themselves in utter disappointment—they understood *nothing*—and I in the tomb-like quiet that followed in the hallowed Rocky Mountain granite of the vast Macky auditorium— burst out laughing uncontrollably. (And I was the only one.) Dirac, normally an endearing bird-like little man, scowled, halted the talk, stepped out from behind the podium, and stared down at me in silence, vulture-like, for a full minute. The rest of the audience looked back and forth between Dirac and me as they coughed and inspected their watches. Then, without a word, after my torturous minute was up, he returned behind the podium and finished his talk as if nothing had happened at all.

13. See Einstein as "Person of the Century," *Time Magazine*, Volume 154, Number 27 (December 31, 1999), http://www.time.com/time/magazine/article/0,9171,99301 7,00.html.

14. See "Can Quantum-Mechanical Description of Reality Be Considered Complete?" by Albert Einstein, Nathan Rosen, and Boris Podolsky in *Physical Review*, Volume 47 (1935), pages 777–780, http://link.aps.org/doi/10.1103/PhysRev.47.777. This title should have probably been Americanized with a definite article as "Can *the* Quantum-Mechanical Description of Physical Reality Be Considered Complete?" Only Nathan Rosen, born in Brooklyn, New York, was a native speaker of English. It is very unusual for the title of a formal scientific article, particularly in *Physical Review*, to end with a question mark. (Typically, the editorial staff frowns on such hypothetical questions in the title.)

15. The EPR paper, with over 4500 citations (over 300 in the year 2010 alone), is *by far* Albert Einstein's most cited work! My source for these numbers is the *Thomas Reuters Web of Science* (formerly known as the *Science Citation Index*) with data pulled on April 5, 2011.

16. See "EINSTEIN ATTACKS QUANTUM THEORY; Scientist and Two Colleagues Find It Is Not 'Complete' Even Though 'Correct'" in the *New York Times* (May 4, 1935), page 11, http://select.nytimes.com/gst/abstract.html?res=F50711FC3D58167A93C 6A9178ED85F418385F9.

17. The Podolsky incident is discussed in *The Historical Development of Quantum Theory* by Jagdish Mehra and Helmut Rechenberg, Volume 6 (Springer, 2001), page 724, http://www.worldcat.org/oclc/7944997, and in *The Philosophy of Quantum Mechanics* by Max Jammer (Wiley, 1974), pages 189–194, http://www.worldcat.org/ oclc/969760. Both of these books have a nice overview of this entire business of Einstein versus quantum theory, but the Jammer book is perhaps more accessible, as it comes in only *one* volume instead of *six*. Perhaps the most accessible is his much smaller book, *The Conceptual Development of Quantum Mechanics* by Max Jammer (McGraw-Hill, 1966), http://www.worldcat.org/oclc/534562, which is out of print, but which can be often found in libraries or used online.

18. For more on radio-controlled atomic wristwatches and clocks, see http://www.nist. gov/pml/div688/grp40/radioclocks.cfm.

19. The radio signal travels farther at night when the Earth's upper atmosphere, the ion-osphere, is less noisy when it is not being buffeted about by radiation from the Sun.

20. Bob's watch may take a licking but Alice's keeps on ticking.

21. For more about these great (and not so great) schisms of quantum interpretations, see my popular article, "Interpreting the Interpretations," in *Physics World*, Volume 14, November 2001, http://physicsworldarchive.iop.org/full/pwa-pdf/14/11/phwv14i11a36.pdf.

22. See "NIST Employees Contaminated by Plutonium" by Ivan Monroe of *The Associated Press*, as reported in *The Denver Post*, July 11, 2008, http://www.denverpost.com/ ci_9726821. The version I heard was that a postdoctoral researcher was tamping a quartz vial full of plutonium powder against a laboratory table when it broke open. He dutifully swept up the plutonium in a dustpan, washed it all down the sink, put the dustpan back in the broom closet, washed his hands, and then went to lunch. Apparently, his supervisor never told him what to do in case of a plutonium spill.

23. This illustration works better if I use the old style pocket watch with analog hands as opposed to the digital pocket watch of the new millennium—the iPhone—which has replaced the wristwatch as the personal timekeeping device du jour—at least among my students. It is telling that in the 20th century, wristwatches supplanted pocket watches, which were popular in the 19th century, only to be supplanted yet again in the 21st century by the iPhone and other such gadgets, which we now again often just use as pocket watches.

24. There is no official subdivision of NIST that is called NIQuIST (but there should be). The joke is that Nyquist noise, first described by Harry Nyquist at Bell Labs in 1928, is a type of electronic noise that plagues the accuracy of the actual NIST trapped-ion atomic clocks.

25. A calcium *ion* is a calcium *atom* that has had its outermost electron stripped off, rendering the atom positively charged, after which it is called an ion instead of an atom. This positive charge allows the atom to be easily trapped using electric fields. There is a David Wineland, a real person and leader of the NIST Ion Storage group (and 2012 Nobel Laureate in physics), who makes some of the world's best atomic clocks using these calcium ions in just such ion traps.

26. For simplicity, we'll confine the ion direction to two dimensions so its north pole is always pointing in the plane of the watch face when measured. In reality, the ion's north pole can point in any direction in three dimensions: north, south, east, west, up, or down. We'll assume here that the Earth is flat, that up and down is prohibited, and that Alice and Bob always hold their pocketwatches horizontal with the ground before popping the lids. This is the fun of thought experiments; we can put in any assumptions we like!

27. For the quantum cognoscenti, I'm proposing that they are prepared in an entangled two-particle spin singlet state of the form $|\uparrow\rangle_A |\downarrow\rangle_B + |\downarrow\rangle_A |\uparrow\rangle_B$. I'm treating the Bohm version of the EPR paradox, in terms of entangled spins rather than the original with entangled momentum and position. I'm not discussing all four possible Bell states. Also, I should really be using a singlet state with a minus sign, instead of a triplet state with a plus sign, but that's life.

28. This is what Einstein referred to, in German, as "*spukhafte Fernwirkung*" or "spooky action at a distance," a phrase that appears in a letter by him to the quantum physicist Max Born, which can be found in *The Born–Einstein Letters, 1916–1955: Friendship, Politics and Physics in Uncertain Times* by Albert Einstein and Max Born (MacMillan, 2005), http://www.worldcat.org/oclc/539202043.

29. See "Experimental Violation of a Bell's Inequality with Efficient Detection" by Mary A. Rowe, David Kielpinski, Volker Meyer, Cass Sackett, Wayne M. Itano, Christopher Monroe, and David J. Wineland in *Nature*, Volume 409 (2001), pages 791–794, http://www.nature.com/nature/journal/v409/n6822/abs/409791a0.html.

30. See "Violation of Bell Inequalities by Photons More Than 10 km Apart" by Wolfgang Tittel, J. Brendel, H. Zbinden, and Nicolas Gisin in *Physical Review A*, Volume 57 (1998), pages 3563–3566, http://link.aps.org/doi/10.1103/PhysRevLett.81.3563.

31. See "Is the Moon There When Nobody Looks? Reality and The Quantum Theory" by N. David Mermin (where N is large) in *Physics Today*, April 1985, page 38, http://dx.doi.org/10.1063/1.880968.

32. Technically, this is called an *anti*-correlation, as Bob's hand is always pointing at the antipodal point in the *opposite* or anti-direction as Alice's.

33. My character Dave is purely fictional and any resemblance to David J. Wineland, 2012 Nobel Laureate and the leader of the NIST Ion Storage Group, is completely coincidental.

34. My character Charlie is purely fictional and any resemblance to Dr. Charles H. Bennett, a fellow in the Quantum Information and Computation Theory group at IBM Research or, for that matter, Charles "Krazy Horse" Bennett, the professional street fighter and member of the Mixed Martial Arts Association, is completely coincidental.

35. See "On the Einstein–Podolsky–Rosen Paradox" by John S. Bell in *Physics*, Volume 1 (1964), pages 403–408; this paper can be found widely online by searching for the title (in quotes) but on somewhat sketchy websites so I dare not to link to them here. Also see the longer version, "On the problem of hidden variables in quantum mechanics," by John S. Bell in *Reviews of Modern Physics*, Volume 38 (1966), page 447, http://link.aps.org/doi/10.1103/RevModPhys.38.447. A more tractable and fanciful discussion can be found in "Bertlmann's Socks and the Nature of Reality" by John S. Bell in the *Journal of Colloquial Physics*, Volume 42 (1981), page C2-41–C2-62, http://hal.archives-ouvertes.fr/jpa-00220688/en/. In the profound words of John Bell, we have the following: "Dr. Bertlmann likes to wear two socks of different colours. Which colour he will have on a given foot on a given day is quite unpredictable. But when you see ... that the first sock is pink you can be already sure that the second sock will not be pink. Observation of the first, and experience of Bertlmann, gives immediate information about the second. There is no accounting for tastes, but apart from that there is no mystery here. And is not the EPR business just the same?"

36. You think I'm making this up? See *Driving Mr. Albert: A Trip Across America with Einstein's Brain* by Michael Paterniti (Delta Publishing Group, New York, 2001), http://www.worldcat.org/oclc/43790771. To quote from the book's synopsis, "Albert Einstein's brain floats in a Tupperware bowl in a gray duffel bag in the trunk of a Buick Skylark barreling across America. Driving the car is journalist Michael Paterniti. Sitting next to him is an eighty-four-year-old pathologist named Thomas Harvey, who performed the autopsy on Einstein in 1955—then simply removed the brain and took it home. And kept it for over forty years." (Random House Digital, Inc., 2001), http://www.randomhouse.com/book/127990.

37. The Scientific Method in a Nutshell: Construct a hypothesis and then test to see if your hypothesis is true or false by conducting an experiment. Is the Moon made of green cheese? No, Buzz Aldrin went to the Moon and reported that it is made of rocks. Is the sky on Mars blue? No, the Jet Propulsion Lab sent a camera to Mars and it observed that the sky on Mars is the color of butterscotch candy.

38. Legend has it that Galileo carried out this experiment from the tower with a one-half-kilogram (one-pound) weight and a five-kilogram (10-pound) weight. Aristotle predicted that the 10-pound weight would fall 10 times faster but instead the two weights hit the ground at nearly the same time. Score: Scientific Method 1; Galileo 1; Aristotle 0, http://www.pbs.org/wgbh/nova/pisa/galileo.html.

39. Twelve-sided dodecahedral dice are used in the board game *Dungeons and Dragons*. In my wasted youth in Texas, I spent many an hour playing *Dungeons and Dragons*, while throwing such curious dice in battles with dungeon master and now video game designer, Richard Garriott, son of the Skylab astronaut, Owen Garriott. In a break from gaming at the Garriott household one evening, in the summer of 1979,

Astronaut Owen Garriott solemnly summoned me into his study, where the walls were festooned with NASA memorabilia, and he confided in me his worries about his son Richard. Owen was concerned that Richard was only interested in fiddling around with computers and playing *Dungeons and Dragons* and was not very serious about his studies in college. Owen asked me to have a little pep talk with Richard on this matter, which I did, and which thankfully had no effect on him. Soon afterward, Richard combined his two interests to develop the wildly popular *Ultima* computer game series (basically *Dungeons and Dragons* on the computer) that quickly made Richard and Owen and the rest of the Garriott family multimillionaires. I, on the other hand, still have some very nice plastic 12-sided dice.

40. See the article "Bohmian Mechanics" in the online *Stanford Encyclopedia of Philosophy* by Edward N. Zalta, principal editor (Metaphysics Research Lab, Stanford University, April 19, 2011), http://plato.stanford.edu/entries/qm-bohm/.

41. See, for example, the science fiction novel, *Time for the Stars*, by Robert A. Heinlein (Scribner's, 1956), http://www.worldcat.org/oclc/471210. The plot revolves around pairs of telepathic twins, separated in their youth, and stationed on Earth interstellar spaceships traveling at near the speed of light. The goal was to evade the upper limit imposed by the speed of light on communications. The twins in the novel communicate telepathically and instantaneously across distances of many light-years, allowing the ships to stay in constant contact with the Earth.

42. Arthur C. Clarke's Third law, as first proposed in the short story *The Sorcerer of Rhiannon* by Arthur C. Clarke in *Astounding Magazine*, February 1942, page 39.

43. "A statue of Le Verrier ... was erected in 1888 in the north court of the Paris Observatory. [The statue] contained on its pedestal a representation of the solar system, including Vulcan. However, Vulcan was subsequently rubbed out." See *In Search of Planet Vulcan: The Ghost in Newton's Clockwork Universe* by Richard Baum and William Sheehan (Basic Books, Oxford, 2003), page 230, http://books.google.com/books?id=PO7yC6BtQ54C.

44. For a popular account of the scientific method for lay people, see Carl Sagan's *The Demon-Haunted World—Science as a Candle in the Dark* (Random House, New York, 1996), http://www.worldcat.org/oclc/32855551. See particularly, starting on page 171, the discussion of Sagan's invisible fire-breathing dragon that lives in his garage and is undetectable by any means; "... what's the difference between an invisible, incorporeal, floating dragon who spits heatless fire and no dragon at all?" In the context of my discussion, the question is: What's the difference between the magic-dice-throwing-telepathic-gremlin interpretation of quantum theory and the Many Worlds interpretation? Neither can be ruled in or ruled out by experiment alone.

45. A more time-stable reference of all the interpretations I give here can be found in *The Philosophy of Quantum Mechanics* by Max Jammer (Wiley, 1974), http://www.worldcat.org/oclc/969760.

46. See "The Hollow Earth: A Maddening Theory That Can't Be Disproved" by Scott Morris in *Omni Magazine* (October 1983), page 128. There do not seem to be any official *Omni Magazine* archived versions of this article online but it's reprinted accurately here, http://www.skepticfiles.org/ufo1/theory.htm.

47. The first of which was "Mirror, Mirror," an episode of *Star Trek: The Original Series*, second-season episode, #33, production #39, broadcast for the first time on October 6, 1967, written by Jerome Bixby and directed by Marc Daniels. In it, Captain Kirk

and Mr. Spock are swapped with their evil twins from a parallel universe. That particular theme appears again in other later episodes of *Star Trek: Deep Space Nine* and *Star Trek: Enterprise*. A different take on the subject altogether, explicitly quoting the "quantum multiverse," appeared in "Parallels," *Star Trek: Next Generation*, season 7, episode 11, originally aired on November 23, 1993, written by Brannon Braga and directed by Robert Wiemer. There is an unforgettable scene when the Starship *Enterprise* encounters hundreds of copies of itself, each copy from a different parallel universe, with a slightly different history and a slightly different crew.

48. In this short story, Bill and Lorraine are separated by death in a traffic accident into two separate parallel universes—only to be eventually and joyfully reunited in one of those universes through the wonders of quantum theory and the power of true love. See "Divided by Infinity" by Robert Charles Wilson in *Starlight 2*, edited by Patrick Nielsen Hayden (Tor Books, 1998), http://www.tor.com/stories/2010/08/divided-by-infinity.

49. Erwin Schrödinger, "Die gegenwärtige Situation in der Quantenmechanik," in *Naturwissenschaftern*, Volume 23 (1935), pages 807–812, 823–823, and 844–849. English translation: "The Present Situation in Quantum Mechanics," by John D. Trimmer, in *The Proceedings of the American Philosophical Society*, Volume 124 (1980), pages 323–338, reprinted in *Quantum Theory and Measurement*, by John Archibald Wheeler and Wojciech Hubert Zurek (1983), page 152, http://www.jstor.org/stable/986572.

50. Adapted from the nursery rhyme "The House That Jack Built" in *The Oxford Dictionary of Nursery Rhymes*, edited by Iona Opie and Peter Opie (Oxford University Press, 1951, 2nd Edition, 1997), pages 229–232, http://www.worldcat.org/oclc/38119972.

51. See "Discussion of Probability Relations between Separated Systems" by Erwin Schrödinger in the *Mathematical Proceedings of the Cambridge Philosophical Society*, Volume 31 (1935), pages 555–563, http://journals.cambridge.org/action/displayAbstract?aid=1737068.

52. Many thanks to my teutonically bilingual friend and colleague, Prof. Dr. Dr. Barbara Höling, at the California State Polytechnic University in Pomona, for pointing out to me this historical inconsistency with Schrödinger's own translation of his word *Verschränkung* into English as entanglement. Barbara Höling, private communication (2009). English not his first language, I can only *surmise* that Schrödinger mistranslated it. Perhaps he just liked the sound of entanglement?

53. A very nice recent book on the history of entanglement in quantum physics that is quite accessible to the layperson is *The Age of Entanglement: When Quantum Physics Was Born* by Louisa Gilder (Knopf, 2008), http://www.worldcat.org/oclc/213765737.

54. Mathematically, the state $|\uparrow\rangle_A |\downarrow\rangle_B$ is unentangled because it can be cleanly separated into Alice's ion state $|\uparrow\rangle_A$, which is definitely pointing up at all times and into Bob's ion state $|\downarrow\rangle_B$, which is definitely pointing down at all times. In Einstein's language, up and down are elements of reality associated with each of the ions. The state $|\uparrow\rangle_A |\downarrow\rangle_B + |\downarrow\rangle_A |\uparrow\rangle_B$ is entangled because it cannot be split cleanly in two in this way. The plus sign implies that, before any observation is made, Alice's ion is pointing up *and* down, while Bob's ion is pointing down *and* up. This encodes what we call the unreality of quantum theory. The ion is pointing in all directions until an observation or measurement is made—that's unreality for you.

Chapter 2

For Whom the Bell Tolls

Now that we have a few more concepts under our belts, I would like to go back and revisit in some more detail just what John Bell proved in 1964 about entangled states. He considered a simple quantum situation like the entangled ions in Alice and Bob's pocket watches and then set about to compare the predictions of quantum theory for such a setup to the predictions of a classical hidden variable theory that had all of Einstein's constraints: *reality, locality,* and *certainty.* Remember—there was already the Bohm hidden variable theory, well known in 1964, which did predict the same outcome as quantum theory, but this Bohm theory doesn't count because it violates Einstein's *locality* condition. (Einstein thumbs his nose at the Bohmians.)

Bell's goal was to see if he could construct a local classical hidden variable theory that would agree with *all* the predictions of quantum theory *and* satisfy Einstein's three conditions. If such a hidden variable theory could be found, then it would be tempting to just replace quantum theory with that hidden variable theory and all the spookiness would just disappear. But in order for

---

* Photo: "The Bell in Chersonesos, Crimea," taken by Dmitry A. Mottl (January 8, 2009).

this project to succeed, the classical hidden variable theory had to obey all three of Einstein's *reality*, *locality*, and *certainty* conditions *and* reproduce *all* the predictions of quantum theory. That was just too tall of an order to ask of any local hidden variable theory, Bell discovered. Something, somewhere, had to give.

Bell proved (at least in this setup similar to the pocket watches) that the predictions of *any* local hidden variable theory *disagreed* with the predictions of quantum theory. This means that a successful local hidden variable theory is not just another interpretation of quantum theory—like the inverted-hollow-bubble earth versus the solid-round-ball earth theories—which could not be distinguished by any means. Bell showed that *any* local hidden variable theory that obeyed Einstein's three conditions was in reality a *different* theory than quantum theory—the two theories gave different predictions of what would happen in an *experiment!* Galileo's cleaving sword of truth and falsehood—the scientific method—now saves us!

For 30 years, from 1935 to 1964, physicists had hoped that the yet-undiscovered local hidden variable theory would be a replacement for quantum theory—and in one stroke John Bell destroyed this hope. Bell had now muted the philosophical debate, but this was immediately replaced by a physics debate—if quantum theory and hidden variable theory gave different predictions, the issue of which was true and which was false could now be settled in the laboratory! Thirty years of philosophizing had come to an end and it was time to get down to building some hardware to uncover the truth.

But before I go on to discuss these experiments, I would like to spend a bit more time focusing on just why any local hidden variable theory cannot keep up with quantum theory. For this, I will return to Alice and Bob and Eve and Evan and the quantum pocket watches once again. We now know that the true National Institute of Quantum Information Standards and Technology (NIQuIST)–certified quantum pocket watches of Alice and Bob contain two ions in an entangled state, $|\uparrow\rangle_A |\downarrow\rangle_B + |\downarrow\rangle_A |\uparrow\rangle_B$.

But this is not the whole story. Entanglement is much richer than just this one state. The same exact state actually can be shown to encode *all* of the following correlations *simultaneously*: $|\uparrow\rangle_A |\downarrow\rangle_B + |\downarrow\rangle_A |\uparrow\rangle_B$ and $|\leftarrow\rangle_A |\rightarrow\rangle_B + |\rightarrow\rangle_A |\leftarrow\rangle_B$ and $|\nearrow\rangle_A |\swarrow\rangle_B + |\swarrow\rangle_A |\nearrow\rangle_B$ and $|\nwarrow\rangle_A |\searrow\rangle_B + |\searrow\rangle_A |\nwarrow\rangle_B$ and ...

The (...) indicates that these are only 6 possibilities out of the 12 possible positions of the hour hand—all 12 positions encoded in a *single* state of two ions. Anton Zeilinger, a quantum physicist at the University of Vienna, likens this idea to that of the short story, "The Library of Babel," by the Argentine writer Jorge Luis Borges.[1] This single library contains all books that have ever been written and all books that will have ever been written simultaneously. In the same way, the single quantum-entangled state described above contains every possible two-ion anti-correlation simultaneously. (We restrict this to just 12 for the pocket watch analogy.) That storage of 12 potential anti-correlations in a single pair of ions is a nugget of quantum weirdness that cannot be reproduced classically. The quantum *unreality* principle ensures that the ions are in *all* of these 12 potential states simultaneously. Only when a measurement is made does this whole collection of uncertain possibilities collapse into a single certain reality of, say, $|\swarrow\rangle_A |\nearrow\rangle_B$, which translates into Alice getting 8:00 and Bob getting 2:00 and then only if Alice and Bob have both chosen to measure the orientation of the ions along the diagonal ($\nearrow$) direction.

The entanglement guarantees that the arrows *always* point in opposite directions, if Alice and Bob choose the same axis along which to measure. But the entanglement also guarantees that Alice and Bob can choose which axis along which to measure *years after the pocket watches have left the shop*, and the siblings still get perfect anti-correlations. How can Eve and Evan know in advance which axes Alice and Bob will randomly choose years later? This waiting until long after the preparation of the state to choose the axis of measurement is called "delayed choice." With delayed choice, the classical and local hidden variable theories of Eve and Evan cannot keep up. Let's see why.

What Eve and Evan are doing with the fraudulent clockwork of the non-quantum pocket watches is to try and produce a local classical hidden variable version of the pocket watches (much easier and cheaper), which matches all the predictions of quantum theory (much harder and more expensive). In this way, they hope to fool Alice and Bob into thinking that they have the genuine quantum items. But, as I said earlier, Bell proved that reproducing quantum theory with local classical hidden variable theory is impossible. What is the best Eve and Evan can do using a half-assed version of quantum theory? They can only give Alice and Bob quantum pocket watches with anti-correlated quantum states of ions in them and just skip the (expensive) step of entangling the ions.

Using their random number generator, they can prepare many copies of the pairs of anti-correlated rigged pocket watches with ion correlations that look like this: $|\uparrow\rangle_A |\downarrow\rangle_B$ or $|\downarrow\rangle_A |\uparrow\rangle_B$ or $|\leftarrow\rangle_A |\rightarrow\rangle_B$ or $|\rightarrow\rangle_A |\leftarrow\rangle_B$ or $|\nearrow\rangle_A |\swarrow\rangle_B$ or $|\swarrow\rangle_A |\nearrow\rangle_B$ or ...

Such a semi-quantum mockup can be explained with a local hidden variable theory—nontelepathic gremlins with quasi-magical dice. These unentangled states are called separable, because Alice's state can be cleanly separated from Bob's state. Separable quantum states still have elements of *unreality* and *uncertainty* associated with them, but because they are not entangled, the long tenuous tether of *nonlocal* influences at a distance (telepathy) has been severed. Remember—it takes all three of the conditions (nonlocality, unreality, and uncertainty) to fool Charlie's code.

Why are such unentangled separable states uncertain? Let's look first at just Alice. If she gets the state $|\rightarrow\rangle_A$, but chooses to measure in the vertical direction ($\updownarrow$), then with a 50–50 chance she will measure up ($\uparrow$) and with a 50–50 chance she will get down ($\downarrow$). The outcome is completely random, which is the condition for *uncertainty*. The state $|\rightarrow\rangle_A$ is also *unreal*, as it can be expressed as a cat-like superposition of $|\uparrow\rangle_A + |\downarrow\rangle_A$. In this notation, $|\uparrow\rangle_A$ is the live cat, $|\downarrow\rangle_A$ is the dead cat, and $|\rightarrow\rangle_A$, which is the same as $|\uparrow\rangle_A + |\downarrow\rangle_A$, is the cat both alive and dead. Being both alive and dead is the condition for *unreality*.

So these separable states above, $|\uparrow\rangle_A |\downarrow\rangle_B$ or $|\downarrow\rangle_A |\uparrow\rangle_B$ or unpaired and without the plus sign, as is found in the entangled state, $|\uparrow\rangle_A |\downarrow\rangle_B + |\downarrow\rangle_A |\uparrow\rangle_B$, are no longer entangled. They are *local*. They also lack the efficiency of Borges' *Library of Babel*. *With* entanglement, *one* pair of ions stores all 12 anti-correlated states *simultaneously*. *Without* entanglement, 12 pairs of ions are required to store 12 pairs of anti-correlated states *one pair at a time*. Eve and Evan are trying to do this without entanglement and thus without complete access to Borges' vast library. The best that Eve and Evan can do, in their nefarious half-assed local hidden variable scheme, is to use a random number to choose from some pair of anti-correlated ions out of 12 to send to Alice and Bob. The random number between 1 and 12, that only Eve and Evan know, is the hidden variable. This means that if in a single experiment Alice and Bob happen to both choose the $\updownarrow$ axis to measure in—and Eve and Evan happen to have randomly sent $|\uparrow\rangle_A |\downarrow\rangle_B$ or $|\downarrow\rangle_A |\uparrow\rangle_B$—then Alice and Bob, when they compare notes over the cell phone, will see perfect anti-correlation—hour hands pointing 180° opposite across the clock face.

But what dooms Eve and Evan to being caught is the business of *delayed choice*. Suppose Alice and Bob randomly both choose in one run to measure on the $\updownarrow$ axis but in that same run Eve and Evan have randomly sent $|\rightarrow\rangle_A |\leftarrow\rangle_B$. Because the state Eve and Evan sent is aligned on the horizontal axis ($\leftrightarrow$), and Alice is measuring along the vertical axis ($\updownarrow$), as we saw above, the quantum *uncertainty* principle kicks in, and Alice's measurement will randomly collapse Eve and Evan's state into either $|\uparrow\rangle_A$ or $|\downarrow\rangle_A$ with a 50–50 probability. In *Katzensprache* (the language of cats), $|\downarrow\rangle_A$ is dead, $|\uparrow\rangle_A$ is alive, and $|\rightarrow\rangle_A$ is simultaneously both dead and alive. According to quantum theory, Alice's measurement causes Eve's dead and alive cat $|\rightarrow\rangle_A$ to randomly and uncontrollably collapse into either totally dead $|\downarrow\rangle_A$ or totally alive $|\uparrow\rangle_A$. Because the collapse is random, Eve and Evan cannot predict which way it will go and therefore cannot manipulate the ion directions in advance to steer the statistics of such events in their favor.

Similarly, if in the same run that Eve and Evan sent $|\rightarrow\rangle_A |\leftarrow\rangle_B$, Bob randomly chooses the vertical ($\updownarrow$) axis to measure in, he will get either $|\downarrow\rangle_B$ or $|\uparrow\rangle_B$, also randomly with 50–50 probability and (this is the key point!) Alice and Bob's results will no longer be perfectly anti-correlated in each of such runs. Only entanglement ensures anti-correlation every time!

Approximately 50% of the time for such an event, they will get the perfect quantum theory–predicted anti-correlation $|\uparrow\rangle_A |\downarrow\rangle_B$ or $|\downarrow\rangle_A |\uparrow\rangle_B$, but the other 50% of the time they will get results that disagree with quantum theory, either $|\uparrow\rangle_A |\uparrow\rangle_B$ or $|\downarrow\rangle_A |\downarrow\rangle_B$. Without telepathy and sufficiently magical dice, the result disagrees with quantum theory.

If Alice and Bob knew for sure what state Eve and Evan had sent every time, then a single measurement of $|\uparrow\rangle_A |\uparrow\rangle_B$ would instantly reveal the fraud. The entangled state (the state Alice and Bob are paying for) would never give such a result, whereas the unentangled, separable state would give that result 50% of the time. However, Alice and Bob do not know what Eve and Evan are doing. (The whole point about hidden variables is that they are hidden!) Hence, to reveal the fraud, Alice and Bob must perform many runs and slowly accumulate data; when they have enough to be statistically significant, they can feed the data into Charlie's computer code, which will then spit out either "true" or "false."

As Alice and Bob accumulate enough data, after a sufficiently large number of test runs, the statistics they share will begin to show that, whatever Eve and Evan are doing at the shop, it strongly disagrees with full entangled-state quantum theory. Quantum theory predicts that whenever Alice and Bob randomly pick the same axes to measure along—and if they both see the hour hand—they will always have perfect anti-correlation, every time. As Bell pointed out, any classical hidden variable, such as in the fraudulent pocket watches produced by Eve and Evan, cannot produce this perfect anti-correlation result every time. All Bob and Alice need to do is to call each other up on Skype, list the events for which they chose identical axes, and then do a statistical analysis of the correlations in order to distinguish whether what is going on is quantum theory or classical hidden variable theory. Bell provided this analysis in his paper, and it was expanded and clarified in 1969 in a paper by John Clauser, Michael Horne, Abner Shimony, and Richard A. Holt, who boiled it down to a single formula.[2] (This paper is called the CHSH paper after the authors.)

All Alice and Bob have to do is feed their data into the formula and the formula will spit out a number. If that number is less than or equal to 2, then all the results can definitely be explained with a classical hidden variable theory. If that number is greater than 2, then only full quantum theory can explain their results. This is the test that Charlie from IBM proposed NIQuIST implement to check for fraudulent quantum pocket watches. Less than or equal to 2, it spews out "false," and greater than 2 it spits out "true." Even better, for this particular setup we are considering in this chapter, CHSH point out that quantum theory predicts the code will spit out 2.83, which is significantly greater than the classical breakeven point of 2.0, making the task of distinguishing the two theories much easier. (A way to view these numbers is that the predicted maximum "strength" of the two-particle correlations is 42% stronger in quantum theory than in local hidden variable theory.)

There is one more step to the logic. In our story, we have evil twins, Eve and Evan, trying to mimic expensive, hard-to-do quantum theory with their cheap easy-to-do local hidden variable theory—in order to defraud the government. What Einstein was worried about was not Eve and Evan, but rather Mother Nature herself deploying a local hidden variable theory to defraud us all into thinking we had to use quantum theory to explain all our experiments, when in fact a local hidden variable theory would suffice. Einstein claimed Mother Nature had fooled us into thinking God played dice, and once her true local hidden variable theory was discovered, we could admonish Mother Nature, embrace the local hidden variable theory, and discard quantum theory (as well as that craps-shooting deity) forever.

To belabor the point, Bell showed that any hidden variable that obeyed Einstein's *locality*, *reality*, and *certainty* constraints made predictions that *disagreed* with the predictions of quantum theory. Local hidden variable theory and quantum theory were not interchangeable theories, like solid-ball earth and inverted-hollow earth, but instead were distinctly different theories giving different predictions. Only one of the two theories, quantum or local hidden variable, could be a correct description of nature—but which one?

By 1964, Bell's statistical test, with large numbers of statistically significant pairs of quantum-entangled particles, had never yet been performed by anyone. Before 1964, nobody knew that such a test even existed, and even if they had known, making pairs of quantum-entangled particles and shipping them over long distances (long enough so that Alice and Bob could choose which measurement to make long after the particles were prepared) had been technologically infeasible. Only a few cases of entangled particles were even known in the laboratory, such as in the pair of electrons orbiting the nucleus of the helium atom, but nobody had ever performed such statistical tests on these electrons. Even then, helium's two electrons are only separated by approximately a billionth of a centimeter (inch) or so in the atom, and because Bell's test needs to allow Alice and Bob time to make their choices, the particles need to be separated over many centimeters or meters or even kilometers (inches or yards or even miles).

So in 1964, it was still possible that Einstein was right. Every experiment in quantum mechanics performed up to that time could be explained with a classical hidden variable theory and so there were no data to rule out hidden variable theory or rule in quantum theory, or vice versa. Galileo's cleaving sword of truth and falsehood was still the Sword of Damocles, hanging over all our heads. A new experiment, built specifically to test the two theories against each other, had to be done.

CLANKING CONTRAPTIONS AND CANTANKEROUS CODGERS

Bell's original 1994 paper contained a specific proposal for carrying out a statistical test in a specific experiment to see which was right—local hidden variable theory or quantum theory. The test is now familiar to us. It involves preparing a large number of pairs of quantum particles in an entangled state and shipping them to Alice and Bob, who make randomly chosen measurements on the state of the particles and compare the statistics after the experiment. The test was whittled down and simplified in an improved version proposed by the CHSH paper in 1969. This CHSH paper showed that Bell's test could be carried out with far fewer measurements than Bell had thought was needed, and thus the test became a possibly doable experiment. CHSH also pointed out that for specific settings of the measurement apparatus, the maximum quantum theory violation of Bell's inequality was 2.83, much greater than the hidden variable upper limit of 2.0.

Determining if an experimental result is less than or equal to 2, versus greater than 2, is very tricky, because of the experimental margin of error. Suppose Charlie's code spits out 2.1 with a margin of error of ±0.2. That does not rule out local hidden variable theory because the result could be a number less than 2.0 within the margin of error. However, determining if an experimental result is either less than 2.0 or around 2.83 is much easier. Suppose Charlie's code spits out 2.7 with a margin of error of ±0.2. With that same margin of error, this result much more conclusively supports quantum theory, because 2.7 ± 0.2 is much, much, much more consistent with 2.83 than with any number less than or equal to 2.0. But where to get the quantum-entangled particles? In a word—photons! Are you listening? There's a great future in photons.[3]

Photons are the quantum particles of light whose existence was first proposed by Einstein himself in his Nobel Prize–winning 1905 paper on the photoelectric effect. And then his own beloved and traitorous photons were used in the first experiments to prove that quantum theory was right and that he was wrong. Oh, the irony. Poor Albert Einstein!

For the past 40 years, much of the progress in the foundations of quantum mechanics and its more practical spawn, quantum technology, has taken place in the development of ever-bigger, -better, -brighter, and -faster laboratory sources that shoot pairs of entangled photons out with ever-increasing abundance. The primary motivation for the development of such entangled photon sources was to test quantum theory—either rule it in or rule it out in favor of Einstein's local hidden variable theory. However, once lots of pairs of entangled photons became freely available in the laboratory, researchers started thinking about other applications besides testing the foundations of quantum mechanics. (I will discuss some of these applications later in the book.)

For now, the only game in town was to produce entangled photons in sufficient quantity that we could run Charlie's statistical test on them and see which was right, Einstein's classical local hidden variable theory (Mother Nature fools us to think God plays dice) or quantum theory (God plays dice). The game was afoot! There were several experiments of this kind carried out between 1972 and 1982.

The first (that was widely accepted) was an experiment carried out by Stuart J. Freedman and my friend and colleague, the cantankerous American quantum physicist John F. Clauser, then at the physics department at the University of California at Berkeley.[4] I have personally heard this story myself over many a beer in discussions with John Clauser, and there is a very nice rendition of it in the recent book *The Age of Entanglement* by Louisa Gilder.[5] The source of photons for this experiment is a calcium atom suffering from an identity crisis. When an atom becomes excited (either by whacking it with another atom or by shining a camera flash lamp on it), the electrons, whizzing around the atomic nucleus like planets around the sun, jump up into higher-energy orbits (a.k.a. energy levels). They do not stay in these higher energy levels for long but rapidly (typically around a billionth of a second) descend back down to where they came from, emitting a photon or two or three in the process. This is the same process by which atoms in neon signs in Las Vegas emit light. The electric current from the wall plug excites "up" the electrons in the neon atoms, and then the electrons come back down again while emitting pretty red photons. (Other atoms besides neon emit photons of different colors.)

The calcium atom is peculiar in that, when it is excited, the outermost electron sometimes jumps up *two* energy levels, and then when the electron jumps back down those same two levels, it emits *two* bluish green photons back to back, firing them out 180° apart in opposite directions. These two bluish green photons are *entangled*.

Photons, like ions, have an internal sense of spatial orientation called polarization. A photon coming straight at you can, say, be polarized either vertically $|\updownarrow\rangle$ or horizontally $|\leftrightarrow\rangle$. This is the direction the electric field inside the photon wiggles as it shoots toward you. Detectors can be made with glass plates and electronic photon detectors (like those in your digital camera) that can distinguish between these two polarizations, photon by photon. The photon detector system will signal "V" for vertical whenever it gets a vertically polarized photon $|\updownarrow\rangle$ and "H" whenever it gets a horizontally polarized photon $|\leftrightarrow\rangle$.

The calcium identity crisis occurs when the electron jumps back down those two levels. When it emits these two bluish green photons back to back, it is impossible to tell—even in principle—if the photon traveling to the left toward Alice's detector is horizontal and the photon traveling to the right toward Bob's detector is vertical—or vice versa. It is a rule of quantum theory that when two processes cannot be distinguished by any means, the state describing the two processes must then account for *both* potentialities *simultaneously*. (The tea-spoon is in my coffee cup in my office *and* in the bathroom sink across the hall.)

In other words, the quantum mechanical state describing the two photons must be written in such a way that the left-moving photon is vertical and the right-moving photon horizontal *and* simultaneously the left-moving photon is horizontal and the right-moving photon is vertical (see Figure 2.1).

We now have a name for such a quantum mechanical state of affairs—*Verschränkung!* The photons are in a polarization state of *quantum entanglement*, and the state may be written as $|\updownarrow\rangle_A |\leftrightarrow\rangle_B + |\leftrightarrow\rangle_A |\updownarrow\rangle_B$ and $|\diagdown\rangle_A |\diagup\rangle_B + |\diagup\rangle_A |\diagdown\rangle_B$ and ...

The notation again implies that the single entangled state encodes a large number of entangled outcomes. If Alice chooses to define $\updownarrow$ as "vertical" and $\leftrightarrow$ as "horizontal" and measures the photon state and gets $|\updownarrow\rangle_A$ (vertical), then if Bob chooses the same definitions of vertical and horizontal, he is sure to get $|\leftrightarrow\rangle_B$ (horizontal). If Alice instead chooses to measure in a coordinate system such that $\diagdown$ is defined to be "vertical" and $\diagup$ is "horizontal" and she gets $|\diagup\rangle_A$ (horizontal), then Bob is sure to get $|\diagdown\rangle_B$ (vertical). Once again, the single entangled state stores all these anti-correlations *simultaneously*. In the language of Schrödinger's cat, $|\updownarrow\rangle_A$ is "atomic nucleus not decayed" and $|\leftrightarrow\rangle_B$ is "cat alive," while $|\leftrightarrow\rangle_A$ is "atomic nucleus decayed" and $|\updownarrow\rangle_B$ is "cat dead."

Local hidden variable theory can produce only unentangled states that can only store these possibilities one at a time: $|\updownarrow\rangle_A |\leftrightarrow\rangle_B$ or $|\updownarrow\rangle_A |\leftrightarrow\rangle_B$ or $|\leftrightarrow\rangle_A |\updownarrow\rangle_B$ or $|\diagdown\rangle_A |\diagup\rangle_B$ or $|\diagup\rangle_A |\diagdown\rangle_B$. Such states, without the plus sign, are called *separable*

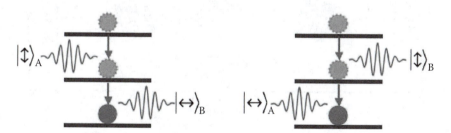

Figure 2.1 A doubly excited calcium atom has an excited electron (fuzzy ball) that rapidly makes two jumps down to its ground state (solid ball) while emitting two greenish blue photons back to back (wiggles). In order to conserve angular momentum, one photon is always vertically polarized $\updownarrow$ and the other is horizontally polarized $\leftrightarrow$. Calcium suffers an identity crisis in that it is impossible to tell—even in principle—if the vertically polarized photon is launched to the left toward Alice $|\updownarrow\rangle_A$ and the vertically polarized photon is launched to the right toward Bob $|\leftrightarrow\rangle_B$ (left-hand-side picture) or the reverse (right-hand-side picture). When such a quantum identity crisis occurs, quantum theory requires that both outcomes must be included simultaneously, giving rise to the polarization-entangled state $|\updownarrow\rangle_A |\leftrightarrow\rangle_B + |\leftrightarrow\rangle_A |\updownarrow\rangle_B$.

states, because the Alice state can be cleanly separated from the Bob state. That extra storage space, provided by quantum-entangled states, is the primary reason why quantum theory beats out classical hidden variable theory in Bell's test—the classical hidden variable theory just cannot compete with this storage power of quantum entanglement. (This extra storage space in quantum-entangled states will also be a key notion in the next chapter for understanding the power of a quantum computer.)

In Figure 2.2, I show a schematic of the Freedman and Clauser experiment. The excited calcium atoms are in a glass tube in the center and Alice and Bob are two separate detector systems on opposite sides of the room. Each detector system is composed of a polarization analyzer and a photon detector. The polarization analyzer in this experiment was a bunch of windowpanes strapped together into a big heavy stack that could be rotated on a semiautomatic ratcheting mechanism into any of a large number of possible definitions of "vertical" ($\updownarrow$, $\nwarrow$, ...) and "horizontal" ($\leftrightarrow$, $\nearrow$, ...). Such a large number of orientations were required to implement Charlie's test—the statistical processing code that checks the Bell inequality. Recall that the IBM-Charlie-improved quantum atomic pocket watches have rotating cover slits that ratchet to any clockface

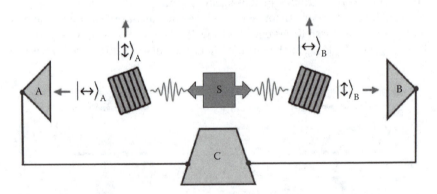

Figure 2.2 The Freedman–Clauser single-channel experiment. A doubly excited calcium atom emits two polarization-entangled bluish green photons back to back. Every so often, these two photons are fired in the direction of Alice's detector to the left (A) and Bob's to the right (B). Two polarization analyzers, made from rotatable stacks of windowpanes (striped boxes), can each sort the horizontally from the vertically polarized photons and transmit one type to the detector system (triangles)—rejecting the other type of polarization upward where it is lost. The computer (C) then records those events when both Alice and Bob's detectors fire simultaneously and it also records the orientation of the two polarization analyzers when such an event occurs. Shown here is a single event where Alice gets horizontal and Bob gets vertical for a particular orientation of the analyzers.

hour position and then are opened to reveal if the hour hand is actually there or not. To reproduce that thought experiment in the Freedman–Clauser experiment, they had to ratchet the polarization analyzer into 16 different "clockface" positions, each separated from the next by 22.5° to make a complete circle. The contraption was connected to a bunch of pulleys and gears to ratchet the heavy stack of windowpanes (the polarization router), clanking around in a step-by-step process.

In Alice's detector system, the stack of windowpanes sits at one polarization orientation for defining horizontal and vertical and routes a large number of photons to the photon detector. (Photon is detected as vertical.) Simultaneously, it routes out of the experiment altogether an equally large number of photons. (Photon is assumed to be horizontal but is not detected.) After collecting a sufficient number of data clicks at that orientation setting, the clanking semiautomatic gear mechanism moves the windowpanes onto the next step of polarization orientation (where the definitions of vertical and horizontal have been rotated by 22.5°), and the windowpanes route another bunch of photons into the photon detector (photon is detected as vertical in the new orientation) or the other photon detector (photon is assumed to be horizontal in the new orientation but is not detected). Bob's detector system is identical.

A subtle point needs to be made here. In the Freedman–Clauser experiment, called a *single-channel* experiment, Alice and Bob have only one photon counter each (not two). Hence, each photon counter has to do double duty—in half the runs, Alice's photon detector looks only for the horizontals (while assuming any verticals went the other way to be lost), and in the other half of the runs, it looks only for the verticals (while assuming any horizontals went the other way to be lost). This assumption slightly weakens the strength of Charlie's statistical test, which was actually designed with the assumption that both horizontals and verticals are caught each time. A true IBM-Charlie *dual-channel* experiment would require Alice and Bob to have two photon detectors each. (I will discuss such dual-channel experiments at the end of the next chapter.)

One reason the Freedman–Clauser experiment was notable is that it was the very first (well-accepted) experiment to test Bell's theorem, and it had enough statistics with the data to rule quantum theory in or out with some degree of accuracy. The second reason it was notable is that Clauser performed this experiment fully believing that he would confirm Einstein's local hidden variable theory as right and quantum theory as wrong. To his own astonishment, he proved just the opposite: Spooky *nonlocal*, *unreal*, and *uncertain* quantum theory was right. Einstein's commonsensical, *local*, *real*, and *certain* hidden variable theory was wrong. Clauser couldn't believe his own experiment!

He kept checking and rechecking and re-rechecking the contraption, looking for hidden flaws, and kept collecting more and more data, improving his statistics, until he came face to face with Galileo's blazing bifurcating blade—the scientific method. The power of the method is shiningly clear—it does not matter what you believe, even if it is your own experiment. It only matters what the data reveal. Clauser, by his own contraption, was forced to capitulate and embrace the quantum theory. (What's the difference between science and politics? In science, when the data disprove your hypothesis, you throw out the hypothesis. In politics, when the data disprove your hypothesis, you throw out the *data*. This is why science produces useful things, like a polio vaccine, and politics produces useless things, like a government shutdown.)

This tension between hypothesis and data is also the reason I trust the Freedman–Clauser experiment. Had there been an experimental flaw (and there were other experiments around the same time that had them), such as a systematic error that skewed the data to erroneously *confirm* Clauser's belief in a local hidden variable theory, there would be less impetus for him to double and triple check everything. Hypothesis confirmed: I believe in hidden variables; my data support hidden variables; publish immediately. That is a much less compelling story than this: Hypothesis denied: I believe in hidden variables; instead my data confirm quantum theory; therefore, I must redo the experiment over and over and over again, and double and triple check the equipment, and when my very own data show that I'm still wrong, I must begrudgingly publish that inconvenient truth and get on with my life. Damn! (Score one for the scientific method!)

In this context, let me talk a bit about experimental statistics, or as my students like to call it, "sadistics," or as the American physicist and quantum cowboy, Marlan Scully, likes to repeat, quoting Mark Twain, "There's lies; there's damn lies; and then there's *statistics!*" In physics experiments, we typically have to deal with systematic errors and random errors. I will explain these types of errors with another precious little thought experiment.

In November of 2012, The "Galumph" Organization conducts an exit poll on the night of the US presidential election to predict which of two hypothetical candidates, Alice or Bob, has won the presidency. A fly-by-night polling organization like Galumph doesn't have enough resources to poll too many people, so Galumph decides to ask only gun-toting Cadillac owners whom they voted for. This is a *systematic error* that incorrectly skews the poll in favor of the conservative candidate, *Alice*. When an independent sadistican, Marky DeSaab, points out the error in their ways (with a whip), Galumph decides instead to poll only arugula-eating Volvo owners. This is a *systematic error* that incorrectly skews the poll in favor of the liberal candidate, *Bob*.

Indeed, in 1948, the very real Gallop Organization predicted "DEWEY DEFEATS TRUMAN"[6] in that presidential election. Gallop polled people by telephone and did not compensate for the fact that, in 1948, well-off urbanites (Dewey voters) were more likely to own phones than poor rural folk (Truman voters). That skewed the poll in favor of Dewey, who actually lost. This was a *systematic error*—an error in the experimental setup—something that can be anticipated and minimized if you are very careful.

Back to the future—in 2012, Galumph Polls decides to randomly exit poll only 10 people nationwide—and predicts Bob wins with 53% of the vote and Alice loses with 47% of the vote. Do you believe it? No. That's because polling only 10 voters out of 100 million puts Galumph's percent margin of error close to 100%, much bigger than the predicted point spread between Alice and Bob. This is a *random error*, owing to polling too few people, an error that can only be minimized by polling more people—taking more data—a bigger sample size.

At the other extreme, if Galumph polls all 10 million people who voted (assuming no mistakes and nobody lies), then the margin of error would be zero percentage points. That poll would be equivalent to running the election itself! What Galumph needs to poll is a number between 10 voters and 100 million voters. But which number? The margin of error reduces only very slowly—quadratically[7] slowly—with increasing poll or sample size (see Figure 2.3).

Suppose instead Galumph exit polls *100* voters at random and predicts Bob wins over Alice, 53% to 47%, with a percent margin of error of 10%. Is it time to celebrate (or to weep)? No. The point spread between Bob and Alice is 6%, which

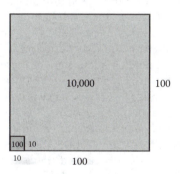

Figure 2.3 How random errors decrease with increasing sample size—the quadratic scaling law. In mathematics, the word *quadratic* refers to a square. The area of the square is the sample size and the length of the side of the square is the error. For a sample size of 100 voters (100 data points), 100 is the area of the small square. The margin of error is the length of the side of the small square, namely, 10. The percent margin of error is 10/100 or 10%. If we increase the sample size to 10,000 voters (10,000 data points), then 10,000 is the area of the large square. The new margin of error is the length of the side of that square, namely, 100. The new percent margin of error is 100/10,000, that is, 1/100 or 1%. Hence, we must increase the sample size by two orders of magnitude (100 to 10,000) to decrease the percent margin of error by only one order of magnitude (10% to 1%). It is very, very expensive to reduce the margin of error, which reduces only quadratically slowly with increasing sample size.

is smaller than the percent margin of error of 10%. This poll predicts nothing. You cannot tell who will win with any certainty at all. To reduce the margin of error from 10% to 1%, all other things being equal, Galumph would have to poll *10,000* voters instead of *100* voters. Then, you have Bob with 54%, Alice with 47%, a spread of 6%, and a margin of error of 1%—Bob most likely wins! (And that is only if Galumph first removed all the systematic errors owing to the Cadillacs and Volvos and telephones.) It is very, very expensive in time, money, and manpower to reduce the margin of error by just a few percentage points, because of a slow quadratic scaling law for how the margin of error decreases as the sample size increases.

So too in a science experiment: You have to collect more and more data to knock down your random error margin by just a few percentage points, which is a slow, expensive, and time-consuming process. Often the experimental equipment breaks down before you get the accuracy you want, particularly in the Freedman–Clauser experiment, where a pair of photons was counted once every 10 seconds. For a 1% margin of random error, they would need 10,000 pairs of photon counts, at one pair per 10 seconds, which would be approximately 3 hours of data collecting for *each* of the 16 rotation settings, or approximately 45 hours of data collecting *per experimental run*. The actual Freedman–Clauser experiment collected data for a total of approximately 200 hours before the contraption expired from Clauser's incessant hammering. The margin of error quoted in polls and physics experiments is usually a combination of both the random error and an estimate of the systematic error.

Let me add a historical note on a few more of these experimental tests of quantum theory that were also carried out in the 1970s, before we move on to the 1980s.[8] This note illustrates just how science is done and also underscores the need to control these experimental errors. The Freedman–Clauser paper appeared in 1972 and ruled quantum theory *in* and local hidden variable theory *out*—within their margin of error. In 1973, Richard A. Holt[9] and Francis M. Pipkin from Harvard University distributed Holt's (unpublished) PhD thesis, which contained results from their new Bell test experiment (using mercury atoms instead of calcium atoms), and which ruled quantum theory *out* and local hidden variable theory *in*—within *their* margin of error. Sometimes this happens, and more experiments are needed to straighten things out.

In 1976, in response to the Holt–Pipkin result, Clauser duplicated their experiment for himself at Berkeley, and *after 400 hours of data collecting*, Clauser's new experiment ruled quantum theory *back in* and local hidden variable theory *back out*.[10] Clauser suggested that the erroneous Holt–Pipkin result arose from an unknown *systematic error* that skewed the data to hidden variable theory. (Cadillacs, Volvos, and telephones all over again!) Finally, also in 1976, and also motivated by the discrepancy between the Freedman–Clauser and Holt–Pipkin results, Edward S. Fry and Randall C. Thompson at Texas A&M University performed a greatly improved version of the Holt–Pipkin mercury experiment and published results with the smallest margin of error up to that point—with a data collection time of only approximately 80 minutes. The Fry–Thompson experiment also ruled quantum theory *in* and hidden variable theory *out*.[11]

By 1976, the score was three experiments for quantum theory and one against (and nobody really believed the Holt–Pipkin experiment, probably not even Holt and Pipkin, which is why they never published it). So it looked like— finally—quantum theory had been ruled in and local hidden variable theory ruled out. And yet, experiments are still being conducted on this issue, even up to this very day. Why?

The primary reason for these ongoing experiments lies in the bowels of Bell's theorem. Bell's theorem was so general that it compared the well-known *single* quantum theory to an *infinite number* of possible unknown local hidden variable theories. In our quantum pocket watch thought experiment, Eve's original hidden variable mockup sent only pairs of ions that were always anti-correlated. This is a very simple-minded hidden variable theory, which is very easy to rule out in a Bell experiment. Evan, on the other hand, replaced Eve's hidden variable scheme with the local hidden variable clockwork of nontelepathic fairies throwing nonmagic 12-sided dice. A much more elaborate experiment is required to rule out hidden variable fairy theory. So with each new experiment, an ever-larger set of local hidden variable theories is ruled out, and the surviving hidden variable theories are painted into an ever-smaller corner. As I write these words, *no experiment has yet ruled out all possible local hidden variable theories*. However, we are getting really close—or at least close enough for government work.

FREEDOM OF CHOICE, FRENCH FINESSE, AND LOOPHOLES

French quantum physicist Alain Aspect, as part of his PhD thesis research, carried out the next Bell test experiments that I will now discuss. His three experiments were performed with his collaborators, Philippe Grangier and Gérard Roger, at the University of Paris from 1980 to 1982, but I will only discuss the last and most notable of these from 1982.

As a PhD student at the University of Paris (where I imagine he was drinking less expensive vintages), Aspect became very excited about the works of John Bell and the foundations of quantum mechanics, so he decided to make a pilgrimage from Paris to the giant atom smasher at CERN and visit John Bell himself. There, Aspect confessed to Bell that his dream in life was to perform even better experiments on tests of the foundations of quantum mechanics and Bell's theorem. What was John Bell's sage response to Aspect's dream? "Don't give up your day job."

In the 1980s, working on the foundations of quantum mechanics was considered a somewhat nutty and career-killing choice for a young scientist. Even Bell himself worked mostly on the practical theoretical models and designs of atom smashers, the work he was actually paid to do. (Bell did his work on the foundations of quantum mechanics somewhat secretly at first and always at night or on weekends.)

Not derailed by Bell's warning, Aspect persisted and pressed him further for advice. Bell focused on the matter of *delayed choice.* In all the experiments of the 1970s, the matter of delayed choice was not implemented properly. As we recall from IBM-Charlie's test of the quantum pocket watches, Alice and Bob should, *completely randomly,* choose which hour hand slit position to dial up, *long after* the pocket watches left the NIQuIST factory floor. This is delayed choice. In the 1970s experiments, the angles to which the polarization analyzers were moved were controlled by a gearbox and preordained in advance; thus, they were not a true randomized delayed choice. Bell pointed out to Aspect that by implementing a true random delayed choice, a much larger class of local hidden variable theories could be ruled out.

Remember, we are testing a *single* quantum theory against *all possible* local hidden variable theories. (It seems unfair to quantum theory.) With each new and improved experiment, the number of surviving local hidden variable theories is painted into an ever smaller and smaller corner whose triangular boundaries are demarcated by Einstein's *locality, reality,* and *certainty* conditions. Bell's advice to Aspect was that an experiment with true delayed choice would make that corner very small indeed.

Bell's concern about the 1970s experiments can be most easily explained with our local hidden variable fairy theory. Without delayed choice, our two invisible local hidden variable fairies, Alfreeda and Breena, could be flying back and forth between Alice and Bob's detector systems and the photon source and messing with the innards of the experiment. If the experimenter knows in advance which polarizer settings Alice and Bob will systematically go through in a particular order, programmed into the gearbox, then Alfreeda and Breena would know this too, and they could then mischievously fiddle with the detectors and the data-collecting process and even the flying photons themselves to make it appear that quantum theory was correct, when in truth local hidden variable fairy theory was true. While mischievous invisible fairies meddling with your experiment may seem implausible, they could not be ruled out based on any of the 1970s experiments. (Remember the inverted Earthers?)

Delayed choice, as in IBM-Charlie's pocket watch test, requires Alice and Bob to choose which polarization setting to use *randomly* and *quickly*—after the photons have left the source. The choice should be made fast enough so that Alfreeda at Alice's detector system cannot observe which random setting Alice chooses and then fly over to Breena at Bob's detector system and conspire with her to doctor Bob's results (putting correlations into the data that mimic those of quantum theory but which could be explained by fairy hidden variable theory).

Aspect and his colleagues, Philippe Grangier and Gérard Roger, performed three experiments in the 1980s, but last and most famous was the one published in a 1982 paper, which implemented delayed choice.[12] A schematic of the setup is in Figure 2.4.

The Aspect experiment, using the calcium two-photon source, had a number of improvements. It was a two-channel experiment, with two detectors in each of Alice and Bob's detector systems, so no photons were purposely discarded. (Compare this to the single-channel setup in Figure 2.2.) Keeping more photons improves the statistics and lowers the run time. In addition, the stacks of windowpanes hooked to gearboxes are now replaced with small glass polarization sorters, called polarizing beam splitters. The polarization sorters are hooked to a very fast electronic switch, which quickly aligns them in any of the 16 possible polarization detection orientations. Finally, the electronic switch is flipped at random, which means the polarization orientation is chosen at random, after the photons leave the source, implementing the sought-after *delayed choice*. The improvements took the run time for the experiment from tens of hours to only tens of minutes (see Figure 2.5). The result?

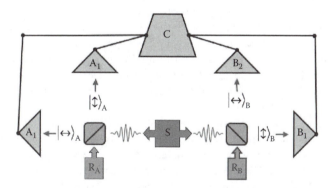

Figure 2.4   A schematic of the two-channel Bell test experiment first implemented by Aspect and his colleagues. The stacks of windowpanes attached to a gearbox have been replaced with small blocks of glass (polarizing beam splitters) that now sort the polarized photons. In the two-channel experiment, there are four detectors that catch all possible photons that exit the polarizing beam splitter, improving the statistics. The polarizing beam splitter is attached to a fast electronic switch, which is controlled by the random number generators R, which allows for the choice of polarization axis to be chosen quickly and randomly for each detector system—long after the photons have left the source—thus implementing delayed choice. All the data are sent to the computer (C), where IBM-Charlie's code tests for violations of Bell's inequality in order to rule quantum theory in or out.

Figure 2.5   Here is a 1981 photo of the experiment by Aspect and colleagues. (Courtesy of Alain Aspect and the Institut d'Optique in Orsay, France.) I have added the label C for the calcium entangled photon source and D is Bob's polarization analyzer and his two detectors. (Alice's are outside the photo to the lower right.) The photons traveled down the long cylindrical tube that had the air pumped out of it to reduce photon losses. (Fairies are not shown.)

When the statistical analysis was run, the code spit out 2.697 with a margin of error of ±0.015. Recall that Bell's inequality puts a strong upper bound on this number. If this number is less than 2, then a local hidden variable theory must be true. If this number is greater than 2, then the quantum theory is true. In this setup, the largest possible violation of Bell's inequality was the number 2.83, as predicted by the quantum theory. The experimental result did not quite hit that maximum violation, owing to remaining experimental errors, but it still was far, far more consistent with quantum theory than hidden variable theory. This time, quantum theory had been ruled *way* in, and hidden variable theory had been ruled *way* out.

The Aspect experiment, for reasons that are unclear to me, garnered a huge amount of attention in the popular media—so much so that it eclipsed the experiments done in the 1970s to the point that they were, for a time, seldom mentioned anymore. When I am at a conference where some well-meaning but undereducated scientist calls Aspect's experiment the *first* to test Bell's theorem, I politely advise the scientist not to make that claim (particularly if John Clauser is in the audience armed with his hammer). Aspect's experiment was certainly the best up to that point, the first to use the two-channel detector scheme, and the first to implement delayed choice. Was this enough to rule out all possible hidden variable theories? No. The Aspect paper itself mentions two remaining loopholes, which I will now explain with our local hidden variable fairy theory.

The first is the so-called *locality loophole*, which I will call the *speed-of-light loophole*. Recall that I have been describing the two fairies, Alfreeda and Breena, flitting back and forth between Alice and Bob's detectors and mucking around with their data. There are actually two fairy attacks possible. In the *stationary fairy attack*, Alfreeda and Breena imbed local hidden variable correlations in the source and detectors *before* the experiment starts, and then just sit back and relax. The source and detectors are skewed in advance to make us think quantum theory is true when in fact local hidden variable theory is true. Aspect's delayed choice clearly rules out such a stationary fairy attack in that Alfreeda and Breena would have to know in advance which polarizer settings Alice and Bob will choose. But because Alice and Bob choose those settings randomly after the experiment starts, Alfreeda and Breena would have to foretell the future, a power that is inconsistent with a local hidden variable fairy theory.

In the *flying fairy attack*, Alfreeda and Breena introduce the faked correlations while the experiment is underway. In particular, Alfreeda observes Alice's random detector setting (and the result obtained by the detector) and then flies over to Breena and tells her this information. Then, Breena can mischievously fiddle with Bob's detector system to skew the data to make it look like quantum theory is true when local hidden variable fairy theory is in fact true.

In order to prevent such mischief from either attack, we must make the *delayed choice*, which certainly eliminates any effect from correlations embedded from the get-go, the stationary fairy attack. However, to also eliminate the flying fairy attack, in addition we must make the delayed choice and collect the data fast enough so that it all occurs before Alfreeda can fly to Breena with this critical information required to install those skewed-up correlations. The Aspect experiment definitively rules out only the stationary fairy attack. Less definitively, the Aspect experiment only rules out the flying fairy attack given certain assumptions about the terminal airspeed velocity of an unladen fairy. The maximum flight velocity of fairies (as far as we know) is not a fundamental speed limit in nature. The true speed limit in nature is the speed of light—299,792,458 meters per second (670,616,629 miles per hour)! (It's not only a good idea—it's the law.[13]) Aspect's delayed choice only rules out the flying fairy attack if the choice is made fast enough—so fast that Alfreeda does not have the chance to reach Breena in time with the data. The delayed choice in the Aspect experiment is too slow to rule out a sufficiently speedy fairy.

Recall that when we first introduced Alfreeda and Breena, they had ham radios, which represent a new and improved and more lethal local hidden variable fairy theory—they don't have to fly back and forth anymore. Alfreeda waits until Alice's polarizer setting is chosen at random, then she records that setting and the polarization result that Alice got—horizontal or vertical—and radios up Breena with this information. With this information in hand, Breena then goes about her mischief of manipulating Bob's detector result to make it look like quantum theory is true. (Radio signals move at the speed of light so this is a much better attack than just flying over there.)

The Aspect experiment did not rule out local hidden variable ham radio fairy theory. To do so, the delayed choice must be made *very* fast and Alice and Bob must be very far apart—fast enough and far enough so that Bob's choice is made and his datum collected *before* any ham radio signal has time to travel from Alfreeda to Breena. I am not sure of the dimensions of the Aspect experiment, but I do know it was carried out in a couple of laboratory rooms (see Figure 2.5) so I'll estimate 7-meters (24-feet) maximum, give or take. The speed of light (and therefore radio waves) is 299,792,458 meters per second (983,571,056 feet per second), so the time it takes radio waves to travel 7-meters is approximately 24 billionths of a second. In order to rule out the local hidden variable ham radio fairy theory, the random choice generator and the electronic switch and the polarizer would have to be moved into their delayed choice position faster than that. Such gizmos in the Aspect experiment were a heck of a lot slower than this and so that loophole remained. That loophole was closed only in 1998 when the research group of the Austrian quantum physicist Anton Zeilinger carried out just such a fast and long-distance delayed choice experiment on the campus of the University of Innsbruck.[14] In this experiment, Alice and Bob were located 400 meters (437 yards) apart on the university campus, and ultrafast electronics implemented the random delayed choice of the polarizer orientation, ruling out local hidden variable ham radio fairy theory for the first time. The experiment also used a newfangled ultra-bright source of entangled photon pairs that put out many more photons per second than the old calcium and mercury atom sources.[15] This and other improvements gave the statistical test result of 2.736, with a margin of error of ±0.02 for 14,700 coincidence events collected in only 10 seconds. It was still not quite the maximum violation of 2.83, owing to remaining experimental errors, but it almost conclusively ruled out the local hidden variable ham radio fairy theory prediction of 2.0 or less.

This Austrian experimental result demonstrated that the odds that quantum theory is true (and hidden variable theory is false) are 1,000,000,000,000, 000,000,000,000,000,000,000,000,000,000,000,000,000,000,000,000,000, 000,000,000,000,000,000,000,000,000,000,000,000,000,000,000,000,000, 000,000,000,000,000,000,000,000,000,000,000,000,000,000,000,000,000, 000,000,000,000,000,000,000 to 1. Local hidden variable theory had now been painted into a very small corner indeed.

There was one remaining popular loophole, the so-called *detection loophole*, which I can explain again through the local hidden variable fairy theory. In all of these photon-based experiments, the detectors sucked. And by this, I mean they really sucked. Even in Zeilinger's 1998 tour de force experiment, the detectors only detected approximately 5% of the photons that were actually present. In all the previous experiments, it was even worse. How could Alfreeda and Breena exploit photon detectors that suck? No flitting about or ham radios required—Alfreeda simply infests Alice's detector system and Breena infests Bob's. As the photons come in, they steer clicks that would (on average) confirm quantum theory into the detector system and block events whose clicks (on average) would confirm local hidden variable theory. ("There's lies, there's damn lies, and *then* there's statistics.")

Alfreeda and Breena are careful to always make the average number of photons detected appear to be approximately 5%, so as not to cause any suspicion, but they skew the statistics so badly that IBM-Charlie's computer code spits out "true" (quantum theory is ruled in) when in reality the answer is "false" (local hidden variable theory is ruled in). The experimenter believes that he has confirmed quantum theory, but in fact the opposite is true. In such experiments, with photon detectors that suck, the experimenter is forced to make a *fair sampling* assumption. The fair sampling assumption posits that the statistics for the 5% of the photons that are actually detected are the same statistics for the 95% of the photons that are lost. This seems like a reasonable assumption, but not if you have mischievous fairies infesting your detectors and skewing around with your data.

Within the constraints of locality, reality, and certainty—the local hidden variable assumptions—the fairies can get away with this ruse as long as the detector efficiency is less than 82%. Above that number, they can no longer skew enough of the data enough of the time to fake quantum theory in favor of hidden variable theory (at least not without getting caught). All Bell test experiments with photons to this date have used detectors with at most 50% efficiency, and so the detection loophole still looms over the photon experiments.

The detection efficiency loophole has been closed in another setup, one that contains our old friends the ions in an electromagnetic trap. The ion experiments have a much better detection efficiency. We can detect the state of the ion—spin up ($\uparrow$) versus spin down ($\downarrow$)—with an efficiency that is now greater than 99%. This is more than enough to exclude any hidden variable theory where a couple of fairies are skewing around with our data. The first such detection loophole–free experiment was indeed carried out by our old friends at the National Institute of Standards and Technologies (NIST) in the Ion Storage group of the American quantum physicist, David Wineland. However, the ions in the trap were separated by only the tiniest fraction of a centimeter (an inch), too close together to also rule out the speed-of-light loophole.

A race is underway, as I write, to be the first group in the world to close both the detection and the speed-of-light loopholes together in a single experiment. One group, currently at the NIST, plans to do this with photons. In addition to the ion traps, the NIST also makes the world's most efficient photon detectors in Sae-Woo Nam's section of the Quantum Devices Group at the NIST laboratory in Boulder, Colorado. Nam's photon detectors, last I heard, exceed 98% efficiency, more than enough to close the detection loophole. So all they need is a good source of lots of entangled photons and some fast electronics and some room to stretch it all out like Zeilinger did in Austria. The NIST laboratory in Gaithersburg, Maryland, where Alan Migdall's laboratory in the Optical Thermometry and Spectral Methods group has been developing entangled photon sources for years, also has the source of entangled photons. The missing piece, the room to run the photons out so the fast electronics can do their thing (before any light or radio signal can be transmitted from Alfreeda to Breena) is to be found in the activity of Joshua Bienfang in the Quantum Telecommunications group, where he has a line-of-sight photon relay set up from one side of the NIST facility to the other via a command station in the attic of the tall administration building that sits in the middle of the NIST campus. Fast electronics can be now bought commercially, off the shelf. (I expect the NIST team to achieve conclusive results in a year or two, if somebody else doesn't beat them to it.)

The runner-up in the competition is Boris Blinov and his Trapped Ion Quantum Computing group at the University of Washington in Seattle, who takes the approach of using entangled ions, with their high detection efficiency, but separating the two entangled ions in their traps by many yards across the campus. The idea is to use a photon system to first entangle the distant ions and then run the Bell test on the ions themselves. This University of Washington ion experiment is more challenging, in my opinion, than the NIST photon experiment, so I predict they will come in second. The delightful thing is that there could be a dark horse candidate that I know nothing about lurking in the wings. However, I do hear that Anton Zeilinger—the purveyor of the first speed-of-light loophole-free experiment in Austria—is just *itching* to get his hands on a couple of those NIST super-efficient photon detectors. I also hear that those photon detectors are, curiously, in short supply. What fun!

I will end with a discussion about loopholes in the loopholes, as there are some loopholes we can never guard against. Here, for illustration, are two imaginary loopholes:

> *The post-experimental computer infestation loophole:* In this loophole, the two fairies, Alfreeda and Breena, just sit back and lie in wait and do nothing but watch as our experimentalist, who we'll call Xavier, carries out the entire experiment. Critical for the Bell test, IBM-Charlie's code must take data from *both* Alice and Bob's detector systems and analyze it for statistical correlations. Alice cannot run the test without Bob, and Bob cannot run the test without Alice. At some point, the data must be combined into a single computer labeled C (as in Figures 2.2 and 2.4). So Alfreeda just waits, chugging her beer, and Breena just waits, sipping her wine, until Xavier has completed the entire experiment and all that remains is the statistical test on the computer. When Xavier steps

out for a cigarette, then Alfreeda and Breena fly into the computer with glee and dance about the transistors, merrily flipping digital zeros into ones and ones into zeros, until the infested computer spits out "true" (quantum theory is true) when in fact the original data supported "false" (local hidden variable theory is true). Even if Xavier suspected foul play and runs quality control tests on his computer, Alfreeda and Breena are smart enough (and hopefully sober enough) not to muck with the computer when those tests are running, but only when Xavier is processing his data. Xavier would have no way in any experimental test to rule this out. So the scientific method fails us and we have no choice but to ignore such a possibility.

The post-apocalyptic supercomputer simulation loophole: In this loophole, Xavier only thinks he is running an experiment in his laboratory. He is actually dreaming about running an experiment in his laboratory while floating about in a clear plastic tube of semitransparent goop with a supercomputer jacked into the back of his head (like in the film *The Matrix*[16]). After World War III took place in 1964, aliens scooped up all the remaining humans and hooked them up like this to their highly advanced (but classical) supercomputers. In the *real* world where the aliens and the supercomputers reside, local hidden variable theory is true. However, for nefarious reasons known only to the aliens (they're evil after all), they program Xavier's virtual laboratory with virtual photon sources and virtual detector systems in such a way that—in his dreamlike state—his virtual data collected on his virtual laptop tell him that quantum theory is true. Because that is Xavier's reality, this loophole too cannot be ruled out by the scientific method, at least by Xavier, so he just ignores this possibility as well. It's the Inverted Earth Society all over again—no way to tell!

The point of these last two loopholes is that it's not worth worrying about any more loopholes at all after some point. Either God does play dice, as quantum theory predicts, or else he is going through some extraordinary measures to hide this fact and to make it look like he plays dice (when he really doesn't).

I NEVER METAPHYSICS I DIDN'T LIKE

For those gentle readers whom I have not yet sent into a coma, and who might now be wondering why they purchased a book on quantum computing only to be subjected to a very long-winded chapter on quantum foundations and fairies and gremlins, I assure you that what comes next will be worth the wait. In the next chapter, I will state that quantum computers are very much different from classical computers, for the simple reason that quantum mechanics (what quantum computers run on) is very much different from classical mechanics (what classical computers run on). I don't want you to just *believe* that is true, taking my word for it, I want you to *know* it is true in *your* gut, in the same way that I know it is true in *my* gut. How do *I* know it is true in my gut?

For 85 years, since the first skirmishes between Einstein and Bohr in the 1920s, the foundations and meanings of quantum theory have been debated heatedly. As we have seen, this feud led to an all-out war with the 1935 publications of the EPR paper and the Schrödinger cat paper, making it clear that quantum theory was fundamentally different and stranger than classical theory, but not yet making it clear that quantum theory was true.

In 1964, Bell changed all that by showing that quantum theory and local hidden variable theories could be tested against each other, *mano-a-mano*, in the laboratory. Those experiments have been carried out from the 1970s to this day, and each and every one of them supports the ever increasingly certain conclusion that quantum theory—with all of its strange features of nonlocality, unreality, and uncertainty—is true. John Clauser himself had to embrace quantum theory when his own experiment and his own data told him that classical local hidden variable theories are false.

And so the strange features of quantum theory are no longer open for debate. Pending one or two last loopholes, those weird features are here to stay. We have seen them and they are not going away. It is time to face the facts: Einstein was wrong and quantum theory is right. It is time to stop asking why quantum theory is so strange; we must embrace that strangeness and get on with our lives and—even better—put that strangeness to work for us. As the infamous gonzo journalist Hunter S. Thompson once said, "As the going gets weird, the weird turn pro."

Richard Feynman, the famous American physicist, bongo player, Casanova, and Nobel Laureate, once said, "Nobody understands quantum theory." I am not sure I agree with or even understand that statement. I think I understand quantum theory. I can use it and calculate with it and make predictions with it that agree with what other people see in their laboratories, so I am content to say I understand it. Maybe Feynman meant that nobody understands why quantum theory is weird the way it is. If that is the case, then maybe I don't understand any theory. I ask my students, why is it in electricity theory we have opposite types of charges (positive and negative) that can either repel or attract each other, and in magnetism theory we have opposite types of poles (north and south) that can either repel or attract each other, but in gravity theory we only have one kind of mass that only attracts and never repels? Why is it that when I drop my chalk it always falls down and never up? Well that's just the way it is. No experiment ever demonstrates the chalk falling up. I don't understand *why* my chalk never falls up; I just accept that. (I also don't run international workshops where all the participants agonize for days about the chalk not falling up.) I also don't understand *why* quantum theory has these spooky nonlocal, unreal, uncertain properties; I just accept it. That's what the theory predicts; that's what the experiments show in the laboratory; that's what I understand to be true; and that's the end of the discussion.

The strangeness and the power of quantum theory—and particularly quantum entanglement—is the engine for a whole host of potential quantum technologies.[17] The quantum computer is perhaps the most famous of these technologies, but there are more. Quantum cryptography has been demonstrated over distances of over 50 miles and used to encrypt banking data, Swiss national election votes, and government secrets. Quantum metrology promises to provide super accurate atomic clocks and novel sensors. Quantum imaging has opened up a whole new ball game in the field of remote sensing and microscopy.

I will touch on these things later in this book. But for now, it is time to stop wailing and gnashing our teeth over the strangeness of quantum theory and just suck it up. It is instead time to exploit that very strangeness of quantum theory for the betterment of humankind. Enough of this incessant whining—let's roll!

NOTES

1. Translated into English by Donald A. Yeats and James East Irby, in the collection *Labyrinths: Selected Stories and Other Writings* by Jorge Luis Borges (New Directions, New York, 2007), http://www.worldcat.org/oclc/86115639.
2. See John F. Clauser, Michael A. Horne, Abner Shimony, and Richard A. Holt, "Proposed Experiment to Test Local Hidden-Variable Theories," *Physical Review Letters*, Volume 23 (1969), pages 880–884, http://link.aps.org/doi/10.1103/PhysRevLett.23.880.
3. Adapted from the conversation between Benjamin Braddock (Dustin Hoffman) and Mr. McGuire (Walter Brooke) in the 1967 film *The Graduate*, where here the word *plastics* has here been replaced with the word *photons*.
4. See "Experimental Test of Local Hidden-Variable Theories" by Stuart F. Freedman and John F. Clauser in *Physical Review Letters*, Volume 28 (1972), pages 938–941, http://link.aps.org/doi/10.1103/PhysRevLett.28.938.
5. See *The Age of Entanglement: When Quantum Physics Was Born* by Louisa Gilder (Knopf, 2008), http://www.worldcat.org/oclc/213765737. See also my light-hearted review of the book "Entanglement with a Twist" by Jonathan P. Dowling in *Science*, Volume 325 (July 17, 2009), page 269, http://www.sciencemag.org/content/325/5938/269.1.summary and http://ageofentanglement.com/science.
6. This was the infamous front-page headline of the *Chicago Tribune* on November 3, 1948, the morning after the election, which erroneously declared that Republican challenger, Thomas E. Dewey, had defeated Democratic incumbent, Harry S

Truman. (Note that this is not a typo. Harry Truman's middle name was the single letter "S," which was not an abbreviation for anything, and so it is properly typeset without the period. Truman had two grandfathers with different given names beginning with the letter "S" so by saddling Harry with no middle name at all except the letter "S," his parents could, with deniable plausibility, claim to each grandfather that Harry was named for *him*.)

7. In mathematics, the term *quadratic* is used to describe something that has to do with the geometric shape of a square, which is a special case of a quadrangle, a two-dimensional object having four sides.

8. These notes are reconstructed from a conversation that took place in March of 2011 with Edward Fry, a quantum experimentalist at Texas A&M University.

9. This is the same Richard A. Holt from the CHSH paper. (See note 2.)

10. See "Experimental Investigation of a Polarization Correlation Anomaly" by John F. Clauser in *Physical Review Letters*, Volume 36 (1976), pages 1223–1226, http://link.aps.org/doi/10.1103/PhysRevLett.36.1223.

11. See "Experimental Test of Local Hidden Variable Theories" by Edward S. Fry and Randall C. Thompson in *Physical Review Letters*, Volume 37 (1976), pages 465–468, http://link.aps.org/doi/10.1103/PhysRevLett.37.465.

12. See "Experimental Realization of Einstein–Podolsky–Rosen–Bohm Gedankenexperiment—A New Violation of Bell Inequalities" by Alain Aspect, Philippe Grangier, and Gérard Roger in *Physical Review Letters*, Volume 49 (1982), pages 91–94, http://link.aps.org/doi/10.1103/PhysRevLett.49.91. (*Gedankenexperiment* is the German compound word for thought experiment.)

13. Adapted from the 1970s US government public service announcements on television and billboards that encouraged adherence to the new, wildly unpopular, but gas-saving *National Maximum Speed Law*, which prohibited states from setting speed limits higher than 55 miles per hour (90 kilometers per hour); "55 MPH—IT'S NOT ONLY A GOOD IDEA, IT'S THE LAW!"

14. See "Violation of Bell's Inequality under Strict Einstein Locality Conditions" by Gregor Weihs, Thomas Jennewein, Christoph Simon, Harald Weinfurter, and Anton Zeilinger in *Physical Review Letters*, Volume 81 (1998), pages 5039–5043, http://link.aps.org/doi/10.1103/PhysRevLett.81.5039.

15. From about 1990 onward, experimentalists replaced the calcium and mercury atom sources for entangled photons with a device called a spontaneous parametric down-converter (SPDC). The problem with the calcium and mercury atoms is that not many pairs of photons came out per second and they came out back to back in random directions so it was unlikely the pair would line up with the detectors. This was bad for the statistics and required lots of data taking. The SPDC works by "down-converting" photons from a bright source of ultraviolet light from a laser into a bright source of entangled pairs of visible or infrared photons that lined up with the detectors. With many more pairs of photons to work with, you can collect data more quickly and greatly reduce your margin of error.

16. *The post-apocalyptic supercomputer simulation loophole* and the film *The Matrix* are both philosophical adaptations of Hilary Putnam's *Brain in a Vat* hypothesis, which is, in turn, an adaptation of the *Cartesian Evil Genius hypothesis*. See "Skepticism

and Content Externalism" by Tony Brueckner in the *Stanford Encyclopedia of Philosophy*, edited by Edward N. Zalta (The Metaphysics Research Lab, Stanford, 2008), http://plato.stanford.edu/entries/brain-vat/.

17. See "Quantum Technology: The Second Quantum Revolution" by Jonathan P. Dowling and Gerard J. Milburn in the *Philosophical Transactions of the Royal Society of London, Series A*, Volume 361 (2003), pages 1655–1674, http://www.jstor.org/stable/3559215. This paper actually began its life as a NASA Technical Brief. As Richard J. Doyle, my former supervisor at the NASA Jet Propulsion Laboratory noted, "That must have been the first time a NASA Tech Brief made it into the *Proceedings of the Royal Society of London!*" I believe that was a compliment.

Chapter 3

The Quantum Codebreaker

The American physicist, Nobel Laureate, Richard Feynman, first proposed the idea of a quantum computer in 1982. In the 1980s, Feynman was thinking a lot about computers and their design and he participated on the development of the Connection Machine, a new type of computer with a million parallel processors.[1] That no one before had ever thought about the ramifications of quantum mechanics on computing was remarkable in that quantum mechanics had been around since about 1925. In contrast, the army completed the construction of the first American programmable electronic digital computer (the "ENIAC" or "Electronic Numerical Integrator and Computer") in 1946, but nobody before Feynman had ever conjectured what quantum theory might have to do with computers. Feynman had a track record of coming up with new ideas and inventing new technologies. Feynman's Nobel Prize was for his work

---

* Photo: "Jorge in Wonderland," by Guglielmo (June 16, 2007). The artist describes the work thusly, "I was sitting in the park when suddenly Jorge Luis Borges came out of nowhere and strolled through the lawn."

on a quantum theory of how light interacts with matter. If Einstein's paper in 1905 on the photoelectric effect was the first word on the matter, Feynman's paper in 1949 was the last word. Feynman constructed a theory of how photons interact with electrons. This theory of quantum electrodynamics was the first such theory to obey Einstein's theory of relativity and give sensible answers. Previous attempts at unifying electrons and photons either did not obey relativity or else predicted nonsensical results such as the charge and the mass of an electron are both infinite. Feynman, along with American physicist Julian Swinger and Japanese physicist Shin'ichirō Tomonaga, swept these infinities under a theoretical rug while leaving behind a sensible theory that gave predictions that agree with high-precision experiments.[2] All three of them shared the 1965 Nobel Prize for this work.[3]

Feynman is also widely credited with inventing the field of nanotechnology with a famous lecture, "There's Plenty of Room at the Bottom," delivered at the 1959 meeting of the American Physical Society held at the California Institute of Technology in Pasadena. The "nano" in nanotechnology comes from the Greek word for "dwarf" and we use it in the scientific unit of length—the nanometer—which is 1 billionth of a meter or 40 billionths of an inch. An atom of silicon, the stuff Intel makes computer chips from, is approximately a quarter of a nanometer in diameter. In his lecture, Feynman speculated on new technologies that would evolve from man's ability to manipulate matter on such nanometer-sized atomic scales.[4] Feynman declared, "And it turns out that all of the information that man has carefully accumulated in all the books in the world can be written in this form in a cube of material one two-hundredth of an inch wide—which is the barest piece of dust that can be made out by the human eye. So there is *plenty* of room at the bottom!"

While Feynman's result is an impressive number, it pales in comparison to the information processing capacity of a quantum computer. In a quantum machine, through the power of entanglement, you can process all the information contained in all the classical computers on Earth in a quantum computer register composed of just 70 silicon atoms, which would form a cube of silicon just about a nanometer on a side. While there may be plenty of room at the bottom, there is plenty more room in the quantum. I will discuss the ramifications of this type of exponential storage capacity of quantum technology in the last chapter of this book, but now back to Feynman's issues with classical computers and how he thought up the quantum computer.

The classical computers we know and love, from our laptops to our iPhones, can do many wonderful things. One of those wonderful things that computers can do is to simulate other wonderful things.

Financed by the US Army, the University of Pennsylvania built the ENIAC during World War II for computing artillery-firing tables. By World War II, the height of artillery science was the development of techniques of "indirect fire" in which the target cannot be seen by the soldier firing the mortar or other big guns. Indirect fire revolutionized warfare in that you could hit targets on the other side of hills or just too far away to see. Artillery guns were powerful enough to shoot such distances by the 1940s but what the army needed was a way to compute where the mortar shell would actually land. In the fateful words of Harvard mathematician Tom Leherer, "'Once the rockets are up, who cares where they come down. That's not my department,' says Wernher von Braun." Indirect fire does you no good if you cannot hit anything. That is where the ENIAC comes in. The artillery firing tables are derived from the science of external ballistics, the science of an unpowered projectile in flight, which accounts for such factors as angle, distance, wind velocity, and other atmospheric conditions, and the Coriolis effect where Earth rotates beneath the shell as it flies. Such calculations are notoriously difficult to carry out accurately by hand and their complexity stumped traditional computational methods.

Before the ENIAC, a bank of 200 female undergraduate students (as well as female military personnel) carried out the army firing table calculations at the University of Pennsylvania. They were equipped with only paper and pencil and adding machines, and did all the calculations by hand. The calculations were broken down into subroutines so that each young woman computed just one small part, and then assemblers collected the answers and assembled them into the firing tables. The army was using these young ladies as one big human computer.[5] Richard Feynman also exploited such female computer calculational strategies in the development of the atomic bomb for the *Manhattan Project*.[6]

Such human computers have a long history. In 1757, the French mathematician Alexis Claude Clairaut and two assistants, a male astronomer (and former Jesuit seminarian) and the wife of the royal clockmaker, using only parchment paper, ink, and goose-quill pens, took on the arduous calculation for predicting the exact time and place of the return of Halley's comet in 1758. By 1794, the director of the French Bureau of Land Registry, Gaspard Clair François Marie Riche de Prony, had assembled an array of nearly 100 human computers from a pool of laid-off servants and hairdressers (gender unknown but likely both male and female) for computing the trigonometry tables used in land surveying. de Prony broke down the calculations into sub-calculations and sub-sub-calculations and so on until he handed each hairdresser one paper form at his desk with numbers already inputted onto it. Her job was to add or multiply the numbers, according to the instructions given to her, and then write his results into little boxes and then hand the form back up the computational pyramid. There at the top of that pyramid, de Prony himself cobbled together all the lower-level computations into a single astronomical result.

In 1821, the English mechanical engineer Charles Babbage adopted this human computer approach of de Prony. Along with the English astronomer John Herschel, Babbage was calculating astronomical navigation tables for the British Navy. Frustrated with errors made by the human computers (as well as the incessant sound from the organ grinders that besieged his home), Babbage designed and prototyped the first programmable, steam-powered, mechanical computer that he christened the *Difference Engine*. (When metalsmiths of the time could not build the parts with sufficient accuracy to construct a working device, the project ran out of steam.) The project did inspire Babbage's patroness, Augusta Ada King, the Countess of Lovelace, to write a computer program for Babbage's engine—the world's first computer programmer was a woman. By the time of the re-return of Halley's comet in 1835 (the year of Mark Twain's birth), the British Admiralty was ready for it. In 1833, they appointed Lieutenant William Samuel Stratford to oversee the calculations for the British Nautical Almanac, and Stratford set up in London a "celestial factory" of human computers to toil away at the comet orbit calculation.[7] Such parallel hand computations are at the extreme limit of human capacity (and also riddled with errors).

It was in order to break such a human computational logjam of the army firing tables that two University of Pennsylvania professors, the American physicist John W. Mauchly and the American electrical engineer J. Presper Eckert, built the ENIAC. The ENIAC was remarkable for a number of reasons, not the least of which was its size. The machine contained over 17,000 vacuum tubes, 5 million hand-soldered connections, weighed nearly 30 tons (metric or Imperial), and at over 167 square meters (1800 square feet), it filled an entire warehouse (see Figure 3.1). That size prompted the infamous prediction that "Computers of the future may have only 1000 vacuum tubes and weigh only 1.5 tons," in a 1949 issue of *Popular Mechanics*. As American baseball player Yogi Berra and Danish quantum physicist Niels Bohr were both fond of saying, "Predictions are very difficult to make, especially about the future." (My iPhone has more memory and processing power than did the ENIAC.) The University of Pennsylvania completed the ENIAC in 1946 and then they transferred it to the army's Aberdeen Proving Ground in Maryland in 1947, where it remained in operation until 1955. In a curious twist of fate, the ENIAC never did the job the army designed it to do—calculate those blasted artillery firing tables.

By 1947, World War II was over and the United States had built, tested, and used the atomic bomb to end that war. By 1947, it was World War III that was on everybody's mind, and the Hungarian mathematician John Von Neumann, working at the Los Alamos weapons laboratory in New Mexico, harnessed the power of the ENIAC and bent it to his own will. The first calculations done by the ENIAC were simulations of hydrogen bomb explosions. Computer simulations of nuclear explosions began with the ENIAC and continue to this day. At

Figure 3.1   This is 1946 photo of the ENIAC computer at the University of Pennsylvania. Physicist J. Presper Ekert and electrical engineer John Mauchly (in suits and ties, center, left to right) pretend to program the ENIAC that they designed and built. The true programmers are Betty Jean Jennings (back right) and Ruth Licterman (front right). Also shown are army electrical engineer PFC Homer Spence (in uniform, back left) and Army Ballistic Research Laboratory liaison, Captain Herman Goldstine (in uniform, back right). (This is a 1946 publicity photo of the ENIAC, "Bird's Eye View," described as "Black and white image of the ENIAC computer": [left to right] PFC Homer Spence, Chief Engineer Presper Eckert, Consulting Engineer Dr. John Mauchly, Betty Jean Jennings, Ballistic Research Laboratory—University of Pennsylvania Liaison Officer Captain Herman Goldstine, and Ruth Licterman. Photo courtesy of the Computer History Museum, http://www.computerhistory.org/collections/accession/102622385. This work is in the public domain in the United States because it is a work prepared by an officer or employee of the United States Government as part of that person's official duties under the terms of Title 17, Chapter 1, Section 105 of the US Code.)

Los Alamos, there are banks and banks of the fastest supercomputers in the world, whose sole job is to simulate nuclear explosions, so that we do not have to test our aging nuclear arsenal by exploding live bombs. Since the United Nations adopted the Comprehensive Nuclear-Test-Ban Treaty in 1996, neither the United States nor Russia has detonated a single nuclear weapon. They do not have to—with their supercomputers, they can just simulate the detonations and we all can (literally) breathe easier. And so it was that the first digital computer was a physics simulator, and in 1982, Feynman, who worked on the atomic bomb, pointed out that there were quantum limits to the simulation capacity of any such classical computer. The classical computer age began with simulations on the ENIAC and the quantum computer age began with Feynman's conjecture about simulations on a quantum computer.

THE TROUBLE WITH THULIUM

In 1982, Feynman pointed out that there was a problem with classical computers simulating complex quantum physical chemistry problems.[8] Particularly, for very specific problems, classical computers set to a quantum simulation suffered an exponential slowdown in performance. Feynman in his brilliance turned this observation around and predicted that perhaps a computer running on quantum principles—a quantum computer—should experience an exponential speedup set to simulating classical problems. In order to understand what that all means, we need to understand what a simulation might be and what Feynman's problem was (see Box 3.1).

BOX 3.1   SIMULATION OF A QUANTUM SYSTEM ON A CLASSICAL COMPUTER

Alice and Bob are playing a game of classical "heads or tails" with a *pair* of ordinary coins. New rules: Alice flips one coin and Bob flips the other. If both get heads then Alice wins. If both get tails then Bob wins. If the outcome is a head-and-tail, or a tail-and-head, then it is a tie and they play again. Using our same notation from Chapter 1, we take ($\uparrow$) to be heads and ($\downarrow$) to be tails.[9] This is all classical, mind you. What are the possible outcomes in any one play? There are four and they are $(\uparrow)_A (\uparrow)_B$ or $(\uparrow)_A (\downarrow)_B$ or $(\downarrow)_A (\uparrow)_B$ or $(\downarrow)_A (\downarrow)_B$. In English, Alice and Bob both get heads, or Alice gets heads and Bob gets tails, or Alice gets tails and Bob gets heads, or Alice and Bob both get tails. There is a 25% chance that Alice will win, a 50% chance of a tie, and a 25% chance that Bob will win. After playing this for some hours, their hands get tired and Alice suggests that they program a special-purpose analog computer on her workbench to simulate the game. Using some simple electronic gizmos, Alice solders up a little black box with two circuits: one circuit to simulate her coin and another circuit to simulate Bob's coin. There are two coins so she needs two circuits. Each circuit has a classical logical bit, represented by a binary one or a zero, to represent the coins: a one is a head and a zero is a tail. Each circuit contains a pseudo-random number generator. When either of them presses the button on the device, Alice's number generator randomly (with a 50–50 probability) puts Alice's bit to zero (tails) or heads and Bob's number generator does the same with his bit. The display has windows with glowing red digits that display the four possible outcomes $(1)_A (1)_B$ or $(1)_A (0)_B$ or $(0)_A (1)_B$ or $(0)_A (0)_B$. Again, in English, this reads heads–heads, heads–tails, tails–heads, or tails–tails. This is a perfectly efficient simulation of the game in that the two circuits replace the two coins. The simulation would be less efficient if we would require say four circuits to simulate the two coins.

After hours of this fun game, Alice and Bob decide to whip out their quantum atomic pocket watches and crack into them to devise a quantum version of the game. Now, the ions are the coins and they are prepared in a two-particle entangled state of the form $|\rightarrow\rangle_A |\leftarrow\rangle_B + |\leftarrow\rangle_A |\rightarrow\rangle_B$, or in English, ions are in a quantum superposition 3:00–9:00 and 9:00–3:00. Here, Alice and Bob agree that 12:00 is heads and 6:00 is tails and they only slide the viewing slits to look at those two numbers; that is, they make measurements only in the up–down direction ($\updownarrow$). Therefore, when Alice measures the cat state $|\rightarrow\rangle_A$ (alive and dead) in the up–down direction here, ion state randomly collapses to either $(\uparrow)_A$ (alive) or to $(\downarrow)_A$ (dead) with 50–50 odds, and similarly with Bob. To make the game more interesting, Alice in Aurora and Bob in Brockley decide to randomly each choose 12:00 or 6:00, then look at their watches, and play the game remotely by Skype. So sometimes Alice and Bob are both on 12:00 (heads–heads), or sometimes Alice is on 12:00 and Bob is on 6:00 (heads–tails), or sometimes Alice is on 6:00 and Bob on 12:00 (tails–heads), and finally sometimes Alice and Bob are both on 6:00 (tails–tails). What do they see? Well, they sometimes see $(\uparrow)_A (\uparrow)_B$ or $(\uparrow)_A (\downarrow)_B$ or $(\downarrow)_A (\uparrow)_B$ or $(\downarrow)_A (\downarrow)_B$ just as in the classical coin toss, provided they force the lasers to measure the ion spin direction only in the up–down direction ($\updownarrow$). However, because of quantum uncertainty and unreality, other times they see other things like $(\varnothing)_A (\uparrow)_B$ or $(\uparrow)_A (\varnothing)_B$ or $(\downarrow)_A (\varnothing)_B$ or $(\downarrow)_A (\varnothing)_B$ or even $(\varnothing)_A (\varnothing)_B$. The symbol ($\varnothing$) means they see nothing through their slit. How could they see $(\varnothing)_A (\varnothing)_B$? Because they both choose the up–down direction ($\updownarrow$) to measure in, it is possible that Alice looks at 12:00 but her ion collapses to 6:00, and she sees nothing, and Bob looks at 6:00 but his ion collapses to 12:00 and he sees nothing. These all occur with various probabilities, but Alice and Bob decide that $(\uparrow)_A (\downarrow)_B$ or $(\downarrow)_A (\uparrow)_B$ or $(\varnothing)_A (\uparrow)_B$ or $(\uparrow)_A (\varnothing)_B$ or $(\downarrow)_A (\varnothing)_B$ or $(\downarrow)_A (\varnothing)_B$ or $(\varnothing)_A (\varnothing)_B$ should all be called a tie and then they play again.

So we have not changed the game much, replaced two classical coins with two entangled quantum ions, but there are many more outcomes in the quantum game. What is going on? Well, in classical heads or tails, with classical coins, the game obeys Einstein's locality, reality, and certainty conditions. In quantum heads or tails, with quantum-entangled ions, the game obeys the nonlocality, unreality, and uncertainty conditions. We can explain classical heads or tails with a local hidden variable theory. Quantum heads or tails requires full quantum theory. Therein lies the rub.

As before, Alice and Bob get tired of playing quantum heads or tails and Alice decides to build a classical analog computer on her workbench to simulate the game. Alice discovers that in order to get all the new outcomes, with all their various probabilities, just right, her classical circuit simulator board requires at least four circuits instead of the classical two. That is double the number of circuits: 2×2, which is 4. It is the infinite library of Borges all over again; we need more computational resources to access the full quantum library. The scaling is very bad.

If Alice and Bob played quantum heads or tails with three ions, the classical simulator would need $2 \times 2 \times 2$, which is 8 classical circuits. If Alice and Bob play with four ions, the classical simulator would need $2 \times 2 \times 2 \times 2$ or 16 classical circuits. The number of classical simulator circuits grows exponentially with the number of quantum ions. This was Feynman's observation. Simulating a quantum system on a classical computer is always a losing game. So no more games, let us go to quantum chemistry.

The hydrogen atom, with one positively charged proton in the center and one electron, has an energy level structure (a chemistry) that is easy to simulate on a laptop and even easy to work out by hand. How about helium? Helium has a nucleus that contains two positively charged protons that is orbited by *two* negatively charged electrons. The electrons, like the ions, have an internal spin degree of freedom and their quantum spin direction can point up, down, left, right, and so on. While hydrogen is easy to solve, helium is every physics students' nightmare. Why is that? The two helium electrons are entangled. The two helium electrons can be found in any of the four quantum states, $|\uparrow\rangle_A |\downarrow\rangle_B + |\downarrow\rangle_A |\uparrow\rangle_B$, $|\uparrow\rangle_A |\uparrow\rangle_B + |\downarrow\rangle_A |\downarrow\rangle_B$, $|\uparrow\rangle_A |\uparrow\rangle_B$, and $|\downarrow\rangle_A |\downarrow\rangle_B$. The first two of these are entangled. This gives my students headaches. The energy levels for the first state are different than the last three. If I were to build a simple analog computer circuit board, like Alice with the two ions, I would be forced to build four circuits to simulate all the properties of the chemistry of the two electrons of helium. Four circuits for two electrons, this does not look good.

Fortunately for most atoms with huge numbers of electrons, there is very little entanglement between the electrons, and their chemical properties can be simulated reasonably well on classical computers. If the number of electrons is large, we may have to use supercomputers, but still the chemistry problem is tractable. But then there is thulium. Thulium is a chemical element that has an atomic number of 69, which means it has a nucleolus with 69 positive charged protons surrounded by a cloud of 69 orbiting negative charged electrons.[10] In addition, as bad luck would have it, every electron is entangled with every other electron. So there are only 69 electrons but a single generic quantum state of the thulium electrons looks like this:[11]

$$|\uparrow\rangle|\uparrow\rangle|\uparrow\rangle|\uparrow\rangle|\uparrow\rangle|\uparrow\rangle|\uparrow\rangle|\uparrow\rangle|\uparrow\rangle|\uparrow\rangle|\uparrow\rangle|\uparrow\rangle|\uparrow\rangle|\uparrow\rangle|\uparrow\rangle|\uparrow\rangle|\uparrow\rangle|\uparrow\rangle|\uparrow\rangle|\uparrow\rangle\cdots|\uparrow\rangle|\uparrow\rangle|\uparrow\rangle.$$

Although I cannot fit them on one line, there are 69 arrows named Alice, Bob, Charlie, Dave, Eve, ... and Zardoz. (I will have to go through the alphabet almost three times for all the names.) That already looks complicated but that is just one possible state. There are many, many, many more and they look like this:

$$|\uparrow\rangle|\uparrow\rangle|\uparrow\rangle|\uparrow\rangle|\uparrow\rangle|\uparrow\rangle|\uparrow\rangle|\uparrow\rangle|\uparrow\rangle|\uparrow\rangle|\uparrow\rangle|\uparrow\rangle|\uparrow\rangle|\uparrow\rangle|\uparrow\rangle|\uparrow\rangle|\uparrow\rangle|\uparrow\rangle|\uparrow\rangle|\uparrow\rangle\cdots|\uparrow\rangle|\uparrow\rangle|\uparrow\rangle+$$
$$|\downarrow\rangle|\uparrow\rangle|\uparrow\rangle|\uparrow\rangle|\uparrow\rangle|\uparrow\rangle|\uparrow\rangle|\uparrow\rangle|\uparrow\rangle|\uparrow\rangle|\uparrow\rangle|\uparrow\rangle|\uparrow\rangle|\uparrow\rangle|\uparrow\rangle|\uparrow\rangle|\uparrow\rangle|\uparrow\rangle|\uparrow\rangle|\uparrow\rangle\cdots|\uparrow\rangle|\uparrow\rangle|\uparrow\rangle+$$
$$|\uparrow\rangle|\downarrow\rangle|\uparrow\rangle|\uparrow\rangle|\uparrow\rangle|\uparrow\rangle|\uparrow\rangle|\uparrow\rangle|\uparrow\rangle|\uparrow\rangle|\uparrow\rangle|\uparrow\rangle|\uparrow\rangle|\uparrow\rangle|\uparrow\rangle|\uparrow\rangle|\uparrow\rangle|\uparrow\rangle|\uparrow\rangle|\uparrow\rangle\cdots|\uparrow\rangle|\uparrow\rangle|\uparrow\rangle+$$
$$|\uparrow\rangle|\uparrow\rangle|\downarrow\rangle|\uparrow\rangle|\uparrow\rangle|\uparrow\rangle|\uparrow\rangle|\uparrow\rangle|\uparrow\rangle|\uparrow\rangle|\uparrow\rangle|\uparrow\rangle|\uparrow\rangle|\uparrow\rangle|\uparrow\rangle|\uparrow\rangle|\uparrow\rangle|\uparrow\rangle|\uparrow\rangle|\uparrow\rangle\cdots|\uparrow\rangle|\uparrow\rangle|\uparrow\rangle+$$

$$\cdot$$
$$\cdot$$
$$\cdot$$

$$|\downarrow\rangle|\uparrow\rangle|\uparrow\rangle|\uparrow\rangle|\uparrow\rangle|\uparrow\rangle|\uparrow\rangle|\uparrow\rangle|\uparrow\rangle|\uparrow\rangle|\uparrow\rangle|\uparrow\rangle|\uparrow\rangle|\uparrow\rangle|\uparrow\rangle|\uparrow\rangle|\uparrow\rangle|\uparrow\rangle|\uparrow\rangle|\uparrow\rangle\cdots|\uparrow\rangle|\uparrow\rangle|\uparrow\rangle+$$
$$|\downarrow\rangle|\downarrow\rangle|\uparrow\rangle|\uparrow\rangle|\uparrow\rangle|\uparrow\rangle|\uparrow\rangle|\uparrow\rangle|\uparrow\rangle|\uparrow\rangle|\uparrow\rangle|\uparrow\rangle|\uparrow\rangle|\uparrow\rangle|\uparrow\rangle|\uparrow\rangle|\uparrow\rangle|\uparrow\rangle|\uparrow\rangle|\uparrow\rangle\cdots|\uparrow\rangle|\uparrow\rangle|\uparrow\rangle+$$
$$|\downarrow\rangle|\uparrow\rangle|\downarrow\rangle|\uparrow\rangle|\uparrow\rangle|\uparrow\rangle|\uparrow\rangle|\uparrow\rangle|\uparrow\rangle|\uparrow\rangle|\uparrow\rangle|\uparrow\rangle|\uparrow\rangle|\uparrow\rangle|\uparrow\rangle|\uparrow\rangle|\uparrow\rangle|\uparrow\rangle|\uparrow\rangle|\uparrow\rangle\cdots|\uparrow\rangle|\uparrow\rangle|\uparrow\rangle+$$
$$|\downarrow\rangle|\uparrow\rangle|\uparrow\rangle|\downarrow\rangle|\uparrow\rangle|\uparrow\rangle|\uparrow\rangle|\uparrow\rangle|\uparrow\rangle|\uparrow\rangle|\uparrow\rangle|\uparrow\rangle|\uparrow\rangle|\uparrow\rangle|\uparrow\rangle|\uparrow\rangle|\uparrow\rangle|\uparrow\rangle|\uparrow\rangle|\uparrow\rangle\cdots|\uparrow\rangle|\uparrow\rangle|\uparrow\rangle+$$

$$\cdot$$
$$\cdot$$
$$\cdot$$

$$|\downarrow\rangle|\downarrow\rangle|\downarrow\rangle|\downarrow\rangle|\downarrow\rangle|\downarrow\rangle|\downarrow\rangle|\downarrow\rangle|\downarrow\rangle|\downarrow\rangle|\downarrow\rangle|\downarrow\rangle|\downarrow\rangle|\downarrow\rangle|\downarrow\rangle|\downarrow\rangle|\downarrow\rangle|\downarrow\rangle|\downarrow\rangle|\downarrow\rangle\cdots|\uparrow\rangle|\downarrow\rangle|\downarrow\rangle+$$
$$|\downarrow\rangle|\downarrow\rangle|\downarrow\rangle|\downarrow\rangle|\downarrow\rangle|\downarrow\rangle|\downarrow\rangle|\downarrow\rangle|\downarrow\rangle|\downarrow\rangle|\downarrow\rangle|\downarrow\rangle|\downarrow\rangle|\downarrow\rangle|\downarrow\rangle|\downarrow\rangle|\downarrow\rangle|\downarrow\rangle|\downarrow\rangle|\downarrow\rangle\cdots|\downarrow\rangle|\uparrow\rangle|\downarrow\rangle+$$
$$|\downarrow\rangle|\downarrow\rangle|\downarrow\rangle|\downarrow\rangle|\downarrow\rangle|\downarrow\rangle|\downarrow\rangle|\downarrow\rangle|\downarrow\rangle|\downarrow\rangle|\downarrow\rangle|\downarrow\rangle|\downarrow\rangle|\downarrow\rangle|\downarrow\rangle|\downarrow\rangle|\downarrow\rangle|\downarrow\rangle|\downarrow\rangle|\downarrow\rangle\cdots|\downarrow\rangle|\downarrow\rangle|\downarrow\rangle.$$

Now that looks bad and it is bad. I cannot write out all the possible states for you because even writing at a speed of one state per second (in the Microsoft Word equation editor), it would take me 20 trillion years, and that's without taking any potty breaks. Because the universe is only 14 billion years old, working out the chemical properties of thulium on a handheld calculator at one calculation per second would take 1336 universe lifetimes—long after my retirement savings have run out. That is time, what about space? Each state takes up approximately 0.5 inches by 6 inches on the page, so 3 square inches per state. The page has a writing area of 6 inches by 9 inches, or 54 square inches per page. Therefore, that is approximately 18 quantum states per page. To write down the entire entangled electron spin state of the thulium atom, I would need 30,000,000,000,000,000,000 pages. The diameter of the known universe is 5,000,000,000,000,000,000,000,000,000 inches, so those are enough pages to stretch across the universe and back 15 million times. Where I am getting these whopping big numbers? Simple. Each electron has two possible states, $|\uparrow\rangle$ or $|\downarrow\rangle$ (up or down). There are 69 electrons. The total number of states then are $2 \times 2 \times 2 \ldots \times 2$, where there are 69 twos in there, which is 600,000,000,000,000,000,000, or 600 quintillion possible states (Figure 3.2). The entanglement ensures that the thulium atom is in all of these states simultaneously.

Not fair, you say. I do not have to work out the properties of thulium on scraps of paper or a handheld calculator; I can use a computer. How big of a computer will I need? I will take a bunch of Intel's "monster chips," the 2010 Itanium 9300 series microprocessor (codenamed *Tukwila*) (Figure 3.3), with 2 billion transistors each, and hook them all up into one big supercomputer.[12] How many Tukwila chips will I need? Well at a minimum, I will need at least one transistor per quantum state. (That is an extremely conservative estimate. I will likely need several transistors per quantum state so this calculation will be a low-ball estimate.) Okay so that's 600 quintillion states, divided by 2 billion transistors ... carry the quadrillion ... and so I need 300 billion Tukwila chips to simulate the chemistry of a single atom of thulium.

Intel does not publicly release its yearly production data, but one industry analyst estimated a production of around 200,000 for the year 2007. Therefore,

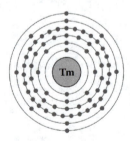

Figure 3.2   A schematic of the 69 electrons arranged in the six energy levels of the thulium atom. Every electron is highly entangled with every other electron, making classical simulations of the properties of thulium exponentially inefficient. The 69 electrons occupy 600 quintillion quantum states at once. (The image is "Electron Shell Diagram for Thulium, The 69th Element in the Periodic Table of Elements" and original work by Greg Robson.)

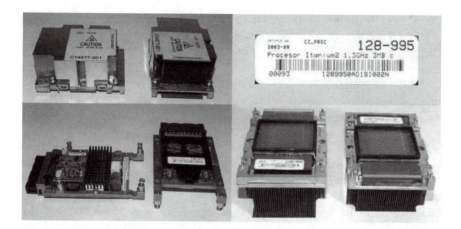

Figure 3.3   Intel Itanium 2 processor. Because of the exponential overhead involved in simulating a quantum system on a classical computer, we need more than 300 billion such chips to simulate the chemistry of a single thulium atom. (Photos of the Intel Itanium 2 Processor 1.3 GHz 3 MB Cache, taken by Piast [October 22, 2007].)

that is 300 billion chips, divided by 200,000 chips, per year. To simulate the properties of a single thulium atom, I need to buy up the entire worldwide production of Intel's Tukwila chips for the next 1.5 million years. Once I have my 300 billion chips, which at $2000 each will cost me a total of $600 trillion (which is 20 times the yearly gross national product of the entire Earth in 2010, but thankfully I have 1.5 million years to save it up), then I just have to hook all the chips together into a supercomputer. I'll assume that the chips will need a chassis (server) to hold them in and maybe after 1.5 million years we'll be able to put 100,000 of the chips in a single server and if each server weighs 455 kilograms (1000 pounds)—got to include power and cooling and all that—we get a machine that weighs approximately 1.5 tons (metric or Imperial). Well, that 1949 issue of *Popular Mechanics* was right all along, "Computers in the future may weigh no more than 1.5 tons." Now, I trust you are beginning to see what the trouble is with thulium.

The computer science lingo for this startling complicated computational situation is that a classical computer cannot simulate the chemistry of thulium "efficiently." Simulating thulium carries an exponential overhead in size and power and time, and this exponential overhead makes the process inefficient. What we seek is a way to simulate the chemistry of thulium efficiently. The hint lies with thulium itself.

Suppose I take a bare thulium nucleus with 69 positively charged protons and drop a handful of 69 negatively charged electrons onto it. How long does it take those electrons to settle into the entangled ground state of thulium? Does it take 1.5 million years? Nope. It takes less than a billionth of a second for all the electrons to fall into their happily entangled ground state. Now I can take a bunch of laser beams, read out the energy levels of thulium in a few hours, and deliver to you the chemical structure of thulium. Problem solved. There is an exponential overhead in the thulium atom itself. I get the energy levels out with very little time or effort. What is going on? The thulium atom is an analog computer—a quantum computer—that efficiently simulates itself.

Not fair, you say. We cannot allow computers to simulate themselves. Well we can, but that is another story. Fair enough; so instead, we take a bare nucleus of ytterbium (with 70 positively charge protons in the nucleus) and drop 69 electrons onto that.[13] The result, also in a billionth of a second, is a singly ionized ytterbium atom whose chemical properties are very nearly the same as those of thulium. We read out the ionized ytterbium energy levels with laser beams, apply a correction for the systematic errors related to this not being thulium (the charge of the nucleus is larger), and we get the energy levels of thulium again, or at least close enough for government work. Singly ionized ytterbium is an analog quantum computer that efficiently simulates thulium. This observation was Feynman's first breakthrough. Classical computers seem to experience an exponential slowdown when put to simulating entangled quantum systems. Yet, that same entangled quantum system shows no exponential slowdown when simulating itself. The entangled quantum system acts like a computer that is exponentially more powerful than any classical computer.

Where is this exponential overhead coming from? If we treated the thulium atom completely classically, with the electrons as balls orbiting the nucleus, what physical input would we need to simulate that? Well, the electrons are flying around in three-dimensional space, so I would need at least three spatial pieces of information. In spherical coordinates, I would have to specify the initial latitude, longitude, and altitude of each electron with respect to the nucleus, like in my GPS (global positioning system). So maybe three numbers corresponding to the three dimensions of space is enough. That is not enough to predict the electron's future position. For that, I also need to specify the initial three components of the electron's velocity in the north–south, east–west, and up–down directions. That brings up to six numbers, three for the space and three for the velocity. This would completely predict the electron's location in the future classically. We call the six-dimensional representation classical phase space, where we treat space and velocity on equal footing. That is all we need to track each electron, six numbers. There are 69 electrons, so a classical simulation of classical thulium would require 6 × 69 or 414 numbers total. Compare that to the 600 quintillion numbers that are required for the classical simulation of the quantum thulium. This is Borges infinite Library of Babel all over again.

The quantum simulation of the 69 electrons must specify all possible 600 quintillion states simultaneously. Entanglement is to blame. If none of the electrons were entangled with any of the others, then I would have to just specify 1 of the 600 quintillion states. Accounting only for electron spin (up or down) (and not the electron position or velocity), that would be 2×69 or 138 numbers needed to specify just 1 of the 600 quintillion electron spin states. (If I wanted to specify position and velocity too, then I would need to multiply by six numbers, three for position and three for velocity, to get $6 \times 2 \times 69$ or 828 numbers needed to specify just one of the unentangled states.) However, with entanglement in the picture, I cannot get away with just specifying 1 of the 600 quintillion states, I need to specify all of them simultaneously. Where do these 600 quintillion states live? Well, they are not in three-dimensional position space or three-dimensional velocity space (or in two-dimensional spin space); they are not in any space we have encountered before. The 600 quintillion states live in an abstract mathematical space called Hilbert space (after the German mathematician David Hilbert).

In the Many-Worlds interpretation of quantum mechanics, the entire universe has split into 600 quintillion parallel universes, each containing a different state. When we make a measurement, the 600 quintillion possibilities collapse at random into just one reality. Before we make the measurement, if we are careful, we can steer the thulium atom (or the ytterbium ion) through these 600 quintillion parallel universes to efficiently extract the chemical properties of thulium. For example, we could take the ytterbium ion and apply a slowly changing electric field and watch with laser beams how its energy levels shift about. Then using this result, we could predict how the energy levels of thulium shift about in the same changing electric field. If we tried to make that prediction by direct classical computation, we would be back to 1.5 million years of 300 billion Tukwilas. However, somehow in just a few hours, we get the answer out of the entangled state of a single ytterbium ion. In the Many-Worlds interpretation, ytterbium is an analog computer with 600 quintillion parallel processors. The shocking point is that all but 1 of those 600 quintillion parallel processors are in parallel universes other than our own; parallel universes where we don't have to buy the chips or pay the electric bill. With the brick and mortar of quantum entanglement, we have built an inter-dimensional portal into Hilbert space—where all those parallel universes lie—and by carefully grasping through this portal, we can manipulate all 600 quintillion of these quantum states and exploit them for our own ends.[14]

In computer science, or the closely related science of information theory, we say that information is "fungible." This is a word most often used in the context of money. If you give your middle school kid $5 lunch money, there is nothing to stop her from spending it all on video games; the money is fungible. Cash is fungible. If I give you a $10 bill and you give me back a different $10 bill, we're still even because the value of $10 is independent of which bill it is sitting on; $10 is fungible. If I ask you to divide 1024 by 16, you may decide to use paper and pencil or an abacus or a calculator. The steps required and the information you use are the same in each case but the type of computer is different in each case. In the first case, the computer is a combination of your brain and your fingers and the paper and pencil. In the second case, the computer is your brain and your fingers and the beads on the abacus. In the third case, the computer is your brain and your fingers and the calculator. The information required to divide 1024 by 16 is fungible—it is independent of the hardware it is stored on and processed with. The recipe for baking a cake is fungible. It does not matter if the two-liter mixing bowl is glass or ceramic, if the quarter of a kilogram of butter is soft or hard, or if you frost with a one- or two-sided spatula; you get the same cake. The 69 entangled electrons in ytterbium contain the same quantum information of the atomic structure of thulium, as do the 69 entangled electrons in the original thulium atom; quantum information is fungible. So next, we can take a radical step.

What if, instead of 69 electrons in an atom, NIQuIST (National Institute of Quantum Information Standards and Technology) builds us a trap with 69 positively charged ions? The ions also can be prepared in the same entangled state, but now getting them to simulate thulium is more complicated. I cannot just drop them on a nucleus with 69 negative charges because nuclei are only positively charged. (If I could, I might actually get a good simulation.) Therefore, I have to leave the ions in the trap and program them somehow. To program them, I need a quantum programming language and a design for the ions in a quantum computing architecture. Simulating thulium with an analog quantum computer made from ytterbium is fine but we want something more. The ENIAC was famous not because it was the first computer but because it was the first programmable digital computer that we could reprogram to handle any classical computational problem. We need that, a general digital programmable quantum computer. Because quantum information is also fungible, by definition, it should not matter whether we use ytterbium or an ion trap to simulate thulium, but ytterbium is only good for simulating thulium (and itself). We could reprogram a programmable quantum computer to simulate, say, hafnium.

There are problems in condensed matter physics, the physics of solids, where atoms with their spins sit on regular sites on a crystal lattice. In certain magnetic materials, the spin of each electron is highly entangled with all the other spins of all the other electrons—the electron spins are in an identity crisis and cannot decide which way to point.[15] It is classically computationally inefficient to simulate 69 spins in such a magnetic spin lattice, just as it is computationally inefficient to simulate thulium. A programmable quantum computer would be able to tackle both jobs. Studies of magnetic materials in the past 20 years have led to breakthroughs in classical computer memory. This is why I can today buy a USB (universal serial bus) flash drive with 32 gigabytes of memory, whereas in 1981, Bill Gates thought 640 kilobytes was all anybody would need for a long time.[16] There is a vast number of interesting simulation problems that are intractable on a classical computer; these would benefit immensely from a quantum machine. A reprogrammable quantum simulator would open up entire new research directions in material science research and nanotechnologies. How do we design, build, and program a quantum computer?

Feynman made the first pitch for this idea in his 1982 paper where he proposed building a "universal quantum simulator," a quantum computer that would be able to simulate any complex quantum system like thulium or singling ionized ytterbium or frustrated spins in a crystal lattice. Feynman even proposed a simple quantum computer architecture. This paper launched the race to find a blueprint for a general-purpose universal quantum computer. To understand what a universal quantum computer would be, we should spend some time to understand what a universal classical computer is so we can compare the two concepts.

TURING MACHINES AND A DEUTSCH TREAT

In 1936, 10 years before the army built the ENIAC, the English mathematician Alan Turing was thinking up a thought experiment for a general-purpose programmable (classical) computer. Turing imagined a machine, now called a "Turing Machine," that consists of three parts: a paper tape with numbers and symbols on it (typically zeros and ones), a tape head that can move the tape back and forth and read and write and erase the symbols on it, and an instruction table that tells the tape head what to do (Figure 3.4). Any detailed description of a Turing machine is too elaborate for this book, but suffice it to say it was a simple imaginary model of a computer.[17] The tape is the memory of the machine, keeping track of what the head has done, and part of the program of the machine, determining what the head will do at each number. The instruction table is the rest of the program and tells the head what to do on each tape position: "If you see a zero on the tape at this position erase it and replace it with a one."

Turing was able to show that this machine, while a simple device, captured the essence of a general-purpose, digital, programmable computer. That is, he proved that a Turing machine efficiently simulates any digital general-purpose programmable classical computer. We know what that word "efficiently" means—it means without an exponential overhead—we do not need 1.5 million years and 300 billion Tukwilas. It turns out that all classical, general-purpose, programmable classical computers can efficiently simulate themselves, each other, and a Turing machine. In this sense, they are all called universal classical Turing machines or universal classical computers. To clarify, if you design a new model of a classical computer, to join the club, you must first show that

Figure 3.4 This is a fanciful depiction of a Turing Machine. The tape has all the information stored in the binary code of zeros and ones. The machine can read, write, or erase the numbers and move the tape back and forth. The machine has an instruction set that along with the tape acts like the computer program. A Turing machine can efficiently simulate any universal classical computer and vice versa.

it can efficiently simulate an arbitrary classical Turing machine. Once done, by abstraction, your new model joins the entire set of universal classical computers. In this set, any machine can efficiently simulate any other machine. Building on Turing's work, Ekert and Mauchly designed the ENIAC as a Turing machine in this universal class. Even though my iPhone is more powerful than the ENIAC, it is not exponentially more powerful, and therefore I can use my iPhone to efficiently simulate the ENIAC and the ENIAC to efficiently simulate my iPhone. We can relate these notions to the computer science term *computational complexity*. The computational problem of one classical computer simulating another is in the same computational complexity class; that is, the overhead in the resources required are not exponential.

This takes us up to the quantum computer. We took great pains to show that a classical computer cannot efficiently simulate a quantum computer. The classical supercomputer made up of 300 billion Tukwilas cannot efficiently simulate a quantum computer made up of a thulium atom simulating itself, or the ytterbium atom simulating the thulium atom, or a bunch of entangled ions in the trap simulating either. Whatever a quantum computer is, it is not a classical Turing machine. There is that exponential overhead in simulating it. Therefore, we need a new Turing machine, a quantum one, and that is where the Israeli–British physicist David Deutsch comes in. In 1985, Deutsch constructed a model of a quantum Turing machine and then showed that no classical Turing machine could efficiently simulate it.[18]

Nevertheless, thanks to David Deutsch, we now have a way to think about programming and reading and writing information out of our ion trap quantum computer to do quantum simulations and quantum chemistry. We must search for computer designs in this new universal quantum computing class. In summary, a universal classical computer can efficiently simulate any other classical computer; a universal quantum computer can efficiently simulate any other quantum computer; a universal classical computer *cannot* efficiently simulate a quantum computer; a universal quantum computer *can* efficiently simulate a classical computer. This hierarchy gives us a way to quantify just in what way a quantum computer is more powerful than a classical one. In terms of computational complexity, the mathematical problem of simulating a quantum computer is in a different (exponentially harder) complexity class than the problem of simulating a classical computer.

These Turing machines are the touchstones of computer science and we use them to tell a universal computer from one that is not. If I can show your machine is just as good as a Turing machine, then we accept it as a legitimate computer in the universal class. All the computers in the classical set are digital; the information is stored as zeros or ones in typical binary code. Any base-10 number or in fact any ASCII (American Standard Code for Information Interchange) character, number, or text can be converted into binary (and back), which is the language of the classical computer. The Turing machine only has zeros and ones (and maybe a few special commands like HALT) on its paper tape. There is some overhead in the conversion to binary. For example, "Alice" becomes "010000010110110001 10100101100 0110110010100001101000001010," and "Bob" becomes "010000100110111 1011000100 000110100001010," and the number "1984" is rewritten as "11111000000."[19] However, this overhead is not exponential, and so the conversion to binary is "efficient." The language of binary is tough on humans but marvelously liberating for calculating machines. We put all text and numbers into a common base-2 language. Even better, computer designers have to only manipulate "0" and "1" instead of all the base-10 digits (0, 1, 2, 3 4, 5, 6, 7, 8, 9), all the letters of the alphabet, all the special punctuation marks, and so on. This simple binary language means then a computer design that can be boiled down to just the rearrangement of a few elementary widgets or "gates" that manipulate the zeros or ones.

Nobody these days builds a classical computer using the Turing machine model. They instead use a circuit model that is easier to lay out on a computer chip. You feed your information into the circuit (encoded in binary) and the circuit manipulates the input and then outputs your answer. It may be the energy levels of hydrogen or your income tax data. Your laptop converts your tax data into binary code, TurboTax then runs it through the circuit, and then your laptop converts the result back into base 10 and fills out your Form 1040 for you. A remarkable result is that you just need three different classical gates to build an arbitrary universal classical computer. These gates are the AND, the OR, and the NOT gates. These gates are the computer designer equivalent of Lego "bricks." With just these three, you can build any computer worth building. Because different combinations of these three gates can build any universal classical computer, we call them a universal classical gate set. I am going to spend a bit of time now showing how they work so we can compare them to the universal quantum gate set that will come.

In the computer circuit, we draw horizontal lines that represent the binary bits moving through the circuit from left to right as we run the calculation. In classical computers, a bit is a transistor that can be in two possible distinct states or positions that we will call up or down. If the transistor is up, then the bit is in the state "1" and if the transistor is down, then the bit is in the state "0." The gate is made of more transistors and you should think of it as a black box that takes in one or two bits and outputs one or two bits. The key to the gate operation is that the output depends on the input. Usually, the gate operation is represented in table form with lists of zeros and ones that is either scary or boring so I'm going to use traffic lights instead. In Figure 3.5, we depict the operation of the NOT gate. I will represent a bit by a traffic light with just the two colors red and green. "Green" is "down" is "0" and "red" is "up" is "1." Critical for classical bits, the bit is always either "0" or "1" but never both and never somewhere in between (like yellow). This will not be true for quantum bits so I point that out now. We will start with the NOT gate as it is the easiest with just one bit in and one bit out. The NOT gate just flips the bit. If the bit is red, then the NOT gate flips it to green. If the bit is green, then the NOT gate flips it to red. (Remember that information is fungible, so it does not matter if I use one–zero or red–green or down–up encoding as they all encode the same bit of information.) In terms of traffic light language, the NOT gate is a Mobile

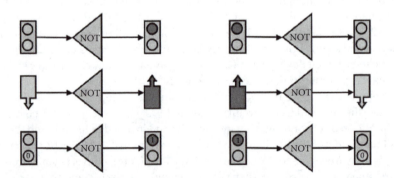

Figure 3.5   A classical NOT gate flips the state of the binary bit. On the left, the NOT gate flips the green-down-zero bit state to a red-up-one bit state. On the right, the reverse happens. Note that two applications of the NOT gate bring you back to where you started.

Infrared Transmitter for Emergency Vehicles that the police use to turn their light from red to green (and consequently the light on the cross street from green to red).

Next, we consider the OR gate shown in Figure 3.6. This gate has two input bits and one output bit. Now that I have set up the key, we will illustrate the gate's operation with the red–green and zero–one pictogram in Figure 3.6. The figure shows all possible two-bit inputs and all possible one-bit outputs, which completely specifies the operation of the gate. The OR gate can be computed by binary modular addition. Imagine you are a cautious driver from Kansas on a four-lane road and you reach an intersection. If both lights for the two lanes in your direction of travel are green, you go (green or green equals green $[0 + 0 = 0]$). If both lights are red, you stop (red or red equals red $[1 + 1 = 1]$). However, if there is a malfunction, and one light is red and the other is green, because you're a cautious driver from Kansas, you stop anyway (red or green equals red $[1 + 0 = 1]$ and green or red equals red $[0 + 1 = 1]$).[20]

Finally, we take on the AND gate in Figure 3.7. In the language of stoplights, imagine you are a reckless driver from New York on a four-lane road. If you come to an intersection and the lights over your lane and the lane next to yours (moving in the same direction) are both green, you should go (in binary modular arithmetic $0 \times 0 = 0$). If the lights over both of your lanes are red, you should stop ($1 \times 1 = 1$). However, if there is a malfunction and the light over your lane is green but the light over the neighboring lane is red, or vice versa, then (because you're the reckless driver from New York) you shout, "What the hell?" and then blow your horn and gun it through the intersection anyway as if both

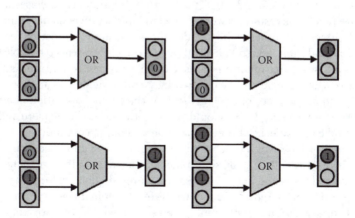

Figure 3.6 A classical OR gate is a two-bit input, one-bit output gate. The four possible inputs and outputs shown here completely specify the gate. The OR gate implements the cautious Kansan's rule—only go if both lights are green.

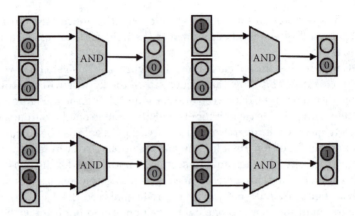

Figure 3.7   A classical AND gate is a two-bit input, one-bit output gate. The four possible inputs and outputs shown here completely specify the gate. The AND gate implements the reckless New Yorker's rule—only stop if both lights are red.

were green ($1 \times 0 = 0$ and $0 \times 1 = 0$). The AND gate implements the reckless New Yorker's rule via binary multiplication.

These three gates are universal. Any universal classical computer, equivalent to a universal Turing machine, can be constructed from just the NOT, AND, and OR gates strung together in different ways. The ENIAC implemented the bits and the gates all with vacuum tubes. The Tukwila implements them with billions of tiny transistors on the chip. Classical information is fungible; you can even build a universal computer with Tinkertoys.[21]

Now that we understand what it takes to build a universal classical computer, we tackle the resources we will need to build a universal quantum computer. Deutsch and others whittled down the required number of resources over the period from 1985 to 1995 until we have just this: we need reliable quantum bits and just three quantum gates to build a universal quantum computer. Recall that the classical bit like our two-light traffic signal has just two states, either green or red (zero or one). Critical for the correct operation of a traffic light (or a classical bit), the bit should not be both red *and* green (zero and one). That would cause wrecks at the intersection and faulty computation in a classical computer. However, a la Schrödinger's cat, we must allow the quantum bit (or "qubit" as it is called) to be simultaneously dead and alive and hence both green and red (zero and one). We have encountered qubits before. The ions in our trap can be prepared in a spin down state $|\downarrow\rangle$, a spin up state $|\uparrow\rangle$, or a quantum cat superposition of both spin up and spin down $|\rightarrow\rangle$ that we can rewrite as $|\uparrow\rangle + |\downarrow\rangle$. Hence, the ions in the trap are a perfectly respectable qubit, as are the spinning electrons in the thulium atom, as are the polarized photons in the EPR experiments. Quantum information is also fungible. It is just a matter of labels. For the ions and electrons, just like in our binary stoplight notation, $|\downarrow\rangle$ is $|0\rangle$ and $|\uparrow\rangle$ is $|1\rangle$. For the photons, we can use $|H\rangle$, which is $|\leftrightarrow\rangle$, which is $|0\rangle$ for the horizontally polarized photon and $|V\rangle$, which is $|\updownarrow\rangle$, which is $|1\rangle$ for the horizontally polarized photon.

Any two-state quantum system that can be prepared reliably in a down, up, or a cat state is potentially a usable qubit, and qubits have been made with ions, electrons, nuclei in molecules, nuclei in semiconductors, semiconductor quantum dots, and little superconducting rings of electrical current. The quantum information is fungible; it does not matter what the hardware looks like. In practice, it is a whole other matter. The key is that, unlike the bit, we can put the qubit into a cat-state superposition of zero and one. Qubits are typically represented as an arrow inside a little sphere as shown in Figure 3.8. In our ion trap quantum pocket watch thought experiment from Chapter 1, I suppressed that ion can point anywhere on a sphere to simplify things a bit, and I will continue to do that mostly, but for now, we should see the qubit in its full glory. We require two numbers, latitude and longitude, to specify the state of the qubit on the sphere. The $|\downarrow\rangle$ or $|0\rangle$ is on the south pole (green light), the $|\uparrow\rangle$ or $|1\rangle$ is on the north pole (red light), and the $|\rightarrow\rangle$ or $|0\rangle + |1\rangle$ is on the equator (orange light) (Figure 3.8). The state $|\uparrow\rangle + |\downarrow\rangle$ is just one possible state on the equator. To get the most general state, we have to write $n|\uparrow\rangle + m|\downarrow\rangle$, where n and m are two numbers that contain the latitude and longitude information. To do this right, I would have to introduce complex numbers and trigonometry, so I will not do it right. Instead, I will revert to my stoplight notation and just represent one state on the equator $|\uparrow\rangle + |\downarrow\rangle$ as an orange light with latitude zero and longitude zero. Other cat states look like this at different points on the globe.

Now that we have our quantum bits, we need our quantum gates. A complete universal set of gates is the CAT gate, the RAT gate, and the ENT or entangling gate.[22] The CAT and the RAT gates are both single qubit gates (one qubit in and one qubit out) and are easiest to describe in terms of what they do to an

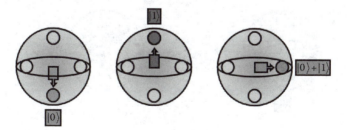

Figure 3.8 This figure depicts three states of the quantum bit or qubit as points on a sphere. The down-green state at is $|0\rangle$ at the south pole, the up-red state is $|1\rangle$ at the north pole, and the right-orange state $|\rightarrow\rangle$ is the cat state $|0\rangle + |1\rangle$ on the equator. Unlike the classical bit, the qubit can be simultaneously both zero and one.

input of either $|0\rangle$ or $|1\rangle$. The CAT gate is, as its name suggests, the Schrödinger cat maker. It takes the $|0\rangle$ state into a positive cat $|0\rangle + |1\rangle$ and the $|1\rangle$ state into a negative cat $|0\rangle - |1\rangle$ as shown in Figure 3.9. A negative cat is just on the opposite side of the equator from the positive cat. For ions in a trap, the CAT gate can easily be implemented by firing a laser pulse of a certain length and brightness at the ion. In fact, by adjusting the length and brightness, you can prepare any cat state you want on the sphere and not just those on the equator.

Next, we have the gate that Rotates around the vertical Axis by Twenty-two and a half degrees, or the "RAT" gate. This gate is needed to cover the entire equator with states that are each separated by 22.5° so we use up all the equatorial space. It does nothing to the $|0\rangle$ or $|1\rangle$ states at the poles but shifts the cat states around on the equator. Because $8 \times 22.5° = 180°$, applying the RAT gate eight times to the positive cat state takes it half way around the equator and makes a negative cat, and applying the RAT gate eight more times to the negative cat brings it back to the starting point as a positive cat. We depict the action of the RAT gate in Figure 3.10 as viewed looking down at the cats from the north pole. As a mnemonic device, you can think of RAT as short for "ratchet" because it ratchets the cat state around the equator in steps of 22.5°. (The Babbage difference engine implements this gate with compressed air and steam and so it can be a "pneumatic" device.)

We will need one more gate, a two-bit quantum gate that generates entanglement between qubits. As you know, it takes two to tangle. The two-qubit gate,

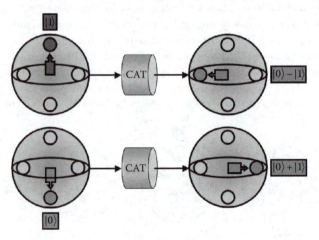

Figure 3.9 This figure depicts the operation of the CAT gate, "the cat maker." In the top figure, the CAT gate converts the red-north-pole state $|1\rangle$ into the orange negative cat state $|0\rangle - |1\rangle$. In the bottom figure, the CAT gate converts the green-south-pole state $|0\rangle$ into the orange positive cat state $|0\rangle + |1\rangle$. The positive cat is just on the other side of the sphere on the equator from the negative cat.

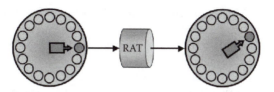

Figure 3.10 This figure depicts the operation of the single-qubit RAT gate as viewed from the north pole looking down toward the south pole. The RAT gate rotates any cat state (on the equator) by 22.5° around the north–south axis of the sphere.

which here I will call the ENT gate for "entangling gate," is more commonly known as the CNOT (pronounced "see not") gate, which is short for controlled-NOT or conditional-NOT gate. As with the classical NOT gate, a quantum NOT gate is a single-qubit gate that flips the state from $|0\rangle$ to $|1\rangle$ or from $|1\rangle$ to $|0\rangle$. The two-qubit ENT gate is a conditional NOT in that the flip of the "target" qubit does or does not take place depending on the state of the "control" qubit. That is, the state of one qubit determines if the gate flips the other qubit. We can completely specify the ENT gate by its operation in Figure 3.11.

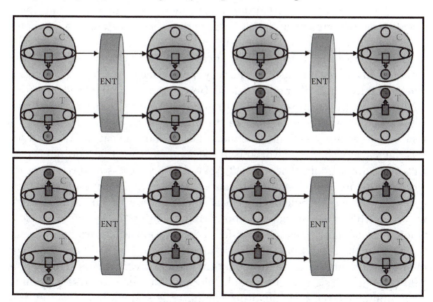

Figure 3.11 This figure depicts the operation of the two-qubit ENT gate. The qubit labeled "C" is the "control" and the qubit labeled "T" is the "target." If the control qubit is in the green-south-pole state $|0\rangle$, then the ENT gate does not change the target qubit (upper left and right boxes). If the control qubit is in the red-north-pole state $|1\rangle$, then the ENT gate flips the target qubit (lower right and left boxes).

While Figure 3.11 may appear a bit daunting, the diagram shows that the state of the target qubit is flipped dependent on the state of the control qubit. In a strained traffic light analogy, imagine you have a box on your dashboard labeled "ENT" with a single button on it and you come to the same possibly faulty two-lane traffic signal. If your light is green and you push the ENT button, the light in the other lane does nothing. If your light is red, then when you push the ENT button it flips the light in the other lane from green to red (if it was green) or from red to green (if it was red). How do we implement the ENT gate in the laboratory? For any ENT gate, it is critical that there is a coupling or interaction between the two qubits. This coupling mediates the operation of the gate. In the ion trap quantum computer, we achieve the coupling through the interaction between the ions via their electric fields. Each ion is positively charged. Benjamin Franklin's rule for electric charges is, "Like charges repel and opposite charges attract." (This rule is the opposite of the business model for online dating service eHarmony, which is based on the idea that like personalities attract and opposites repel.) This mutual repulsion of the ions in the trap causes the energy levels of the remaining electrons to shift around in a way that the state of one ion depends on the state of another. The ions talk to each other! We can then exploit that dependence with carefully tuned and timed laser beams that implement the ENT gate between a pair of ions. Box 3.2 shows how the ENT gate generates entanglement.

There are two final caveats. The first caveat is that we must implement the gates quickly. Large objects like cats very rapidly decay from being both alive and dead into being either alive or dead. However, even small objects like ions do decay on a time scale of seconds to minutes. We need to carry out all our gates and computations long before the minute is up, so the gates must be very fast. The gate speed must be much faster than the cat decay speed for this to have any chance of working. The second caveat is that nothing is perfect. We cannot implement the CAT, RAT, and ENT gates in the laboratory with infinite precision. Even in classical computers, there are sometimes mistakes and there is a whole field of classical computer error correction, where we accept that mistakes happen and correct them on the fly. There is also a whole field of "fault tolerant" classical computation where we build architectures for the computers that are resilient to mistakes. The same holds for quantum computers. There are quantum-error-correction codes and quantum computer designs that are fault tolerant. The quantum ENIAC, a fully programmable large-scale quantum computer, is still many years away, but it will likely require millions of qubits to simulate thulium with most of those qubits devoted to error correction and related "housekeeping" chores to keep the thing running properly.

Now that we have worked out all these gates for designing a universal quantum computer, what is it good for? Feynman made two conjectures in his 1982

BOX 3.2   GENERATION OF ENTANGLEMENT WITH THE ENT GATE

Here in Louisiana, in any cookbook, the first step of every recipe is "first you make a roux."[23] In this book, the first step in any recipe is "first you make a cat." We start with two qubits with the control qubit in the zero state and the target qubit in the one state, $|0\rangle_C |1\rangle_T$. I label them "C" for control and "T" for target. How do we make a cat? We apply the CAT gate to the control qubit (CAT $\Rightarrow |0\rangle_C$) $|1\rangle_T$, which gives $(|0\rangle_C + |1\rangle_C) |1\rangle_T$, which we write through as $(|0\rangle_C |1\rangle_T + |1\rangle_C |1\rangle_T)$. Now, we apply the ENT gate to these two qubits and the quantum identity crisis kicks in. If the control is zero, then the ENT does nothing to the target. However, if the control is one, then ENT flips the target. Because the control qubit is in a cat-like superposition state of zero and one, the ENT gate does a quantum superposition of simultaneously not flipping and flipping the target. We can write this out as ENT $\Rightarrow (|0\rangle_C |1\rangle_T + |1\rangle_C |1\rangle_T)$, which we can expand out as (ENT $\Rightarrow |0\rangle_C |1\rangle_T$) + (ENT $\Rightarrow |1\rangle_C |1\rangle_T$), which gives $|0\rangle_C |1\rangle_T + |1\rangle_C |0\rangle_T$. If we relabel "control" as Alice and "target" as Bob and switch back to the up–down arrow notation, we have $|\downarrow\rangle_A |\uparrow\rangle_B + |\uparrow\rangle_A |\downarrow\rangle_B$, which is our old friend, the two-ion entangled state. If there are 69 ions in the trap, we can concatenate this procedure. We entangle Alice with Bob in step 1. We entangle Alice with Charlie in step 2, which also entangles Charlie with Bob. We entangle Alice with Doug in step 3, which also entangles Doug with Bob and Charlie. In this fashion, we work our way all the way up to the 69th qubit named Zardoz. At the end, we have 69 entangled qubits and we are well on our way to simulating the chemistry of thulium in our ion trap quantum computer.

paper: that a quantum computer would be exponentially faster for quantum chemistry simulations and that there might be abstract math problems that would acquire an exponential speedup if run on a quantum computer. The simulation problem conjecture was proved to be true in general by Seth Lloyd in a 1996 paper, "Universal Quantum Simulators," which had the succinct one-sentence abstract, "Feynman's 1982 conjecture, that quantum computers can be programmed to simulate any local quantum system, is shown to be correct."[24] The final conjecture was that there might be hard math problems that would experience an exponential speedup on a quantum over a classical computer. The first such problem to be discovered was the so-called Deutsch–Jozsa algorithm published in 1992 by David Deutsch and the Australian physicist Richard Jozsa. While not particularly an important problem in mathematics, it was the first to show this exponential speedup and heralded future exponential speedups to come, so we will spend a bit of time on it.[25] I will illustrate the problem with another game (see Box 3.3).

BOX 3.3   QUANTUM COMPUTERS AND EXPONENTIAL SPEEDUP

Alice hands Bob a bit-flipping box as shown in Figure 3.12. The box takes a bit of information in and outputs a bit of information either flipping or not flipping the input bit. Because there are two possible inputs (zero or one) and two possible outputs (zero or one), there are two times two or four possible bit-flipping boxes, labeled in the figure as U_1, U_2, B_1, and B_2. Their action on a bit is also shown in the figure. The boxes are divided into two categories. The "unbalanced" boxes are labeled "U" and output either always zeros or always ones regardless of the input. The "balanced" boxes labeled "B" always output either zero and one or one and zero. The question Alice asks Bob is to decide by experiment which kind of box he has, balanced or unbalanced. To answer that question, Bob has to use the box twice first feeding in a zero bit, observing the output, and then feeding in a one bit, observing the result. If he gets zeros out both times or ones out both times, he knows it is unbalanced (Figure 3.12, top row).

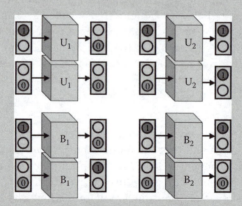

Figure 3.12   This figure depicts the operation of Alice's black box on all possible classical input bits. There are four possible box functions: U_1 and U_2 (top row) as well as B_1 and B_2 (bottom row). The U boxes are unbalanced, outputting always zero or always one. The B boxes are balanced, outputting either zero and one or one and zero. Bob must decide if Alice has given him a U or a B box. Classically, he must input at least two bits (a zero bit and then a one bit) to tell them apart.

If he gets a zero and then a one or a one and then a zero, he knows it is balanced (Figure 3.12, bottom row). He cannot tell from feeding in just one bit whether it is unbalanced or balanced. In the lingo, he must make two "calls" to Alice's box to decide.

In the quantum version of this game, we replace the box with a quantum box and replace the bits with qubits. We call this particular version of the game with one qubit in and one qubit out the Deutsch problem or the Deutsch algorithm that was first worked out by Deutsch in 1985 (see note 25). This quantum game is shown in Figure 3.13. Here, we send in the two qubits at once so there is just one call to the box. The details are messy but we apply the CAT gate to a $|0\rangle$ state qubit and a CAT gate to a $|1\rangle$ state and then launch them into Alice's box at once. Then, we apply two more CAT gates to the two qubits that come out and throw one away in the trash while measuring the state of the other one. If we measure in the up–down

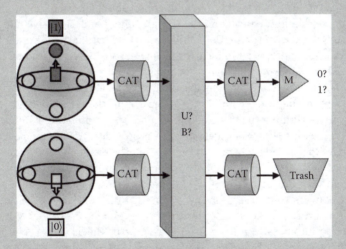

Figure 3.13 This figure depicts the operation of the quantum version of Alice's black box. The goal is still for Bob to decide if the box is unbalanced (U) or balanced (B). But instead of sending in one classical bit after another in two calls to the box, he sends in two qubits at once. Bob applies a CAT gate to each and sends them in simultaneously. He then applies a CAT gate to each qubit that comes out and throws the bottom qubit away. He measures the top qubit in the up–down (one-zero) direction and if he gets zero, the box is balanced and if he gets one, the box is unbalanced. There is a speedup by a factor of 2 in the processing as quantum mechanically he needs to call the box only once instead of the two times required in the classical game.

or zero–one direction, the state of the measured qubit will collapse from some strange cat state into zero or one. If it is zero, the box is balanced, and if it is one, the box is unbalanced. So, in the quantum version of the game, we only have to make one "call" to the box instead of two. There is a quantum speedup by a factor of 2 in figuring out balanced versus unbalanced. The idea is that in the Many-Worlds interpretation, there is one universe checking the zero input and another universe checking the one input and then the answer is combined by constructive interference into a single qubit that we measure. Constructive interference is the term we use to describe waves interacting in a pond. Where two waves add peak to peak, you get a wave twice as high; that is constructive interference. If two waves add peak to trough, they completely cancel each other out; this is destructive interference. The ability of the qubits to interfere constructively or destructively is called "coherence." In the wave–particle duality of quantum theory, the qubits obey a wavelike theory. In the analogy with water waves, the two peaks cohere in constructive interference to give a bigger wave, and the peak and the trough cohere in destructive interference to give no wave at all. Coherence is a critical property of a qubit that allows it to be prepared in a cat state, a superposition of dead and alive instead of a superposition of a peak and a trough.

This constructive interference is one of the tricks that you need for all speedups in a quantum computer. It is fine to start the calculation by thinking you are running the thing on parallel processors in parallel universes, and the access to those parallel universes is provided by entanglement, but you have to carefully construct things so that the result ends up all in your universe at the end. This is not always possible. In most cases, a measurement in your universe only gives you information about the computation being done on your processor and you are cut off from the rest of Borges' infinite library; the library shatters into an exponential number of books scattered into parallel universes never to return. This is where coherence and constructive interference comes in. For the quantum speedup, the library books have to conspire to produce the answer in the one book that you have access to in your universe or there is no quantum advantage at all. This can be tricky or difficult or impossible to arrange. Some hard problems get no speedup on a quantum computer because the parallel processors can never all be corralled back into our universe (via constructive interference) at the end of the calculation to provide the single complete answer.

What Deutsch and Jozsa did in 1992 was to extend the game to an arbitrary number of bits in and one bit out, and this problem shows an exponential speedup in the quantum version of the game. It is a bit too complicated to explain in full detail but the idea is an extension of the one in versus one out (see Box 3.3). Suppose in the classical Deutsch–Jozsa case we have 69 bits in and 1 bit out. The total number of input states then are $2 \times 2 \times 2 \ldots \times 2$, where there are 69 twos in there, which is 600,000,000,000,000,000,000, or 600 quintillion possible states. It's beginning to look a lot like thulium. With two output states (zero or one), then there are $2 \times 2 \times 2 \ldots \times 2$ or 70 twos (an extra two for the output) or 1,200,000,000,000,000,000,000 or 12 hundred quintillion possible combinations of inputs and outputs. To be sure that the box is balanced versus unbalanced (in the classical scenario), Bob has to try out just a bit more than half of these combinations. In the worst-case scenario, Bob has to send in 600,000,000,000,000,000,001 classical bits to know for sure if the box is balanced or unbalanced. At one bit per second, this is again 20 trillion years. Once again, this would take longer than many lifetimes of our universe for Bob to check.

The quantum version of the Deutsch–Jozsa game takes just 69 qubits, applies 69 CAT gates, sends the whole armada of 69 qubits into the box at once, applies 69 more CAT gates to the 69 qubits that come out of the box, and then throws the 69th qubit in the trash and makes 68 measurements on those that remain. If all 68 measurements yield "zero," the box is unbalanced, and if any one measurement (or more) yields "one," then the box is balanced. The qubits can be ions and the CAT gates can be implemented easily in a trap and out pops the answer in a few minutes. There is the exponential speedup. The quantum algorithm requires 600,000,000,000,000,000,001 parallel processors, but only 69 of them are in this universe where we have to pay the electric bill. Once again, the algorithm is carefully constructed so that the answer flows from all 600,000,000,000,000,000,001 parallel processors in parallel universes back into the 69 processors in our universe where we can read out the answer.

There are two features to this problem that bear emphasizing again. While I do not go into the detail of the operation of the box in the quantum case, it is actually designed to be a modified version of the ENT gate. That is, the box is entangling the 69 input qubits so that we access all 600 quintillion possible states simultaneously. That access is necessary but not sufficient for the exponential speedup. The parallel processors are checking all 600,000,000,000,000,000,001 inputs simultaneously, but the program must be carefully engineered so that the answer "unbalanced" or "balanced" is steered back into the 69 qubits that we finally measure (or throw out) in our universe. This steering is related to the coherence of the qubits. Coherence is the wavelike phenomenon that allows us to steer the answer to constructively appear in our 69 qubits in our universe and destructively disappear in the remaining states in the other universes. If this steering is not done carefully, then the measurement shatters the answer all over the 600 quintillion–dimensional Hilbert space and the probability that the correct answer in our universe is only 600 quintillion to 1, and we are stuck repeating the experiment 600 quintillion times and we lose just as badly as we did in the classical case. While coherence (cat states) is required to generate entanglement (see the ENT gate above), you can have coherence without entanglement. Water waves and radio waves are coherent but not entangled. Both coherence and entanglement are needed for the exponential speedup. Entanglement provides the exponentially large number of processors in other parallel universes but coherence steers the answer into our processors so we can extract the answer in one shot.

In summary, in 1992, Deutsch and Jozsa had proved conclusively the second of Feynman's 1982 conjectures: For some mathematical problems, there is an exponential speedup solving the problem on a quantum computer. The important question remained: Was there any useful problem that people cared about for which a quantum computer provided an exponential speedup? The answer came in 1994 and the problem was cracking secret codes on the Internet. To set the stage for this 1994 result, I now digress into a bit of history of secret code cracking with modern computers.

YOUR PAD OR MINE?

The title of this chapter is 'The Quantum Codebreaker,' but I have yet to discuss code breaking at all. It is time to get cracking. During World War II, while the Americans were fussing over computers for simulating artillery-firing tables, the British had a much more pressing issue, the Battle of Britain and subsequently the Allied invasion of Normandy. While the Americans were simulating firing tables, the British were breaking Nazi secret codes. In a curious twist of fate, during World War II, the Germans were using a secret code transmission system developed in World War I by a US Army Major named Joseph Mauborgne. Mauborgne, along with the AT&T Bell Labs engineer Gilbert Vernam, co-invented a cryptography system now known as the Vernam cipher or the one-time pad. In my newspaper, in the comic pages, there appears everyday a puzzle called "The Cryptoquote." There you find each day some famous quote, such as "ALL GOOD THINGS MUST END." However, puzzle makers rewrite the quote in encrypted form as "bmm hppd uijoh nvtu foe," and the newspaper reader must crack it. (I will use uppercase for the unencrypted letter and lowercase for the encrypted.) The Cryptoquote is a simple substitution cipher. Here, I have used A goes to b, B goes to c, C goes to d, … Z goes to a, in a very simple and obvious substitution. Usually, they give you a couple of hints like "t is S and e is D." The Cryptoquote is not hard to figure out because *they reuse the same letters.* Once you have figured out "e is D," then everywhere in the quote, you replace e with D to get, "bmm hppd uijoh nvtu foD." Once you figure out "p is O," then you do it again to get, "bmm hOOd uijoh nvtu foD." This is the feeling you get watching *Wheel of Fortune* when with one good guess you gain a large number of letters. With some guesswork and knowledge that the most common English letters are remembered mnemonically as "ETAOIN SHRDLU,"[26] good players can crack these in under a minute. Again, the trick is that once you know "p is O" in one place, then you know "p is O" everywhere. What Vernam and Mauborgne figured out was that you could make a crypto system where that assignment of letters is random from shot to shot. That is, the first time it appears, "O is p," but then the second time it appears, "O is w," and the third time, "O is e," where the p, w, and e are chosen at random. This makes cracking the quote much harder because the letters never repeat—there is no pattern. In 1949, Claude Shannon at Bell Labs proved such a secret code was unbreakable, but even in World War I and II, the cryptographers realized that it was very good and very likely unbreakable.

Cryptographers also called this the "one-time-pad" cipher since in the early days there were actual pads of paper with the random key on it and workspace to keep track of the letter replacements. The pads had a huge collection of random letters and characters on them with the important word here being "random." That randomness makes it unbreakable. Because the notion of one-time-pad cryptography appears here and in the section on quantum cryptography later in this book, I will explain how it works in some detail. The key is to use modular arithmetic, which is how we add times on a clock. If it is 11:00 a.m. and I tell you that you have an appointment in 2 hours, you know that the appointment is at 1:00 p.m. and not 13:00 p.m. Even if you do use the 24-hour clock, you should be able to convert 13:00 hours to 1:00 p.m. just by subtracting 12. (See the encoder in Figure 3.14.) Now, if I use the full English alphabet for the pad, then I will need to work with a 52-hour clock, which is confusing. It will be easier to just use a language, like Hawaiian, which only has 12 letters in its alphabet for the example, so the math is just that of ordinary clocks. Two star-crossed Hawaiian lovers, 'A'ala from Ahukini Landing on the island of Kauai and Pa'ahana from Puuohala Village on Maui are madly in love. However 'A'ala's father, a former naval intelligence officer, does not approve of her beau Pa'ahana, who is a surfer. So, 'A'ala and Pa'ahana decide to secretly elope and move in together after the wedding. They have to plan everything secretly, such as who will move in with whom (your pad or mine), and so they decide to use encrypted text messages on their cell phones. 'A'ala knows that her dad can break any code so she resorts to the one-time pad, which she knows is unbreakable. We show a sample test communication in Figure 3.14.

I generated the pad using a random number generator in Mathematica. Notice that some letters like "O" appear several times and one letter "W" does not appear at all. That is randomness for you. In the figure, 'A'ala encrypts the text message "ALOHA" into "NEOUL." Note that, unlike the Cryptoquote, the first letter "A" is encrypted as "N" but the last letter "A" is encrypted as "L." There is no pattern; the letter "A" is assigned a different random code letter

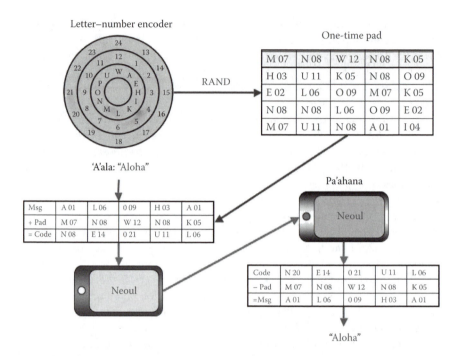

Figure 3.14 Here, we show two star-crossed Hawaiian lovers, 'A'ala and Pa'ahana communicating via encrypted smart phone text messages using a one-time-pad cipher. The number–letter encoder assigns each letter of the Hawaiian alphabet to two numbers on the 24-hour clock (either 1 or 13 is "A" and so forth). 'A'ala generates (RAND) a table of 25 random numbers between 1 and 12 and assigns the appropriate letter from the clock to construct the pad. She makes one copy of the pad and gives it to Pa'ahana. In the first test run, she texts Pa'ahana and tells him to use the first row of the pad. She then takes her message "ALOHA" and encodes it by adding the numbers corresponding to each letter in her message to the number of the letter on the pad. Reading the numbers off the clock, she constructs the encoded message "NEOUL" and sends it. Pa'ahana receives the message, assigns the clock numbers 1–12, and subtracts the pad numbers. (If he gets a negative result, he goes back and chooses the clock numbers 13–24.) He then reads off the message by comparing the final list of numbers to the clock.

each time it appears in the message. That is what makes it unbreakable. What can go wrong with this protocol? Reusing the pad is bad. Some letter assignments will start to repeat and that repetition pattern, which we used to break the Cryptoquote, can be used by 'A'ala's father to break the code. The South American revolutionary, Che Guevara, used a one-time pad to communicate with Fidel Castro. Rumor has it that, running low on pads, Guevara started reusing them, which allowed the CIA to crack and read his communications and locate him for the Bolivian Army, which then executed him.[27] 'A'ala and Pa'ahana must also be sure that they have the *only* two copies of the pad. If 'A'ala's dad sneaks into her bedroom one night and makes a Xerox copy of her pad, then he can read everything.

We will end this thought experiment with a little crypto lingo, which we'll need later. The process or system for encoding and decoding the secret messages—the clock plus the pad—is called a "cipher." (Think of the opposite word, which is "decipher.") This particular system is called the one-time pad cipher or the Vernam cipher. The two pads are called the "private key" as only 'A'ala and Pa'ahana should have copies and they must keep them private for the scheme to work. The message to be sent, "ALOHA," is called the "plain text." The encrypted message transmitted, "NEOUL," is called the "cipher text."

As it became clear that the one-time pad was very likely unbreakable, its use took off. In the 1920s, the pads were actual pads of paper that the diplomatic corps transported to embassies in diplomatic pouches. Around the same time, commercial electrical–mechanical versions, which were automated electronic pads that were plugged into Teletype machines or Morse code systems, became available. These electronic pads had cylinders set up like the old-fashioned odometer on a car with cogged wheels that the operator adjusted to different rotational settings to choose one out of perhaps billions of possible pads. If Alice wanted to communicate with Bob, they would first set it up so they had identical machines. Then, Alice would transmit a short message telling Bob what setting to use, what pad to choose, and then they would begin encrypting and decrypting. The procedure was automated but still identical to the set up in Figure 3.14. Alice still had to ship Bob a machine and initiate the pad settings. As before, reusing the pad was bad. The German military quickly realized the usefulness of such machines and developed military versions with three and four rotor wheels that we call Enigma machines. While not completely random, the pad eventually repeated, the scrambler in the machine allowed you to quickly switch pads, and the pads were so long, much longer than any message, that they effectively did not repeat.

In 1928, the Germans rolled out a three-rotor machine for use in the army, navy, and air force. The Americans, British, and French were unable to crack it, but a group of Polish mathematicians at the Polish Cipher Bureau in Warsaw broke the code for the three-rotor machine in 1932. The Poles were able to do this using an instruction book for the machine and a copy of the keys used for 2 months of that year provided to them by French Military Intelligence, information that had been obtained by old-fashioned spying. Remember that one of the rules that make the one-time pad unbreakable was that there should only be two copies of the pad. The French leaked a third copy of the pad for those months to the Poles. With that and the instruction booklet, it was enough for the Polish mathematicians to figure out how the machine worked, and using mathematical group theory, they reduced the number of possible pads that had to be guessed from billions to just a hundred thousand. All that remained was to systematically go through these hundred thousand pads and find the right one. That task was first carried out by human computers using index cards but it took days or weeks to decrypt a message, making the information of limited value. It does no Pole good to decrypt the message "INVADE POLAND" a week after the event. So immediately, they began automating the process with an ever more sophisticated set of calculating machines, cumulating in 1938 in an electromechanical device called a cryptologic "bombe."[28] In essence, the Poles were efficiently simulating the Enigma machine on their bombe to rapidly search through every possible combination of thousands of possible pads for the specific one used for a given intercepted message (or a day's worth of messages).

The project came to a quick end with the 1939 German invasion of Poland, but in the nick of time, the Poles managed to transfer the entire cryptology program from Warsaw to Bletchley Park, near London, including a few Polish bombes, where the British and the Poles collaborated on deciphering the Nazi Enigma code for the rest of the war. Immediately, British mathematician and inventor of computer science, Alan Turing, designed and built a new and improved bombe that was even faster and more general than the Polish version. The Turing bombe came online in 1940. The Turing bombe was 2.2 meters wide (7 feet), 1.83 meters tall (6 feet), 0.6 meters deep (2 feet), and weighed a ton (metric or Imperial). It was a special-purpose computer, really an electromechanical calculating machine, which would quickly find the daily one-time pad being used by the Germans. (The standard protocol was to change the settings on all the three-rotor Enigma machines each day, based on a predetermined system, to produce new pads.) The Turing bombe provided the bulk of the Enigma decryption from 1940 through 1942. This classified program at Bletchley Park was called "Ultra" and the information they provided to the Allied Powers shortened World War II by 2 years.[29]

In early 1942, the German U-boat traffic (codenamed "Shark") arriving at Bletchley Park suddenly became much different from the rest of the Enigma encryptions (codenamed "Dolphin") to and from the German High Command. The Germans had upgraded the naval Enigma from three to four rotors, drastically increasing the number of possible pads that needed to be searched, and hence vastly increasing the security. Breaking the Shark Enigma code would have taken 50 to 100 times more time on the Turing bombe, and in addition, the bombes were all saturated decrypting the Dolphin traffic, which consisted primarily of German Air Force and Army messages. The scientists and engineers went quickly to work to develop a new class of programmable and fully electronic computers called the Colossus machines. They were very similar to the ENIAC in that they were vacuum tube devices. Although not universal classical computers in the Turing sense, they were special-purpose codebreakers and they did that task quite well. The Colossus Mark I came online in 1944 and the Mark II on June 1, 1945—just in time for the D-Day Allied invasion at Normandy. The Colossus was critical to that invasion; on D-Day, the Allied forces knew the location of all the German U-boats. The British provided this intelligence to the Americans (who were developing their own code-breaking machinery) under the code name Ultra, and it was likened to manna from heaven. Ultra dispatches were rarely issued and for only the most dire events as the cryptographers knew if too much was released too often the Germans would wise up and change the code. By 1945, all German Enigma traffic could be routinely decrypted in a day or two—the Turing bombes handling the Dolphin code and the Colossuses on the Shark.

Yet, the Germans remained convinced of the Enigma's absolute security. However, there were lapses in the German security protocol that allowed the Allied cryptographers to break all the Enigma codes. Occasionally, a German communication would not get through and the exact message would be retransmitted. The rule is you must not reuse the pad. German Teletype operators caught reusing the pad, like Che Guevara, were taken out and shot. However, a few times, the German radio operator would retransmit the same message or an abbreviated version with the same pad setting. Such mistakes allowed for the optimization of the Colossus. Additional information came from captured German radio operators, captured U-boat Enigma operation manuals, and two captured Enigma machines that were taken from U-505 in June of 1944. (The intelligence value of the captured machines was limited. By June of 1944, Bletchley Park cryptographers had worked out the properties of the machines by other methods and with other captured intelligence information.) By the end of the war, there were 10 Colossus machines in operation at Bletchley Park. The entire Ultra program was kept secret until the 1970s and all the Colossus machines were destroyed. (A 1943 photo of the Colossus in operation is shown in Figure 3.15.)

The Enigma machines were the height of one-time pad technology by the end of World War II. The story of the arms race between the German Enigma machines and the American and British code-breaking machines illustrates both the advantages and drawbacks of the system. This story illustrates the importance of advances in computer technology for code breaking, which will be a theme for the quantum computer coming up next. The one-time pad is a private key; Alice and Bob must keep their pads private. The quantum computer attacks a type of public key, which we will discuss in the next section. The story of Allied Ultra versus Axis Enigma illustrates a number of advantages and disadvantages of the one-time pad. Let us list them here.

Figure 3.15 Here, we show two programmers with the Colossus computer at Bletchley Park, Buckinghamshire, England, circa 1943. Funding for this code-breaking machine came from the Ultra project, an Allied intelligence program to break Nazi secret codes during World War II.

The one-time pad is unbreakable. If the pad is never reused and nobody but the intended users has copies of the pad, the secretly coded messages can never be read.

The one-time pad is secret key intensive. The code is unbreakable only if each character in the message corresponds to one character in the pad. If you are sending a lot of messages, you need a lot of pad.

The one-time pad must never be copied. If Alice and Bob are communicating secretly, they must be sure that they have the only copy of the pad. If Eve manages to get a third copy, then Alice and Bob are doomed.

The one-time pad requires a secret transmission system. In order for Alice and Bob to be sure that Eve does not have a third copy, Alice must arrange a way to get Bob his copy in a way that is secret so that Eve cannot copy the pad en route.

The Germans did not have proof that the one-time pad was unbreakable. That proof came in 1949. However, because nobody had ever broken it, they were lulled into assuming it was. It would have been had there not been violations of the rules. German radio operators frequently resent transmissions without changing the Enigma pad settings, thus often reusing the pad. Reusing the pad produces patterns that, a la Cryptoquote, allow the message to be guessed. Politics is immune to this pitfall; carelessly reusing the pad brought down both Che Guevara and the Nazis. The problem that a one-time pad is key intensive (requires vast numbers of pads) the Germans solved by constructing the Enigma machine itself—the device could store billions of pads in the different combination of rotor settings. The secret pad transmission requirement exposes a paradox common to all private-key encryption systems. If Alice has a communications channel that allows her to send the pads to Bob in utter secrecy, with no possibility of copying, then why does she not just use that channel to send the secret message and do away with the pads? The pads have the same amount of data on them as the messages! The answer is time. You transmit the pad by slow U-boat at your leisure in a diplomatic pouch in secret in advance and then use the pad to transmit your secret message in real time by a public channel such as the radio, email, or text message. To handle the two requirements of no copying and the secret transmission system, the German High Command physically shipped the Enigma machines and the paper tables of daily rotor settings (pad choices) to all military end users in the equivalent of diplomatic pouches—locked and sealed briefcases handcuffed to Teutonic couriers. That was a slow process to set up, but once in place, they had all the pads they needed and could communicate quickly and publicly over the radio. Nevertheless, some of the daily rotor setting lists were indeed copied or stolen by Allied spies, which also weakened their system by allowing the computers to guess the daily pad settings in the Enigma simulator running on the Colossus.

THE PEOPLE'S KEY

The cryptography system that the Internet currently uses to encrypt all your banking transactions is not a one-time-pad private-key cipher system primarily for the reasons elaborated above. In the Internet age of instant gratification, we will not wait weeks or months for Internet Nazis to arrive at our front doors, handcuffed to briefcases packed with one-time pads and Enigma machines, before we can set up our online banking or gambling account. And given the inherent laziness of the average consumer, there would be too much incentive to start reusing the pads, and the average consumer is resistant to being taken out and shot. The Internet uses a system developed in the 1970s called "public-key encryption" (the people's key), and this invention single-handedly launched the age of Internet commerce. The security of the public-key cipher system relies on *the assumption* that breaking the public-key cipher is inefficient on a classical computer. Breaking a public key is harder than simulating the thulium atom—it takes 300 billion Tukwila chips running for billions of years. However, just like the thulium atom, a quantum computer can crack the Internet public-key cipher in just a few minutes. The security of the public-key cipher relies on the assumed hardness of a mathematical problem from the field of number theory and your third grade arithmetic lesson—that division is harder than multiplication. The public key uses very large numbers in the encryption system. These numbers are the multiplicative product of two smaller numbers called prime numbers. Prime numbers are numbers that are evenly divisible by the whole number 1 and the prime number itself. The numbers 2, 3, 5, 7, and 11 are all prime because they cannot be divided evenly by any whole number without a remainder. The integer 12 is not a prime number; it is divisible by 4, giving 3 with no remainder. The integer 11 is a prime; it is not divisible by any other whole number without a remainder. The process of dividing a large number into its primes is called factoring. Hence, 12 can be factored into 3×4, 2×6, or 1×12. Because there are other options beside 1×12, 12 is not a prime. However, the only option for 11 is 1×11. Eleven is a prime.

In 1994, an American computer scientist, Peter Shor at Bell Labs, devised an algorithm for a quantum computer—if one could be built—that could factor large numbers into smaller primes exponentially faster than any known classical computer algorithm. An arcane property of number theory (factoring) turns into a critical problem in the cracking of public-key encryption—which is the basis for most of the secure data transfer on the Internet. The discovery of a *classical*, exponentially fast factoring algorithm, enabling you to hack the Internet, was the plot of the 1992 movie, *Sneakers*, starring the improbable cast of Robert Redford, Sidney Poitier, River Phoenix, Dan Aykroyd, and Ben Kingsley. Who can forget Poitier's finest moment, when he dashes away from the Scrabble game and runs up to Redford and Phoenix and Aykroyd, all hunched over a clanking and chattering Braille-enabled computer terminal, and shouts "Shut it down! Shut it down now!" as theses hackers successfully and sequentially break into the US Federal Reserve Bank's Communications and Records Center in Culpepper, Virginia—"Anybody want to shut down the Federal Reserve?"—the US Electrical Power Grid—"Anybody want to blackout New England?"—and the US Air Traffic Control System—"Anybody want to crash a couple of jumbo jets?" No wonder the US Government has spent tens of millions of dollars trying to build such a thing.

One would not think that a breakthrough in the mathematics of classical number theory would be a reasonable plot for a Hollywood blockbuster, but there you have it. (In this movie, the codebreaker is a mathematician who "… specializes in large number theory, prime numbers, factoring …" and who finally gets "whacked" by Ben Kingsley.) In *Sneakers*, they are hacking public-key cryptography, also called public-key encryption, which is the most widely used crypto system in the world and the backbone of nearly all Internet security. Whenever you log on to your account on *Macy's* and purchase something with your *Bank of America World MasterCard Credit Card* (with *WorldPoints Rewards* and *PayPass*), the image of a little padlock suddenly appears on the bottom of your browser screen. What does this padlock mean?

It means (we hope) that nobody can steal your account and credit card information as it is being transferred from your laptop to *Macy's* for the purchase of your "Martha Stewart Collection Three-Piece Sieve Set—$24.99."[30] Why can nobody steal your secrets? In the words of the *Sneakers* character, the notorious Carl Arbogast (the equally notorious River Phoenix), "You won't get in. It's encrypted!" While a complete overview of prime number theory, mathematical trapdoor functions, and Diffie–Hellman key exchange is—I hope—beyond the scope of this book, I'll instead present here a highly abridged explanation of public-key cryptography (see Box 3.4). The point is, as you all learned in third

---

BOX 3.4   PUBLIC-KEY CRYPTOGRAPHY

Typical public-key cryptography such as that on *Amazon.com* uses 512-bit encryption, which involves a number that is 512 binary bits (010110100101 … .) long (base 2). We'll take a more famous example of a 155-digital base-10 bit scheme (equal to 512 binary digits) called RSA-155 that was hacked in the year 1999 using an array of computers hooked together worldwide on the Internet. ("RSA" stands for the names of the inventors of the public-key system: Rivest, Shamir, and Alderman.) Converted to base 10, these are very large numbers. Let's go through the steps. Suppose Alice at *Amazon.com* is going to set up a secure link with Bob in Bletchley Park. Alice takes two large, prime, 70-digit integer numbers, oh let's say for concreteness, 10263959282974110577205419657399 1675900716567808038066803341933521790711307779 (my lucky number) and 10660348838016845482092722036001287867920795857598929152222 70608237193062808643, which for short we will call p and q, respectively. (In this game of large numbers, we typically drop the commas so that one thousand [1,000] becomes just 1000.) She keeps these numbers secret on her computer or on a yellow sticky note in her desk drawer. These two large numbers constitute Alice's private key. She then uses her computer to multiply these numbers together, which takes just a fraction of a second on this *iMac* I'm typing on, even though the numbers are very long, because *multiplication is easy!* She immediately gets the result 1094173 8641570527421809707322040357612003729454492059909138421314763 4998428893478471799725789126733249762575289978183379707653724 4 02714674353159335433897, which we'll call c for "short." (Here, p and q are prime whole numbers and c is a composite whole number.) There is no need for Alice to keep the composite number c secret, and the whole point is that she is now free to post it to the world on her website at *Amazon. com*, in an email, or to send out by carrier pigeon (which is how I think some of my online orders are often processed). The single huge number c is Alice's public key. Then, Bob in Bletchley Park takes the publicly available and very long number c and uses it (and the public-key encryption algorithm) to encrypt his credit card information and send it back to Alice at *Amazon.com*. "You won't get in. It's encrypted!"

The point is that Bob does not know Alice's short prime numbers p and q separately, her private key, only their multiplicative product, the long composite number c, her public key. Public-key encryption uses the long number c to encrypt but the short numbers p and q to decrypt. This is why the system is sometimes called an asymmetric encryption scheme— Alice and Bob do not have the same information; they do not share identical copies of the private key. Unlike the one-time-pad, the encryption key and the decryption key are different. Only Alice knows p and q separately, so only Alice can decrypt Bob's message. At this point, not even Bob can even read his own encrypted credit card number back to himself! The encryption process is complicated but it is all carried out in the background in the software of your web browser. This is how Bob sends secret data to Alice. What about the reverse? Bob does the same thing. He chooses at random two large prime number of his own, p-Bob and q-Bob, and then multiplies them out to get his own composite number c-Bob. Just like Alice, he posts c-Bob on his web page (Bob's public key) but keeps his knowledge of p-Bob and q-Bob under lock and key on his laptop (Bob's private key). The public-key cipher solves the biggest problem in cryptography—there is no need to ever transmit the secret keys p and q. What is transmitted is the public key c but that is the whole point, everybody can see that; it is public. In the private-key one-time-pad cipher, the paradox is that if Alice and Bob have a top secret channel (that nobody can eavesdrop on or copy information from) over which they can transmit their private-key pad in real time, why not just do away with the private-key pad altogether and simply transmit their data over this top secret channel? In a classical world, such a channel can never exist.

grade, that multiplication with paper and pencil is much easier than longhand division. That's it. Everything you ever needed to know about public-key cryptography and world domination via the Internet you learned in third grade. This third-grade observation holds regardless of how big the numbers get— division is always much, much, much harder than multiplication. How much harder? Well, let us see.

Okay, so what's the big deal? Shouldn't public-key crypto systems be easy to crack? Can't eavesdropper Eve just divide out (factor) Alice's public key, the big number c on her web page, and figure out p and q for herself and then steal Bob's MasterCard number? As a test, in Box 3.4 when I multiplied the two big numbers p and q together to get c on my iMac, I also at the same time told my iMac to factor c back into p and q. The multiplication of p and q to get c took place in the blink of my left eye. The factoring command is still running after about an hour. I imagine it will still be running long after I finish this book! I imagine it will be running long after I'm dead. I imagine it will be running long after the universe dissolves away into cold nothingness. Why is that? Recall everything you need to know you learned in third grade. Multiplication is easy! It is non-exponentially easy. Multiplication can be *efficiently* simulated on any classical computer. Long division is very, very, very hard! It is exponentially hard. Long division cannot be efficiently simulated on any classical computer. Long division is so much harder my iMac, running Mac OS X Version 10.5.8 on a 2.66 GHz Intel Core 2 Duo core processor with 2 GB of RAM, is still struggling with the problem of dividing out c into p and q. What's taking so long!? (Still running. Still running.)

My iMac is running a variant of an ancient 2210-year-old Greek factoring algorithm known as the *Sieve of Eratosthenes*.[31] After 2210 years, the sieve (or some variant) is the fastest known algorithm for factoring a large composite integer number c into its primes p and q. How does the sieve work? Well, just as a sieve is a sifter or strainer or net for separating unwanted things (hot water) from wanted things (*farfalle* pasta), so is the *Sieve of Eratosthenes* a strainer to filter out unwanted things (numbers that are not the factors p or q, of which there are a lot) from wanted things (numbers that are either p or q, of which there are precisely two). So, the factoring algorithm on my iMac, running on Mathematica, is carrying out the following procedure. Divide the number c by the whole number 2 and see if you get another whole number with no remainder. Nope. Divide the number c by the whole number 3 and see if you get another whole number with no remainder. Nope. Divide the number c by the whole number 5 and see if you get another whole number with no remainder. Nope. Divide the number c by the number 7 and see if you get … you get the idea. Continue this process with every possible prime until you hit either the unknown prime factor p or the unknown prime factor q and then stop. For practice—the nuns who taught me long division in third grade would be proud—try this last one on your own using longhand division. Divide 10941738641570527421809707322 0403576120037329454492059909138421314763499842889347847179972578912 67332497625752899781833797076537244027146743531593354333897 by 7 and show there is (or is not) a remainder. (Be sure to show your work or I'll smack you upside the head with a ruler.)

My iMac will continue to do this testing 2, 3, 5, 7, 11, and all the prime numbers[32] up to p, which is 10263959282974110577205419657399167590071656780 8038066803341933521790711307779, when only then it will find that c divided by p equals q, which recall is 1066034883801684548209272203600128786792 07958575989291522270608237193062808643. Around there it will stop. (Still running … . Still running … .) Now, you see why it is taking so long. I have to carry out a lot of long, long, long divisions! Long division is very hard. It is taking a very long time. The security of the Internet rests on this observation. How long is it going to take? Well, that depends on how many prime numbers have to be tried. In this case, we would like to know how many prime numbers there are between 2 and the smaller prime number p, which is again, 10263 959282974110577205419657399167590071656780803806680334193352179071 1307. This can be estimated using the *Prime Number Theorem*.[33] (In order to do this, I'm going to abort the factoring calculation, which after 2 hours has gotten nowhere: "$Aborted.") There is a rule, as explained to English physicist Stephen Hawking by the Bantam Dell Publishing Group in 1988, when he submitted his book *A Brief History of Time*, that for each equation in the book, the number of sales would drop by an order of magnitude. This is why I am going out of my way to avoid writing anything with an equals sign in it or some mathematical function like the logarithm function. The formula for computing the number of prime numbers less than p, which have to be sieved, was discovered by the German mathematician and lightning mental calculator, Carl Friedrich Gauss.[34] The formula states that the approximate number of primes less than or equal to prime p is the number p divided by the natural logarithm of the number p. This my iMac's now aborted Mathematica program can do easily and the answer is, roughly, 60228400000000000000000000000000000000 00000000000000000000000000000000000000, or more accurately, in scientific notation, there are approximately 6.02284×10^{71} primes less than p that you have to test. This is *much* worse than computing the ground state of thulium. That is, you would have to carry out approximately 600 duovigintillion[35] long divisions of primes before you hit the first prime factor p of the composite number c. (Upon hearing this news, right above me in heaven, the ghost of Sister Mary Adeline, who taught me long division, is singing the "Hallelujah" chorus from Handel's *Messiah* and dancing a fandango.) This is why my iMac was taking so long; it would have taken many trillions of universe lifetimes to find the answer. This is why Bob can't extract his own encrypted credit card number. This is why the Internet is secure. The best-known classical algorithm for factoring large numbers, that *Sieve of Eratosthenes*, runs exponentially slowly on any classical computer in terms of the number of digits in the public key.

In the film *Sneakers*, the plot revolves around the mathematician Dr. Gunter Janek, who has found an efficient *classical* factoring algorithm that is exponentially faster than the *Sieve of Eratosthenes* for factoring numbers, and Redford and his team use that classical algorithm to hack the Internet while the bad guys are trying to kill them. While most mathematicians think that such an efficient classical factoring algorithm is unlikely to exist, nobody has proved that it does not. The logic goes something like this: We have spent over 100 years looking for an efficient classical factoring algorithm with no success, and we are smart mathematicians, and so therefore it does not exist. The security of the entire Internet hinges on this one hope of those egotistical mathematicians that nobody ever finds an efficient classical factoring algorithm. You want to know what keeps me up nights? This keeps me up nights. This hope is also why the most paranoid people on Earth use the one-time pad instead of the public-key encryption. Sure, the one-time pad has all these drawbacks about transmitting the pad, copying it, and so forth, but at least we have proof it cannot be hacked. However, the difficulty of distributing the pads by carrier pigeon is incompatible with the need for speed on the Internet. The public-key encryption—where no secret or private key is ever transmitted—is a much faster and more efficient way to go. That is why I call the public key the people's key. It is perfectly secure, unless Dr. Gunter Janek (or his real-world counterpart) finds an efficient classical factoring algorithm and posts it to his blog, which would cause the collapse of the world's entire economy in 3 days. What are the odds of this? We do not know.

In a strange twist of fact imitating fiction, there was recently a close call. A related problem to factoring large numbers is called the PRIMES problem. I hand you a very large number, say, 10263959282974110577205419657399167590071656780803806680334193352179071307 and ask you to tell me if it is a prime number or not. Up until 2002, there was no efficient algorithm to tell this. All classical algorithms for deciding if a number n is a prime or not were inefficient—they required an exponential amount of time to run on a classical computer. It was thought there was no efficient classical algorithm. The logic was identical to factoring. We smart mathematicians have been looking for a hundred years for an efficient algorithm to test if n is prime or not, and we have not found one, so such an efficient algorithm is unlikely to exist.

Then, in 2002, Manindra Agrawal, an Indian professor of computer science at the Indian Institute of Technology in Kanpur, along with his graduate student Neeraj Kayal and his *undergraduate* student Nitin Saxen, published an *efficient* classical algorithm for testing if n is prime or not. It ran exponentially faster than any previously known algorithm and shattered the widely accepted belief that no such efficient algorithm existed. This number theory result made the headlines of a story in the *New York Times* on August 8, 2002.[36] Thank goodness the security of the entire Internet did not hinge on *that* problem. But unlike the fictional Dr. Janek, what are the odds that three Indian scientists, working with only paper, pencil, and a crate of erasers, working in an un-air-conditioned attic somewhere in Kanpur, will announce tomorrow the proof of an equally shocking and efficient classical factoring algorithm? If they immediately post the result to their blog, then hackers can crack the public-key encryption system and bring down the Internet in 3 days—collapsing the entire economy of Earth. (*This* is what keeps *me* up nights.) The odds of this happening are completely unknown. Just because nobody has found one so far does not mean that somebody will not find one tomorrow. But while the Indian computer scientists are stewing over this classical factoring problem in their steaming attic, I have a story to tell. It is about an efficient factoring engine, but it is a quantum factoring engine and not a classical machine.

THE BOLT FROM THE BLUE

In 1994, the American mathematician and Bell Labs scientist, Peter Shor, published a *quantum* computer algorithm that efficiently factors a large composite number c into the two primes p and q—and all hell broke loose. The only thing standing between Peter Shor and the hacking of the entire Internet was the small detail that no quantum computer had yet been built to run his algorithm on. So we were safe (and still are safe) for the time being. To factor the 38-digit numbers that we currently use in public-key encryption on most web browsers, you would need a quantum computer that had about a million entangled qubits. The current record for a quantum computer is approximately 10 entangled qubits. If you ask the American physicist and Nobel Laureate, Bill Phillips, what are the odds of building a quantum computer, he will tell you, "Fifty–fifty. And by that I mean a fifty percent chance of building one in fifty years." Some of us are more optimistic. In Chapter 4 ("You're in the Army Now"), I will give you a respite from the mathematics with a bit of the history of how we went from 0 to 10 qubits in the past 17 years, but right now, while visions of secret codes and prime numbers are still dancing in your heads, we will polish off Shor's algorithm.

When I was a graduate student at the University of Colorado, I took two semesters of the subject called *Number Theory* in the math department. My

---

* "Multiple Cloud-to-Ground and Cloud-to-Cloud Lightning Strokes during Night-Time. Observed during Night-Time Thunderstorm," by C. Clark, from NOAA Photo Library, NOAA Central Library; OAR/ERL/National Severe Storms Laboratory (NSSL).

fellow physics students all laughed and laughed at this *Number Theory* course with the retort, "Number theory! What self-respecting physicist would ever need to know number theory? You should take something useful." I found number theory fun and ignored them, and I had the last laugh. Shor's factoring algorithm is a large, dry, stale, and weighty fruitcake recipe of classical number theory—embedded with the hard candied citron of Euclid's Greatest Common Divisor Algorithm and the bitter Brazil nuts of the Chinese Remainder Theory—that is then delightfully iced with a bit of light and fluffy quantum frosting. I was a US Army research physicist in 1994 when I first alerted the Army Research Office to the importance of Shor's algorithm for cryptanalysis. I was tasked to give lectures on the algorithm to army scientists and brass. At one memorable lecture, a two-star army general stopped me when I got to the *Chinese Remainder Theorem* and exclaimed, "How long have the Chinese been working on this?" I replied with a completely straight face, "General, when it comes to the remainder theorem the Chinese are years ahead of us." The general turned to his aide and barked, "The Chinese are ahead! Write that down." (The Chinese mathematician Sun Tzu first reported on the remainder theorem in his book, *The Mathematical Classic*, which was published in the third century AD.)

So as to not spoil your dinner, I will skip the bulk of the fruitcake and focus instead on the fluffy quantum icing. By now, you have gotten used to my clock arithmetic from the one-time pad example. The trick is to use the quantum part of the algorithm to find the period of particular number theoretical function related to the clock arithmetic. This clock arithmetic is called modular arithmetic, and the Chinese were writing about it over a thousand years ago. You know the drill. Any number on the clock that goes over 12:00 we subtract off 12 or 24 or 36 until we get a number between 1:00 and 12:00. If it is 11:00 and we are to meet in 2 hours, that is 1:00 and not 13:00. (Subtract 12 from 13:00.) If it is 11:00 and we are to meet in 22 hours, that is 9:00 and not 33:00. (Subtract 12 twice or 24 once from 33:00.) I don't keep track of the a.m. or p.m. or the day, just the time. That is clock arithmetic. That is modular arithmetic. That is what the Chinese were working on in the third century AD.

So now, we worry about finding the period of a function. This is done in engineering and science with something called a Fourier transform, after the French mathematician, Jean Baptiste Joseph Fourier who, when he was not assisting Napoleon Bonaparte with the 1798 invasion of Egypt, was thinking about periodic functions. The Earth spins on its axis. The motion is periodic. If you make enough observations on the time it takes to spin around once, say by timing when the sun is overhead each day, you will find that the period of rotation is 24 hours. The Moon orbits the Earth. If you make observations of the Moon's location in the sky and correlate that to the time, you will eventually figure out that the Moon's position repeats, and it repeats with a period of

1 month. There is a close relationship between the timing of an event and the period of the event. Fourier provided a mathematical transform that allowed you to look at your data, taken in time, and lift out the period (if there is one). The period has units of frequency that is one over the time: once per day or once per month. Fourier's transform is used all the time in signal analysis to particularly lift periodic signals out of the noise. Down the road a piece from here at Louisiana State University, there sits the Laser Interferometer Gravitational-Wave Observatory (LIGO), which is a giant optical antenna that scans the heavens for evidence of the gravity waves predicted by Einstein's theory of general relativity. A particular signal they look for is a periodic one produced by distant binary astrophysical objects such as a rapidly rotating pair of neutron stars. These pairs orbit each other like a tumbling dumbbell with a period of a thousandth of a second, which is expected to be the period of the gravity wave. Some of these compact binaries are known from astronomical observation and the hope is to see at least gravity waves from them. A standard technique for analyzing the temporal data at LIGO is to use a Fourier transform and look for peaks in the transformed signal that would correspond to one revolution per thousandth of a second.

There are classical Fourier transforms and classical computing fast Fourier transforms but nothing compares to the speed of the quantum computing Fourier transform, which is exponentially faster than even the classical "fast" Fourier transform. To implement the quantum Fourier transform on 10 qubits requires only about a hundred ENT, CAT, and RAT gates. This scaling is quadratic in the number of qubits and so is considered efficient. David Deutsch first worked out a type of efficient quantum Fourier transform as early as 1985 (see note 22). What Peter Shor did nearly 10 years later was to modify the Deutsch quantum Fourier transform to handle certain periodic functions of integer numbers. That is the fluffy quantum icing on the classical number theory cake. Once you have an efficient way to find the period of these integer functions, that period can be used to extract the prime factors p and q of Alice's composite number c and hence to hack her private key from her public key. The secret is in the periods of the clock arithmetic of numbers. Let us make a list of numbers following the pattern: $2, 2 \times 2, 2 \times 2 \times 2, \ldots$ and we get, 1, 2, 4, 8, 16, 32, 64, 128, 256, 512, 1024 … . I now apply the 12-hour clock arithmetic rule and reduce the numbers by subtracting 12 or 24 or 36 or … until we get something back on the 12-hour clock: 2, 4, 8, 4, 8, 4, 8, 4, 8, 4 … . It repeats! After settling down, the period is two: 4, 8, then 4, 8 again and so forth. The goal now is to connect this period to the prime factors.

Twelve-hour clocks are problematic as 12 is not the simple product of only two primes. That is, 12 can be written as $2 \times 2 \times 3$, which is the product of three primes. (The number 1 is not considered a prime.) The public-key encryption assumes that Alice's public-key composite number c is the product of two and only two private-key primes p and q. To handle this, let us move to a 15-hour clock. The number 15 is the product of 3×5 and no other primes and so is perfect. Alice publishes a public key that is a c of 15 on her web page and we must figure out that the two private-key primes p and q are 3 and 5. We go back to our sequences of numbers from the two-times list: 1, 2, 4, 8, 16, 32, 64, 128, 256, 512, 1024 … . Then we rewrite them using 15-hour clock arithmetic as 2, 4, 8, 1, 2, 4, 8, 1, 2, 4 … . Once again, the list of numbers repeats and the period is four: 2, 4, 8, 1 repeat 2, 4, 8, 1 repeat … . Okay then, but what good does this do us?

We dive into the molassified rum extract of the classical core of the Shor algorithm fruitcake and emerge (all sticky) with a theorem proved in the 1760s by the Swiss mathematician Leonhard Euler (pronounced "oiler"). Euler was the most prolific mathematician of all time and at one point was writing one scientific paper a week. His complete collection of publications totals nearly 900 and fills 90 volumes of books. To quote the French mathematician François Jean Dominique Arago, "Euler calculated without apparent effort, as men breathe, or as eagles sustain themselves in the wind."[37] This prolific output continued unabated even though he had 13 children (a prime number) and calculated with one hand while bouncing a baby on his knee with the other. Even going nearly totally blind in his old age barely slowed Euler down—he dictated his results. The particular result of Euler we need has to do with these periodic lists of numbers. Euler's theorem states that the product of $(p - 1) \times (q - 1)$ is *very often* evenly divisible by the period of the list. Let us check. Our composite number is 15. We use that to generate the two-times list and apply the 15-hour clock arithmetic to get 2, 4, 8, 1, 2, 4, 8, 1, 2, 4 … . The period of that list is four. The prime factor p is 3 and the prime factor q is 5, so $(p - 1) \times (q - 1)$ is $(3 - 1) \times (5 - 1)$ is 2×4, which is 8. The number 8 is evenly divisible by the period 4 to give 2 with no remainder. Check.

So let us parse this. The protocol to find the prime factors of 15 is to construct the two-times list of numbers in 15-hour clock arithmetic. Read off this from this table the period. Euler tells us that the period very often divides evenly into $(p - 1) \times (q - 1)$. Guess p and q. We can improve the speed of the guessing by constructing a three-times list and then a four-times list and each will have its very own period on the 15-hour clock, and very often all those periods divide evenly into $(p - 1) \times (q - 1)$.[38] Very rapidly, we can extract p and q and learn Alice's private key and hack her encrypted bank card data. This seems like a very long-winded process. Why don't we just divide 15 out longhand by 2 then by 3 then by 5 to guess p and q that way? Well, we could, but remember what you learned in third grade; long division is hard. And Alice is never going to use 15 as her public key, she is going to use instead a number like 10941738 641570527421809707073220403576120037329454492059909138421314763499842

88934784717997257891267332497625752899781833797076537244027146743531593354333897. Remember that dividing that out longhand will take trillions of times the lifetime of the universe. Instead, we resort to this periodic function generation but now using a 10941738641570527421809707322040357612003732945449205990913842131476349984288934784717997257891267332497625752899781833797076537244027146743531593354333897-hour clock.

If we stick to classical Fourier transforms, this does us no good. It would take trillions of times the lifetime of the universe for the two-times table on this clock arithmetic to start repeating and so extracting the period even with the classical fast Fourier transform is not nearly fast enough. However, I have argued that the *quantum* Fourier transform is exponentially more efficient than the classical one, and using it we get the period and then guess p and q in a second and hack Alice's private key.[39] The Internet is doomed. Or is it?

The quantum Fourier transform is the only quantum part of Shor's factoring algorithm. It is the fluffy quantum icing on the leaden core of fruitcake number theory. I will illustrate how it works for factoring 15 so long as we keep in mind the actual public-key number we are trying to factor is the very much longer public key c. So back to the two-times list on the 15-hour clock. It looks like: 1, 2, 4, 8, 16, 32, 64, 128, 256, 512, 1024 …. This grows exponentially. To cut it down to size, we apply the 15-hour clock arithmetic to get 2, 4, 8, 1, 2, 4, 8, 1, 2, 4 …. On the 15-hour clock, the numbers are never bigger than 15. We now these convert the numbers into binary: 00010, 00100, 01000, 00001, 00010, 00100, 01000, 00001, 00010, 00100 …. Here, I have padded the numbers with extra zeros in the front for ease of converting into the quantum computer register: $|0\rangle|0\rangle|0\rangle|1\rangle|0\rangle$, $|0\rangle|0\rangle|1\rangle|0\rangle|0\rangle$, $|0\rangle|1\rangle|0\rangle|0\rangle|0\rangle$, $|0\rangle|0\rangle|0\rangle|0\rangle|1\rangle$, $|0\rangle|0\rangle|0\rangle|1\rangle|0\rangle$, $|0\rangle|0\rangle|1\rangle|0\rangle|0\rangle$, $|0\rangle|1\rangle|0\rangle|0\rangle|0\rangle$, $|0\rangle|0\rangle|0\rangle|0\rangle|1\rangle$, $|0\rangle|0\rangle|0\rangle|1\rangle|0\rangle$, $|0\rangle|0\rangle|1\rangle|0\rangle|0\rangle$ …. So now the bits are qubits. Next, we use a sequence of CAT, RAT, and ENT gates to entangle these. The result looks something like $|0\rangle|0\rangle|0\rangle|1\rangle|0\rangle$ + $|0\rangle|0\rangle|1\rangle|0\rangle|0\rangle$ + $|0\rangle|1\rangle|0\rangle|0\rangle|0\rangle$ + $|0\rangle|0\rangle|0\rangle|0\rangle|1\rangle$ + $|0\rangle|0\rangle|0\rangle|1\rangle|0\rangle$ + $|0\rangle|0\rangle|1\rangle|0\rangle|0\rangle$ + $|0\rangle|1\rangle|0\rangle|0\rangle|0\rangle$ + $|0\rangle|0\rangle|0\rangle|0\rangle|1\rangle$ + $|0\rangle|0\rangle|0\rangle|1\rangle|0\rangle$ + $|0\rangle|0\rangle|1\rangle|0\rangle|0\rangle$ …. Notice here that only four qubits are now used to represent what was before an exponentially growing list 1, 2, 4, 8, 16, 32, 64, 128, 256, 512, 1024 …. That is true in general. The quantum Fourier transform requires only around the number of qubits in the original number 15. The classical Fourier transform would act on an exponentially larger number of bits. The quantum entanglement buys us the exponential processing power. Most of the information is stored in an exponential number of quantum registers in parallel universes but only four qubits are in our universe.

But remember there are two requirements for an exponential speedup. We need an exponential number of quantum processors in parallel universes, like an exponential number of ships in parallel oceans, but we need a quantum tiller that controls all the ships and steers the answer back into our universe. That tiller is the quantum Fourier transform, which we apply next, and is yet another sequence of CAT, RAT, and ENT gates. The trick is to find just what sequence to use and that was Shor's genius. When the quantum Fourier transform is complete, we measure just the four qubits in our universe in a particular way and out pops the number four, which is the period. Part of the secret of Shor's algorithm is that we only need to know this period, which is not any single number in the exponentially large list but a global property of all the numbers in the list. We then guess p and q and we are done.[40] This procedure works even if we replace a public key c of 15 with public key c of 10941738641570527421809707 32204035761200373294544920599091384213147634998428893478471799725789 1267332497625752899781833797076537244027146743531593354333897. We'll now need maybe a million qubits to factor that, using Shor's algorithm, but that is not so bad, there are 300 billion classical bits (transistors) on the Tukwila chip. If I can build a quantum computer using the same technology as I build a Tukwila, a million qubits should be easy. If it were not for the quantum exponential speedup, we would be in serious trouble. To carry out this algorithm without quantum entanglement, we would need many more classical bits than there are atoms in the known universe. With quantum entanglement, I just need approximately a million atoms or Tukwila-like quantum transistors.

So the big question looming over everybody in 1994, after Shor published his result, was just how hard or easy is it to build a quantum computer with a million qubits? That is where I enter the story, and we'll discuss this in the next chapter.

NOTES

1. See "Richard Feynman and the Connection Machine" by W. Daniel Hills, in *Physics Today*, Volume 42 (February 1989), http://dx.doi.org/10.1063/1.881196. This text is also available with video of Danny Hill's reminiscences at http://longnow.org/essays/richard-feynman-connection-machine/.
2. See *QED and the Men Who Made It: Dyson, Feynman, Schwinger, and Tomonaga* by Silvan S. Schweber (Princeton University Press, Princeton, NJ, 1994), http://www.worldcat.org/oclc/28966591. Freeman Dyson made key (but somewhat later) contributions to the theory but did not share the Nobel Prize with the others, since the prize in physics can be awarded to at most three people in any given year.
3. See *QED: The Strange Theory of Light and Matter* by Richard P. Feynman (Princeton University Press, Princeton, NJ, 1985), http://www.worldcat.org/oclc/12053221.
4. See "There's Plenty of Room at the Bottom: An Invitation to Enter a New Field of Physics" by Richard P. Feynman, in *Engineering and Science* (February 1960), which is reprinted online here: http://www.zyvex.com/nanotech/feynman.html.
5. See "Before the ENIAC—Weapons Firing Calculations" by Harry Polachek, in the *IEEE Annals of the History of Computing*, Volume 19 (1997), pages 25–30, http://ieeexplore.ieee.org/xpl/freeabs_all.jsp?arnumber=586069. For a feminist perspective,

see also "When Computers Were Woman," in *Technology and Culture*, Volume 40 (1999), pages 455–483, http://muse.jhu.edu/journals/technology_and_culture/v040/40.3light.html.

6. See "Last of the Human Computers" by Swati Pandi in *The Daily* (March 30, 2011), http://www.thedaily.com/page/2011/03/30/033011-opinions-history-human-computer-pandey-1-2/.

7. See *When Computers Were Human* by David A. Grier (Princeton University Press, Princeton NJ, 2007), http://www.worldcat.org/oclc/635305974.

8. See "Simulating Physics with Computers" by Richard P. Feynman, in the *International Journal of Theoretical Physics*, Volume 21 (1982), pages 467–488, http://www.springerlink.com/content/t2x8115127841630/.

9. As in Chapter 1, we'll denote classical states with parentheses ($\uparrow$) and quantum states with the funny angular brackets $|\uparrow\rangle$. The angular bracket is actually a quantum notation introduced by the British quantum physicist Paul Adrien Maurice "P.A.M." Dirac and has a special meaning, but here we just use it as a bookkeeping device to track quantum versus classical.

10. Thulium was named in 1879 for the city of Thule in Greenland. (The city has since been renamed Qaanaaq in the Eskimo-Aleut language of Greenlandic.) In classical literature and on European maps, Thule is any region in the far north (from the Greek Θούλη).

11. The story of the computational complexity of thulium comes from recollections of discussions from 1986 to 1989 that I had with members of Ingvar Lindgren's Atomic Theory group from the University of Göteborg in Sweden. It is possible that I misremember which element they were having trouble with, but I'm pretty sure it was thulium. I was sure quantum entanglement was behind all their computational complexity problems. I therefore may have exaggerated the degree of the entanglement in thulium, as well as the computational complexity, to make this point.

12. The Tukwila was named for a town in Washington State with the same name that means "land where the hazelnuts grow" in a Native American dialect. The chip was originally codenamed "Tanglewood," but after complaints from the world-famous Tanglewood Music Festival, Intel changed the name to "Tukwila" in 2003. See "Intel's Tukwila Slips Yet Again," *CNET News* (CBS Interactive, May 21, 2009), http://news.cnet.com/8301-13556_3-10246293-61.html.

13. The rare-earth elements erbium, terbium, ytterbium, and yttrium were all named after the otherwise nondescript Swedish village of Ytterby after all four were discovered in a mine nearby.

14. For Dilbert's amusing take on quantum computers, click on the following URL: http://dilbert.com/strips/comic/1997-03-22/.

15. See "Quantum criticality" by Subir Sachdev and Bernhard Keimer in *Physics Today*, Volume 64 (February 28, 2011), http://dx.doi.org/10.1063/1.3554314.

16. A byte is eight binary bits. A kilobyte is a thousand bytes and a gigabyte is a trillion bytes. Quote from Gate's 1989 talk, http://csclub.uwaterloo.ca/media/1989%20Bill%20Gates%20Talk%20on%20Microsoft.

17. See the article "Turing Machines" in the online *Stanford Encyclopedia of Philosophy* by Edward N. Zalta, principal editor (Metaphysics Research Lab, Stanford University, April 22, 2011), http://plato.stanford.edu/entries/turing-machine/.

18. See "The Church-Turing Principle and The Universal Quantum Computer" by David Deutsch, in the *Proceedings of the Royal Society of London A*, Volume 400, pages 97–117, http://rspa.royalsocietypublishing.org/content/400/1818/97.

19. See the webpage "Exploring Binary" by Rick Regen (April 25, 2011), http://www.exploringbinary.com/.

20. In binary addition, 0 + 0 = 0, 0 + 1 = 1 + 0 = 1, but 1 + 1 = 10, which is the binary for the number two. In this type of binary *modular* arithmetic, we don't carry that zero in the one's place, so 1 + 1 = 1 instead of 10. Hence, the OR gate implements binary modular addition. This is the same as addition on a clockface modulo 12:00 where I don't care about a.m. or p.m. and the time wraps around: 0:00 + 0:00 = 0:00, 0:00 + 12:00 = 12:00, 12:00 + 0:00 = 12:00, but 12:00 + 12:00 = 24:00 = 12:00 again.

21. See *The Tinkertoy Computer and other Machinations: Computer Recreations from the Pages of Scientific American and Algorithm* by A.K. Dewdney (W.H. Freeman, New York, 1993). The particular article on the universal Tinkertoy computer is "A Tinkertoy Computer that Plays Tic-Tac-Toe" by A.K. Dewdney in *Scientific American* (October 1989) found reprinted here: http://www.rci.rutgers.edu/~cfs/472_html/ Intro/TinkertoyComputer/TinkerToy.html. Undergraduate students at MIT built the tic-tac-toe playing Tinkertoy computer and it is in a museum there. A photo of it can be found at "The Tinkertoy Computer" by James Grahame in *Retro Thing* (April 24, 2011): http://www.retrothing.com/2006/12/the_tinkertoy_c.html.

22. The standard but not particularly illuminating names for these gates are the Hadamard (CAT), $\pi/8$ (RAT), and the CNOT (ENT) gates. I am using an $O(3)$ representation of the Bloch sphere rather than the usual $SU(2)$ in order to avoid confusing statements such as "The $\pi/8$ gate is a rotation about the z-axis by $\pi/4$."

23. See "Making a Roux" in Cooking Louisiana, http://www.cookinglouisiana.com/ Cooking/Making-a-Roux.htm.

24. See "Universal Quantum Simulators" by Seth Lloyd, in *Science*, Volume 273 (1996), pages 1073–1078, http://www.sciencemag.org/content/273/5278/1073.

25. See "Rapid Solutions of Problems by Quantum Computation" by David Deutsch and Richard Jozsa in the *Proceedings of the Royal Society of London A*, Volume 439 (1992), pages 553–558, http://rspa.royalsocietypublishing.org/ content/439/1907/553.

26. The Linotype ("line of type") was a typesetting machine for setting metal "hot type" letters for a printing press. Each keystroke injected molten metal into a mold and the Linotype would produce an entire line of metal type at a time in a single metal ingot. The Linotype letters were arranged in order of frequency of use in the English alphabet, and so ETAOIN SHRDLU were the letters of the first two vertical columns on the left side of the keyboard and approximate the most common letters in English from most common to the left "E" and less common to the right "U." This should be compared to the QWERTY typing standard in English keyboards, which was purposely designed with some of the *least* common letters on the primary row in order to slow typists down to keep them from typing so fast that they jammed the keys of the first mechanical typewriters. *Etaoin Shrdlu* was the title of a short story by Fredric Brown about a sentient Linotype machine of the same name. SHRDLU was the name of an artificial intelligence program developed in 1972 in the programming language LISP. (Thanks to Tony Schneider for this information.) The last

issue of the *New York Times* composed using the Linotype machine appeared in July of 1978, and the typesetting of that last issue was captured in the documentary film *Farewell Etaoin Shrdlu*.

27. See *RSA and Public-Key Cryptography* by Richard A. Mollin (Chapman & Hall, CRC Press, 2003), page 10, http://www.worldcat.org/oclc/50339535.

28. I cannot find any clear reason why this was called a "bombe." The Polish word *bombe* translates as "bomb." I suspect that it was named not after the explosive device, nor for the explosive device–shaped dessert, but instead for a type of 18th century furniture design such as a chest or a commode with an outwardly bulging shape, which is a somewhat apt description of the shape of the machine.

29. In return for nearly single-handedly winning the war, the British government rewarded Turing, who was a homosexual, by convicting him of the crime of "gross indecency." They revoked his security clearance and sentenced him to a choice of prison or chemical castration. In 1954, Turing chose instead to take his own life by eating an apple that he had injected with cyanide. In 2009, British Prime Minister Gordon Brown, goaded into action by an Internet petition drive, issued Turing a public government apology.

30. It's a good thing.

31. Eratosthenes of Cyrene was a Greek mathematician who lived around 200 BC.

32. A prime number p is an integer that evenly divides with no remainder (is evenly divisible) by only two numbers, the number 1 and the prime number p itself. Examples are 2, 3, 5, 7, and 127.

33. "In number theory, the prime number theorem ... describes the asymptotic distribution of the prime numbers. The prime number theorem gives a rough description of how the primes are distributed." Gauss guessed the prime number theorem. He was a lightning mental calculator, and he calculated the first few thousand primes and plotted them looking for a pattern. He fit the plot with this function, which tells us for any given number, say 100, approximately how many primes there are between 1 and 100.

34. The formula uses the natural log function log (x) and states that the number of primes less than or equal to the prime p is given by $p/\log(p)$. The number of primes less than p grows a bit slower than linearly due to the logarithm. This should be called Gauss's prime number *conjecture* because he did not prove it. He just ran out the first few thousand prime numbers in his head, plotted them, and then fitted the plot to a curve that he guessed was of the form $x/\log(x)$. Gauss's conjecture was made around 1800. The proof of the conjecture was not made until around a hundred years later, only then promoting the *conjecture* to the Prime Number *Theorem*.

35. Yes, this is a real word. A "duovigintillion" is written as the number 1 followed by 69 zeros. See "Duovigintillion" in *Googology Wiki* (Wikia, June 16, 2011), http://googology.wikia.com/wiki/Duovigintillion.

36. See "New Method Said to Solve Key Problem in Math" by Sara Robinson in the *New York Times* (August 2, 2002), http://www.nytimes.com/2002/08/08/us/new-method-said-to-solve-key-problem-in-math.html?scp=2&sq=Manindra+Agrawal&st=nyt. A longer essay on the problem appeared in the science essay "From Here to Infinity: Obsessing With the Magic of Primes" by George Johnson in the *New York Times* (September 3, 2002), http://www.nytimes.com/2002/09/03/science/essay-from-here-to-infinity-obsessing-with-the-magic-of-primes.html.

37. See *Men of Mathematics* by Eric T. Bell (Paw Prints, 2008), page 139, http://www.worldcat.org/oclc/227033601.
38. Typically, what we do is try several different x-times lists where x is a whole number chosen at random. This gives us different divisors of p and q and makes them progressively easier to guess.
39. This description of the period-finding algorithm is a simplified version of the "Explanation for the Man on the Street" by American computer scientist Scott Aaronson. See "Shor, I'll Do It" by Scott Aaronson in *Shtetl-Optimized: The Blog of Scott Aaronson* (May 12, 2011), http://www.scottaaronson.com/blog/?p=208. Scott assumes that the man in the street understands modular clock arithmetic and exponentiation notation, so I have taken it down a notch. I suppose then mine is the explanation for the man in the gutter. ...
40. Or practically done; there is some small chance Shor's algorithm spits out the wrong answer. However, once we guess the private key numbers p and q, it is easy (on my laptop) to multiply them together and see that we do indeed get the original composite number c. If we don't, we just run Shor's algorithm again.

Chapter 4

You're in the Army Now

The summer of 1994 was a particularly eventful year in the history of quantum information processing. In 1994 at the International Quantum Electronics Conference in Anaheim, I heard a talk by the English physicist John Rarity (now at the University of Bristol), who gave a presentation on his experimental demonstration of quantum cryptography—more accurately quantum key distribution—over 10 kilometers of fiber. The idea of quantum cryptography had been around since the 1980s and had hitherto been demonstrated over a few meters (yards) of fiber or empty space. I remember thinking, "Well that's pretty useless; a crypto system that works over three yards." Rarity's talk changed my mind. The promise of quantum cryptography, which we'll discuss in great detail in

---

* Photo: Library of Congress Prints and Photographs Division, Washington, D.C. 20540, USA. This image is a work of a US military or Department of Defense employee, taken or made during the course of an employee's official duties.

the next chapter, is that by exploiting the quantum principles of unreality, uncertainty, and nonlocality, one could produce an unbreakable quantum key to be used in an untappable quantum communications system. The idea is to use the Vernam or one-time-pad cipher, where two parties, typically our old friends Alice and Bob, share a sequence of random bits they use to encode and decode messages.[1]

Rarity's talk triggered me to write up a two-page assessment of this technology and fax it to Robert Guenther, who was then the head of the physics division at the Army Research Office (ARO) in Durham, North Carolina. I laid out the history of quantum cryptography from the mid-1980s and the potential use of a system with a 6- to 10-mile radius to the army in the battlefield. The ARO does little research on its own but provides funding to outside entities, principally universities.

The response to the crypto fax came quickly. Henry Everitt, a newly hired program manager and experimental physicist in the ARO physics division, called me up to discuss the cryptography results and plan (with our group supervisor, Charles Bowden) an ARO-sponsored workshop on that topic for the spring of 1995. If the ARO is interested in a new area, they will typically hold a workshop on the topic, solicit opinions from experts in the field, and then, if they decide to run with it, issue a call for proposals.

There were some concerns at the ARO that having a program in cryptography might infringe on the turf of the National Security Agency (NSA). The NSA was founded in 1952 within the Department of Defense (DoD) and its primary mission is related to making and breaking secret codes. The mere existence of the NSA was classified until around 1990, and before then, even the initials NSA were not allowed to be used in unclassified communications (leading to the joke that the acronym stood for "No Such Agency"). We decided to have the workshop and invite the NSA to see how they would react.

THE GREAT QUANTUM DIASPORA

At the Fourteenth International Conference on Atomic Physics, held the summer of 1994 at my alma mater, the University of Colorado at Boulder, the Polish physicist Artur Ekert (now a professor at Oxford University) gave a talk entitled "Quantum Computation." It was a revelation because very few of us in the audience had ever heard of quantum computation, and it was the first time any of us had ever heard of Shor's factoring algorithm, which had just recently been widely circulated as a preprint. Ekert gave a wonderful overview of quantum computation, quantum cryptography, and Shor's algorithm, and then finished with a number of insightful ideas about how we could build a quantum computer using atoms or quantum dots (artificial atoms). This talk was for physicists, not computer scientists, and Ekert succeeded in convincing most of us that quantum computation was the next new big thing in physics. For most of us, Shor's algorithm and Ekert's talk came out of nowhere. We were electrified. It was a game changer and we all knew it. Quantum computation had gone from a theoretical backwater to the hottest new thing in just under an hour. Quantum entanglement was actually useful for something practical!

Ekert's talk triggered the great quantum-computing diaspora of quantum physics. At this lecture were some of the key players for the next 10 years of the field's development. David Wineland from the National Institute of Standards and Technology (NIST), Boulder, went back to his laboratory to look at making quantum-computing gates in his ion traps, work for which he was awarded the 2012 Nobel. H. Jeffery Kimble, an American physicist at the California Institute of Technology (Caltech), went back to his laboratory to look at making ENT gates between photonic qubits in optical cavities. Austrian physicist Peter Zoller and his collaborator, Spanish physicist Ignacio Cirac, went back to their desks and began cranking out theory papers on just how to implement quantum gates in an ion trap and other systems. I returned to construct another two-page fax for the ARO.

The ARO response to the quantum-computing fax was much more enthusiastic than their response to the one about cryptography. While there was kvetching at the ARO about cryptography being on the NSA's turf, nobody could doubt that building a new type of computer was solidly in the army domain—after all, the army had built the ENIAC (Electronic Numerical Integrator and Computer). The fact that the primary application of the quantum computer seemed, in 1994, to be for code breaking was set aside. The ARO liked the idea that it could do physics simulations, just as the ENIAC simulated the explosions of H-bombs, and it was also hoped there would be new and better exponentially fast quantum algorithms just around the corner with applications to army-hard problems like logistics calculations. Henry Everitt, Charles Bowden, and I were instructed to expand the planned ARO workshop to include quantum computing (in addition to quantum cryptography). The army allocated the funds, and we set a venue and began rounding up participants.

With the help of faculty at the University of Arizona, including American physicist Hyatt Gibbs, we organized our workshop to be held at the university in Tucson in February 1995. The first US government workshop on quantum information processing, "NIST Workshop on Quantum Computing and Communication," had taken place the summer of 1994. Our "Army Research Office Workshop on Quantum Computing and Cryptography" was pitched as specific to US Army and DoD applications. We also mined NIST's list of invited speakers (see Figure 4.1).[2]

Figure 4.1   Chance meeting in Singapore of four alumni of the 1994 NIST Workshop on Quantum Computing and Communication. From left: Charles Clark, then Chief of the Electron and Optical Physics Division at NIST, now Chief of the Electron and Optical Physics Division at NIST; Ignacio Cirac, then a graduate student, now a Director of the Max Planck Institute of Quantum Optics; Artur Ekert, then a postdoc, now Director of the Centre for Quantum Technologies at the National University of Singapore; Andrew Chi-Chih Yao, then a professor of computer science at Princeton, now Dean of the Institute for Interdisciplinary Information Sciences, Tsinghua University. (The photo was taken in Singapore by Keith Burnett at the request of Charles Clark on January 28, 2011, and it is the property of Charles Clark who has given his kind permission to use it here.)

THE NOTEBOOK, THE SPY, AND THE WORKSHOP

Participants of the two-day workshop included scientists from universities (e.g., American physics Nobel Laureate Willis Lamb from the University of Arizona), as well as participants from government laboratories (physicist Richard Hughes from the Los Alamos National Laboratory and electrical engineer John Rarity from the British Defence Evaluation and Research Agency), industry (computer scientist Charles Bennett from IBM), and other DoD funding agencies. In addition, Henry Everitt told me that they were expecting two guys from the NSA, but he did not know their names. Because I knew most of the invited participants, I was quickly able to rule them out and zoom in on two suspiciously ordinary looking guys. Then, one of them reached into his backpack and slowly pulled out a vomit lime green–colored laboratory notebook. Bingo! These hideously colored notebooks are issued, even to this day, to all government scientists. They are made at the Government Printing Office and distributed by the Federal Supply Service free to any government employee.

These two gentlemen clearly did not want it to be known for *which* branch of the DoD they worked, but it was clear to me that they were the ones from the NSA.

A final workshop report contained four overviews on the discussion topics: *Future Issues in Theoretical Quantum Cryptography*, by Pierre Meystre; *Future Experimental Work in Quantum Cryptography*, by James Franson; *Theoretical Quantum Computing*, by Jonathan P. Dowling; and *Future Experimental Work in Quantum Computing*, by Michael Littman. I will focus here on the quantum-computing aspects of the report. In re-reading this after some years, I am struck not by how much we knew in 1995 but how much we did not know.

An example of what we did not know is the answer to the question, what is the origin of the exponential speedup of a quantum computer over a classical computer? This question has been debated over the past 15 years and even now there is not a unanimous consensus in the community. However, the majority of scientists in the field now believe, and it has been argued, that an exponential speedup requires two ingredients: quantum coherence and quantum entanglement. Quantum coherence is, in this book's lingo, the ability to create a Schrödinger cat state of a single qubit by applying the single-qubit CAT gate. Quantum entanglement is a stronger requirement: the ability to make a two-qubit entangled state of the EPR type by applying a two-qubit ENT gate. Coherence is required for entanglement. The first step in generating a pair of entangled particles is to apply the CAT gate to one of them and then the ENT gate. Without the CAT, the ENT does not produce multi-qubit entanglement. However, entanglement is not required for coherence. You can make a single-qubit cat state without reference to another qubit. Only entanglement gives us the exponentially large computational space, the so-called Hilbert space, containing an exponential number of parallel processors in parallel universes. That is what most of us in the field now believe. But in 1995, there was a cadre of physicists who either did not understand the need for entanglement or did not believe it, and at the workshop, there were many discussions about why simple coherence (the making and maintaining of cats) was not enough. In particular, some participants suggested that Shor's factoring algorithm might be run on a coherent system that would not need entanglement, greatly simplifying the technological requirements for building a quantum computer.

Let me discuss a bit of the language here to get us all up to speed on the quantum speedup. The ability to produce a cat state requires coherence. That is, we must be able to make the qubit in a superposition of up and down or alive and dead or 0 and 1. Coherence at heart is a wave phenomenon, and even the word is suggestive of wave behavior. Quantum systems like ions or photons or other physical realizations of qubits inherit their coherence (ability to make a cat) from the wave–particle duality of the quantum theory. Photons, electrons, and ions are both waves and particles. The wave nature imparts upon them the coherence. Water waves are coherent as are sound waves, earthquake waves, and radio waves. There is nothing or little quantum mechanical about them. Coherence is a term to describe how well the waves interfere. If two water wave peaks meet and produce a wave twice as high, the water waves are said to be perfectly coherent. If a water-wave peak and water-wave trough meet and completely cancel to produce no wave at all, they are also perfectly coherent. The waves "cohere" to add up in phase or cancel out of phase. If the waves do not add up to precisely twice the height (or do not cancel out to exactly zero), they are called partially coherent. If the waves do not add or cancel at all, they are called incoherent. If you make very nice clean water waves by dropping two stones in a pond that are perfectly coherent, they may lose their coherence as they propagate from one side of the pond to the other, say by interacting with gusts of wind, so by the time they reach the other side of the pond from where you drop the rocks, they are partially coherent and do not add up to double or cancel out to zero anymore. The process by which the waves lose coherence (the interaction with the wind in this example) is called decoherence. The random motion of the wind decoheres the waves. If the wind is strong enough (or the distance is far enough), the waves are totally decohered and no interference (doubling up or zeroing out) is seen anymore.

Quantum particles inherit their coherence properties from the wave properties of their wave–particle nature and the words describing their coherence are all the same. If you can make a perfect cat state, dead and alive, a superposition of zero and one, $|0\rangle + |1\rangle$, the qubit is perfectly coherent. Interaction with the environment (like the wind on the pond) can cause decoherence. If the decoherence is strong or you wait a long time, the cat will completely decohere and collapse to either dead $|0\rangle$ or alive $|1\rangle$. Decoherence is the origin of the collapse of the cat. For large objects like cats, there is a lot of interaction with the environment and the decoherence-induced collapse occurs on a time so short it cannot be measured by any technology. However, for small objects like a single ion qubit, it is much easier to protect the ion from the environment, and the decoherence-induced collapse may take place on time scales of seconds or minutes. At very short times, a good ion cat state $|0\rangle + |1\rangle$ is perfectly coherent, and at long times, the collapse state $|0\rangle$ or $|1\rangle$ is completely decoherent or has completely decohered. At intermediate times, the state is somewhere between alive and dead ($|0\rangle + |1\rangle$) and dead or alive ($|0\rangle$ or $|1\rangle$), and in these intermediate time zones, the cat state is partially coherent. For ions in a trap, decoherence is caused by random thermal fluctuations of the electromagnetic field that emanate from the walls of the ion trap. Just like the random buffeting of the water waves by the gusts of wind, the random buffeting of the ions by random puffs of electromagnetic radiation decoheres their state.

Decoherence is the curse of Schrödinger's cat and the bane of the quantum computer. For an exponential speedup, we need to make many entangled states. This is done by many applications of the ENT gate to cat states first created by the CAT gate. If, immediately after the application of the cat-creating

CAT gate, the cat decoheres from dead and alive to dead or alive, before we have time to apply the two-qubit entangling ENT gate, then it is not only the cat who is dead meat. This leads us to a metric of goodness for building a quantum computer, the ratio of the gate speed to decoherence time. In the ion trap quantum computer, the gates can be implemented in a millionth of a second. The decoherence time is at least a second. Thus, the rule of thumb is that you can do approximately a million quantum gate operations before your cat states have all decayed away. Coherence in the form of the cat is needed to make an entangled state, but coherence is also needed to maintain an entangled state. Entanglement requires coherence. Two ions in a trap can be prepared in a general entangled state of the form $|\uparrow\rangle_A |\uparrow\rangle_B + |\downarrow\rangle_A |\uparrow\rangle_B + |\uparrow\rangle_A |\downarrow\rangle_B + |\downarrow\rangle_A |\downarrow\rangle_B$. The State is a superposition of all four possible vertical orientations of the ions: up–up and down–up and up–down, and down–down. Notice that there are only two ions but we require four states. The exponential speedup is to take the two ions and construct the 2×2 or 4 pairs of states.[3] Upon measurement, this state of the two ions collapses into one of the four possible outcomes, up–up *or* down–up *or* up–down *or* down–down, each with a 25% probability. Measurement destroys the coherences (the "and" in the superposition state) and produces one outcome that is the "or" in the collapsed state. The direct destruction of the coherence, by measurement or, equivalently, interaction with the environment, results in the indirect destruction of the entanglement. You can have coherence without entanglement but you cannot have entanglement without coherence.

It is critical to note that "measurement" is really just a rapid decoherence by your measuring device, a laser beam in the ion trap, which rapidly causes the state to go from a four-state system to a one-state result. The buffeting of the ions in the trap is equivalent to a measurement, a measurement by the environment, and it also causes decoherence of the four-state system into one of the four outcomes. Uncontrolled measurement by the environment (decoherence) destroys our cat just as surely as a controlled measurement by our laser beam. Without decoherence, there can be no entanglement, and without entanglement, we have no access to the exponentially large number of states—the Infinite Library of Borges is lost to us. But do we really need entanglement and coherence, or is coherence alone enough to view the exponential vastness of The Library of Babel?

In *Katzensprache* (the language of cats), let us recall that the atom in the box in the Schrödinger cat thought experiment is in a superposition of decayed and not decayed; that atom superposition alone is provided by coherence. The cat is also in a superposition of dead and alive; that cat superposition alone is also provided by coherence. However, the final ingredient is that state of the cat is correlated or entangled with the state of the atom (via the action of the hammer either smashing or not smashing the flask of cyanide); this is an additional feature above and beyond just coherence. You have to have coherence to get the atom or the cat into a superposition state, but entanglement—the atom–cat coupling via the bipolar mallet and deadly flask—is what strongly correlates the state of the atom to that of the cat. You could have the atom in a superposition of decayed and not decayed in a state completely independent from the cat in a state of dead and alive. In such a case, the atom and cat both separately possess coherence but there is no entanglement, because a measurement we make on the atom has no effect on the cat (or vice versa). Only when a measurement on the atom causes the state of the cat to collapse does the power of entanglement come into play.

In summary, you can have coherence without entanglement but you can never have entanglement without coherence. The question? Is coherence alone good enough to build a quantum computer? The answer turns out to be "maybe" and depends a great deal on what type of quantum computer you would like to build. It seems clear that entanglement is required for a full universal quantum computer but there is some wiggle room for a special-purpose "quantum" computer that runs on coherence alone. Coherence alone is not enough to build a universal quantum computer, a quantum computer capable of simulating any other quantum computer and doing any quantum Turing computable task. Coherence alone is probably not enough to run Shor's algorithm efficiently and hack the Internet. However, coherence alone is sufficient for a special-purpose quantum computer that, while not universal, does one thing very well. We will cover this in the upcoming section on the Grover search algorithm, Needle in a Haystack, where we will see that coherence alone can give a quadratic but not an exponential speedup. Remember our discussion on quadratic errors. If a classical computer can do the calculation in 100 seconds, then a quadratic speedup means the pseudo quantum computer can do it in 10 seconds, because 10 times 10 is 10^2, which is 100. It is almost now universally believed that only entanglement, and only entanglement of just the right kind, gives the exponential speedup. But for some important problems, a nonexponential quadratic speedup on a nonuniversal quantum computer may be just fine.

This point of whether or not quantum entanglement (or coherence alone) was required for the exponential speedup of the quantum computer was curiously (in hindsight) the subject of much debate at the ARO workshop. To attendees with a background in quantum foundations, it seemed obvious that entanglement with its exponentially large Hilbert space (exponential number of parallel processors in an exponential number of parallel universes) was needed for an exponential speedup. In fact, many of us view this exponentially large Hilbert space as the true resource for universal quantum computing and entanglement as a tool or portal to access it. However, there were a number of participants, many of whom had little experience with the notion of quantum entanglement, who argued that Shor's factoring algorithm only required coherence and not additionally entanglement. This minority argued that a quantum computer then could be built using classical light waves, radio waves, or water waves, all of which are coherent—that is, they all display constructive and destructive wave interference.

That debate went on for a remarkable number of years and while now almost everyone agrees entanglement is required for the exponential speedup, there are some uses of coherence alone for a smaller quadratic speedup such as in the Grover search algorithm, which we will discuss in an upcoming section. However, it took a few years to stamp out the coherence adherents who even submitted proposals to the ARO and the NSA on such coherence-only schemes for factoring. One in particular was from an experimental group that claimed that they could build a universal quantum computer to implement the Shor factoring algorithm by manipulating a single electron in a hydrogen-like atom. (I think this proposal came in 1996.) This proposal came to the NSA and ARO review panel (which I was on at the time) and actually got quite high scores from some of the panel members, but I smelled a rat. There was no entanglement in the scheme at all, only coherence. I then showed that their scheme would actually require an exponential amount of time to implement—much, much, longer than the lifetime of the universe. With only coherence, the exponential speedup comes at a cost of some exponentially large resource, in this case, time. I blackballed it and gave the proposal a 0 out of 100 by checking the box "nonresponsive to the call for proposals." That was enough to drop its average so low it did not qualify for funding. Not daunted by this, the investigators resubmitted the same proposal the following year, acknowledged my complaint, but then claimed to have a new "encoding" scheme that got around the exponential blowup in time. I then showed that their encoding scheme was in an exponentially hard class of mathematical problems and the new encoding scheme would also take an exponentially large amount of time to implement. This killed it off once and for all. They were able to show that the scheme could implement the Grover search algorithm, discussed in the next section, but that does not give an exponential speedup. The question of entanglement or not entanglement can be restated in terms of our quantum computer programming language this way. Is a quantum computer consisting of only RAT and CAT capable of creating coherence alone and capable of running Shor's algorithm, or is the ENT gate required as well? It has been mathematically proven that the three ruling gates, RAT, CAT, and ENT, are required to make a universal quantum computer but the dangling marsupial in the room is whether or not RAT or CAT alone is sufficient to make a special-purpose (nonuniversal) factoring engine for running Shor's algorithm. Most quantum computer scientists suspect they are not, but there is no proof. Hell, there is no proof that there is not an efficient classical computing algorithm for factoring. (That was the premise of *Sneakers*.) Hence, without a general proof of impossibility, we are left with attacking individual proposals for using RAT and CAT alone for factoring one at a time, each time looking for the exponential blowup in some resource or another.

John Clauser and I even constructed our own design of a factoring engine that used only coherence in a paper we published in 1996 called "Factoring Integers with Young's *N*-Slit Interferometer," which was published in the journal *Physical Review A*. The idea was to use coherent classical light beams in an optical device called an interferometer that, like with the water waves, produces wave patterns that either constructively or destructively interfere with each other. This paper was instructive (to me at least) because from it I learned just how far you could push coherence without entanglement and what explicitly goes to hell in a hand basket if entanglement is not in play. An N-slit interferometer is basically a black screen with a bunch of parallel slits carved into it to allow the light to pass. If N is 100, then there are 100 such transparent slits in the screen. Coherent light, like from a laser, goes through the slits and illuminates a second screen, making a wavy interference pattern, peaks where the waves add destructively and troughs where the waves subtract destructively. The waves of the laser light coming from different slits have to travel slightly different distances depending on the location of the slit. If we take a pair of slits, then if the light from slit one travels a full wavelength longer to hit the screen, there will be constructive addition—peak will meet up with peak. If it travels a half of a wavelength, then there will be destructive cancellation—peak meets with trough. This type of interaction has to be considered by adding up the contributions from all the slits at each point on the screen, and the math gets hairy, but Clauser's theory handles it all nicely. However, the pattern is very complicated because interference from light arriving at the screen from each of the 100 slits needs to be tracked. The curious and fun part about the mathematics, which is what gave Clauser and I the idea, is that the primary mathematical expression for determining this pattern of peaks and troughs looks nearly identical to the primary expression for finding the prime factors of N in the Shor factoring algorithm. If the math is the same, then the physics must be the same!

Clauser was working with these interferometers and atom waves (not light waves) to try and make a sensitive gravity-measuring device for use in looking for oil pockets near oil wells. (The idea was to lower the gravity wave detector down the oil well bore hole, and when the contraption passed the less dense oil pockets, the interference pattern would shift in such a way to reveal the pocket.) In 1996, I was instead deeply obsessed with coherence versus entanglement and the factoring problem. When Clauser gave a talk on this at the ski resort in Snowbird, I had a "eureka" moment when I saw his primary mathematical result closely resembled that of Shor's quantum Fourier transform. I followed Clauser out to the snow bank at the coffee break where I began excitedly drawing equations in the snowdrift with my finger.[4] That was how the paper got started (see Figure 4.2).

The idea is that we have coherent laser light with an adjustable wavelength (the distance from peak to peak in the light-field oscillation). We then construct an interference grating (black paper with parallel slits) with an arbitrary number of N slits in it, where N is the integer to be factored. We scale the wavelength of the light in such a way that it is always also an integer number n less than N. We

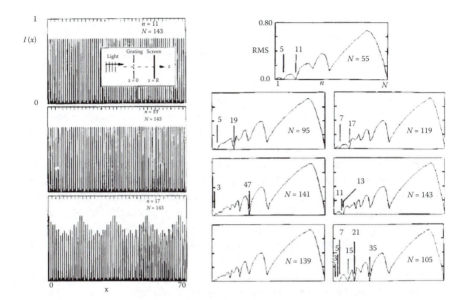

Figure 4.2 Factoring integers with an interferometer. On the upper left (inset), we see coherent laser light coming from the left going through a black screen grating that has an integer number N slits carved into it to let the light pass. The light then falls on the screen where the intensity $I(x)$ is recorded as a function of the vertical position x on the screen. The light interferes to generally produce constructive interference peaks and destructive interference troughs seen in the bottom left. The three leftmost graphs correspond to patterns from gratings with numbers of slits N equal to 143. What changes is the wavelength of the light, which is indicated (in units of wavelengths) as n, which is also an integer. When n evenly divides into N, the pattern goes flat as in the top left (n is 11 divides evenly into N is 143 to give 13) and the middle (n is 13 divides evenly into N is 143 to give 11). However, the bottom left pattern is not flat (n is 17 divides into 143 to give 8 with a remainder of 7/17). On the right, we plot a "flatness function" that goes to zero whenever we have a factor that evenly divides N. By just noting the dips in the curve from the data, we can find all the factors of N by just adjusting the laser light. (Figure courtesy of John Clauser.)

count our way up to N one wavelength at a time: one wavelength then n is 1, two wavelengths then n is 2, and so on. We then ratchet the wavelength from n is 1 up to n is N (say N is 100) and monitor the interference pattern (pattern of constructive wave peaks or destructive wave troughs) on the far screen to the right of the slits (Figure 4.2, top left and middle left). The beauty of the mathematics underlying the setup is that it mimics that of the Shor algorithm in a concrete and visible way on the screen—whenever we hit upon an integer n (adjustable) that evenly divides into our target N (fixed), the interference pattern goes flat and we are sure that the n evenly divides into N with no remainder and that proves n is a factor of N.

Thus, for example, in Figure 4.2 (top and middle left), the integers n equal to 11 or 13 divide evenly into N equal to 143 to give 13 or 11 and the pattern goes flat. At the bottom left, we have n is 17 that does not divide evenly into 143 but gives a result of 8 with a remainder of 7/17. (To "divide evenly" means there should be no remainder or the remainder is zero.) In this case, the pattern is wavy and not flat, so 17 does not divide into 143 evenly. The bumps in the pattern come from the remainder. So we just step n from 1 to 143 and monitor when the pattern goes flat to find the factors 11 and 13 of 143, and because the pattern goes flat only twice at these two integers, they are the only two factors. Voila! The number 143 has only two factors: 11 and 13. To speed up the process, in Figure 4.2 (right), we can scan the wavelength quickly and plot a "flatness" function taken from the pattern. (This is computed by taking the height of the peaks *minus* the height of the troughs and then dividing by the height of the peaks *plus* the height of the troughs, which always gives a flatness number between zero and one.) Simply recording the data and sending into a computer program can carry it all out. Whenever n is a divisor of N, the flatness function dives to zero (the height of the peaks is the same as the height of the troughs), and from the graph, we can read off all the factors of any integer N. Hence, for example, in the topmost right of Figure 4.2, we see that N equals 55, which has two and only two factors, namely, 5 and 11.

Thus, this setup neatly explains how coherence can be used to factor integers in a special-purpose factoring machine that uses coherent laser light. What goes wrong? Well, Clauser and I estimated that by using this technique, we could factor integers N that were about four or five digits long in a realistic laboratory experiment. Thus, we could set up a laser system and factor 5183 into the two primes 71 and 73. Recall that N is the number of slits so we need a big diffraction grating to make so many slits on it. But the NSA is not interested in factoring a four-digit number like 5183. Recall from Chapter 3 that a typical 512-bit public key has approximately 155 decimal digits, say, RSA-155 given by a composite number c equal to 10941738641570527421809707322040 35761200373294544920599091384213147634998428893478471799725789126733 2497625752899781833797076537244027146743531593354333897. That means we need 10^{155} or 1000000000000000000000000000000000000000000000000 0000000000000000000000000000000000000000000000000000000000000000000000 00000000000000000000000000000000000000000000 slits. Okay, so in a typical setup, each slit is separated from the other by approximately a micrometer (0.0004 inches). Hence, our grating would have to be 0.0004 inches per slit multiplied by 10^{155} slits or a grating 1200000000000000000000000000000000000000 0000000000000000000000000000000000000000000000000000000000000000000000 00000000000000000000000000000000000000000000 kilometer across, which is 10^{134} or 100000000000000000000000000000000000000000000000000 0000000000000000000000000000000000000000000000000000000000000000000000 00000000000000000000000 light-years across. The known universe is only approximately 10^{11} light-years in diameter so the grating would need to be 123 *orders of magnitude* larger than the universe. That is a whopping big interferometer! Worse is that the laser needed to power it, that is, illuminate all the slits equally, would be just as huge. In order to get a good signal-to-noise ratio, let us assume we need around a microwatt of power or 0.0000001 watts of laser power *per slit.* That is, 0.0000001 watts times 10^{155} slits is 10^{139} gigawatts of laser power, enough to power 10^{139} flux capacitors on the same number of DeLorean time machines.[5] The total power output of the Sun is only 10^{25} watts, so this is the power of 10^{114} Suns. Even better, the power output of the Big Bang that created the entire universe was a measly 10^{52} watts, so we'd need 10^{87} Big Bang explosions to power our laser to illuminate the giant grating vastly bigger than the entire universe to be of any use to the NSA. Now that's what you learned in third grade—factoring is hard.

Once again and most clearly, here we see the penalty we pay if there is coherence alone with no entanglement. Without entanglement, the door to the Library of Babel is slammed shut. Entanglement is our only vehicle through that portal. Without it, we are forced to produce 10^{155} parallel slits that are parallel processors not in parallel universes—but in our own universe where we must pay the full electric bill and pay the space contractors to build us a grating that stretches to the next universe and beyond. Another way to say it is, without entanglement (and only coherence), the quantum computer is exponentially large in the real space in which we live. With entanglement, the computer is exponentially large in some abstract Hilbert space where we don't live and the qubits we control are polynomially small in our real space.

Attempting factoring using coherence alone is tempting at face value, because of the similarity of the math with Shor's algorithm, but one must check those pesky scaling laws. Without entanglement, there seems always to be Feynman's conjectured exponential slowdown of the classical computer or, equivalently, this exponential blowup in real physical resources. Only entanglement seems to avoid that blowup. Clauser and I had worked this all out with the slits in 1996, and when I got the proposal to implement Shor's algorithm in a single hydrogen-like atom in that same year, I was ready. Their proposal was nearly identical to our proposal in the mathematics with the photons in our laser beam replaced by the electrons in their atom and our N parallel slits replaced by N energy levels the electron has access to in the atom. There are an infinite number of levels in any hydrogen-like atom, but they become exponentially closely spaced in energy as the number you are trying to factor grows larger. From the Heisenberg energy–time uncertainty relation, it was easy to show that to read and write information into exponentially small spaced energy levels (energy is small), you would need an exponentially large amount of time to do it (time is big). The hydrogen-like quantum computer could factor 155-digit numbers but no more efficiently than Clauser's slit. It would take much, much, much longer than the lifetime of the universe to do it! That is why I rejected their proposal (twice).

Despite the lack of exponential speedup in such a coherence-without-entanglement factoring scheme, there has arisen a small cottage industry of theory and experiment surrounding this idea, viewed now as an interesting interplay between physics and number theory instead of a public-key cracking application. The champion of this industry has been my friend and colleague, German physicist Wolfgang Schleich, who has vastly extended the idea of factoring integers using various number theory–inspired coherence-based interference strategies and who has collaborated on the demonstration of the effect experimentally in coherent light waves, atom waves, and nuclear spin coherent states.[6]

This detour into coherence versus entanglement leads us directly into the debates about coherence versus entanglement as quantum-computing requirements. These debates took place in the 1995 ARO Tucson meeting in the quantum-computing theory breakout sessions, which I was assigned to document and report. I'm going to paraphrase them here. There will be some repetition of ideas discussed previously but if you don't know them well, now is a good time for a review. I will demark report text with an italicized subheading and text and follow each report with some clarification.

Preamble: *This theoretical quantum-computing session contained mostly theoretical physicists, mathematicians, and computer scientists. Originally, we intended to discuss all aspects of quantum computing. However, it was decided to discuss predominantly mathematical, computational, and complexity theoretic[7] material as a separate group—and then to rejoin the experimental quantum-computing session to discuss theoretical aspects of the actual physical implementation of a quantum computer with the experimentalists. Participants in the theory session included Willis Lamb, Pierre Meystre, and Ewan Wright, University of Arizona; Artur Ekert, University of Oxford; Umesh Vazirani, University of California at Berkeley; Charles Bennett, IBM-Watson; Bill Wooters, Williams College; Peter Milonni, Los Alamos National Laboratory; Jon Dowling and Charles Bowden, Army Missile Command; Jagdish Chandra, ARO; Howard Brandt and John Pellegringo, Army Research Labs; Hersch Pilloff, Office of Naval Research (ONR); and Keith IIIIII, NSA.*

Introduction and Background: *It was Feynman who first pointed out in the early 1980s that the simulation of a quantum process on a classical computer implies an exponential slowdown in computer time. In a series of papers in the late 1980s and early 1990s, Deutsch elaborated on the inverse idea: that classical calculations might be performed on a quantum device with an exponential speedup. Although preliminary work showed that an exponential increase in computation was possible for at least some problems (Deutsch–Jozsa algorithm), the real breakthrough came in 1994 when Shor announced that he had constructed* *a quantum algorithm that could factor large composite integers on a quantum computer with an exponential increase in speed over a classical algorithm. The rapid factoring of large numbers has tremendous consequences for the security of public-key encryption schemes, as illustrated in an entertaining fashion in the movie Sneakers. The hope is that other exponentially intractable problems might be solved—perhaps even the famous NP-completeness question. Hence, quantum computers hold the promise of a nearly unimaginable breakthrough in computing power.*

We now know that quantum computers can provide only a quadratic speedup on the NP-completeness problem, but in 1995, that was still an open question. I will give a detailed definition of the NP problem in the endnotes, but suffice it to say that the NP problems are very important problems such as the traveling salesman problem, logistics problems, scheduling problems, and so forth that the best-known classical algorithms can solve only with an exponential slowdown. Because "NP" or "nondeterministic polynomial" is math jargon, I will instead typically call these "traveling salesman–type" problems. Any breakthrough on any one traveling salesman–type problem implies a breakthrough on all of them. The traveling salesman problem posits a salesman who must drive to a number of different cities on his sales route, say Atlanta, Barstow, Chicago, Dallas, . . . and Zigzag. The goal of the salesman is to find the route that minimizes his driving time and gas expense, that is, a route that has minimum distance and passes through each city once and only once. This problem is in a certain large class (called the NP class) of classical computing exponentially hard problems. Like factoring, the best-known classical algorithm takes an exponential amount of time to find the optimal route, but once the route is found, it is easy to check that it is optimal. (This "easy to check" feature is where the term "nondeterministic polynomial" of "NP" comes from. The solution is exponentially hard, but once you have it, the check is polynomially easy.) The particular charm of the NP or traveling salesman–like problem class is that it contains a large number of very important decision problems such as the traveling salesman, the scheduling problem, and other logistics problems related to finding an optimal solution that minimizing resources—say the shortest route an email—should take over the Internet to get from Beijing to Baton Rouge, given the exponentially large number of routes it could take. This class of problems is called NP complete and has the important property that if you can solve any one problem in the class in polynomial time, you can solve all the problems in polynomial time. Any crack in this edifice of super-hard problems brings down the entire computational façade. For many years, classical computer scientists struggled to find such a crack, with little success, when along came quantum computer science, which offered new hope and quickly provided a new result—a quantum computer algorithm can solve any problem in the NP class *quadratically* faster than the best classical algorithm.

Not an exponential speedup like Shor but still potentially useful. The Indian–American computer scientist Lov Grover invented this algorithm in 1996 and it will be discussed in Needle in a Haystack.

Recall also that, in factoring the best-known classical algorithm, it takes an exponential amount of time (hard) to factor c into the two primes p and q, but once you have p and q, it is very fast (easy) to multiply them back together to check that they give back c. The factoring problem is not in the traveling salesman (NP) class of hard problems but has properties like them. (Hard to solve but easy to check.)[8] Also, like factoring, it is unknown if there is an efficient polynomial speed classical algorithm to solve the traveling salesman problem. The best-known solutions are exponentially slow, but there is no proof that there is not a classically exponentially faster algorithm out there waiting to be constructed on a legal pad in an un-air-conditioned attic in Kanpur. Once again, mathematicians have searched for years for such a classical algorithm and never found one and so in a fit of sour grapes declared that there is likely no fast polynomial algorithm (P) that solves the hard traveling salesman nondeterministic polynomial problem (NP), leading to the Zen-like unproved mathematical conjecture that NP ≠ P.[9]

Search for Problems: *A consensus seemed to be reached that the most important task for quantum computational theory is to identify problems that are only known to be solvable in exponential time on a classical computer (hard) and to see if they can be transformed into a form that can be solved in polynomial time on a quantum device (easy). Of course, such a research thrust will depend on what types of quantum computation schemata are envisioned—say the Deutsch methodology. The prime example of a problem such as this is the quantum prime factorization algorithm of Peter Shor of AT&T Bell Laboratories. Vazirani illustrated another example of a potential problem: Investigating the calculation of the reliability of certain multi-nodal networks against failure in one of the links. This problem is currently thought to be classically un-computable (in polynomial time), and it is not known if this could be made polynomially tractable on a quantum machine. Applications would include the optimization of telephone, highway, and microchip networks.*

For the term "exponential" we have given plenty of examples, above. The term "polynomial" is more math jargon but in this context means "easy" or "not exponential."[10] The term "polynomial" refers to an algebraic scaling and the most common one encountered in this book is a quadratic scaling where the polynomial is x^2. In all polynomial expressions like this, the x is in the "downstairs." If x is the size of the problem, then x^2 is the time it takes to compute the answer to a problem; quadratic scaling means if you double the size of the problem, then you quadruple the time it takes to solve it. Take x as originally 10, then x^2 is $10^2 = 10 \times 10 = 100$. Doubling the problem size gives x as 2×10, which is 20, and then the time to solve it is $20^2 = 20 \times 20 = 400$, which is quadruple the original time of 100, which is $4 \times 100 = 400$. This may seem like a lot of extra time, but let us compare it to an exponential scaling law, which can be written 2^x. This is an *exponential* expression and not polynomial (the x is upstairs in the *exponent*). If the original problem size is x is 10, the original run time is 2^{10} is $2 \times 2 \times 2 \dots$ (10 times), which is 1024. If we double the problem size to x is 20, then the run time is $2^{20} = 1048576$, which is much, much bigger than 1024. Exponential scaling is—in this way—much, much worse than quadratic scaling. In this sense, any polynomial scaling x^3, x^4, and so on is still considered "easy" when compared to exponential scaling 3^x, 4^x, and so on, which is considered "hard." It is that pesky exponential scaling that gets us quickly into lasers requiring more power than the Big Bang to factor a 155-digit number comprising a 512-binary-bit public key.

Quantum-Computing Engines: Quantum computers at first will not be versatile, programmable devices—but rather hardwired "engines" that are designed to solve specific types of problems. For example, one can envision that there would be a prime factorization engine, a traveling salesman problem engine, and a network reliability engine—each completely devoted to a specific problem. An actual programmable general machine seems incredibly long term.

I wrote the above paragraph and what I had in mind at the time (I think) was that the goal of building a *universal* quantum computer, a reprogrammable machine that could do any task a quantum computer might be set to do, seemed far off, but perhaps some special-purpose *nonuniversal* quantum "engine" could be designed to attack particular problems like factoring or searching a database. Historically, we can think of the Polish–British bombe and even the Colossus, which were not universal classical computers but rather special-purpose engines for cracking the German Enigma code. Cracking the Enigma was all they could do. The ENIAC in contrast was a universal classical computer, originally designed to calculate army artillery tables but then reprogrammed to simulate H-bomb explosions—quite a different problem. The ENIAC was a more difficult machine to build but the British were not worried about all classical problems; they only needed to solve one—the location of the German U-boats on D-Day. As I will discuss below, the D-Wave Corporation has constructed a commercially available special-purpose coherence-based "quantum" computer for running database searches. The D-Wave machine has no entanglement and is therefore not universal, but it is good at doing just one thing, searching a database with a quadratic speedup.

Quantum versus Classical Optical Computers: There was a discussion initiated by Bennett on the advantage of a quantum computer over a classical, optical, interferometric computer. For example, the physical ingredients required to implement Shor's algorithm for prime factorization are the ability to make coherent superpositions (CAT gates), unitary transforms (RAT and ENT gates), and interferometric measurements. It is apparent that all these ingredients are *available in a classical, optical, linear, interferometric device. One could imagine a classical optical computer that carries out all the steps in Shor's algorithm—utilizing linear optical elements, such as phase plates. So why bother with a quantum computer? The difference is that the EPR-type entanglements found in a quantum computer are not found in a classical optical device, and this is crucial in order to achieve exponential-time speedup while maintaining polynomial-space storage and power requirements. For example, let's say one wanted to store all 8 of the 3-bit binary numbers {0, 1, 10, ... , 111} as a single superposition state $a_1|000\rangle + a_2|001\rangle + a_3|010\rangle + ... + a_8|111\rangle$. In a quantum computer, this quantum superposition of $2^3 = 8$ states could be made using one single 3-qubit register. On a classical optical computer, one would have to superpose $2^3 = 8$ separate classical beams to achieve the same classical superposition state. Hence, on the classical machine, storage and power requirements grow exponentially with the number of bits in the register—to encode a 100-bit superposition state would require 2^{100} separate optical beams classically. In the quantum register, all 2^{100} possible states are superposed in one single 100-qubit register. Each term in the classical superposition has a real existence, whereas in the quantum machine, each term has only a quantum probability of existence that is not actualized until a measurement is made. In the context of the many-worlds interpretation of quantum mechanics, Deutsch has pointed out that it is as if each of the 8 potentialities embodied in $a_1|000\rangle + a_2|001\rangle + a_3|010\rangle + ... + a_8|111\rangle$ are stored in 8 separate computers in parallel, with each computer in a separate universe.*

This is a rehash of what we have covered before. The exponential speedup requires the exponentially large Hilbert space of an exponentially large number of parallel processor in parallel universes. Coherence is necessary but not sufficient. One must have entanglement in addition to coherence. Entanglement requires coherence but coherence does not require entanglement.

Decoherence Time versus Error Correction Schemes: *Bennett pointed out that an estimate of the decoherence times alone is not sufficient to determine the practicality of an actual quantum device. This is because the advancement in error correction schemes might be able to compensate for more and more of the decoherence. As an example for this point of view, it was noted that at the dawn of the age of classical computers in the 1940s and 1950s, it was thought that innate limitations to the accuracy of calculation from physical considerations alone would prevent the type of computations that are carried out today without a second thought. What was not foreseen at the time was the advance of powerful error correction algorithms that could allow highly accurate computer operation even at today's standards of super high-bit rates and giga- or teraflop processing speeds, even though the physical computing process generated errors at a fairly high rate. A similar effect can be hoped for in the future progression of quantum computers, and error correction schemes should develop in tandem with attempts to quantify and reduce the decoherence from physical external processes.*

This was a very prescient observation by Charles Bennett (IBM Charlie). That is, in particular, the Shor algorithm assumed that the quantum computer, if it existed, would carry out the calculation with perfect accuracy and precision. However, nothing is ever perfect in the real world. Particularly, as we have discussed before, the coherence of the qubits (ability to make a cat state) must be long lived when compared to the speed at which one implements the gate. Recall that the first step in preparing an entangled pair of qubits is to apply a CAT gate to one of them (to prepare it in a cat state of dead and alive) and then the ENT gate to the pair to generate the entanglement, which is the portal to our exponentially larger Hilbert space. It is critical that the cat state must survive for a long time in the quantum split identity of dead and alive so that there is time for the ENT gate to be applied to that state. If the cat collapses to dead *or* alive *before* the application of the ENT gate, the gate just kills the dead cat (or resurrects the live one) but no entanglement is generated and the portal to the exponentially large Hilbert space closes abruptly. Ideally, the coherent cat state should survive many, many gate operations to be useful. A quantum computer capable of cracking a standard 512-bit public key (the composite number public key is about 155 digits long as we have seen) would require approximately a million qubits and millions of gate operations. The best-known coherent system for qubits is ions in the ion trap that have coherence times of seconds and gate speeds of a thousandth of a second. That means we can do approximately a thousand gates before the qubit decoheres or becomes not coherent (collapses to dead or alive). In practice, we need the qubit to be much closer to dead and alive than alive or dead, which restricts the number of gates per qubit to maybe a few tens not a few tens of millions.

This problem of decoherence of the qubits limiting the size of the calculation led some researchers, such as French physicists Serge Haroche (2012 Nobel Laureate) and Jean-Michel Raimond, to publicly declare in a 1996 article fetchingly entitled "Quantum Computing: Dream or Nightmare?" that it would be impossible to ever build a "practical" quantum computer.[11] Decoherence is impossible to stop entirely, but as in the early classical computers (such as the ENIAC), there are quantum error correction schemes capable of mitigating the effects of decoherence. In early 1995, no such quantum error correction schemes had yet been found, but at the ARO workshop, Bennett predicted confidently that they would be found, and soon, and the road to quantum computing would be clear. In their 1996 article, Haroche and Raimond bet in the pessimistic direction that no useful quantum error correction scheme would ever be found and that quantum computers would never be built. Haroche and Raimond both forgot British science fiction writer Arthur C. Clarke's *First* law: "When a distinguished but elderly scientist states that something is possible, he is almost certainly right. When he states that something is impossible, he is very probably wrong." In fact, Peter Shor had already published a quantum error correction code for protecting against decoherence in a Bell Laboratories report in May of 1995 but word was slow to get around. Bennett was the distinguished scientist to predict quantum error correction and hence that quantum computing was possible and Haroche was the distinguished scientist to predict that quantum error correction and quantum computing were impossible. (Neither were they then—nor are they now—*that* elderly.)

What threw Haroche and Raimond and other pessimists (such as German physicist Martin Plenio and English physicist Sir Peter Knight who also published a paper in 1997 on the inability to do large computations in ion traps[12]) was that classical error correction usually entails making multiple copies of a bit and that redundancy ensures protection against errors. However, in quantum mechanics, it is not possible to copy a qubit in an unknown quantum state. The uncertainty principle prohibits it. To copy a state, one must measure it to see what the state is, but the measurement destroys the state. Imagine taking an ion trap quantum computer and setting it on a Xerox machine to take a picture of all the states of all the ions. As in Heisenberg's microscope, in Heisenberg's Xerox machine, the photons from the copying machine, required to take such a photo, uncontrollably destroy the states of all the ions, and the copy is not of the state of the quantum computer but random useless noise. This noncopying idea is formalized by the quantum "no-cloning" theorem: *it is impossible to make a copy of an unknown quantum state.* Since 1995, on the basis of the understanding of how classical error correction works (by copying classical bits for redundancy), it has seemed reasonable to some quantum physicists that copying a quantum state would be required to correct quantum errors. People ruled out (too quickly) the possibility of quantum error correction if modeled directly after classical error correction. What they missed was that there is a way around the no-clone zone.

For example, a simple classical correction scheme is called a bit-flip protection scheme and works by majority rule. If the physical bit (transistor) is in the state (0), then you make two copies of it (000), and these three physical bits (three transistors) now become what is known as the logical bit that I'll write "(0)," which is just a new name for physical (000). Similarly, we copy (1) twice to get (111), which we call logical "(1)." We will represent each logical bit with three physical bits and work instead with the logical ones. This is a waste of two physical bits, but if it protects against error, it may be worth the overhead so long as the number of physical bits needed to correct does not grow exponentially with the number of errors you are correcting. Three physical bits make up the one logical bit. Now, if a cosmic ray, voltage surge, or other disturbance flips one of the physical bits, say the second one, that is an error, giving (010), which is no longer the logical bit (000) or "(0)" or the logical bit (111) or "(1)" but something else that will be construed as an error. But it is an error that is easy to fix. We measure all three bits and use a majority opinion. Because the first and the last physical bits are both (0), then it is likely that the middle physical bit is in error and we just flip it back to (0) as well to recover (000), which is "(0)," our logical bit. The error is fixed. This works well if the probability of a single bit flipping is small, say 1/10. If the probability of one bit flipping is independent of any other, then the probability of two bits independently flipping is 1/10 × 1/10 = 1/100, or 1% of the time, which is very small. As long as we check and correct often, we can always catch when one bit has flipped by accident and fix it before a second one also goes bad. If we wait too long for two bits to go bad, for example, (000) becomes (110), then the majority rule will produce (111) and we'll have a real error on our hands; "(0)" will be converted into "(1)." One way to prevent this is to check often. Another way is to increase the encoding size to five physical bits per logical bit; thus, (00000) is "(0)" and (11111) is "(1)." Now, three bits would have to flip for us to mistakenly correct a "(0)" into a "(1)"; the majority rule can still handle two physical bit errors, say (00000) suf-fers two random bit flips (01010), then the majority rule says there are more zeros in that bit than ones, and so we fix it by flipping the ones back into zeros to get (00000) or "(0)" again. Three bits would now have to flip in order for us to mistake a "(0)" for a "(1)." In that case, (00000) becomes (01011) and major-ity rule turns this mistakenly into a (11111). But three independent random bit-flip errors would occur with a probability of 1/10 × 1/10 × 1/10 = 1/1000, which is 0.1% of the time, which is very small. Quick measurements and large encodings can handle most bit-flip error scenarios so long as the errors don't occur too often.

What goes wrong with this procedure quantum mechanically? Recall that, in the quantum computer, we require qubits not only in the states $|0\rangle$ *or* $|1\rangle$ but also in cat states such as $|0\rangle + |1\rangle$, that is $|0\rangle$ *and* $|1\rangle$. Any attempt to *measure* the state of the cat $|0\rangle + |1\rangle$ uncontrollably collapses into $|0\rangle$ *or* $|1\rangle$, giving no infor-mation that the original state was even a cat. A measurement of the state $|0\rangle$ gives back the state $|0\rangle$ with 100% probability, but measurement of a cat $|0\rangle + |1\rangle$ gives $|0\rangle$ with a 50% probability. From the outcome $|0\rangle$, you cannot reliably con-clude that you had a $|0\rangle$ or $|0\rangle + |1\rangle$ to begin with. This is the uncertainty and the unreality principles at work—the no-cloning theorem in action. Triplicating $|0\rangle + |1\rangle$ to get $[|0\rangle + |1\rangle, |0\rangle + |1\rangle, |0\rangle + |1\rangle]$ cannot work, as the triplication pro-cess requires a measurement of the original qubit to see what it is that we are triplicating.[13]

Shor's way around this was not to copy the state to be protected but to instead entangle it with nine other qubits and then perform the correction on the nine-qubit logical qubit without making a destructive measurement that could destroy the quantum cat-like information. This code was eventually improved by British physicist Andrew Steane and others to a five-qubit encod-ing (five physical qubits makes one logical qubit) that protects against all types of errors, including decoherence, provided the decoherence is not too fast. The reason that it is an odd number like five or nine is similar to the classical case—the same reason there are nine judges and not eight on the US Supreme Court—majority rule (even in the entangled version) works best if there can never be a tie vote.

Quantum Fidelity versus Quantum Tuning: *Vazirani initiated a discussion of what the group decided to call "quantum fidelity" versus "quantum tuning." By quantum fidelity, we mean the physical tendency of the qubit elements of the quantum-computing device to resist decoherence. Hence, a device that decoheres only slowly has a high quantum fidelity. Vazirani pointed out that there is another source of error introduced in the quantum processor. If, say, the Deutsch scheme of quantum computing is implemented, then a key ability required of the quantum processor is its ability to perform highly accurate unitary quantum transformations or "rotations in Hilbert space" on the quantum state of the qubit registers—independent of how fast these registers decohere. For example, if it is required that a qubit unitary transformation operation performs a rotation in Hilbert space by an abstract rotation angle θ, then how much error $\pm\Delta\theta$ in the actual transformation—carried out on the machine—can be tolerated? It is clear that this is a separate problem from decoherence. It was suggested that the ability to define these unitary transformations accurately should be called the "quantum tuning" of the device. Hence, a "well-tuned" quantum computer carries out these transforms very accurately. Thus, the ideal quantum machine would need to be both well tuned and of high fidelity.*

This "fidelity" versus "tuning" is a subtle point and, in the end, with the advent of quantum error correction in May of 1995, turned out not to be that critical of a distinction. The distinction is important physically but not computationally. One way a qubit goes "bad" is by decoherence. It interacts with the environment, stray heat fluctuations, cosmic rays, or vibrations from a passing troupe of Irish tap dancers and decoheres from "dead and alive" to "dead or alive." That is, the desired state $|0\rangle + |1\rangle$ quickly becomes $|0\rangle$ *or* $|1\rangle$. This is an error that must be corrected for the quantum computer to work, particularly because long-lived cat states such as $|0\rangle + |1\rangle$ are a precursor to entanglement generation, but decoherence is an error that is mostly out of the hands of the experimenter or at least not caused by her. She can try to protect the qubit from the environment by shielding it from electromagnetic fields, evacuating all the air, cooling it, and so on, as they do in the ion trap quantum computers, but still some decoherence takes place (from stray unshielded fields and remaining thermal heat radiation) and this is why the ion decoherence time is seconds or minutes but not days or years. This is what Vazirani meant by fidelity. Perfect fidelity means no decoherence.

Another type of error is in the uncertainty of the qubit preparation and particularly the gate operations, which is in some sense a fault owing to the experimenter's apparatus and not the environment. The CAT gate must be applied to the qubit $|0\rangle$ to generate an exactly equal 50–50 superposition cat state of $|0\rangle + |1\rangle$. But the CAT gate in the ion trap is implemented with laser pulses of carefully controlled intensity and time duration tailored to give that precise 50–50 setup. The experimenter cannot specify the intensity of the pulse or its length in time to infinite precision. This is what Vazirani meant by tuning. Sometimes, the laser pulse will yield a superposition cat state that is a bit more dead ($|0\rangle$ with a 51% probability) than alive ($|1\rangle$ with a 49% probability). That lack of precise tuning is also an error and such errors will accumulate as you apply hundreds or thousands or millions of such gates. As it turned out, the quantum error correction codes can fix both the fidelity-type errors and the tuning-type errors with equal ease, and without much care about where the errors came from, and thus this distinction has blurred over the years.

The American physicist Michael Littman was the scribe and composer responsible for writing the section in the ARO final report entitled "Future Experimental Work," which consists of background material already covered, but most interesting was this working group's list of potential physical hardware platforms the quantum computer might be built upon. In 1995, no one had yet demonstrated an ENT gate, although the equivalent of CAT and RAT gates had been demonstrated with qubits made from photons, atoms, ions, and collections of nuclear spins in molecules. The ENT is the trickiest gate because you must have long coherence and the ability to make a CAT before applying the ENT that produces entanglement. In 1995, the making of CAT gates had been considered a challenge and the follow-up ENT gates were even harder.

What is curious to me about this list of potential quantum-computing hardware platforms from the "future experimental work" group is that it is a list of almost entirely systems where quantum coherence (CAT gates) had been observed or was likely to be observed soon, but there was almost no mention about the need for generating entanglement, that is, the making of a two-qubit ENT gate. You have to have a CAT gate before you can make an ENT—perhaps this was the logic of the group in making a list of places to look at. However, I suspect there was still confusion here about the need for ENT gates at all. This goes back to the discussion of whether coherence alone—wavelike phenomenon—is sufficient to build a quantum computer.

Here is that 1995 list again in italics with my own comments following.

> ***What are possible devices for exploration?*** *Any quantum system is a possible candidate for quantum logic. Listed below are systems or experiments that display coherent quantum effects.*
>   1. *Electron spin in atoms or lattices—quantum beat spectroscopy, electron paramagnetic resonance, or photon echoes.*

This approach is currently being explored in a number of systems particularly the electrons of ions embedded in a solid host. Photon echoes have evolved into a technique by which quantum information can be read in and read out of the electron spins for a type of quantum-computing memory as well as single-qubit CAT and RAT gates and a two-qubit ENT gate.[14]

2. *Nuclear spin in atoms or lattices—hyperfine beat spectroscopy, nuclear magnetic resonance, spin echoes. Here, nuclear magnetic spin orientation could code a QUBIT 0 or a QUBIT 1.*

Nuclear magnetic resonance (NMR) of large numbers of molecules in a bulk liquid had a heyday in the late 1990s (see the Section on Quantum Computing in a Coffee Cup, below) but eventually crawled into a dead end where it there died a slow and painful death. Another approach, proposed in 1998 by American physicist Bruce Kane, exploited individual phosphorus ions that are embedded in the lattice of a silicon host material. The nuclear spins of the ions are the qubits. This scheme has proved challenging in that entirely new fields of nanotechnology had to be invented to place the phosphorous ions at single atom sites in the silicon lattice. Perhaps more promising is the recent work on single nuclear spins of nitrogen ions embedded in a diamond host structure, which, unlike most proposals for solid-state quantum computing, can in principle be operated at room temperatures (instead of near absolute zero). The field of manipulating the single electron or nuclear spins of atoms is now called "spintronics," a word that did not exist in 1995, but which has become a research field in its own right—a "spin-off" from quantum-computing research in some sense.[15]

3. *Photon polarization—single quanta interferometer. Here, two possible polarization states of a photon (e.g., vertical linear and horizontal linear) code a QUBIT 0 and a QUBIT 1. Devices based on single-photon propagation may be able to be constructed using electro-optic elements or materials with large nonlinear polarizabilities.*

Photonic quantum computers are currently an area of active research. Texan physicist H. Jeffery Kimble demonstrated a photon–photon ENT gate in a 1995 experiment (part of the outcome of the Great Quantum Diaspora) at Caltech where the nonlinear polarizability of a single atom was radically enhanced in an optical microcavity that briefly trapped the two photons.[16] Another proposed approach is to couple the two photons coherently to a large number of coherently prepared atoms in a gas cloud. The large number of atoms boosts the nonlinear polarizability effect by the number of atoms N and the coherence gives you an extra quadratic boost to N^2.[17] The field really took off particularly after the 2001 work of American physicist Emanuel Knill, Canadian physicist Raymond Laflamme, and Australian physicist Gerard Milburn, which showed that the photon–photon coupling could be implemented without materials with large nonlinear polarizabilities but only with a source of single photons and simple optical devices such as mirrors and slabs of transparent glass as well as good photon detectors.[18] Called linear optical quantum computing, it is a field I dabbled in during the period 2000–2010, when in 2010 the bulk of DoD funding for this scheme was not renewed. (More about this later.) I had a good 10-year run with it in any case. In a photonic quantum computer, typically the encoding is in the photon polarization, as discussed in Chapter 2. Hence, the horizontally polarized photon is a $|0\rangle$ and the vertically polarized photon is a $|1\rangle$. Given such photons, the single-qubit CAT and RAT gates are easy and can be implemented with simple optical elements such as mirrors and slabs of glass. The two-qubit entangling ENT gate is much harder as it requires a pair of photons to "talk" to each other via some interaction through the electrons in an optical material and such two-photon interactions are typically very weak. For a chunk of off-the-shelf two-photon nonlinear material needed to make an ENT gate, the gate would work only about one time in 10^{20}. To boost this interaction probability, you need to have the photon interacting with a lot of atoms—say by bouncing it back and forth millions of times between a pair of mirrors with the atoms in between. The linear optical approach does not require such real photon–photon interactions but somewhat mysteriously generates them in a virtual environment. To make a two-photon ENT gate, you mix your control and target qubit photon with each other and some additional photon qubits in a separate circuit called the "ancilla" or "slave" circuit (see Figure 4.3). Then, you make a measurement on the ancilla circuit, and if the measurement outcome is a particular one of many, then you know that the ENT gate on the control and target was successful.

What is curious about this type of gate is that it does not require "large nonlinear polarizabilities," properties of specific chemical molecules or atoms that are hard to come by. Typically years of chemical experimentation produce designer molecules that still have only very small nonlinear polarizabilities. The idea is that when the first control photon passes by one of these nonlinear molecules, the photon's electric field causes the molecule to stretch

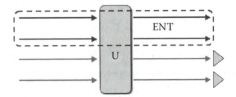

Figure 4.3 How to make an ENT gate without really trying. The control and target qubit photons enter in the two circuit lines (top left). Two additional qubits enter on the two ancilla circuit lines (bottom left). All the photons are combined in the box labeled "U" in the middle, which consists of simple optical elements such as lenses, mirrors, beam splitters, and other chunks of glass found in ordinary optics laboratories. A measurement on the two ancilla circuit lines is made using photon number counting detectors (bottom right). For a specific set of detector outcomes, say both detectors get a $|0\rangle$ (a horizontally polarized photon), then we are sure that ENT gate has successfully been implemented on the top two circuits and out comes a target entangled state (top right). Monitoring the detectors for the correct answer allows us to know when the ENT gate was successful and use the target output state in future computations.

like a spring that is pulled so hard it bends out of shape in such a way that the second passing target photon senses the presence of the first, and if things are set up just right, the second photon's polarization flips (or does not flip) depending on the polarization of the first. In this way, the control photon "talks" to the target and tells it when to flip (or not). Without lots of work, this happens only about one time in 10^{21} or one time in a sextillion for the best non-linear materials available; thus, it is often not useful. The nonlinear response, the stretching and bending of the molecular spring, is just too weak at the single- or two-photon level. This nonlinear response can be boosted either by putting the molecule or atom between two mirrors so that the photons interact with it millions of times or by rigging it so that millions of molecules talk to the same photon at once, but this is technologically challenging.

In contrast, the whole gate in Figure 4.3 is "linear" in that the response of the system is proportional to the inputs and there is nothing like the bending spring. This is the easiest type of optical system to build—light passing through window glass on a sunny day is using a linear response of the glass molecules. What is strange is that in Figure 4.3, which is called a linear optical gate, the detection process (combined with the opening and closing of the shutter based on the detector outcomes) produces an "effective" strong nonlinear polarizability—that is, the system behaves like it is filled with these stretching, bending, springy molecules, but in reality, there are none there. Somehow, the information of the firing order of the detectors on the ancillas at the bottom right of Figure 4.3 propagates to the left—backward in time—and then up the box "U" filled with lenses and slabs of glass—and then out the top right to the target and control qubit photons giving rise to the conditional ENT gate. Even stranger, this whole thing works 1 out of 16 times. But remember, the "regular" optical ENT gate using "large" nonlinear polarizabilities works only one out of sextillion times, so in some sense, the linear gate works 20 orders of magnitude better than the nonlinear one.[19] This idea then means that you can essentially design strong nonlinear optical polarizabilities not by spending years growing newfangled and exotic molecules in a chemistry laboratory but instead just by rearranging lenses and beam splitters and other bits of glass on a laboratory table, along with the photon detectors, to mimic the strong nonlinearities. That approach, linear optics plus photon detectors, has given a boost to the entire field known as nonlinear optics. Now, building the required two-photon nonlinear optical element is much less like dabbling in alchemy and much more like playing with Legos.

4. *Molecular polarization—coherences in rotational alignment, vibrational alignment, or electronic alignment (magnetic or electric dipole); birefringence echoes; and photo-induced electron transfer in polymers. Here, a QUBIT 1 might correspond to the location of a electron in a polymer—this concept was suggested by John Hopfield of Caltech.*

A simple two-qubit polar molecule–based quantum computer was recently used in 2009 to execute the Deutsch–Jozsa algorithm.[20] Other than that, I do not know of much work in this area, perhaps because it is difficult to cool and trap molecules using lasers and electric fields, as opposed to atoms and ions, where this is comparatively simpler. In addition, the idea is not really scalable to millions of qubits. You might be able to make two qubits on a single molecule, as demonstrated here, but not likely millions. For that, you would have to entangle the quantum states of millions of molecules by manufacturing them in some sort of regular array and then arranging for ENT gates between molecules. That would be very hard.

5. *Self-induced transparency in atomic vapors—the ability to traverse optically opaque vapors has been shown by S. Harris of Stanford to be caused by a quantum coherence induced by a strong laser field. A QUBIT 0 might correspond to the absorber being in a ground state, whereas a QUBIT 1 might correspond to being in an excited*

continuum state. The laser driving field could serve as the controlling signal.

There is nothing that I am aware of (even a proposal) that implements a quantum computer in this way. "Self-Induced Transparency," now universally called "Electromagnetically Induced Transparency" (EIT), has been proposed as a photonic quantum-computing medium, but now here the qubits are photons and the vapor is the medium that couples the photons together to make the ENT gate. Here, it sounds like the atoms (not the photons) are meant to be the qubits. There are, however, proposals to exploit the storage capability of the EIT system for a quantum memory. In this way, a photon in some quantum state, say a photon polarization cat state $|0\rangle + |1\rangle$, is written into a collection of atoms that then behave as if they are in a collective atomic state $|0\rangle + |1\rangle$ where now this is a superposition of the two collective atomic electron energy levels. (All the electrons in the cloud store the state information collectively.) Then, this information is stored for milliseconds and then read back out into the photons for later use in the computation or for long-distance quantum communications applications we'll discuss in an upcoming chapter.[21]

6. *Coherences in atomic ions in traps—Zoller (theory) and Wineland (experiment). Here, the ion is coupled to an electromagnetic trap. The excitation is shared between the trapping field and the internal state of excitation of the ion. A QUBIT 0 might correspond to the ion being excited, whereas a QUBIT 1 might correspond to the field being excited. This same idea would apply to electrons in traps following the work of Hans Dehmelt of the University of Washington* (Figure 4.4).

The "future experimental work" group at this ARO workshop had an edge up on this prediction. Ignacio Cirac and Peter Zoller were part of the Great Quantum Diaspora and were working on a theory preprint in early 1995, "Quantum Computation with Cold Trapped Ions," which was submitted for publication in November of 1994 and published in May of 1995. Some details of this theory were already known in early 1995. David Wineland and his group at NIST published the experimental results of the first working ENT gate between two ions in a paper published in December of 1995 but submitted for publication in July of 1995. Everybody knew in early 1995 that such an experiment was underway. Ion trap quantum computing to this day remains one of the strongest candidates for a future quantum-computing hardware platform, with entanglement generated between 14 qubits recently demonstrated by German physicist Rainer Blatt's group in Innsbruck, Austria.[22]

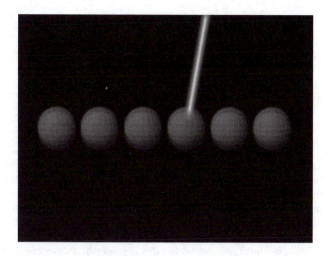

Figure 4.4   A schematic of an ion trap quantum computer. The ions are the spheres and their internal spin states are the qubits. The ions are positively charged and hence have their own electric field as well, and it is through this field that the requisite two-qubit interaction is made needed for the ENT gate. The laser beam is responsible for cooling the ions and for implementing the single-qubit CAT and RAT gates as well as the two-qubit ENT gate. Via the electric field, an ENT gate can be performed between any two ions in the trap—they don't have to be next to each other. In 2011, 14 such atoms in a single trap were prepared in a large 14-qubit entangled state, the record as I write this. (The image is taken from "NIST Physicists Coax Six Atoms into Quantum 'Cat' State" by Laura Ost [NIST Press Release, November 30, 2005]. The image was produced by Bill Pietsch for NIST and is the property of the US Government and hence in the public domain http://www.nist.gov/pml/div688/cat_states.cfm.)

7. *Quantum dots. Here, a QUBIT 0 might be an electron on one quantum dot, whereas a QUBIT 1 might correspond to it being on a nearby neighboring dot (Figure 4.5). This concept follows the work of Lent et al. of Notre Dame. Alternatively, a single electron transistor such as that described by K. Likharev of the State University of New York at Stony Brook might be useful in building such devices.*

This prediction was close to the mark. In 1997, Swiss physicist Daniel Loss and American physicist David DiVincenzo proposed a scheme for quantum dot–based quantum computing. In a slight twist of the above proposal, Loss and DiVincenzo proposed that the spin of the electron would represent the qubit, not the energy level of the electron in the dot. A quantum dot, called an artificial atom, is typically a chunk of semiconductor material, like silicon, that is fashioned roughly into the shape of a nanometer-sized

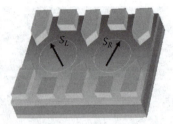

Figure 4.5 A schematic of a double quantum dot circuit made from quantum dots. S_R and S_L are the spins of the right and left electron held inside each of the two dots (circles). The pyramidal and square blocks are electrodes in the circuit. The electrodes can be turned on and off in such a way to make single-qubit CAT or RAT gates on either dot and to make a two-qubit ENT gate between the two dots. (This is an image created by Vitaly Golovach [December 18, 2006].) He has granted permission to publish the picture without restrictions in electronic form or in print. This work has been released into the public domain by the copyright holder. This applies worldwide. In case this is not legally possible, the copyright holder grants any entity the right to use this work for any purpose, without any conditions, unless such conditions are required by law.

box.[23] A free electron in the box behaves like an electron in a hydrogen atom and has energy levels. A proposal for using two of these energy levels are one way to make a qubit, but using the spin of the trapped electron is thought to be more immune to decoherence. This field is today somewhat advanced in the direction of making one- and two-qubit gates but progress is hampered by the short coherence times and the difficulty in making regular repeatable quantum dot arrays at the nanometer scale. Every rubidium ion in an ion trap has exactly the same levels because every rubidium atom is identical. It is hard to make the quantum dots all identical and so the energy levels tend to change randomly from dot to dot making the scale up to large quantum circuits difficult.

8. *Quantum wells. Here, a QUBIT 0 might correspond to an electron being in one well, while a QUBIT 1 might correspond to it being in another. A shaped optical pulse from a mode-locked laser might serve as a control signal. The work of Jagdeep Shah of AT&T Laboratories has observed quantum beats owing to coherences in optically excited two-well devices.*

A quantum dot is a small three-dimensional box, typically made of semiconductor material, that behaves like an artificial atom with electrons confined in the box instead of the potential well around an atomic nucleus. A quantum well is an analogous device that is a two-dimensional planer structure where the electrons are confined in the plane between two layers of semiconductor material. It is a theme of this list from the ARO Workshop, as I have mentioned above, that only coherence or the ability to make CAT states seems to be given as a requirement for quantum computer with no mention of entanglement. To make an ENT gate, you would have to somehow entangle different electrons in the same quantum well or different electrons on different quantum wells. I do not know of any way to do that. Once again, coherence of single electrons is easy, but entanglement of pairs of electrons (or more) is hard. Perhaps the most hype associated with quantum wells comes from the proposal of topological quantum computing. The idea of topological quantum computing exploits the fact that in a quantum well, electrons tend to form quasiparticles (not real particles at all but collections of electrons) that have the curious property that their spin angular momentum (in quantum units) can be anything. Such quasiparticles with any spin are called "anyons." All fundamental particles are divided into two classes, bosons and fermions, on the basis of their spin angular momentum (the speed at which the particle spins on its axis). Fermions have half-integer spin angular momenta of $1/2$, $3/2$, $5/2$, and so on in units of Planck's constant. Electrons, neutrons, and protons all have a spin angular momentum of $1/2$ and are hence fermions. Bosons on the other hand have integer spin angular momenta such as 0, 1, 2, 3, and so on. Photons have a spin angular momentum of 1 and are hence bosons. Every atom, depending on the total number of neutrons, protons, and electrons making up the atom, is either a boson or a fermion. However, these collections of electrons as quasiparticles in the two-dimensional quantum well have a spin angular momentum that violates this rule. You might have a quasiparticle with say spin $2/3$. In any case, this strange feature could in principle be exploited in topological quantum computing to make a type of computing circuit that is immune to noise and decoherence. This would be great, as you would not need to deploy error correction or other techniques to run the quantum computer. However, working the topological quantum computer would need the use of a very specific quasiparticle, a nonAbelian anyon, which has never been observed in any experiment in quantum wells or anywhere else. Hence, for now, topological quantum computing is a theorist's pipe dream—all dressed up with beautiful mathematics but with no place to go. Topological quantum computing is the super-string theory of quantum computing, and I'm afraid I don't mean that to be a complement. Super-string theory is also a physical theory that is adored by theoreticians for its beautiful mathematics but shunned by experimenters, as it makes few predictions that can be tested in any laboratory. As you know from my diatribes in Chapters 1 and 2 about the scientific method, I feel very strongly, to paraphrase a quote I have always attributed to the great German–American theoretical physicist Hans Bethe, that at the end of the day, it is the job of every theoretical physicist to produce a number that can be compared to experiment. If you are not doing that, then you are not doing your job.

9. *Atomic polarization—quantum beats, coherence in Rydberg states. Here, a QUBIT 0 might correspond to excitation of one Rydberg state, while a QUBIT 1 might correspond to another state. The control signal might be a shaped microwave field pulse.*

   Once again, the focus is on coherence without much thought to entanglement. Here, we have the idea of the Rydberg state, a hydrogen-like atom, being discussed as an avenue for quantum computing. As discussed in detail above, there is only one electron in such an atom that is being manipulated. A hydrogen-like atom has a core of strongly bound electrons around the nucleus and a single loosely bound electron orbiting outside the pack. Its energy levels are similar to those of hydrogen in structure. A Rydberg atom is a hydrogen-like atom where the outermost electron has been excited to a very high orbit. This proposal is clearly not scalable into a large quantum computer as it explicitly labels the states of a single electron as the qubit states zero and one. This is just ordinary wavelike coherence with no entanglement. Again, it is possible to make CAT and RAT gates with such a thing, but to make proper ENT gates, one would have to entangle electrons *between* Rydberg atoms. That is a much taller order than just manipulating a single electron in a single atom. There is no entanglement in a single electron, regardless of how many orbits it has access to, and without entanglement, the portal to the exponentially large Hilbert space—the Library of Babel—spirals close upon us. Some simple computations that do not require entanglement, such as a simplified Deutsch–Jozsa algorithm or a slimmed down Grover search algorithm, can indeed be performed on coherent but not entangled systems. I'll discuss this further later.

10. *Mössbauer effect—recoil coherences.*

    I'm not going to spend much time on this one. This idea makes it clear that many of these proposals were simply a laundry list of physical systems where wavelike coherence had been demonstrated but not much thought about entanglement had been given.

The Mössbauer effect is the experimental generation and detection of gamma rays from nuclear decay emission and coherent nuclear reabsorption in large crystals of materials. Other than the process has the word "coherent" in it, I see no way this could be used to make a quantum computer, and I doubt that whoever proposed it had any idea either.

11. *Ramsauer effect—coherences in electron resonance scattering.*

This is another item on the coherence laundry list. The Ramsauer effect is the coherent wavelike scattering of electrons by gas atoms. Other than it is a coherent process, there is no known connection to quantum computing and I suspect there never will be. At this stage, it certainly does seem that the workshop participants were just rattling off every coherent physical effect they could think of.

12. *Neutron interferometer.*

One more for the laundry list—neutron interferometers are devices where single neutrons emitted from nuclear reactors undergo wavelike scattering and coherent interference in interferometers—interfering devices—made from solid crystals of metal. The effect relies on wave–particle duality and is coherent but no neutron in the interferometer ever is entangled with another, nor likely could be, so other than coherent, it is roadkill on the quantum-computing road map.

13. *Electron interferometer—electron coherences are routine in Bohm–Aharonov-type devices and resonant tunneling GaAs structures. In addition, the Princeton URI wavepacket encoding/decoding study (H. Rabitz, S. Lyon, M. Sheyagan, and M. Littman) is exploring principles of quantum interference in semiconductor devices.*

Like the neutron interferometer, the electron interferometer uses wave–particle duality to make interference patterns that exploit the coherence of electron waves. Unlike neutron interferometers, there is at least a plausible chance that the electrons in many such interferometers could be made to interact with each other, as in photonic quantum computing, and make a large-scale quantum computer. Neutrons, because they are electrically neutral, don't interact easily with much of anything. Something very much like an electron interferometer-based quantum computer was recently proposed in 2010, but to my knowledge, there has never been any demonstration of anything like an ENT gate in this system in the laboratory, although at least in contrast to neutrons, there is a concrete proposal here on how to make one.[24]

14. *Coherent effects in multi-photon ionization. Here, the superposition of intermediate states in a multi-photon chain can determine the probability of ionization. The use of many lasers tuned to different resonances in the chain of ionization may give the ability to vary several control signals at once—this idea follows the work of William Cooke of USC.*

This is another item on the coherence laundry list. The process involves the coherence of pairs or triplets or more of photons that dislodge the electrons from an atom. Other than that the process requires wavelike coherence, there is nothing here to suggest a road to entanglement or a quantum computer.

15. *Optically induced coherences in atoms and molecules—(Brummer of Toronto and Shapiro of Weitzmann Institute) pump-probe studies involving interferences owing to coherent excitation of states by different excitation channels.*

Again, coherence but no quantum computer here—but this effect could be used as a photonic qubit memory similar to topic (1) above. The photons are coherently stored in the atoms and molecules and then read back out—all coherence but no entanglement.

16. *Atom interferometer—coherent scattering of atoms by masks and by standing wave optical fields. Examples of atom interference include the work of Dave Pritchard of the Massachusetts Institute of Technology.*

This scheme is in the same ballpark as the neutron interferometer; lots of coherence but no entanglement. Neutral atoms can be made to interfere with themselves in an interferometer from their wavelike properties. However, if the atoms are flying one at a time though an interferometer, there is no mechanism to entangle them. Much more promising are schemes where the atoms are held more or less fixed in optical traps and the internal states of the atom are made to interfere, giving a very plausible road to quantum computing. Neutral atoms trapped in light fields were not technologically ready for prime time in 1995, but now they are a competitive approach for the hardware platform of the general quantum computer.[25]

17. *Atoms in cavities—second quantization (i.e., quantizing the field and quantizing the atoms). Here, exchange of excitation between field states and atoms states (e.g., the work of H. Walther of the Max Planck Institute) or the effects of energy exchange in quartz microspheres with atoms attached (e.g., the work of Jeff Kimble of Caltech).*

This was discussed a bit in topic (2) above on photonic quantum computing. In one scenario, the photons are the qubits and

Figure 4.6 Atoms (glowing balls) are shuttled by the wiggly diagonal laser beam in and out of the optical cavity (two cone-shaped cylinders) where they interact with photons (wiggly beam bouncing back and forth between mirrors) that are bouncing back and forth between the cavity mirrors. Once in a while, a photon emerges to the right (wiggly arrow) carrying information about the state of the atoms. The non-wiggly diagonal laser beam helps trap and cool the atoms into their ground state. Two schemes are possible. In one scheme, the photons are the qubits and the atoms mediate the photon–photon interaction to make a two-photon-qubit ENT gate. In the second, the atoms are the qubits and the photons take the role of mediator of the qubit–qubit interaction to make a two-atom ENT gate. (Graphic courtesy of Gerhard Rempe, Max Planck Institute of Quantum Optics, Garching, Germany.)

the atoms mediate the ENT gate between a pair of photonic qubits. A variant is to make two atoms the qubits and use the photons to implement the ENT gate between a pair of atoms. The "cavity" is often just a pair of mirrors that are placed close to each other so that the atom sees many reflections of itself in the pair of mirrors going off to infinity like the effect you see sometimes in a well-mirrored barbershop (see Figure 4.6).

Thus, this litany of potential quantum-computing hardware platforms had some hits and misses. I've clearly identified the hits in the discussion above but not all the misses (just the unimportant ones). Two current front-runner approaches not prayed for in this above litany are quantum computing with neutral atoms in optical lattices and the superconducting qubits. Somewhat magnanimously, I have credited "atom interferometry" in the above list as anticipating quantum computing with neutral atoms in an optical lattice but this is really too generous. Most of the techniques needed to carry out the neutral atom idea were not in place or not even fully worked out in theory in 1995. The theory and experiments began coming online in around 2000. The idea is a bit like ions, and let us recall that an ion is an atom with an electron stripped off and is thus positively charged and not neutral. There are a couple of approaches, with the most popular being placing single neutral atoms in the potential wells formed by interference maxima and minima made by crossing six laser beams. Such a laser configuration forms an "eggcrate" potential with an atom (instead of an egg) sitting at each pocket in the crate (Figure 4.7). The combination of magnetic (not electric) and light fields is enough to trap, cool, and manipulate the neutral atoms—even though they do not have that positive charge that gives the ion trap its handle. In the case of the neutrals, the combination of light and magnetic fields, interacting with the internal electron orbital structure, is enough to trap and, even better, make quantum gates. This technology is well advanced in that the atoms, lattices, magnetic fields, laser beams, and control systems for them all have been reduced in size until they have all been put on a quantum computer chip—called an atom chip—that is about the size of a penny. Difficulties remain in the process of getting exactly one atom per egg-crate site. (Most of the time, there are none and often there are two.)

The other big miss was the superconducting approach to the quantum computer, which was not mentioned at all at this 1995 ARO workshop, but which

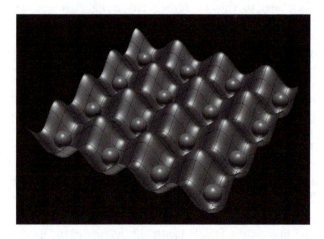

Figure 4.7   Schematic of neutral atom quantum computer: Atoms (balls) are trapped by the "eggcrate" potential of the laser field. The qubits are the atoms or more precisely the internal electron orbital states inside each atom. The light field traps, cools, and manipulates the atomic qubits. Additional laser beams can be brought in to address individual atoms to implement single-qubit CAT and RAT gates. In a popular scheme, a pair of atoms is made to collide off each other, much like two pool balls, which creates the ENT gate. (The image is from a NIST press release entitled "Physicists Find Way to Control Individual Bits in Quantum Computers" by Chad Boutin [July 14, 2009], and as it was created by a US Government employee for their work, it is not subject to copyright, http://www.nist.gov/pml/div685/qbits_071409.cfm.)

is now considered a front-runner technology. Once again, there was just not enough in place in 1995 to make that prediction, and perhaps the fact that the workshop was weighted more to atomic and optical approaches and persons with that kind of experience caused us to miss it. Superconductors are materials that, when cooled to a few degrees above absolute zero, lose all electrical resistance. They become ideal conductors of electricity and hence are superconductors. Superconductors have the advantage that the electrons in the circuit do move altogether as in a wave and are hence coherent, which we know is necessary (but not sufficient) for entanglement. They also have the advantage that by 1995 there was a great deal of expertise worldwide in building large superconducting chips with thousands of superconducting transistors. The problem was that these were all classical transistors—not quantum. The idea was to use such large superconducting circuits to build very fast classical computers, and the National Aeronautics and Space Administration (NASA) also exploited these circuits to make sensitive electromagnetic sensors for use in outer space. In 1995, however, despite about 10 years of effort, nobody had ever succeeded in making even a single CAT gate in such a circuit, much less an ENT gate. This track record led one luminary from the quantum optics community to declare in 1996 that, "These guys have been trying to make a cat state for ten years with no luck, and if you give them another ten years they still won't have one." But let's not forget Arthur C. Clarke's First law, "When a distinguished but elderly scientist states that something is possible, he is almost certainly right. When he states that something is impossible, he is very probably wrong." The superconducting guys had a single-qubit CAT gate working by 1999 and a two-qubit ENT by 2002. Both were demonstrated in the group of the Japanese physicists Yasunobu Nakamura and Tsuyoshi Yamamoto, as well as Chinese physicist Jaw-Shen Tsai, all working at the NEC Corporation and the Institute of Physical and Chemical Research in Tokyo.[26] These two devices used a type of superconducting transistor called a Cooper pair box or charge qubit.

Much like a quantum dot, the qubit is a two-level system, but instead of the energy level or spin state of a single electron forming the qubit state, the electrons in superconductors pair up. This is what in fact makes superconductors superconducting—each electron has spin 1/2 and is a fermion but the pair bond loosely in the metal to form a $1/2 + 1/2 = 1$ spin particle, a boson, named a Cooper pair after the Nobel Prize–winning American physicist Leon Cooper, who first postulated their existence in 1956 to explain superconductivity.

In an ordinary conductor, like most metals at room temperature, the electrons move unpaired and collide with other electrons and nuclei doing a random walk through the regular lattice of atoms of the metal, like a drunk staggering through a parking lot full of cars. When the temperature drops, in some metals and other special materials, the electrons pair up and move through the metal crystal without resistance. It would be like if you strapped two drunks to a long bamboo pole by their necks and made the pole just longer than the length of a car and launched them into the parking lot with one drunk in one row of cars and the other drunk in the neighboring row. The bamboo pole pairing keeps the drunks from staggering and keeps them walking in a straight line and the pole length ensures they walk in parallel straight lines always between the cars and never hit them. In a Cooper pair box qubit, the $|0\rangle$ state is when the electron pair is off a metallic island and the $|1\rangle$ state is when the pair is on the island and the cat state $|0\rangle + |1\rangle$ is a superposition of the pair being simultaneously on and off. The island and the metal "sea" are separated by a few nanometers and so it is exactly like the spoon being simultaneously in my coffee cup in my office and in the sink in the bathroom down the hall (except that is a few meters). Indeed, the breakthrough in superconducting quantum-computing hardware development occurred (to my mind) in 1997 when Shnirman, Schön, and Hermon published a theory paper laying out just how small superconducting circuits could be made to exhibit CAT, RAT, and ENT states. As I recall, the roadblock that held back the field for 10 years was the inability to make the circuits small enough. Recall that Schrödinger cat states decay with a speed that scales exponentially with the size of the cat or in this case the distance. Hence, if you are trying to make a cat over a distance of a micron (0.00004 inches), the cat is too big and quickly collapses to dead or alive before you can make a superposition. However, advances in nano-technology allowed for the fabrication of structures on the order of nanometers (0.00000004 inches) that was small enough to see this effect.[27] A closely related approach to this charge qubit is the superconducting flux qubit. In this device, a square-loop circuit a micron or so on a side and etched into a superconducting metal substrate, the electrical current can flow (without resistance) either clock-wise ($|0\rangle$ state) or counterclockwise ($|1\rangle$ state) around the loop. The cat state $|0\rangle + |1\rangle$ is made by making the same electrical current flow simultaneously clock-wise and counterclockwise round the loop. It does not get any more cat like than this—there are millions of electrons all in Cooper pairs undergoing a quantum identity crisis circulating at the same time clockwise and counterclockwise. This is a very large system and remember that one of the important things about the Schrödinger cat thought experiment is that we seldom find in macroscopic objects like cats the strange quantum behavior that we attribute in microscopic objects like electrons. But a micron-sized circuit is almost visible to the eye (see Figure 4.8)![28]

There are a few more approaches to universal quantum computing not anticipated at the 1995 conference; the saga of bulk-liquid state NMR quantum computing will be discussed in the Section on Quantum Computing in a Coffee Cup, below. My favorite of the nonpredicted platforms was a scheme utilizing electrons floating in a pool of liquid helium. The negatively charged electron floating over the surface of liquid helium in a shallow bath sees a positively charged image of itself in the helium, like in a mirror, and binds with the image

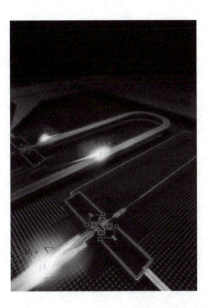

Figure 4.8   Artist's rendition of a superconducting cable connecting two superconducting qubits (rectangular circuits). (Image created by Michael Kemper and was taken from NIST press release entitled "Digital Cable Goes Quantum: NIST Debuts Superconducting Quantum Computing Cable" by Laura Ost [NIST, September 26, 2007]. The image is the property of the US Government and not subject to copyright, http://www.nist.gov/public_affairs/releases/quantum_cable.cfm.)

charge to form a structure that has the energy levels of a one-dimensional hydrogen atom. You can use the ground and excited energy levels for the states 0 and 1, or instead take the spin of the electron up to be 0 and down to be 1 as in an ion trap. Electrodes submerged beneath the liquid helium bath then manipulate (via electric or magnetic fields) individual electrons to implement single-qubit RAT or CAT gates and pairs of electrons to implement two-qubit ENT gates.[29] The properties of liquid helium keep the electron floating just above the surface (rather than just diving into the pool). The American physicist John Goodkind at the University of California at San Diego had several years worth of DoD funding to investigate this scheme, but when I visited him and his laboratory in the fall of 2007, he told me that the funding (and hence the liquid helium) had all dried up. He indeed had made a great deal of progress in the design and testing of one- and two-qubit gates, but I think the project suffered from too many politically incorrect jokes such as "Oops, I spilled the quantum computer again" or "All of our results just went down the drain." DoD funding agencies do not typically appreciate such jokes about their programs. Plus, it is widely believed that the DoD agencies are currently in a phase where they are no longer entertaining new ideas for quantum-computing platforms but rather down selecting every few years with the goal of arriving at one or two hardware platforms that are scalable and practicable.

This issue of coherence alone versus coherence plus entanglement was somewhat put to rest with the dissemination in the late 1990s of five criteria for the development of a scalable universal quantum computer, put forth by American physicist David DiVincenzo, and now called the DiVincenzo criteria.[30] For a system to be a candidate hardware platform for a *universal* quantum computer, it should

1. Be a scalable physical system with well-defined qubits
2. Be able to be initialized to a simple reference ground state such as $|000\ldots\rangle$
3. Have qubits with much longer decoherence times than gate-speed operation times
4. Have a universal set of quantum gates
5. Permit efficient, qubit-specific measurements that rapidly collapse a cat state to $|0\rangle$ or $|1\rangle$

Sometimes called the DiVincenzo commandments, these five rules have become a guide for the development of quantum computing for over the past 10 years. Recall that the CAT, RAT, and ENT gates are a *universal* set of gates. The proposal for implementing Shor's factoring algorithm in a single hydrogen-like atom lacked the equivalent of an ENT gate. There is no entanglement in such a system and hence it cannot be used to give an exponential speedup in factoring and more generally cannot be used to build a universal quantum computer. Criterion number one is a bit subjective. Well-defined qubits are certainly obvious, but what about scalable? Many in the community have taken scalable to mean that you can fabricate many of them on a small integrated circuit–like chip. But remember the classical computer was thought in the 1970s and 1980s to be *not* scalable precisely because you could fabricate lots of small wires and transistors on an integrated circuit silicon chip. There was the electromagnetic cross talk to worry about. Classical computer chip scalability came only when Mead and Conway and others showed that, by using careful design rules, the electromagnetic cross talk could always be eliminated. Putting lots of little gizmos on a chip is necessary but not sufficient for a scheme to be scalable, but this nuance continues to be overlooked by my colleagues in the quantum-computing community. For example, the ion trappers furiously design and build small chips with the plan to put hundreds of ion qubits on these and shuttle them about like beads on a quantum abacus. They then declare with glee, "See! I have all my qubits on a little chip—the ion trap system is scalable!" But where are the design rules? Where is the proof that as you scale down in size, something else does not blow up? There is no proof, and to my mind, the jury is out on whether the ion trap approach to quantum computing is truly scalable, absent any design rules to make their case.

Another caveat about the DiVincenzo criteria is that they are for constructing a *universal* quantum computer, a computer capable of any quantum calculation. What if you don't want to do any calculation but one specific one over and over again? One could imagine developing a quantum-factoring engine that was not universal but factored numbers efficiently—exponentially fast—and that is all it could do. The NSA would be very happy with that as their primary interest is in using factoring for code breaking. It is unlikely that you could build such a factoring engine without DiVincenzo's criteria but it was not proven to be impossible. As we'll discuss in an upcoming section, there is a company, D-Wave, which makes a type of "quantum" computer that does not meet all the DiVincenzo criteria, but they market it as an engine to execute a quantum search algorithm that requires only coherence and not entanglement for its operation. The improvement is only quadratic and not exponential, but there seems to be a market for that.

THE UNLYING LANDS

From 2006 through 2010, I participated on a large, $1.5-million-a-year Quantum Computing Concept Maturation (QCCM) in optical quantum computing that was funded by the Intelligence Advanced Research Projects Activity (IARPA), which was formerly known as the Disruptive Technology Office, which was formerly known as the Advanced Research and Development Activity (ARDA), which was formerly known as the NSA, which were all funding agencies for the US intelligence community. The changes in names, acronyms, and, more importantly, the logos took place at a frightening pace that made it hard for the research scientists to keep up. I personally had funding for optical quantum computing from 2000 to 2010, which came under the umbrella of each of these agencies in sequence and there were even two separate logos for IARPA in use at the same time. When the acronym IARPA showed up in 2007, all my colleagues would ask me, what the heck is IARPA? To this I would respond, it is the Defense Advanced Research Projects Agency (DARPA) for spies (see Figure 4.9).[31]

But in any case, the photonic QCCM, led by American physicist Paul Kwiat, had collaborators that stretched from Austria (Anton Zeilinger) to Australia (physicist Andrew White). We had what we all thought were great results; we submitted in 2010 an essentially renewal proposal and we were not funded and neither was anybody else in photonic quantum computing.

Figure 4.9 A composite study of the logos of ARDA throughout the ages. From 2000 to 2010, I had continuous funding for research in quantum information processing, which as far as I could tell came from one place, but for which I had to change logos five times, starting with the NSA logo on the left (2000) and ending with the second IARPA logo on the right (2010). The penultimate IARPA logo on the right had a life span of only 2 weeks and you can see that it is the logo for the Director of Central Intelligence with the letters IARPA badly and hastily photoshopped across it. In the background is a spoof of a composite map of the lands of Arda from the fictional works of J.R.R. Tolkien. (The sea monster and sailing ship are taken from ancient manuscripts and no longer subject to copyright. The map is based on "A Map of Middle Earth and the Undying Lands: A Composite Study of the Lands of ARDA," author unknown [The Tolkien Gateway, July 22, 2011], http://tolkiengateway.net/w/images/a/a4/A_Map_of_ Middle-earth_and_the_Undying_Lands_%28color%29.jpg. Explanation of the jokes: The Unlying Lands should be the Undying Lands in Tolkien's works. Nimanrø should be Numenor, Mittledöd should be Middle Earth, Odinaiä is Ekkaia, and Darpagar is Belegaer. ODNI is the Office of the Director of National Intelligence, NIMA is the National Imaging and Mapping Agency [now the National Geospatial-Intelligence Agency], NRO is the National Reconnaissance Office, DARPA is the Defense Advanced Research Projects Agency [parts of which were carved out into ARDA], and DoD is the Department of Defense.)

In the case of photonic qubits, this dropping of optical quantum computing by IARPA was a bit hasty in my opinion. While the photonic quantum computer may be a bit of a long shot for the scalable quantum computer, all hardware platforms are a long shot, and photonics is the only technology that would allow us to build the scalable quantum Internet. There is a good analogy. In the 1970s and 1980s, there were predictions that silicon chip technology was coming to an end, and there was a great DoD-funded push to develop scalable *classical* optical computers. The thought was that as we put more and more circuits closer and closer together on the silicon chips, the electromagnetic cross talk between the wires and the transistors would grow without bound limiting the number of processors on a chip. What was not foreseen was the development of good integrated circuit design rules, developed by American computer scientists Caver Mead, Lynn Conway, and others, which showed that the cross talk could be completely eliminated. But until that was understood, the funding for the competing optical computing rose and ran for a while and then collapsed in the mid-1980s when it became clear that the Intel silicon chips were not going anywhere and that predictions of their demise were overrated. The optical classical computer program was viewed as a colossal failure and to say you were working on optical classical computing became the kiss of death. But it was not a failure at all. The optical switches and transistors developed for the scalable optical classical computer found their way into the switches and routers and hubs for the fiber-optic-based classical Internet. The future quantum Internet will also require the manipulations of photons at the quantum level—a quantum repeater is a device for transmitting quantum information over long distances. The quantum repeater is a small, special-purpose, optical, quantum computer that executes a particular error correction protocol. The future of the quantum Internet is in photons and the short circuiting of the development of optical quantum information processors in the United States means that the future quantum Internet will have "Made in China" stamped all over it.

Many of the above approaches to the hardware development for the quantum computer were crystallized in time from 2002 to 2009 in the ARDA quantum-computing road map project, which lists most of these technologies with a great deal of detail and discussion on their relative merits. This road-mapping project was led by English–American physicist Richard Hughes at the Los Alamos National Laboratory in New Mexico (the same place they made the atomic bomb). In fact, the race to build the quantum computer has a lot in common with the US Manhattan Project to build the atomic bomb, except that in the case of the quantum computer, you hope in the end that it does *not* blow up. (The joke here is that you do not want the resources to blow up exponentially. Also, in the case of one of Canadian physicist Raymond Laflamme's experiments at Los Alamos on NMR quantum computing, the molecule they were using to construct the qubits for one proposed scheme turned out to be some type of chemical explosive—at least as told to me by Richard Hughes. As I recall, when they realized that there was a chance the quantum computer might actually detonate, and that they would have to fill out a lot of extra safety paperwork to ensure that no animals, graduate students, or postdocs would be harmed in the experiment, they switched to a much more stable and less lethal molecule—caffeine I think—at least the surplus of which could be fed to the graduate students and postdocs instead of blowing them up.) The ARDA quantum-computing road map is available to download from the Los Alamos website.[32] It lists technologies such as NMR quantum computers and optical quantum computers that are no longer funded by the DoD at any substantial level and, as I like to say, are now roadkill on the road map.

Unlike the NIST workshop, which was more of a scientific event, the 1995 Tucson ARO workshop had a particular goal, stated succinctly by US Army physicist and ARO program manager Henry Everett in the letter to attendees in the final workshop report, "In addition to clarifying several issues, a road map for future research was established. This road map will be of help to researchers and funding agencies alike, as the goals of functional quantum cryptographic systems and quantum computers are advanced." It is typical for such funding agencies as the ARO to have these workshops on new research areas. The workshop attendees produce a report and the report is then used to make a decision on whether DoD funding should be allocated to the new research area or not. In the case of quantum computing and cryptography, the decision was to fund research in these two areas. After the workshop, an alliance was formed between the ARO and the NSA. The ARO had a long history of tens of years experience in funding extramural research work, particularly at universities. The ARO had an entire system of issuing a "Broad Agency Announcement" or BAA, which is DoD lingo for a call for research proposals. The "broad agency" in the announcement means that a number of institutions or agencies may submit proposals, such as universities, industry, or government laboratories. Typically, the work is unclassified and of the basic research or applied research category.

There are two types of government laboratories or centers, most of which were set up after World War II to conduct basic and applied research. We'll need this distinction of the two types in what is discussed next. The first is what I will call a "true" government laboratory. In the true government laboratory or center, almost all the employees are US civil servants; that is, they are all employees of the US government. My first job at the Research, Development, and Engineering Center at Redstone Arsenal, Alabama, was of this type. I was a civil servant paid by the federal government. Then, there is the second type of federal laboratory or center, which has a specific name, a Federally Funded Research and Development Center (FFRDC), which sits somewhere between the true government laboratory and the industrial sector. In an FFRDC, almost nobody is a federal government employee but rather a contractor for the federal government. An example of an FFRDC was my second job at the NASA Jet Propulsion Laboratory (JPL), which is run by the Caltech for NASA. Almost all the workers at JPL were employees of Caltech and not the federal government. Compare this to the NASA Ames Research Center in Moffett Field, California, near San Jose. NASA Ames is a true government laboratory where almost all the employees are civil servants. Both a true government laboratory and an FFRDC receive the vast majority of their funding from the government. In this example, NASA funds almost all projects at JPL and NASA Ames. The FFRDC is a bit more flexible in its response to changes in mission and program goals. At JPL, if there is a major shift in NASA priorities, as occurred between the Clinton and G.W. Bush administrations, JPL can (and did) lay off 10% of its workforce and hire new people with skills relevant for the new programmatic goals.[33] This is almost impossible to do at NASA Ames because it is almost impossible to lay off a civil servant. For this reason, the average JPL'er is in her thirties while the average civil servant at NASA Ames is in his fifties. The FFRDC is also able to pay its employees better and more competitively, closer to industry pay standards, where US civil servants are paid under a General Schedule pay scale that is set by congress and often is barely keeping up with inflation. As I tell my students who are job hunting, there is a job security/salary uncertainty principle. The more secure your job is, the less you are paid. Think tenured university professors, who because of tenure have secure jobs but are at the bottom rung of the pay scale for PhD physicists. At the other extreme, PhD physicists in industry earn the most but can be laid off with only 2 weeks' notice if the project they are working on is cancelled. When I moved from the US Army to the NASA JPL in 1998, from a true government laboratory to an FFRDC, I got a big raise, but I was taking a job security risk by going from a very secure job to a less secure one.

The alliance of the NSA and ARO in 1995 had its origins in the 1995 Tucson workshop, as I have said. I like to think that because I had a major role in organizing this workshop, my interests in the foundations of quantum mechanics, and thence quantum computing and cryptography, led to this serendipitous alliance of these two organizations, both agencies in the DoD. Recall that the primary mission of the NSA is code breaking (cryptanalysis) and code making (cryptography). Typically, within the DoD, other agencies, even the Army Intelligence Agency, coordinate their own cryptanalysis and cryptography activities with the NSA, which oversees such activities and makes sure that the systems of the different services, air force, army, and navy, are all compatible with each other and meet NSA levels of standards, down to the NSA providing "crypto boxes" or devices that the services could interface with their existing communications technology. It is typical for the DoD agencies to run two parallel communications systems, a classified one and an unclassified one. Thus, it is typical that when a DoD employee gives you their business card, you see two separate phone and fax numbers, one on the classified system and one run by the ordinary phone company. This parallel communications system eventually embraced emails and other Internet communications. However, in 1995, email and the Internet were still new things and this parallel Internet-based classification system was not up and running in any complete sense. The NSA folks did have a classified email system, but because of their principle of maximal paranoia, in 1995, that system did not connect to the outside world. When I met Keith ||||||| at the Tucson conference, we exchanged business cards; his read "Keith **I I I I I I**, Piano Player" and had his AOL email address. The only way they could communicate with the outside world via email was on their home computers using AOL! This state of affairs lasted until at least the year 2000, when they began developing parallel unclassified email systems akin to their phone and fax systems, classified and unclassified, but until then, if you wanted to communicate with the NSA via email, it was by AOL.

In 1995, the NSA had no mechanism for issuing grants to universities. Up until that time, if they needed physics research work done, they would do it "in-house"; that is, they would hire physicists to work in the NSA laboratories or contract the work out to other DoD laboratories. However, it was clear in 1995 that the bulk of the expertise for developing quantum-computing hardware and software was at universities, industrial laboratories, and non-DoD laboratories such as the Los Alamos National Laboratory, and that most of the work that needed to be done was at this stage unclassified. The teaming would allow NSA funding to flow to the ARO and then out to such external research activities. The ARO, now a branch of the Army Research Laboratory, is "the Army's premier extramural basic research agency in the engineering, physical, information and life sciences." The ARO had this type of extramural (outside the DoD) experience and grant-issuing capability as long back as the 1800s and in its modern form at least as far back as the 1950s.

During World War II, many universities contributed greatly to scientific research related to war the effort, such as at the University of Chicago where Italian–American physicist Enrico Fermi led the development of the first nuclear chain reaction in 1942,[34] and the development of advanced radar systems at the Massachusetts Institute of Technology Radiation Laboratory from 1940 to 1945. Such successes, as well as the arms race with the Soviets, led to a great deal of funding being put into such service agencies as the Air Force Office of Scientific Research (AFOSR), the ARO, and the ONR—all three of which have the mission to fund extramural basic research, primarily at universities. These are sometimes collectively referred to as the "OXR" patterned after "ONR" with the "X" to be substituted with "N" for navy, "AF" for air force, or "A" for army.[35]

The AFOSR funds extramural basic research of interest to the air force, the ARO funds extramural research of interest to the army, and the ONR funds extramural basic research of interest to the navy. These days, a university single-investigator grant from any of these three agencies can run from $100,000 to $150,000 per year for a 3-year grant—enough to fund part of the professor's summer research salary and a few graduate students.[36] In 1958, the DoD created a new overarching agency, the Advanced Research Projects Agency (ARPA), to fund extramural research for the military that was not specific to the air force, army, or navy, but that was intended to provide research that all three agencies (and others) could benefit from. Renamed the Defense Advanced Research Projects Agency (DARPA) in 1972, then renamed ARPA again in 1993, then renamed DARPA again in 1996, this agency (with now a surplus of stationary with the wrong logo on it) was created in direct response to the Soviet launch of Sputnik, and funded (for example) the creation of the precursor to the Internet (called ARPANET) in 1969.[37] Today, DARPA can be viewed as an extramural funding agency that covers military basic research of interest to all three services, air force, army, and navy, but with its own vision and substantially larger pocketbook. A single-investigator grant from DARPA can run starting at $300,000 a year for 3 to 5 years, and for hot topic areas with multiple investigators, it can be substantially more.[38]

This extramural funding capability of the OXRs and DARPA was focused primarily on technologies that were not of a cryptological nature. That was the purview of the NSA and other intelligence agencies, which, because they were so secretive, did not have this extramural funding capability—at least not at this level—in 1995. The advent of quantum computing, cryptography, and related quantum technologies led the intelligence agencies to develop their own parallel funding agencies starting in 1995 and continuing to this day. The culmination of this effort is the IARPA, discussed above, which is the intelligence agencies' version of DARPA.

But in the spring of 1995, none of this intelligence agency–based grant-issuing capability was in place. The NSA wanted to fund extramural research in quantum information processing, the ARO had a mechanism to emit calls for proposals and issue such grants, and so it was a match. In addition to the mechanics of issuing university grants, I suspect that in 1995, the NSA was also pleased if it was not obvious where the money was actually coming from. As far as the outside world was concerned, the ARO suddenly had developed a great interest (and an equally great budget) for funding extramural research work in quantum information processing. Fund transfers between two DoD agencies, in this case the NSA and the ARO, can easily be carried out using a Memorandum of Understanding (MOU and pronounced "moo") and a Military Interagency Procurement Requisition (MIPR and pronounced "mipper"). Motivated by that Tucson workshop and report, the first ARO call for proposals, or BAA, came out in the fall of 1995, with much of the wording for the desired research to be funded looking like stuff from the Tucson report. That is the way of things. There was a focus on new technologies that could be used for the hardware of quantum information processing, as well as funding for the development of theory of such hardware, and finally funding for the research and development into quantum-computing algorithms and quantum communications protocols that would run on such hardware. Quantum information processing had made it into the big leagues!

INTERIOR PANEL SIDING

The quantum information proposals came in the winter of 1995–1996 and a joint panel of around 10 army and NSA scientists reviewed them. I was on this panel as an army scientist. In the National Science Foundation (NSF), a panel of your peers—also from academia—reviews your academic proposal. This has the advantage that the reviewers are likely to be extremely knowledgeable in the particular area your proposal is in. This has the disadvantage that the reviewers are likely to be your direct competitors for funding. The DoD works differently. To avoid such possible conflicts of interest, the reviewers are limited to US government employees, scientists who are funded by intramural or internal DoD funding, and who hence by definition are not in competition with, say, universities for their funding. The disadvantage of this DoD system is that although the DoD scientist reviewers have a scientific background, they may not necessarily be experts on the particular topic being proposed. I found working on the DoD panel really broadened my thinking skills. I had to become an expert on quantum information theory in a hurry just to understand what was being proposed in the theory proposals, and for the experimental proposals, I had to give myself crash courses in areas of physics that I did not work in. My background was in quantum optics, but we had proposals for semiconductor quantum dot quantum computers, superconducting quantum computers, and electrons floating on a sea of liquid helium, all things that I did not know much about.

One thing to understand is that, as mentioned in the previous section, all of us panel reviewers were US government scientists and civil servants, with PhDs in technical fields, but none of us were experts in the fields of quantum computing and cryptography, which, as far as most of us were concerned, were fields of research that we had only become aware of in the past year or so. One example, discussed in earlier sections, was the proposal to carry out Shor's factoring algorithm using the energy levels of a single electron in a hydrogen-like atom. This sounded good on paper, and for panelists who were not aware of the issue of coherence versus entanglement, particularly, they gave this proposal a fairly high score. I instead gave it the kiss of death and checked the box marking it "unresponsive to the BAA" because the BAA clearly stated that the goal was to build a scalable universal quantum computer and I was sure that, with no entanglement, this was not universal. On a panel of only 10 panelists, a single zero score is more than enough to pull something near that red line of unfundedness to well below it to insure taxpayers' dollars were not wasted on it.

Another example for this procedure was a proposal from a large defense contractor.[39] The proposal was to build a quantum computer using quantum dots. That seems reasonable as one of the first proposals for solid-state quantum computing used quantum dots. The proposal itself was a slick affair, literally; it was printed on glossy card stock, bound, professionally done, with pages and pages of colored photos of their laboratory equipment, electron micrographs of the quantum dots, and plots of lovely fresh data. Everybody on the panel (except Marv) was impressed and gave it a fairly high score. Unfortunately, this was not a beauty pageant for snazzily prepared proposals but a call for ideas to build a quantum computer (or the software that would run on it). Past the glossy pages of graphics and reams of plots of lovely data was one simple but wrong idea, "We can build these quantum dots. We can build a computer with the quantum dots. Therefore it must be a quantum computer." What they proposed was to build a classical computer with quantum dots and call that a quantum computer and hope nobody noticed. Nowhere in the proposal was the discussion of coherence or entanglement or two-qubit gates. I suspect that the proposers really had no idea what a quantum computer even was. I gave the proposal a zero and wrote in the comments, "There is no quantum entanglement in this system whatsoever and therefore it cannot be used to build a quantum computer. What is new is not interesting and what is interesting is not new." That score was enough to drop it well below the line of unfundability and it was not funded.

NEEDLE IN A HAYSTACK

In 1995, the primary driving force for the US Government program in quantum computing was Shor's factoring algorithm. However, if cracking public-key encryption was all a quantum computer was good for, then it would not be particularly widely used by society but rather every country's version of the NSA would have a couple of them to hack Internet communications until at least someone came up with a new type of public-key encryption that was not attackable by Shor's algorithm or turned to quantum cryptography, which is advertised as uncrackable by any means, including a quantum computer (see Chapter 5). During the exciting period of 1994–1996, there was hope then that one useful algorithm, Shor's algorithm, had been devised and shown to have an exponential speedup on a quantum computer and then perhaps a whole slew of algorithms will be discovered, some with more practical import than code breaking.

The hope was for a quantum exponential speedup on the class of NP-complete problems discussed above. The NP-complete computational problems all can be mapped onto the traveling salesman problem. Recall that the salesman is driving from city to city peddling his wares and he wishes to minimize his gas costs by taking the shortest route that hits each city but hits it once and only once. As the number of cities the salesman must visit grows, the difficulty of finding the shortest path grows exponentially. Hence, just with tens or hundreds of cities, the best classical algorithm would take trillions of years to find the solution. All the problems in the NP-complete class have the property, similar to Shor's algorithm, that if somebody hands you the potential solution, then you can check if it is a solution quickly in polynomial time; checking a solution is easy but finding the solution is exponentially hard. In the NP-complete class, there are many problems that are of great importance to the DoD and industry, typically logistical problems whose solution could save lots of time and money, such as the traveling salesman problem and the scheduling problem. A telecommunications entity would like to route telephone and Internet traffic between two points on a large network so that the traffic gets from point A to point B as quickly as possible; this is in the traveling salesman class. Electric companies would like to lay out the power grid in a way that the power gets from the generator to all the disparate users in a way that minimizes the amount of transmission line required. Gas companies want to minimize the amount of gas pipe. The army wants to optimize all their supply lines in a battlefield. These are all in the NP-complete class. The scheduling problem is also in this class and was one of the reasons NASA got into the quantum-computing game in 1998 and I moved to the Quantum Computing Technologies group. In any industrial process or on any complex satellite, there are a number of tasks that have to be performed in a specific sequence. For example, in an automobile factory, one has inputs to the process that include labor, raw materials, and electric power, and the outputs are the different types of cars or trucks the factory produces. Things must be done in sequence and optimizing this sequence can save a typical automobile factory millions of dollars a year. In an optimized sequence, the raw material shows up before the workers begin work. In an unoptimized sequence, the workers are paid to be sitting around drinking coffee while waiting for a shipment of steel to arrive. The worker whose job it is to bolt doors onto Fiats had better find a bucket of bolts waiting for him when he starts his shift. Finding the optimum ordering of all the tasks is a problem that is exponentially hard in the number of tasks. For an automobile plant, there can be millions of tasks, and finding the optimal solution—the traveling salesman's path through time—that minimizes the time it takes to produce a single car can take trillions of years on a classical computer. This scheduling problem is exponentially hard. NASA is a bit like an automobile factory but the output is a Mars rover instead of a Fiat. On the spacecraft itself, there are many millions of tasks that have to be performed in a specific sequence, typically by the onboard computers, in order for some complex process like the landing of a spaceship on Mars to take place seamlessly, quickly, and safely. Finding the optimal schedule is hard on a classical computer. Can the quantum computer help? Well, yes and no.

There is one simple way to think about the traveling salesman problem. Let us recall that you have solved all the problems in the NP-complete class. That is because there is a known "easy" (polynomial-time) algorithm to convert a solution to any problem in the NP-complete class to a solution of any other in the class. Thus, the traveling salesman problem can be reduced to a searching problem; of all possible routes, we want the one that is shortest and hits every city once. We can do this in a mockup analog computer. Let's take a ping-pong table and paint onto it a map of the United States. Then, we label each city the salesman must visit: Atlanta, Barstow, Chicago, Dallas, . . . and Zigzag. In this example, there are 26 cities. Now, we take a hammer and a box of nails and hammer a nail into the center of each city about halfway down the nail's length into the ping-pong table (see Figure 4.10).

On each nail, I place a drop of honey. The honey contains a special molecular tag so that each of the nails is painted with honey with a distinctly different tag. Now, I release a hive of genetically modified honeybees on one corner of the table. The honeybees have been genetically programmed to fly to the

Figure 4.10 Solving the traveling salesman problem with bees. A map of the United States is hammered to the top of a ping-pong table with nails located at each city the salesman is to visit, where here A is for Atlanta, B is for Barstow, C is for Chicago, . . . and Z is for Zigzag (Oregon). There are a total of 26 cities to visit, and the salesman, to minimize gas expenditures, must find the shortest path that visits all the 26 cities but each of the cities only once. A genetically modified classical bee (ceebee) is released in Seattle and must find its way to Miami and find the shortest such path that does so. The ceebee is programmed to visit all the cities, to visit each city only once, and to sample each of the 403291461126605635584000000 possible paths one at a time and report back to the salesman which is the shortest. At one path per second, the ceebee will take nine orders of magnitude longer than the age of the universe to find the shortest path. This time can be reduced quadratically by deploying a genetically modified quantum bee (qubee) whose coherence and wavelike nature allows it to sample all 403291461126605635584000000 possible paths simultaneously, and who at a rate of one qubee per second will find the shortest path in only 3 billion years or around one-fourth the life of the universe.

honey-coated nails, visit each nail only once, and keep track of their total flying time. I collect the bees at the other corner of the table and interrogate each about what its total flying time was. (I assume that all the bees fly at the same speed.) Now, I rank the bees from the longest flying time to the shortest. The honeybee with the shortest flying time has the solution—it had to have taken the shortest path! I now peel off the 26 layers of molecularly tagged honey from her feet, read the molecular tags from the nails from each layer, and I have the solution: Moscow (Idaho), Detroit, Birmingham, . . . and Jacksonville! Perfect. My genetically programmed bees have solved the traveling salesman problem! But how many bees will I need? Well, that depends on the total number of paths, which depends on the number of cities. The number of paths scales like an exponential function called the factorial function, which for our example of 26 cities is 26 factorial, which is written in shorthand as 26! and is written in longhand as $26 \times 25 \times \ldots \times 3 \times 2 \times 1$. That does not seem so bad, but let's multiply it out and we get 403291461126605635584000000 possible paths. Now, we see the problem. I'll need 403291461126605635584000000 bees!

I could instead use just one bee over and over again but assuming the entire flight takes 1 second, I'll need 403291461126605635584000000 seconds! In scientific notation, this is either 4×10^{26} bees or an equivalent number of seconds. There are only approximately 10^{19} insects on the Earth at any moment, so we're seven orders of magnitude short on bees. Reusing one bee won't work either since the age of the universe is approximately 10^{17} seconds, so we are nine orders of magnitude short in seconds. The fastest computer on Earth can carry out approximately 10^{16} operations per second, so it would take approximately 10^{10} seconds or around 317 years. It is in this sense that the traveling salesman problem, and hence any problem in the NP-complete class (because they are all equivalently difficult), is computationally hard. What can quantum computers do for us here? There is strong mathematical proof that the quantum computer can provide a quadratic but not an exponential speedup for such problems. This means that running this on a quantum computer could reduce 10^{10} seconds to 10^5 seconds, because 10^{10} is the quadratic square of 10^5 seconds, and 10^5 seconds is only about a day. There might be a market for solving NP-hard problems in a day rather than in a few centuries and perhaps there is.

I'll explain the way the quantum search works with a new breed of quantum bee, the "qubee." In addition to bee-ing a genetically engineered bee, the qubee is also a quantum particle in the sense that it flies like a wave but is detected as a particle. That is, it obeys wave–particle duality and has simultaneously aspects of both. Because the quantum search algorithm is usually stated in terms of a quadratic speedup in time, I will use one qubee over and over and see how much we can shave off that 403291461126605635584000000 seconds it would take the classical bee (or "ceebee") to sample all the paths.

The key quantum feature of the qubee is that, like a wave, it will scatter off the nails in a fashion like waves hitting posts in a pond that create even more waves moving in different directions. If I wall up the sides of the ping-pong table so the qubee does not flow off the edges, the waves will sample every possible path simultaneously, bathing all those 403291461126605635584000000 possible paths in the gentle, undulating, softly buzzing, wavelike ebb and flow of that single qubee. When looked at this way, it would seem we are home free to an exponential speedup—because the qubee samples every possible path, we can just interrogate her when she arrives at the far corner of the ping-pong table and extract from her the shortest path hitting every city only once. But therein lies the bug.

That single qubee holds all the information of the 403291461126605635584 000000 possible paths in her little noggin in a quantum superposition of itty bitty qubee brain cells. There are an exponentially large number of qubees in parallel universes but we have access to the information in only our own. The interrogation requires a measurement, and that measurement causes a collapse of the qubee. In the Shor factoring algorithm, it is always possible to shift all the information about the prime factor into our universe so that when the collapse takes place, we know with near certainty what the prime factor is. However, that ability to shift all the probability into our universe arises from a very specific property of the Shor algorithm; the entire question reduces to finding the period of a function that can be had in just one shot. The information on the shortest possible path on the map does not have this property and so we do not get it out with an exponential improvement. Well, do we get it out with any improvement or is all the information lost? Is using a qubee no better than using a ceebee?

The answer is that there is some improvement, a quadratic improvement, in using the qubee over the ceebee. That is, if we have to send a single ceebee through 403291461126605635584000000 possible paths or approximately 4×10^{26} seconds at one ceebee per path per second, then the qubee does quadratically better and takes only approximately 2×10^{13} seconds, which is the square root of 4×10^{26} seconds. This is some improvement, but it is certainly not exponential improvement. However, translated from ceebees and qubees to classical and quantum computers, we get the solution in 10^5 seconds (about a day) instead of 10^{10} seconds (317 years) because 10^{10} is the quadratic square of 10^5 seconds. While AT&T might not want to wait 317 years to figure out the way to lay down fiber-optic cable between 26 cities to reduce costs, they could afford to wait a day. Hence, in some scenarios, the quadratic improvement is good enough.

The lesson of the qubee and the ceebee in searching for the shortest path is that quantum computers do not give an exponential speedup on all problems, but rather just on some mathematical problems that have a certain structure, like factoring. The traveling salesman problem does not have this same structure and hence it is impossible to guarantee the right answer with exponentially improved efficiency. The qubee does indeed sample all paths simultaneously in its coherent wavelike flight through the nailed up map, but most of that information is lost when a measurement upon the qubee is made at the end of the flight. This is where the quantum uncertainty works against us. To use the parallel universe argument, in the Shor factoring algorithm, owing to the mathematical structure, it is exponentially likely that when we make a measurement at the end of the calculation, we are in the universe that has the right answer, the factor of the large number. The mathematical structure allows us to sweep all that probability of getting the right answer into our universe and not in any of the others.

However, the search-for-the-shortest-path problem does not have this same high degree of structure or symmetry in its mathematical setup as the factoring problem. Thence, we can sweep some of the probability of getting the right shortest path from all the other universes into our universe but not nearly an exponentially large amount of it, as with Shor's algorithm, but only a quadratically greater amount. That is, we still get the answer faster quantum mechanically than we do classically but it is not that much faster only quadratically, not exponentially. To compare: if a problem takes 2^{100} = 1267650600228229401496703205376 seconds (4×10^{22} years or approximately 300000000000000 times the age of the universe) classically, an exponential speedup on a quantum computer could mean it only takes 100 seconds or a little more than a minute. However, a quadratic speedup is not so dramatic. For a quantum computer that gives only a quadratic speedup on such a problem, then 2^{100} = 1267650600228229401496703205376 seconds becomes the square root of that number, which becomes only 2^{50} = 1125899906842624 seconds, which is still about one fourth of the age of the universe, which is still over 3 billion years. Things do become interesting though when the quadratic speedup takes something like 317 years down to 1 day. Hence, for certain problems that are not too big but still big enough to be interesting, then the quadratic speedup can mean a lot. The trick then is to pick a problem where a quadratic speedup provides a practical result.

Okay, enough with the ceebees and the qubees. The quantum search algorithm that provides this quadratic speedup was invented in 1996 by another researcher at Bell Labs, the Indian–American computer scientist Lov Grover. Grover's research paper originally appeared in a preliminary form in 1996 in the *Proceedings of the 28th Annual ACM Symposium on the Theory of Computing* and was entitled "A Fast Quantum Mechanical Algorithm for Database Search," but then it was followed in July of 1997 with an article in *Physical Review Letters* with the more fun title "Quantum Mechanics Helps in Searching for a Needle in a Haystack."[40] This article has over 1400 citations.

I have tried to encapsulate the idea behind Grover's search algorithm in the analogy with the qubees versus the ceebees in the traveling salesman problem, but let's now try to make things a bit more concrete. I once asked Grover how he came up with the algorithm in the first place and his response was, paradoxically, "I had been thinking about classical radar antenna theory." At its core, the Grover algorithm is all about the interference of waves, just as in qubees and in the design of a classical phase array radar system. As I had made somewhat of a career for myself by solving quantum problems using classical antenna theory, I was remarkably partial to this answer.

A classical phase array radar system is a line of radar antennas where the emitted radio waves from each antenna are coherently "phase locked" to all the other antennas in an array. That is, there is a fixed, known, and programmable arrival time at any point in space for the peaks and troughs that make up the radio waves and we have controllable wave interference. Let us consider the case of two antennas first and scale up.

Consider the two antennas in Figure 4.11. The antennas are "phased" so that the electric field of the radio wave leaves each antenna at its peak (and not the trough). If the distance from the two antennas to the target is an integer number of wavelengths, then the waves arrive at the target peak to peak as well. The way to calculate the intensity I of the electric field at the target, if the waves are coherent and the electric field from each antenna has a value of "one," is to write $I^{coherent}_{two\ antennas} = (1 + 1) \times (1 + 1) = (2) \times (2) = 4$. The intensity from either antenna alone is $I^{coherent}_{one\ antenna} = (1) \times (1) = 1$, and so the intensity from two coherently phased antennas at the target is four times the intensity from either antenna alone. This is a result of coherent wave interference. Let us compare this to the case when the antennas are not phase locked, and the peaks and troughs of each antenna vary randomly from each other. We say then that the two electric fields are incoherent and there is in this case no wave interference and the rule for adding the two incoherent waves is $I^{incoherent}_{two\ antennas} = (1 \times 1) + (1 \times 1) = 2$. Hence, the coherent intensity is double the incoherent intensity. This is our old friend, the quadratic scaling law!

If there are three antennas, then the intensity at the target is $I^{coherent}_{three\ antennas} = (1 + 1) \times (1 + 1) \times (1 + 1) = (2) \times (2) \times (2) = 8$ for the coherent phase array system and only $I^{incoherent}_{three\ antennas} = (1 \times 1) + (1 \times 1) + (1 \times 1) = 3$ for the incoherent phase array. Hence, if there is no coherence, no wave interference, the radar power at the target scales as N where N is the number of antennas (three antennas gives three times as much power). If there is coherence, wave interference, then the radar power at the target scales as $N \times N = N^2$ (three antennas gives eight times as much power). Thus, for example, if each antenna alone can place 1 watt of power on the target, and if there are 100 antennas in the array, and if the antennas are coherently phase locked with waves all in phase, then we can place a quadratically enhanced total of $100 \times 100 = 10,000$ watts

Figure 4.11   Two phased radar antennas (triangles) radiating at a single target (star). The antennas are coherently phased so that the electromagnetic radar waves (wiggles) are perfectly in sync. That is, the waves' peaks and troughs line up peak to peak and trough to trough. The rule of coherent addition of radio waves tells us that the radar power at the target, $I^{\text{coherent}}_{\text{two antennas}} = (1 + 1) \times (1 + 1) = (2) \times (2) = 4$, is double than it would befrom the two antennas if there were no coherence, $I^{\text{coherent}}_{\text{two antennas}} = (1+1) \times (1+1) = (2) \times (2) = 4$. In general, for N coherent antennas, the coherent power at the target scales like $N \times N = N^2$, but for N incoherent antennas, the incoherent power scales as N. This is a quadratic improvement in power placement.

on the target. If the 100 antennas are randomly and incoherently phased, then the most we can place on the target is 100 watts. The more watts on the target, the more radar energy returns to the receiver, and the more likely we can see the target. This is why they go through the trouble of building phase array radars in the first place. They are particularly useful against stealth aircraft technology, where the return signal is designed to be very small. A stealth fighter under

ordinary radar can look like a small bird on the return signal; hence, to coun-
teract the technology, more power at the right wavelength is critical.

Lest the "clueful" reader protest that we seem to be creating power out of
nothing, and hence not conserving energy, let me calm her fears. The key word
phrase is "power on the target," and as we can see in Figure 4.12, the power of
the coherent radiators is being swept up—interfered away by destructive interference from directions that point away from the target—and interfered up by
constructive interference and concentrated on the target in a spike-like blast of
energy. Compare this to the uniform radiation pattern in the figure of the incoherent radiators that forms the much lower horizontal line where at any point
the power scales only as the number of antennas and not the square of that as
in the spiked lobe of the coherent array. This is the power of classical coherence
and interference at work. For a well-defined phase array, you can place all of the
radar power from all the antennas on the target and zero it out in any direction
away from the target.

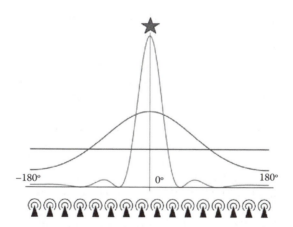

Figure 4.12 Sixteen phased radar antennas (triangles) radiating at a single target
(star). In general, for N coherent antennas, the coherent power at the target scales like
$N \times N = N^2$, but for N incoherent antennas, the incoherent power scales as N. This is a
quadratic improvement in power placement. Here, we plot across from horizon to hori-
zon the average power intensity on the target for 16 incoherent antennas, $I_{16\,\text{antennas}}^{\text{incoherent}} = 16$
(horizontal curve), 16 partially coherent antennas, $I_{16\,\text{antennas}}^{\text{partially coherent}} = 64$ (bell curve), and
finally 16 fully coherent antennas, $I_{16\,\text{antennas}}^{\text{fully coherent}} = 256$ (sharp-peaked curve). The scaling is
quadratic because $16 \times 16 = 256$. The increasing coherence sweeps the antenna power
out of the wings and focuses it on the target. (Plots are not to scale.) In the same way,
Grover's algorithm exploits quantum coherence and sweeps probability from "the
wings" of the phone book onto the target phone number of the phone book search, also
giving a quadratic improvement.

So what do phase array radar systems have to do with Grover's search algorithm? Well, in the algorithm, instead of enhancing and directing radar waves in *space*, we are enhancing and steering qubit quantum probability waves in *time*. As opposed to the needle-in-the-haystack problem, let's take the slightly more ordered problem of finding a name in the Los Angeles phone book. So suppose you are out at a bar and somebody writes down their phone number on a scrap of paper for you, but in your drunken haze, you forget the matching name and are hesitant to call the next day to see if you can line up a date with who's it. No problem, you can just pull out the Los Angeles phone book, which contains on the order of 10 million entries, and start with Aaron Aardvark and end with Zuzanna Zyskowski, checking each name until you find the number on your scrapola of weißbier-infused serviette. This is called an unordered search of an unstructured database, as the numbers are ordered alphabetically by name, which means they are randomly ordered by number. Because the order by number is random, there is no strategy that can speed up your search and the best strategy is the obvious one: start with Aaron and end with old Zuzanna. If you are extremely lucky, the first number you look at will be the right one (Aaron), and if you are extremely unlucky, the 10 millionth number you look at will be the right one (Zuzanna), and on average, if you repeat the search many times with different but still random phone numbers, you will have to search the book halfway through. That is, on average, you have to search 5 million entries out of 10 million. This is mathematically provably the best search strategy, and so, if there are N entries in the book, it takes on average $N/2$ tries to find the one you are looking for.

Searching the phone book can be transformed into finding the shortest path for the traveling salesman. In the traveling salesman problem, you also have a large number of paths to try and you have an alert system that allows you to check (quickly) if the path is the shortest or not. Hence, you can first check path one and then path two and so forth, and on average you'll have to check about half to find the shortest one. In both problems, the alert system is the key. In the phone book problem, your alert system is simply that when you find the right number, you can check it is right by comparing it to the number scrawled on your beer-soaked napkin. In computer science, this alert system is called an "oracle," which is a system in our subroutine that lets you know for sure when you have the right result. Usually, the oracle is programmed into the computer algorithm as a "black box" to which you may input questions and it gives yes or no answers as output. ("Yes," speaketh the oracle, "this is Marylebone Macadangdang's phone number," or "No this is not.") The oracle in the phone book search is just your very act of comparing each number to that on your Guinness-infused napkin. In the traveling salesman problem, it is a small subroutine that runs that "magic test" to see if the path you are checking is indeed the shortest or not.

The idea behind the Grover quantum search is to put the database to be searched (filled with qubits and gates) and the oracle (filled with more qubits and gates) all into a coherent superposition of qubits. The same quadratic improvement in the radar system, improvement of radar beam power in space, transfers neatly to a quadratic improvement in the quantum system, improvement of the qubit joint probability in time. It is the coherence and wavelike properties of the qubits (and not the radar beams) that now provide the quadratic speedup. Back to the phone book. If we are searching the entries at one per second, it would take classically on average 5 million seconds, around 2 months, to find poor ol' Marylebone in the book, who by now has cleanly forgotten ever giving you that phone number in the first place and so that ruins your chance at a date. Running the Grover search, however, improves your social life. The Grover search takes only around 3162 seconds or around 53 minutes—more than enough time to give Mme./M. Macadangdang a ring the next day and follow up with a timely invite to a meet and greet at Starbucks.

When it comes to cracking secret codes, the quadratic improvement of Grover's search algorithm is not nearly as practical as the exponential speedup of Shor's factoring algorithm. But it is not a typical problem in everyday life that the average person on the street is in need of cracking a 512-bit crypto key. The number of practical everyday problems such as search and traveling salesman and the related NP-complete problems is vast and all benefit from the Grover search by this quadratic amount. Recall that the nifty thing about all the problems in the NP-complete class is that once you have a way of tackling any one of them, then you have a way of tackling all of them. Finally, it can be shown that the Grover attack is the best you can do. There is no exponential improvement waiting in the wings—to be found by three Indian mathematicians in an un-air-conditioned attic in Madras—the NP-complete problems suffer only a quadratic improvement from the application of quantum coherence and that is the best you can do on a quantum machine. Where does the improvement come from? Like the radar system, the probability that you get the wrong answer when you make a measurement is destructively interfered away from the target and constructively interfered up on the target answer until a measurement gives you the right answer in just the square root of the time it would take classically.

A bit more on how the Grover search algorithm works. In one scenario, you simply assign one qubit to each of the possible items to be searched, say the 10 million phone numbers, a number we'll call N. (That's a lot of qubits.) The oracle is a separate subroutine that knows which qubit is the target. (It has the beer-soaked napkin and a keen eye.) Each of the coherent qubits is assigned initially an equal probability or, more correctly, a probability amplitude. That is, if a joint measurement is done on all these qubits in this initial superposition of qubits, the probability that the measurement outcome reveals the target qubit is $1/N$, so you would have to repeat this procedure $N/2$ times or 5 million on average to find the target phone number and the associated name. But now, what we do is after preparing the equally weighted superposition of qubits, we run Grover's algorithm, which contains something called "amplitude amplification." Just like the antenna array, where you phase the antennas one by one to quadratically put more radar power on the target (and proportionally less off the target), in Grover's algorithm, you phase the qubits one step at a time to put quadratically more probability on the target phone number (and proportionally less off of it). The oracle provides the steering of the probability because part of the algorithm is to query the oracle if any particular qubit is the target or not, and then if the oracle says yes, the amplitude of the target qubit gets its probability amplified and all the other qubits get theirs deamplified. In the square root of N iterations of the algorithm, you have placed all the probability on the target qubit and none at all on the remaining qubits, and so a measurement at this stage reveals the target with absolute certainty, because the target has a 100% probability now of being revealed in the measurement. Because the square root of N is extracted geometrically by constructing a square of area 10 million and then taking the length of one of the sides, we get a much smaller number. If the square has an area of 100, then the length of a side is just 10 because $10 \times 10 = 100$. Similarly, because $3162 \times 3263 = 10000000$, the square root of 10 million is 3162, and hence, in just 3162 iterations, we have the phone number. (Recall that at one iteration per second, this reduces the search time on the 10-million entry database from 2 months to about an hour.)

We can also show that this quadratic speedup is the best you can do on such a search. Then, the final step is to show that all the NP-hard problems like traveling salesman can be reduced to this type of search and we have proof that a quantum machine provides a quadratic improvement on finding solutions to such problems. But just what kind of quantum machine? Well, at least one that has quantum coherence. But what about entanglement? What about that? The original version of the Grover algorithm did indeed assume all the qubits were entangled and that the database (phone book) was mapped to an exponentially large number of states a few qubits can represent. Thus, in addition to the quadratic speedup in the search, there is an exponential compression of the database. That is, the 10 million entries in the Los Angeles phone book can be stored in only approximately 23 qubits (instead of 10 million qubits), because if you take $2 \times 2 \times 2 \times \ldots \times 2$ a total of 23 times, you get approximately 10 million. The exponential compression of the database looks impressive, but it does not provide any advantage in the classical search of the phone book.

In order to program all the 10 million numbers in the phonebook, I would have to have a classical-to-quantum encoder that takes at least 10 million steps to program the numbers into the quantum computer one at a time. Think of this like entering 10 million contacts into the contact list on your smart phone. Once they are in there, it is very convenient, but putting them all in there is a royal pain. The same is true with the quantum smart phone. And in the process of typing each number one at time into the quantum encoder, you might as well just set that beer-soaked napkin next to the keyboard and oracularly just check each one as it is entered for that euphonious and melliferous of all names, Marylebone Macadangdang. Hence, the exponential database compression aspect of Grover's algorithm only does you some good if the database to be searched is already in this exponentially compressed quantum form in the first place. If it is not, and it is not likely to be for searching a classical database, then you are better off assigning one qubit per phone number and just exploiting the quadratic speedup and leave the entanglement-induced compression out of the picture, because entanglement is so hard to generate anyway.

In the year 2000, American scientist Seth Lloyd first pointed out that Grover's search algorithm does not require quantum entanglement in order to give that quadratic speedup.[41] This result then paves the way for building some sort of a quantum computer—a quantum search engine—that exploits some of the weirdness of quantum mechanics but not all of it. Recall that for a *universal* quantum computer, a quantum computer capable of solving any problem, I need three types of quantum gates: the CAT, the RAT, and the ENT gates. What Lloyd showed is that quadratic speedup in Grover's search can be had without the ENT gate, which requires entanglement, but by using only the single-qubit CAT and the RAT gates, which require quantum coherence but not entanglement. There is still some improvement using quantum coherent systems over, say, trying to carry out the Grover search with classical waves (the radar system). That is, because of the three things Einstein did not like about quantum theory, unreality, uncertainty, and nonlocality, the lack of an ENT gate only precludes nonlocality. A quantum coherent machine of some sort with just CAT and RAT gates would still manifest elements of quantum unreality and uncertainty, which no classical wave machine would have. Because entanglement and the ENT gates are the hardest things to make in a universal quantum computer, a quantum search engine that just got by with CAT and RAT gates would be much easier to build. Such a quantum search engine would not be universal, but it would be good for providing a quadratic speedup via Grover's search algorithm for searching the phone book or solving the traveling salesman or any other of the NP-hard (but often very practical) problems.

For example, such a specialized quantum search engine, without entanglement, would be useless for running the Shor factoring algorithm, where an exponential speedup is critical for the factoring algorithm to be of any use in cracking public-key encryption and hacking the Internet. However, a quadratic speedup in searching is useful for many other noncryptographic things. These days firms on Wall Street make computerized stock trades using elaborate computer programs on lightning-fast supercomputers. In such a cutthroat business, if your quasi-quantum computer trading machine is a few milliseconds faster than the classical computer trading machine of your competitor, you stand to make millions. A quantum quadratic speedup on a 100-millisecond stock trade would be a 10-millisecond stock trade that would leave your competitors to eat the dust of your electron cloud computer.

QUANTUM COMPUTING IN A COFFEE CUP— WHEN THE BUZZ WEARS OFF

In 1996, the NSA proposal call brought in two separate proposals for carrying out quantum computing using molecules, in a liquid, *at room temperature.* One proposal was by American physicists Isaac Chuang and Neil Gershenfeld and the other by American physicists David Corey, Amr Fahmy, and Timothy Havel. The proposals appeared in my crate of proposals to review nearly contemporaneously with publications from both teams, "Bulk Spin Resonance Quantum Computation," by Chuang and Gershenfeld (over 1000 citations), and "Ensemble Quantum Computing by NMR Spectroscopy," by Corey, Fahmy, and Havel (over 500 citations).[42]

NMR stands for "nuclear magnetic resonance" and is the same technology behind the magnetic resonance imaging (MRI) machines used in hospitals. The medical device was originally called a *nuclear* magnetic resonance imaging machine, but the word "nuclear" was dropped because the marketeers assumed, and rightly so, that nobody would want to be placed inside of a large, white, cold, humming, coffin-like tube with the word "nuclear" plastered on its side.[43] NMR works this way: by taking a sample of liquid with molecules floating around in it, which holds for MRI in the approximation that the human body is a bag of water with molecular impurities, and then by subjecting the sample to a strong magnetic field, one could roughly align the nuclear spins of the atoms in some of the molecules with the direction of the magnetic field. Then, the quantum states of those nuclear spins could be manipulated by whacking them with elaborately choreographed pulses of radio waves. The radio waves can be used to write information into the nuclear spins, manipulate it while there, and then read it all back out. In the MRI machine, the readout provides an image of body parts that contain specific molecules, but NMR has been used for years commercially as a chemical analysis tool. You pop a vial of some goop into the hopper, push the run button for a specific sequence of pulses, and read out what is in the goop or what the molecules in the goop are doing.

These features of room-temperature operation and preprogrammed semi-automated operation made the system ideal for a practical quantum computer. Given the commercial state of the art, and the ready availability of off-the-shelf commercial NMR spectrometers, running a quantum-computing experiment was much more like just sitting at a computer terminal and typing in sequences of programming commands than it was like other quantum-computing experiments that involved gluing tiny gizmos together and aligning them for days on end with various sorts of epoxy. It was so easy to do that even theoretical physicists could run these machines and carry out quantum-computing experiments, because much of what we theoretical physicists do consists of sitting and typing away computer commands into keyboards, and so many of them did, and for some of them, the results of these experiments helped jumpstart their careers.[44]

I recall attending the Southwest Quantum Information and Technology Network Annual Meeting (Albuquerque, New Mexico, May 19–21, 2000) where they had a small exhibit hall set up for commercial vendors, and there was this booth where you could order your own personal quantum computer! The company running the booth made commercial NMR spectroscopy machines for chemistry and physics laboratories, and on their flyer and order form, they had a picture of one of their commercially available machines where it looked like they had simply photoshopped the words "QUANTUM COMPUTER" onto a photo of their regular NMR machine. The flyer explained that, unlike the NMR spectrometer, the "quantum computer" came with different software. It appeared that the race to build the quantum computer by the year 2000 was over! Or was it?

The peak of the NMR quantum-computing frenzy came and went in 2001, when Chuang and his collaborators used such a device to run Shor's algorithm and factor the number 15![45] I would say the field has been slowly going downhill ever since.[46] This story is one born out of an interesting idea, which turned out to be not that interesting after all, especially with the confluence of government funding, scientists' egos, and what happens when the speed of hype exceeds the speed of write. Scientists like to think that scientific research is somewhat immune to this kind of thing, that science is disconnected from the personalities and the proclivities of the scientists, but this is not so. As the jolly and rotund and, sadly, now late American physicist Peter Carruthers once explained to me, physics is all about "the-mouth-to-brain ratio," which he illustrated to me in 1998 with the diagram in Figure 4.13 that he drew on a chalkboard at the Max Planck Institute of Quantum Optics.

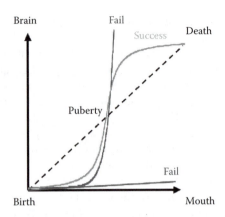

Figure 4.13 The Peter Carruthers "mouth-to-brain ratio" diagram for predicting success in a career in physics. We all start out as mostly mouth at birth, but around puberty, the three curves begin to diverge. Someone who stays too close to the mouth axis, talks a lot but never publishes anything of any consequence, will fail. (See, for example, "God Does Not Play Dice" by Dmitry Chakalov [October 28, 2011], http://www.god-does-not-play-dice. net/#Dowling.) On the other hand, if you stay too close to the brain axis, produce brilliant ideas that you never bother to write up, you will also fail as nobody will know who you are and somebody else will someday just reproduce your results and get all the credit (curves labeled 'fail'). (These days, this includes publishing in English language journals. During the Cold War, Russian scientists would discover great things years before Westerners but not get any credit because none of the Westerners were reading the journals in the original Russian. That reminds me of a story from Israeli physicist Gershon Kurizki, who was lecturing at a workshop in Russia when, after his talk, a Russian scientist came up to him and demanded to know, "Why don't you read my papers?" Western scientists are now well aware of this history of underciting their Russian colleagues and are particularly sensitive to this protestation. Kurizki apologized profusely to the Russian scientist, looking at his name tag, with a name he does not recognize, and promises to look up all of the Russian's publications and cite them in the future. The Russian smiles and nods and walks away, and sheepishly Kurizki turns to one of the workshop organizers and asks, "Who was that guy? Should I know him?" "Nah, don't worry," replies the Russian organizer to Kurizki, "That's just Boris. He does this at all the international conferences. I think the only complete English sentence he knows how to say is, 'Why don't you read my papers?'") The successful trajectory hovers around the diagonal and the optimal is mostly mouth at the beginning (by default) and mostly mouth at the end (by design) with a great deal of brain in the middle to establish your reputation (curve labeled 'sucess').

So how did the NMR quantum computer work and what was the origin of its demise? As I have mentioned, the NMR spectrometer, typically a device the size and shape of a water heater tank, had a place where you could insert a sample of some liquid inside the tank (Figure 4.14). The machine would then expose the liquid to a strong magnetic field, which then aligned or "initialized" or "polarized" some of the spins of the nuclei that made up the molecules that made up the liquid to line up with the magnetic field. Typically, the idea was to dissolve a very specific chemical in ultrapure distilled water and focus on the spins in the molecules of that chemical. One of the originally proposed molecules was caffeine, which led to a news article in January of 1997 in *Science* entitled "Putting a Quantum Computer to Work in a Cup of Coffee," giving credence to the vision that you could make a quantum computer by pouring a cup of coffee into the hopper on this hot water tank–sized NMR spectrometer and then by issuing a sequence of commands to the machine from an attached personal computer (PC). In fact, this simplicity was part of the original appeal of the scheme, but that same simplicity also aroused suspicion that the whole thing seemed too good to be true. In 1997, the most elaborate quantum computers were ion trap quantum computers and photon-atom quantum computers, where experimentalists such as David Wineland and his team at NIST or H. Jeff Kimble at Caltech carried out Herculean efforts to cool and trap ions or atoms in a labyrinthine room–sized maze of lasers and electronics and vacuum

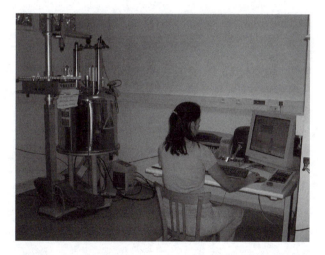

Figure 4.14 The dream of the NMR quantum computer: A hot water tank–looking NMR spectrometer of the type typically used in NMR quantum computing. The vial of molecules in solution to be used as the qubits is loaded into the port at the top. The bulk of the tank consists of strong magnets whose magnetic fields are used to align the spins of the nuclei in the molecules. The other metallic things sticking out of the tank are the radio wave pulse generators that are used to manipulate the qubits via a computer console off to the right. It was so simple it seemed to be too good to be true and it was. (Photograph of an NMR spectrometer by Daniel Alexandre [March 28, 2008].)

pumps to produce even a two-qubit ENT gate. Then, suddenly along came these upstart scientists such as Chuang and Gershenfeld as well as Corey, Fahmy, and Havel, and soon to follow Canadian physicist Raymond Laflamme, some of whom I had thought of, up to this point in time, as theorists (not experimentalists) and all of whom were suddenly magically able to carry out operations on handfuls of qubits, at room temperature, in something as mundane as a vial of coffee, in a store-bought machine that looked like a big water heater.

When I first read the NMR quantum-computing proposals in 1997, I was confused, as were many people, as to just what the qubits were supposed to be. My initial thought was that the qubits were the single, individual ions in, say, the caffeine molecule (see Figure 4.15). The idea then was to use simple radio wave pulses to make the simple single-ion RAT and CAT gates and more complicated radio wave pulses to exploit the spin–spin coupling between the nuclei in the molecules as the mediator for the ENT gates. As Seth Lloyd showed, almost any two-qubit interaction is suitable for making an ENT gate, and the spins of the atoms in the caffeine molecule can interact with each other directly, like two bar magnets causing each other to rotate about, or they can also interact via

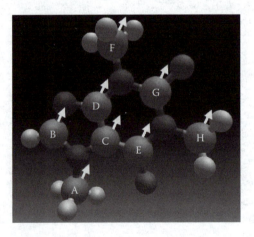

Figure 4.15   A three-dimensional visualization of the caffeine molecule. The balls labeled A, B, C, and so forth, are nitrogen atoms and the other balls are hydrogen, oxygen, or carbon atoms. Each atom has a spinning nucleus that can serve as the qubit for quantum computation, although in the NMR approach, large collections of such molecules and their atomic nuclei form "effective" qubits. It is typical to use just the spins of only the nitrogen nuclei as the qubits as shown here. The qubits are coupled by the ion–ion interactions in the molecules and manipulated with externally applied radio waves. (Image created by Michael Ströck [January 30, 2006].)

the intermediary of the clouds of electrons whizzing around the nuclei. All this can be carried out with radio wave pulses that are finely tuned and choreographed to make the relevant gates. Much like laser pulses are used to manipulate the ions in the ion trap quantum computer, here the radio wave pulses are used instead. The correct orchestration of pulses is programmed into the neighboring PC (or Mac) and the PC then instructs a pulse generator, such as the setup shown in Figure 4.14, which then launches the pulses of the correct energy and temporal duration and sequence to carry out an arbitrary computation such as Grover's search or Shor's factoring algorithms. This on face value seems reasonable.

But that simple picture didn't make any sense to me as there would be around 100 sextillion (10^{23}) caffeine molecules in a single cup of coffee. How would the NMR machine address an individual ion out of all of those? And in the ion trap quantum computer, people like Wineland and his group go through a great deal of effort to cool the ions with additional laser beams until the ions are only a thousandth of a degree above absolute zero, around 0.0001 milli-Kelvin or very close to −273°C (−460°F). The ion trappers do this to prevent decoherence from killing their quantum gates. The NMR quantum computer was operating at room temperature, around 300°K (27°C or 77°F), which was six orders of magnitude hotter than ions in the ion traps. Shouldn't decoherence be killing them? Well, as my friends in the math department are fond of saying, there is safety in numbers.

There were several cute ideas in these first two NMR proposals, but the key idea was that when you place a cup of coffee in a strong magnetic field at room temperature (lukewarm coffee), *some* of the ions in the 10^{23} molecules would find themselves aligned with the magnetic field. The ions inside the molecules are somewhat immune from the frequent molecular collisions in a liquid at room temperature because the ions are cloaked in a shield of electron clouds that protect them from the worst of the incessant buffeting. However, there are thermal photons at work; just as your stove emits red photons from the heating element, so does everything at room temperature emit infrared radiation that would potentially scramble and decohere the state of any particular ion too quickly to be of any use for a quantum computer. However, there are a huge number of ions, 100 sextillion of them, floating around in the coffee cup. From the theory of statistical mechanics and the study of tiny things in thermal equilibrium with their environment, you can estimate that the vast majority of the ions will have their spins scrambled and decohered by the thermal jiggling and wiggling of the molecules in the hot cup of coffee. Hot molecules tend to carry more energy and hot photons tend to do so also. However, some exponentially small fraction of, say, the nuclear spins of the eight carbon atoms in some of these 100 sextillion caffeine molecules would find themselves aligned with the field in such a way that they would obey DiVincenzo's second commandment for a quantum computer: "Be able to be initialized to a simple reference ground state such as |000....⟩." The idea of the NMR quantum computer then would be that of creative bookkeeping: draw a circle around the collection of caffeine molecules that are properly initialized, with all their nitrogen nuclei spins pointing along the magnetic field |00000000⟩ and then manipulate only those and ignore all the other "bad" molecules in the soup that are pointing every which way else. The beauty of the scheme is that the radio wave pulses could select and manipulate just those special, properly initialized molecules.

Thus, with this understanding, we now can see what the qubit is in NMR quantum computing. The qubit labeled A in Figure 4.15 is not one spin in one nitrogen nucleus in one caffeine molecule; the qubit is the collection of all spins on all atoms A in the subset or ensemble of all the molecules that just so happen to point along the direction of the magnetic field. Once in a great while, this happens and you nab it. That particular "qubit" A or actually collections of qubits A can be isolated from all the other "bad" qubits A that point in some other random direction by just looking for the signal from the correctly initialized qubit and ignoring the others. A two-qubit register is the collection of molecules where ions A and B both accidentally point in the same direction as the magnetic field and so forth. Hence, the eight-qubit register is the collection or ensemble of all the caffeine molecules whose eight nitrogen nuclei all accidentally point in the same direction. Once isolated from all the other bad molecules by signal processing, the radio waves then just manipulate these few, these very few, spins on the good molecules. Thence, the collection of all nuclear spins labeled A on all these good molecules is qubit A, the collection of all the nuclear spins labeled B on all these good molecules is qubit B, and all the way up to qubit H. This is why NMR quantum computing is sometimes called "ensemble" quantum computing. Each qubit is not a single nuclear spin, like in the ion trap quantum computer; each qubit is a collection of all the nuclear spins for the ensemble.

This is confusing and so let me give an example. Suppose you have a flotilla of a billion (10^9) sailboats being tossed about in a hurricane far out at sea. The boats are the molecules and the hurricane is the turbulent thermal fluctuating sea of water molecules and thermal photons at room temperature. Now suppose each sailboat has eight magnetic compasses onboard labeled compass A, compass B, ... up to compass G. The compass is the nuclear spin of the nitrogen atom. As the boats are buffeted about by the fluctuating and angry sea, the compass needles will swing wildly about the dial of the compass but, just by accident, occasionally the compass needle will be pulled to the point along the magnetic north by the Earth's magnetic field. Let us take a snapshot of the fleet and then carefully inspect the photograph, ship by ship, and circle each ship whose compass A is pointing magnetic north. This will be a small number of the total ships. We will call qubit A the collection of all compasses A that just so happened to be pointing magnetic north when we took the photo. Let us suppose, just for this thought experiment, that the odds of any compass pointing magnetic north at the time of the snapshot is 1 in 10. Hence, we can now estimate how many boats will have compass A pointing magnetic north! It is just 1 billion divided by 10, and so 100 million (10^8) of the boats. Okay, now out of those 100 million boats, we circle just the boats that have *both* compass A and compass B pointing magnetic north. Well, the odds of B pointing north is independent of A pointing north, but still 1 in 10, and so it is 100 million divided by 10, or 10 million (10^7) ships. All the compasses A and B on those 10 million ships *taken together* form a single two-qubit A–B register. Continuing in this fashion, we have 1 million (10^6) ships with A, B, and C pointing north; then, we have 100,000 (10^5) ships with A, B, C, and D pointing north . . . all the way down to a paltry 10 ships with all A, B, C, D, E, F, G, and H pointing north. Those eight compasses on those 10 ships, taken together, form a single eight-qubit register—or the qubyte.

The game is ingenious in that instead of protecting our qubits from the scrambling thermal environment, we simply ignore the qubits that have been scrambled and manipulate the ones that have not. In the flotilla in the hurricane example, there are only 10 ships that are useful to make our eight-qubit register, which implies that there are 1 billion minus 10 or 999,999,990 useless ships with eight times that or 7,999,999,920 useless compasses. Everything relies on our being able to single out and manipulate the compasses on just those 10 ships, and particularly read out the radio wave–induced signals from those compasses, while being able to ignore the other compasses. Right away, we can see one problem. If I wanted to make a 10-qubit register in the flotilla analogy, there would be $10^9 \div 10^{10}$ or 0.1 ships with all 10 compasses pointing north. That is, on average, there would not even be one single ship with all 10 compasses pointing the right way and we could not then use this method to make a 10-qubit register. The scaling of NMR quantum computing becomes exponentially bad as the number of qubits grows.

In the nitrogen nuclei in the caffeine molecules, the scaling is just as bad. This scaling problem was first and most vociferously pointed out by the American chemist with the wonderfully alliterative name of Warren S. Warren—the name was so nice, they named him that twice![47] Warren pointed this scaling problem out quite early in the game, in an article in the September 12, 1997, issue of *Science* magazine entitled "The Usefulness of NMR Quantum Computing."[48] The scaling laws are different from my simple model with the ships in the hurricane at sea but the idea is the same. Taking the extreme case of making a 100-qubit register in a large molecule, Warren states, "To fully understand the scope of this problem, note that 99.99999999% of the time a generously sized room-temperature sample (10^{22} [molecules]) contains no 100-spin molecules in the ground state. . . ." Taken another way, this means that out of the 10 sextillion (10^{22}) total molecules in the vial, only approximately 1 billion ($10^{22}/10^{10} = 10^{12}$) of them can be circled and called the 100-qubit register. A billion may still seem like a lot, but there is a signal-to-noise ratio issue. It all has to do with the readout, which brings us to DiVincenzo's fifth commandment, "The proposed quantum computer must permit *efficient*, qubit-specific measurements that rapidly collapse a cat state to $|0\rangle$ or $|1\rangle$." Remember, in a billion molecules, the "qubit" A in the hundred-qubit register is not just one spin; it is the billions of spins of A altogether. That gives us an advantage. The way that you detect a spin with radio waves is to send a pulse and whack the qubit and the qubit will respond with its own return pulse indicating what state it is in. For a single nuclear spin, this return pulse is so weak it is undetectable. Even for a billion spins, it might just be barely detectable, but for the noise from the other 10^{22} spins in the "background" molecules, which will also respond to being whacked with their own return pulses. There are tricks to separate out this 10-sextillion-spin background noise from the billion-spin signal, but they can only work so far. The issue is called in engineering the signal-to-noise problem. If 10 people are in a room talking, it is very difficult to understand any one person or locate that one person in the room by his or her voice alone. However, you can do some selective filtering if the person you are trying to find is your spouse as you know what his or her voice sounds like. The NMR signal analysis is something like this; you know what "pitch" you expect from the good molecules, which can help you ignore the rest. But if there are a hundred people in the room, finding your spouse blindfolded and by voice alone becomes harder, and if there are a thousand people in the room, it will become nigh impossible.

The poor scaling in the number of molecules with qubits that are good gets exponentially bad with increasing numbers of qubits in the register, as we saw. That means the signal we get back from the good versus the bad qubits gets also exponentially bad. Realistic estimates of the largest number of qubits that could be made in bulk-liquid room-temperature NMR quantum computers have been on the order of a few tens of qubits, owing to this scaling law, and the record largest number of qubits so manipulated in the NMR approach was 12 qubits, in an experiment with L-histidine molecules, led by Raymond Laflamme's group at the University of Waterloo, in Canada.[49] To my knowledge, this has been the upper limit for this scheme, and the wall they hit is that of the poor signal-to-noise scaling. Recall that to run Shor's factoring algorithm to crack a 512-bit public key, we need a machine with approximately a million entangled qubits. NMR will never get us there and NMR quantum computing violates DiVincenzo's first commandment: "Thou shalt be a *scalable* physical system with well-defined qubits." NMR quantum computing was realized in 1997 to not be *scalable*. However, it was thought still to be *universal*—with a true universal set of CAT, RAT, and most importantly ENT gates—and hence it was sold as a sort of a test bed for quantum computing. While everybody else was struggling to make even two entangled qubits, the NMR folks easily made at least 10 or so entangled qubits, and using this approach, they produced a series of astounding results such as small-scale implementation of Grover's search algorithm and the use of Shor's algorithm to factor the number 15.[50] These publications appeared in the most prestigious journals, over the period of 1998 to 2006, such as *Science, Nature*, and *Physical Review Letters*, garnering thousands of citations. However, in the same time frame, a series of publications, championed particularly by Australian physicist Samuel Braunstein and American physicist Carlton M. Caves, ever more and more conclusively showed that in all of these spectacular experiments, there was no quantum entanglement in any of these experiments whatsoever.

The bulk-liquid NMR quantum computer violated DiVincenzo's fourth commandment: "Thou shalt have a universal set of quantum gates." With no entanglement, the NMR machine had no true ENT gate, and with no true ENT gate, there was no entanglement, and without the nonlocal correlations of quantum entanglement, the portal to Hilbert's exponentially large computational space slammed shut. The exponential number of parallel processors in parallel universes could not be accessed. When it came to NMR quantum computers, what was quantum was not quite a computer, and what was a computer was not quite quantum. And so the whole experimental field went from the stellar heights of scientific adulation to die a slow and agonizing death at the hands of a band of stubborn theorists, such as Australian physicist Samuel Braunstein, the Woody Allen of theoretical physics, who knew in his gut that the whole darn thing was just too good to be true. It is true that there were dark storm clouds—heavy, black, and pendulous—toward which the entire field was driving. Death's cold embrace began with a chilly little air kiss, a virtual peck on the cheek borne on an ill-begotten breeze out of Australia, which arrived in a paper by Braunstein and colleagues, entitled "Separability of Very Noisy Mixed States and Implications for NMR Quantum Computing," a paper that appeared in *Physical Review Letters* in 1999.[51]

As Braunstein recalls the genesis of this work, he and coauthor Jozsa were at a Quantum Information conference at a beach resort in beautiful Heron Island, off the coast of Australia, in September of 1998 where there were two talks on NMR quantum computing: one by Canadian physicist Raymond Laflamme and another by American mechanical engineer Timothy Havel. As Braunstein relates in the discussions over "sumptuous" conference dinners, one of the NMR guys (he does not remember who) claimed he could violate Bell's inequality. Braunstein writes, "I asked how could they do that without entanglement. I then sketched on a napkin a proof that entanglement was needed and surely they didn't have any." Many great calculations have had their origin on a wine-soaked napkin at a sumptuous dinner.[52] Braunstein then worked out the details with Jozsa. In the meantime, Carlton Caves and German mathematician Rüdiger Schack, as well as English mathematician Noah Linden and Romanian physicist Sandu Popescu, were working out their own versions of this proof, all in 1998. Braunstein says, "We decided to combine papers because of the overlap," which was the origin of the 1999 paper, "Separability of Very Noisy Mixed States and Implications for NMR Quantum Computing." The result of that calculation was that there was no entanglement in any bulk-liquid NMR quantum-computing experiment carried out by 1999 and by extension in any bulk-liquid NMR quantum-computing experiment ever carried out to date (2011). The claim by the NMR proponents that their computer was universal but not scalable was wrong. In addition to not being scalable, it was not even universal—no entanglement, no ENT gates, no quantum computer.

To see this, let us go back to the compasses on the ships. Recall that in a snapshot in time, I drew a circle around all the compasses that just happened to be accidentally pointing north, regardless of what ship I was on. This is the initial state $|00000000\rangle$. Now, to carry out my RAT, CAT, and ENT gates, I begin whacking the entire vial of goop with the choreographed sequence of radio wave pulses. The problem is, let us say I send in one pulse per millisecond (one pulse every thousandth of a second), but just because I drew my circle around all the good spins at time zero, it did not stop the compasses and their needles from being buffeted by the wind, the waves, and the rain brought about by the hurricane. Thus, as the pulse sequences go on lockstep in time, the original chosen "good" spins will drift out of alignment at the same rate that the not chosen "bad" spins will drift into alignment and proceed to be whacked as if they were the original good ones. After a long enough pulse sequence, the spins I am manipulating at the end of the elaborate dance are not even the ones I wrote down on my dance card at the beginning. For that reason, the entanglement is an illusion. The apparent appearance of entanglement is an artifact of my bookkeeping—there are no real two (or more)-qubit, strong, nonlocal, unreal, and uncertain quantum correlations in liquid, bulk, room-temperature NMR. What is a computer is not quantum and what is quantum is not a computer. Let us now compare what the NMR machine is with the five DiVincenzo commandments for a quantum computer, all five of which must be satisfied.

1. Be a scalable physical system with well-defined qubits? No. The system is not scalable and the qubits are not well defined.
2. Be able to be initialized to a simple reference ground state such as $|000....\rangle$? No. This implies that the experimenter initializes the qubits herself and not that she waits until they randomly look initialized and then takes a picture of them.

3. Have qubits with much longer decoherence times than gate-speed operation times? No. Their gate speeds are much faster than the decoherence times of the collection of spins being called a qubit, but these are not well-defined qubits.
4. Have a universal set of quantum gates? No. No entanglement implies there are no true ENT gates. The machine is not universal.
5. Permit efficient, qubit-specific measurements that rapidly collapse a cat state to $|0\rangle$ or $|1\rangle$. No. Qubit specific implies there are well-defined qubits, which there are not.

Hence, on the most broadly agreed set of criteria for the existence of a quantum computer, the NMR machine satisfies none of them. Whatever this machine is, it is not a universal quantum computer. Now, I want to be clear that everybody in the community, including me, thought this was a universal quantum computer from about 1997 to 1998. That is because we were all lulled into this position by a combination of the somewhat complex bookkeeping argument for how the qubits were to be identified and wishful thinking. Once the paper by Braunstein and coauthors appeared, it was utterly clear to me that we all had made a mistake, and because it is the job of every theoretical physicist to make as many mistakes as quickly as possible, that is fine, we learn from our mistakes and move on. But moving on was not so easy and less so for some than others.

When I read this paper by Braunstein and his coauthors, I said to myself, "Well that is the end of NMR quantum computing." But to my surprise, it wasn't the end—at least not a quick and painless death, as in a massive stroke—but more like a slow and agonizing death by mad cow disease. Fields of scientific research, particularly experimental fields of research, tend to have an inertia that carries them along well past their expiration date. This is in part due to the expensive investment in equipment and infrastructure that goes with experimental research. When a theorist finds that she has been on the wrong path, she crumples up her piece of paper and tosses it into her recycling bin and gets out a fresh sheet. "The most important tool of a theoretical physicist is the wastebasket."—Albert Einstein. Contrariwise, when an experimentalist realizes that he has been pursuing a dead end, or a dead cat, he stands and surveys the $5 million NMR machine in his $300 million laboratory, which is only good for doing NMR quantum computing if you're a quantum computer scientist, and thinks, well I'm not sure what he thinks, because I am not such a person, but I have observed a tendency for them to just go about their business as if nothing has happened and hope nobody notices, at least for a while. When I tell this story of the birth and death of NMR quantum computing to theorists, they universally are appalled that NMR quantum computing did not end instantly in 1999. When I tell this same story to experimentalists, they are universally sympathetic that the field was allowed to die slowly over the ensuing 10 years. I'm sure this all says something about the psychology of doing science but I'm not a shrink so I'm not going to go there. This is just my observation of the historical facts, tainted no doubt by the fact that I'm a theorist. The funding agencies also have no interest in killing a dead end in research instantaneously. And so the funding spigot is only slowly turned off as the data and theory accumulate, showing that the approach has failed. Failure is important, in science and in finance. Failure is a sign you are willing to take risks. If you are risk adverse in science, you will never fail spectacularly, but you will likely never succeed spectacularly either.

A curious episode to this business took place in February of 1999 in Cambridge, "Our Fair City," Massachusetts, at a DoD-sponsored workshop, "Nuclear Magnetic Resonance and Quantum Computation," held on the campus of Harvard University. The cold kiss of death was replaced with the icy cold winds of a Boston Nor'easter that ne'r blow anybody any good; as the temperatures dropped to −6°C (20°F), I walked with Keith I I I I I I, from the NSA, from the hotel to the conference center while I was wearing only a light jacket. (I had just flown in from Pasadena, California, where the temperature was around 21°C [70°F].) Keith looked at me as my ears turned blue walking across campus and declared, "Dowling—where the hell is your hat!?" I retorted, "Hat? What hat?" To this day, Keith still kids me about arriving in Boston in the dead of winter without bringing a hat. (Keith had on some giant fur-lined hat with fur-lined earflaps that made him look like the Chief of Police, Marge Gunderson, from the 1996 film *Fargo*.[53]) The workshop was a 3-day affair, February 22–24, 1999, filled with lectures on the theory and experiment of quantum computing with bulk-liquid NMR. They saved the best for last and scheduled Carlton Caves to give the very last talk on the very last day, where he tried to succinctly summarize the result of his paper with Braunstein and coworkers that there was no entanglement in any of the experiments presented at the workshop and what they were doing was likely not quantum computing at all. This talk was received with catcalls and outright heckling from a member of the audience—one NMR experimentalist member of the audience in particular.

I seriously doubt this experimenter even understood Cave's theory—he just knew it was bad news for his experimental NMR quantum-computing program. Ever since, it has been hard for me to discuss NMR quantum computing without reliving this experience and to keep a fair and balanced tone.

Over the years since the 1999 paper, I have made my feelings on NMR quantum computing widely known. At conferences, I would often sit in astonishment while NMR experimentalists would lecture about their latest quantum computation carried out on their NMR "quantum computer." I would raise my hand and ask, "What about these proofs that there is no entanglement? How can you call this a quantum computer?" Once called out, they would say to me, "Oh, of course we are aware of those results." My thought was, "Well if you are aware of those results how come we don't hear anything about it in your talk until I bring it up in the question and answer session?" One answer I got was, "Yes the states are not entangled but perhaps the gates are entangling?" I have no idea what that even means, but Caves and collaborators also apparently ruled it out. The final answer I get now when I attend such talks or review such programs or referee such papers is, "The elaborate radio wave pulse sequences we have developed for the NMR machine will be used someday on real quantum computers." I have to agree with that, but is it really justified to spend millions of dollars developing pulse sequences on a machine that is not a quantum computer in order to use them some day on a quantum computer? Why not just give the money to the people building actual scalable universal quantum computers, like the ion trappers, and let them develop their own pulse sequences?

In the end, the paradigm is shifting. Funding for NMR quantum computing has vanished, papers are no longer being published at the same rate, students are no longer graduating in the field, and slowly focus is moving on to other things. However, this initial overselling of the field may have caused many of us to miss some interesting physics that lies at the heart of the NMR machine, and that story may be more interesting than the idea that it was a universal quantum computer! Let me explain.

In 2002, American physicists Nicolas Menicucci and Carlton Caves published a paper entitled "Local Realistic Model for the Dynamics of Bulk-Ensemble NMR Information Processing," which was to be the last nail in the coffin for NMR quantum computing.[54] The kiss of death had become death's cold embrace, well if you call room-temperature computing cold. In this paper, Menicucci and Caves construct a local, realistic, and certain hidden variable model that explains all bulk-liquid room-temperature NMR quantum-computing experiments to date, up to about 12 nuclear spins or qubits. This model ruled out any quantum entanglement in any such experiment and showed that all the NMR experiments are in this sense classical. Recall that it was Einstein who suggested that all of quantum mechanics should be replaceable with such a local, real, certain, hidden variable theory and it was Bell who showed that for quantum systems with maximal strong correlations, quantum entanglement, such a theory disagreed with experiment. Quantum mechanics, or a replacement that agrees with the data, must be nonlocal, uncertain, and unreal. Read this way, NMR quantum computing was classical and hence a 12-qubit NMR quantum computer should be easily simulatable on my iMac, in which case I don't need to go out and buy a $5 million NMR spectroscopy machine.

But there was a subtle and perhaps overlooked caveat in the paper by Menicucci and Caves, the local hidden variable theory they cooked up—something like representing all the quantum mechanical nuclear spins with my classical hurricane-buffeted compass needles—was not efficient! The number of compass needles needed to explain an N-qubit NMR experiment scaled exponentially with the number of spin qubits N. Neither Einstein nor Bell ever discussed the idea of an *efficient* local hidden variable theory. Efficient is a term that comes from computer science and describes the scaling of the resources needed to carry out a particular computation. The best-known classical factoring algorithms are inefficient on a classical computer (exponential overhead in time) but efficient on a quantum computer (exponential speedup in time and polynomial overhead). But the paper by Menicucci and Caves was not discussing a particular computation but rather a physical theory designed to explain the workings of a particular set of experiments. If NMR was truly purely classical, there should exist an efficient local hidden variable theory of it. An efficient model would be in my description 12 compass needles needed to descript 12 nuclear spins. But recall in my compass needle model that there are many compass needles associated with "the qubit" just as in NMR quantum computing each qubit is assigned to a collection of nuclear spins. An inefficient description, what Menicucci and Caves found, implies that the number of compasses grows exponentially, something like 2^N, where N is the number of qubit and where 2^N is the number of required compass needles. Hence, 1 qubit requires 2 needles, 2 qubits requires 4 needles, 3 qubits require 8 needles, . . . , and finally 12 qubits require 4096 needles. What is going on here? Just as Menicucci and Caves drive the last nail into the coffin, signaling the final brain death of NMR quantum computing, they dislodge all the nails and that bloodsucking vampire pops back out again, now undead, to scour the funding agencies for virginal program managers with deep pocketbooks.

Remember Feynman's original logic when he first proposed the idea of a quantum computer. There are atoms like thulium whose properties cannot be efficiently simulated on a classical computer, which suffer an exponential slowdown when you try to simulate it. A thulium atom, with its 69 entangled electrons, can be thought of as a quantum computer that perfectly efficiently simulates other thulium atoms. Hence, a generic quantum computer with 69 logical qubits should also be able to efficiently simulate thulium. Quantum computers show an exponential speedup in simulation capability on the thulium simulation problem. Perhaps there are other math problems that quantum computers also show an exponential speedup? Yes! Shor's factoring algorithm. So with this NMR business, we have come almost full circle except we also know that the NMR experiments do not have entanglement. Maybe they have something else quantum? Something not as strange as full quantum entanglement but something weaker that is still not yet quite classical? Well, yes they do.

In a 2001 article entitled "Quantum Discord: A Measure of the Quantumness of Correlations," Harold Ollivier and Wojciech H. Zurek proposed a new measure of quantum weirdness, quantum discord, that describes quantum correlations that are much weaker than quantum entanglement but that are nevertheless explainable with a purely classical theory. If quantum entanglement is the gold standard of quantum correlations, then quantum discord is the zinc standard—the nickel and dime standard—the loose change of quantum weirdness. Every time a new measure of "quantumness" is proposed, people go to their laboratories and look to see if whatever is in their vat of goop has some of it. NMR quantum machines are no exception, and in 2011, Canadian physicists Gina Passante, Osama Moussa, Denis-Alexandre Trottier, and Raymond Laflamme published a paper entitled "Experimental Detection of NonClassical Correlations in Mixed State Quantum Computation," in which they show in a four-qubit NMR experiment that there is nonvanishing quantum discord.[55] I spoke with Dr. Passante about this experiment after her presentation on the topic at a recent conference in Tokyo. And, after she patiently and hopefully not too uneasily endured my diatribe on the ills of NMR quantum computation, she managed to convince me that in fact—despite my protestations to the contrary—something quantum indeed was going on in her experiment that could not be explained in classical terms. When you couple this existence of quantum discord in such experiments with the result that the experiments cannot be described efficiently with a local hidden variable theory, it is clear something interesting is going on that tells us something about the nature of computing and physics and their interface—something far more interesting than getting large government grants in order to factor the number 15! (And by this, I mean 15 and not 15!) The NMR quantum computer is dead—long live the NMR quantum computer!

So I should be careful to say I have not totally capitulated. The NMR machines are not quantum computers in the ordinary sense of the word, although for 10 years, experimental groups claimed in their talks in publications that they were (not scalable) but universal quantum computers despite ever mounting evidence to the contrary. This well-documented history is the wellspring of my diatribunal. Nevertheless, at the end I think, would this curious and most interesting business of quantum discord and quantum computational efficiency in such NMR machines have ever been uncovered if the field was not overhyped and overfunded to begin with? I think about the funding and hype surrounding optical classical computing in the 1980s that parlayed into the Cisco routers that transmit the bits and bytes of information now at my fingertips. Would the Internet have ever been built had it not been for all those optical computing gizmos that could be immediately harnessed for optical communications? The most important discovery to emerge from the race to build the world's first quantum computer will most certainly *not* be a quantum computer.

D-WAVE, BOSS, D-WAVE

Well, if it is so much easier, using coherence only, to build the quantum search engine than the universal quantum computer, why hasn't anybody done so? Well maybe they have, but then again, maybe they have not. In February of 2007,

D-Wave Systems Inc., a Canadian company that specializes in superconducting technology, revealed in a press conference in Mountain View, California (the heart of Silicon Valley) a prototype superconducting quantum computer with 16 coherently coupled qubits—or so they claimed. The initial reaction of the academic and government quantum-computing community was, frankly, that of disbelief. The record number of qubits claimed in a quantum computer up to that time was a claim in 2006 for an NMR quantum computer, which had 12 qubits, but none of those qubits turned out to be entangled with each other (see 'Quantum Computing in a Coffee,' above). Indeed, in the academic and government communities, by 2007, only ENT gates entangling *two* superconducting qubits had been demonstrated and published in peer-reviewed journals; hence, D-Wave's announcement in an un-peer-reviewed press release of a machine with 16 superconducting qubits seemed preposterous. If one insists on both peer review and true quantum entanglement, then the record by February 2007 for the maximum number of entangled qubits was 8 ion qubits, demonstrated in 2005 conclusively to be entangled in an ion trap by the group of German physicist Rainer Blatt at the University of Innsbruck in Austria. (Herr Prof. Dr. Dr. Blatt has just in 2011 demonstrated 14 entangled ion qubits in his trap.)[56] I recall the eight-qubit milestone quite vividly from one of Rainer Blatt's lectures in 2005 when he proudly announced the production of the world's first "qubyte" or entangled eight-qubit register.[57]

The D-Wave announcements in 2007 of a 16-qubit machine and in 2011 of a 128-qubit machine have been either ignored or attacked by the academic quantum-computing community. Part of the problem is that D-Wave does not release details of what their quantum computer is, exactly, as they claim it is proprietary. Even worse, their announcements and demonstrations of their ever more elaborate "quantum" computers are made in press releases and not in peer-reviewed scientific journals—the gold standard for scientific announcements. Remember what happened to poor old Boris Podolsky when he caused a *New York Times* press release on the Einstein–Podolsky–Rosen paper to be released before the peer-reviewed journal publication? Podolsky was run out of town by a coven of theoretical physicists from the Princeton Institute for Advanced Study who, in an angry mob, carrying torches and pitchforks, chased him to the border of New Jersey where he was forced to swim the Hudson river to New York under the cover of darkness while towing his few earthly belongings behind him in a waterproof haversack.[58]

D-Wave announced their 16-qubit machine in 2007. I got rafts of emails from puzzled technorati and laypersons alike, with subject lines along the lines of "Have you seen this?" and "Do you know anything about this?" and "Have you heard about D-Wave's quantum computer?" and "What the #%$@!?" I read over the press release carefully, which is all I had to go on, and at first I suspected that it was a mistake similar to the quantum dot proposal I had reviewed in

1996; "We have made a computer with transistors made out of superconducting quantum interference devices (SQUIDS). The SQUIDS are clearly quantum, the thing clearly computes stuff, and so it is a quantum computer." Again, D-Wave did not give details of what exactly they had built or how it worked so I, along with everybody else in the community, just had to guess. My guess in 2007 was that D-Wave just did not understand what a quantum computer was. Others were less kind. Some speculated that they knew full well they did not have a quantum computer and the announcements were fraudulent attempts to boost D-Wave's stock prices and to lure investors.

As a sign of the times, the public relations battle on just what D-Wave's quantum computer actually was was fought in the quantum blogosphere, where David Bacon on his blog "The Quantum Pontiff" did battle with Scott Aaronson on his blog, "Shtetl-Optimized."[59] Aaronson promoted the viewpoint of his PhD advisor, Indian–American computer scientist Umesh Vazirani, who suggested that the D-Wave quantum computer was based on D-Wave's misunderstanding of one of Vazirani's own research papers, and Vazirani stated that "A 16-qubit quantum computer has smaller processing power than a cell phone and hardly represents a practical breakthrough."[60] To be fair, the ENIAC had approximately 25,000 vacuum tube and crystal diode processors (transistors had not been invented yet) and my cell phone (a T-Mobile G2 Google smart phone powered by Android) has approximately a billion transistors, so it might be better to compare the first quantum computer to the first classical computer, and by this metric, 16 qubits still sucks, unless the qubits are entangled and then fair comparison would be the size of the Hilbert space spanned by those qubits, which would be $2^{16} = 65{,}536$, which puts us squarely in ENIAC territory. The "D-Wave One" quantum computer, released in 2011, has a 128-qubit superconducting chip, and if those qubits be entangled, then a fair comparison might be to their whopping big $2^{128} = 340 \times 10^{36}$-dimensional Hilbert space—340 undecillion parallel processors in 340 undecillion universes! That dwarfs the processing power of my cell phone and that of all smart phones (or smart toasters) on Earth hooked together. As I write these words, on October 2, 2011, D-Wave has sold precisely one of the D-Wave One quantum computers, to the defense contractor Lockheed–Martin, and I hear rumors a second may soon go to Google.[61] At a cost of $10 million *each*, I suppose they are the only two companies that can afford one of these gizmos during the present economic downturn. The NSA will neither confirm nor deny that they have purchased one but I doubt it. When the first 2007 announcement of the D-Wave Orion prototype appeared, I mentioned this to Keith ⅠⅠⅠⅠⅠⅠ at the NSA with the interrogative, "Have you heard about the D-Wave quantum computer? Does this mean I won our bet?" to which Keith responded, "Don't bet on it." (In May of 1999, I bet Keith that a quantum computer useful to the NSA would be built in 10 years. Keith bet against. The bet was for a pizza and a beer and I lost and I paid up in the summer of 2009.)

The great D-Wave debate raging in the quantum blogosphere comes down then to this: Are the D-Wave qubits coherent qubits, and if so, are they entangled qubits? As we recall from the discussion of Grover's search algorithm, entanglement is not required if all one wants is a quadratic rather than an exponential speedup. That quadratic speedup in search can be immediately applied to the class of NP-hard problems. It is telling then that Google—master of all search engines—bought one of the D-Wave computers but the NSA (I suspect) did not. The running joke is that if a quantum computer is ever built capable of factoring large numbers and cracking 1024-bit public-key encryption, then the NSA will buy exactly one of them. Not exactly a moneymaking business model. The other running joke is that the NSA already has a quantum computer in a basement someplace and its multimillion-dollar development program for quantum computation is just to throw everybody off the scent so people continue to use public-key encryption, which the NSA can hack. Okay, this is more of a conspiracy theory than a joke, but I suspect it is not true or else very good quantum information scientists would have been disappearing over the years into the maw of the NSA, never to be heard from again, and this has not happened. The closest thing the NSA has to a quantum computer development facility is the innocently named Laboratory for Physical Sciences at the University of Maryland, which can't be that secret because it has its own web page.[62]

Coherent or not coherent? Entangled or not entangled? Those are the questions. My views of the D-Wave effort have evolved over the years in an inverse fashion to my views on NMR quantum computing. In the beginning, I believed that the bulk-liquid NMR machines were quantum computers, but now I do not. In the beginning, I believed that the D-Wave machine was not a quantum computer, but now I believe it is. Not an entangled universal machine with exponential speedups, mind you, but something quantum with quadratic speedups nevertheless. Reading Bacon's and Aaronson's blog posts and interpolating the truth is a bit like watching MSNBC and the Fox News Channel and trying to extract a balanced viewpoint, which is like subtracting infinity from infinity and hoping to get a sensible result like –1/120.[63] So in this vein, I turned to CNN, which in this analogy is my old friend, colleague, and collaborator, Welsh computer scientist Colin Williams. Williams has been very closely involved in the development of the D-Wave machines since the very beginning.[64] He is one of the few scientists I know that also has a reasonably good business sense about him. The D-Wave researchers apparently got the idea for their quantum computer after reading the 1998 first edition of Williams' book, *Explorations in Quantum Computing*, and Williams has been a consultant for D-Wave for over 10 years.[65] It is then my opinion, after taking in all this information, that the D-Wave machine is coherent but has no (or very little) entanglement. In our lingo, they can make CAT and RAT but not ENT gates. Hence, the D-Wave machines give no exponential speedups that would be required for running Shor's factoring algorithm efficiently but do give a quadratic speedup similar to that which can be had from Grover's search algorithm even without entanglement. What good is a quadratic speedup in search? Recall that a quadratic speedup reduces the time to find Marylebone Macadangdang's name in the Los Angeles phone book from hundreds of years to about a day. We live in an era where stock trades are made by computers at lightning-fast speeds, and if your company's computer can make trades milliseconds faster than your competitor's computer, then you stand to make billions while they go bankrupt. In such an era, a quadratic speedup may mean everything! In the end, I do not care if the D-Wave quantum computer has entanglement or is universal. In the end, I do care if it useful for anything. I think it probably is.

How does the D-Wave quantum computer work? Well, that is proprietary information, remember, and so what follows is what I like to call a "Wild-Assed Conjecture" (W.A.C.) in my class lectures.[66] Well, it is not totally W.A.C. but guided by a combination of what I have distilled from Bacon and Aaronson, the D-Wave press releases, and information I extracted from Colin Williams after a bottle or two of wine. The D-Wave machine clearly has qubits and those qubits are coherent with each other but not entangled. As discussed above, this coherence alone is sufficient to run Grover's search algorithm, which would give a quadratic speedup in search and all NP-hard problems, but the Grover search algorithm when written out in quantum-computing circuit form has a lot of overhead even in just the CAT and RAT gates and all the calls to the oracle. Coherence, the ability to make a cat state, as discussed was difficult to achieve in this paradigm of CAT gates in superconducting circuits: "They have spent 10 years trying to make a Schrödinger cat state and give them another 10 years and they still won't have one." However, superconducting systems have a built-in coherence, a wavelike description, just from the fact that they are superconducting—all the electrons move coherently like waves in synchronicity through the superconductor. What D-Wave does, instead of a circuit decomposition of the Grover algorithm, is something called adiabatic quantum computing. This type of adiabatic quantum computing is much more like getting the thulium atom to simulate itself by throwing electrons at the thulium nucleus and just waiting for them to all settle rapidly (adiabatically) into the ground state. Now, for something like thulium, the ground state is highly entangled. In fact, Israeli computer scientist Dorit Aharonov and colleagues showed that the adiabatic quantum computer paradigm can have entanglement, and if it does have entanglement, it is equivalent to regular circuit-and-gate quantum computing—that is, that it can also be universal.[67] In the early proposals for adiabatic quantum computing, there was a suggestion that it might be better than ordinary quantum computing—able to solve NP-hard problems in polynomial time with an exponential speedup over a quantum computer.

This paper by Aharonov et al. shows that is not so and full adiabatic quantum computing is equivalent to ordinary quantum computing and that both are universal and able to efficiently simulate each other and solve NP-hard problems with only a quadratic improvement.

My point is that there is a type of watered-down adiabatic quantum computing that is *worse* than ordinary quantum computing—a regime of adiabatic quantum computing with coherence but no entanglement. This is the regime D-Wave operates in. With coherence alone, as we have seen with Grover's algorithm, we can get a quadratic speedup in search (and hence in NP-problem solving) provided we do not need in addition that pesky exponential compression of the database to be searched—which we never do. Hence, the D-Wave machine has coherence without entanglement. It cannot efficiently simulate thulium and it cannot factor large numbers and bring down the Internet by cracking 1024-bit public-key encryption. But in the regime it operates in, with quantum coherence alone, it can (similar to Grover's search algorithm) robustly provide a quadratic speedup in search and consequently a quadratic speedup in solving NP-hard traveling salesman–type problems. That is what I mean by useful—getting Ol' Marylebone on the home phone in days rather than years. Would somebody buy such a thing for $10 million? Well Google did ... best if you are the world's fastest search engine that you continue to stay quadratically better than your competitors who can't afford a $10 million co-processor for their computer server farm.

So why the distain and outright hostility for the D-Wave machine from the academic and government quantum-computing community? Well, there are several issues working against D-Wave: no peer-reviewed journal articles, no exponential speedup, and no universality. The D-Wave press releases did have their fair share of hype: statements that were designed to be vague or ambiguous, making the reader think they had an exponential speedup when they did not. (In their defense, D-Wave never claimed to have entanglement or an exponential speedup—that was just wishful thinking on the part of the reader.) We already discussed the lack of journal articles, and this very lack of detail leaves D-Wave open to accusations of fraud or hype. Quantum complexity theorists, persons who study whether a quantum algorithm or machine gives an exponential speedup versus a quadratic speedup, are typically only impressed with exponential speedups. They tend to lump quadratic speedups (power of two) in with all other polynomial speedups, cubic (power of three), quartic (power of four), quintic (power of five), and so on, as all *sub*-exponential and thus uninteresting. The universality is closely related. Recall that a quantum computer is universal if it can efficiently simulate any other quantum computer with only polynomial overhead. You might just lose that quadratic speedup in the noise when switching from adiabatic quantum computing to circuit-and-gate quantum computing. However, this classification overlooks the question—the only real question—of whether the machine is useful or not—exponential speedups be damned! By my criterion of usefulness, I again and finally conjecture that the D-Wave machine—that infernal, coherent, nonentangled, nonuniversal, nonexponentially sped-up, nonpeer-reviewed machine—is actually useful.

There is a more insidious factor at work here as well. The vast majority of funding for quantum-computing research has come from the intelligence agencies in the DoD and their focus is and always has been on Shor's algorithm and the exponential speedup in factoring. There is a fear, rightly or wrongly, in the academic community that if one appears to be interested in anything other than building or designing a universal quantum computer for factoring, that one's funding will disappear. Frankly, this fear is not supported from my own discussions with program managers, such as American scientist Mark Heiligman at the IARPA, who currently heads a number of intelligence agency DoD programs in quantum information processing.[68] Heiligman very clearly states at the IARPA program meetings that IARPA's interest (at least on the software side) in quantum computing is very much greater than just Shor's factoring algorithm. However, I suspect that many of my colleagues simply do not believe him or assume his voice is a minority at IARPA (especially on the hardware side). This focus on factoring is a particular US belief system, real or imagined, and I think it prevents the US program in quantum information processing from being as broad based as it might otherwise be and as it otherwise is, say, in Australia, Europe, and Japan. Agencies such as the NSF cede too much of the programmatics to IARPA, because IARPA apparently has all the money and focuses instead on much smaller niche subfields.

A specific and personal example: Of the many approaches to building a quantum computer, one that is most versatile, with the most room for potential nonquantum-computing spin-offs, is the photonic quantum computer approach. I had been working on this approach for 10 years, funded by IARPA, as well as a number of other groups from around the world—on the order of 10 of them—stretching from Australia to Austria. By 2006, all these groups had been merged into a single large team, led by American physicist Paul Kwiat at the University of Illinois, with both experimental and theoretical components funded at a rate that exceeded $1 million a year. A typical NSF grant for a single investigator rarely exceeds $100,000 a year, and grants from the ARO, ONR, and AFOSR are at a similar small level. Hence, nobody was funding photonic quantum computing except IARPA, as the NSF and the OXRs were happy to cede the funding and the responsibility to IARPA and focus on other niche applications. Then, in 2010, we were all told that the program would not be renewed and there would be no funding from IARPA for photonic quantum computing at all. The $1-million-a-year program went to a $0-million-a-year program overnight. The other agencies could not possibly pick up the slack and the United States was left in a situation where there was practically no program in photonic quantum information processing. Like the electrons sloshing around on a tray of liquid helium, photonic quantum computing was axed. Why? Well, I am not privy (anymore) to the inner workings of the IARPA program, but I suspect that photonic quantum computing was viewed as a nonscalable road to building generally a universal quantum computer and specifically a quantum factoring engine capable of running Shor's algorithm and breaking secret codes—that is and always has been the main focus of IARPA.

The problem was in marketing, I suspect. In the original photonic quantum-computing scheme proposed in 2001 by Knill, Laflamme, and Milburn, it required approximately 10,000 photonic gizmos (beam splitters, mirrors, phase shifter, detectors) per ENT gate. Given that approximately 1,000,000 ENT gates are needed to crack a 1024-bit public key, this means that the photonic quantum computer capable of factoring large numbers would have to have approximately 10 trillion photonic gizmos. Unlike solid-state and superconducting circuitry, which can be made just a few nanometers in size (a few tens of billionths of an inch), each photonic gizmo can be no smaller than the wavelength of the photon in use, approximately a micron in size or 0.000001 meters (40 millionths of an inch), and so a chip with 10 trillion photonic gizmos on it would be no smaller than 1 square centimeter (approximately 0.16 square inches), which does not seem so bad, now that I think of it. However, I suspect the 10-trillion-photonic-gizmos-per-chip scaling was the number that stuck and even though that number was brought down by an order of magnitude a year between 2000 and 2010, photonic quantum computing was axed. IARPA, I suspect, at least in the hardware development, is under great pressure to winnow down the competing platforms for the quantum computer. First, the electrons floating on liquid helium were flushed down the toilet and then the photons flying in optical microchips were thrown out the window. The hardware folks at IARPA hope to have the one, true, scalable quantum computer platform in the next 10 to 20 years by discarding platforms that the IARPA technical folks view, rightly or wrongly, as un-scalable. In the end, Betamax was a better format for recording videotapes than VHS (video home system), but VHS won out because of marketing.

I'm not bitter. But I am worried. While there was not much I suspect that electrons floating on a refrigerated sea of liquid helium was good for other than quantum computing, certainly having small quantum photonic processors was good for something else, the quantum Internet! Let us consider a classical analog. In the late 1970s and early 1980s, a great fear arose that semiconductor classical computing was coming to an end. The fear was that as you packed more and more transistors onto a chip, the electromagnetic cross talk between them would grow without bound and the chips would cease working, Moore's law would come to an end, and the classical computer revolution along with it. Out of this fear arose a proposed solution: replace the electrons running around wires on a semiconductor chip with light waves running around waveguides on an optical chip. Light waves, unlike electrons, do not experience electromagnetic cross talk and the pitch was that classical optical computers would replace the classical electronic computers and Moore's law (the number of transistors on a chip and, hence, the processing power of a computer chip double every 2 years) would continue on the back of the light waves. The DoD is easily swayed by a slick combination of hype and paranoia. Starting around 1980, the DoD began dumping millions into the development of the all-optical, classical computer. The funding rose rapidly year by year and all the optics folks in the United States were focused on this goal: developing optical switches, transistors, and diodes and figuring out ways to put them all on an optical chip.

Then, by 1990, the sky did not fall. American computer scientists Carver Mead and Lynn Conway showed that Moore's law did not have to end. Using careful design rules, they showed that the cross talk on electronic chips could be made vanishingly small (rather than ravishingly large) with decreasing transistor size if you only carefully designed the chip that way. The semiconductor revolution continued unabated. And the optical classical computer? Well, the funding for the optical classical computer dropped even faster than it had risen, and by 1990, there was virtually no DoD funding for optical classical computers. The entire optical classical computer program was viewed as an overpriced failure and even to whisper that you were working on optical classical computers in the 1990s was the kiss of death for any potential DoD funding.

But the optical classical computer was not a failure at all! It just did not succeed at its intended goal of making chips for a scalable classical computer processor. The same issues arose as in the quantum photonic computer; for example, the light waves were too fat to get too many on a tiny optical chip. But the switches, transistors, diodes, and repeaters developed for the classical optical computer, paid for by the DoD, found their home in every Cisco router, hub, and switch in the optical fiber–based Internet. The realization was that a classical optical computer processor needed trillions of such gizmos (too tall an order) but the classical optical Internet router only needed tens of them (just right). The Internet revolution in the 1990s was driven by optical devices developed for computing, and paid for by the DoD, but whose home was in the relaying of information over long distances. If you want to send lots of information over long distances, you are talking photons.

Meanwhile, back in the 2000s, in quantum land, it was widely whispered among the academics that the quantum optical computer would follow the same trajectory. The DoD would pay for the development of the scalable photonic quantum computer with trillions of quantum photonic gizmos on a quantum optical chip, for the goal of factoring, but when we could only figure out how to make tens of such gizmos on a chip, the spin-off would be the photonic quantum repeaters, routers, switches, and transistors that would power the future quantum Internet and be sold by an as of yet unincorporated corporation named "Quisco." But to speak this out loud was prohibited—lest the DoD cut all our funding once they concluded that we were not true of heart and factoring was not our goal. Maybe they figured it out. Maybe they just did not like those trillions of quantum photonic gizmos, but they cut our funding anyway—too soon!

The quantum optical gizmos for the future quantum Internet (or QuInternet) were not quite ready in 2010 for prime time. All we needed were four more years! The IARPA bulldozer continues to plow inexorably onward to the goal of building the million-qubit universal quantum computer for the singular goal of factoring, and the floating electron qubits were bulldozed down the toilet and the flying photonic qubits were plowed out the window. The volcano of funding for photonic quantum computing has collapsed here. Meanwhile, other countries, whose quantum information programs are much more broad based than in the United States, such as Australia, China, Europe, and Japan, continue to work on the photonic quantum information processors knowing their primary application is to the quantum Internet, which will one day be a reality.

NOTES

1. The American physicist N. David Mermin stated in a public lecture at the 1999 Toronto conference on Algorithms in Quantum Information Processing that, "The introduction of Alice and Bob is the greatest contribution from quantum information theory to physics—I use them in my relativity class. No longer called S and S-$Prime$—they are now called Alice and Bob. My reference frames have gender!" This, of course, was a veiled insult directed at the field of quantum information. Perhaps as penance or perhaps he had a change of heart, but in any case, eight years later he wrote a splendid book on the topic. See "Quantum Computer Science: An Introduction" by N. David Mermin (Cambridge University Press, 2007), http://www.worldcat.org/oclc/137221653.
2. See the email announcement "NIST Workshop on Quantum Computing and Communications," *Interesting People* (May 17, 2011), http://www.interesting-people.org/archives/interesting-people/199407/msg00025.html.
3. For a two-qubit system, the exponential speedup is $2^2 = 4$. (The exponent is two. The dimension of the Hilbert space is four. There are four parallel processors but only two qubits. Two of these processors are in parallel universes.) For a three-qubit system, the exponential speed up is $2^3 = 8$. (The exponent is three. The dimension of the Hilbert space is eight. There are eight parallel processors but only three qubits. Five of the processors are in parallel universes.) The exponential growth is very rapid. For a 10-qubit system, the largest built so far, the exponential speedup is $2^{10} = 1024$. (The exponent is 10. The dimension of the Hilbert space is 1024. There are 1024 parallel processors but only 10 qubits. Hence, 1014 processors are in parallel universes.)

4. Legend has it that when Archimedes was directed by King Heiro II to uncover the mass density of the king's golden crown, he discovered the principle of buoyancy (named the Archimedes Principle) while taking a bath, whereupon he ran naked through the streets of Syracuse yelling, 'Eureka!'. However, the weather is much warmer in ancient Syracuse than in Snowbird, Utah, in the winter.

5. The flux capacitor required 1.21 gigawatts, which is pronounced as "jig-a-what" (as in the movie *Back to the Future*), but is the least commonly used pronunciation, about 10% of the time, but still the recommended pronunciation by the *US National Bureau of Standards Pronunciation Guide for the Metric Prefixes*—your tax dollars at work. Then, there is "gig-a-what" (like "gig 'em Aggies"), which is the most commonly used pronunciation, used about 95% of the time, and then there is a 4% minority who pronounce it "gia-ga-watt," like in the word "giant," and a 1% super-minority that pronounce it "ga-ga-watt" in honor of a female pop star who sometimes dresses in all-meat clothing.

6. See "Factoring Numbers with Waves" by M. Suhail Zubairy in *Science Magazine*, Volume 27 (April 2007) pages 554–555: http://www.sciencemag.org/content/316/5824/554.full.

7. Computational complexity is the study of the level of difficulty of problems that can be run on a computer. We have already alluded to this idea in the above work. If a problem requires an exponentially large number of resources—time, space, energy—to solve, then it is said to be in a difficult complexity class. The notion of complexity depends on the computing model, a point not widely appreciated until the notion of quantum computing was invented. Hence, for example, factoring is a hard problem on a classical computer and is in a difficult classical computer complexity class. Checking the result, by multiplying the two primes back together to see if they give the original number, is easy on even a classical computer, so multiplication is in an easy classical computer complexity class. If we switch to a quantum computer, factoring and multiplication are both easy.

8. A more accurate but yet physical explanation of NP can be found in the abstract of "Virtually-Deterministic Quantum Computing of Nondeterministic Polynomial Problems" by J.D. Brasher, C.F. Hester, and H.J. Caulfield in the *International Journal of Theoretical Physics*, Volume 30 (1991), pages 973–977: "It is common to measure the computational complexity of an algorithm or process in terms of how the computational resources (time, space, energy) must scale with some linear measure N of problem size. In optical processing, N might be the number of input beam resolution cells, output detectors, interconnections, etc. Concentrating on the resource-dominate term, we find that many calculations scale as N^p, where p is some small (often integer) number. We call these polynomial problems or algorithms. Other problems scale in a nonpolynomial way, e.g., p^N, and we call these exponential problems. It is a peculiar feature of exponential problems that they can often be solved by decision-tree algorithms. In such cases, if we magically knew what paths to take (a nondeterministic situation), then we could solve the problem with polynomial resources. Such problems are called nondeterministic polynomial (NP). NP-complete (NPC) problems, a subset of NP, are particularly interesting because each of these problems can be transformed into any other NPC problem with polynomial resources. Thus, if we could ever find a polynomial solution to any one of the several thousand NPC problems, we could solve any other one also with polynomial resources."

9. As a teaching assistant in a large office in the math department at the University of Colorado in the early 1980s, I recall somebody writing on the chalkboard one day P = NP? that was followed in quick succession by more mathematical graffiti, "P ≠ NP!" and "PP = NP!" and "∀∃∀!" (for all backwards E's, there exists uniquely at least one upside-down A) and so forth. At the time, I had little idea what the joke was. In fact, even now, I have little idea what the joke is. Those hilarious mathematicians.

10. The term *polynomial* refers to an algebraic expression of the form $a_n x^n + a_{n-1} x^{n-1} + \ldots + a_2 x^2 + a_1 x^1 + a_0$ where the a's are constants and x is a variable. The expression $x2 + x + 1$ is a quadratic polynomial and $x + 1$ is a linear polynomial.

11. See "Quantum Computing: Dream or Nightmare?" by Serge Haroche and Jean-Michel Raimond in *Physics Today*, Volume 49 (August 1996), page 51: http://dx.doi.org/10.1063/1.881512. I was able to spoof this paper in my winning design for the Southwest Quantum Information and Technology Network logo, which consists of a Native American dream catcher and two kokopelli fertility gods (named Alice and Bob). The dream catcher is a woven web that you hang over your bed to let in dreams but block out nightmares: http://www.squint.org/WhatSQuInT.html.

12. See "Decoherence Limits to Quantum Computation using Trapped Ions" by Martin B. Plenio and Peter L. Knight in the *Proceedings of the Royal Society of London, Series A*, Volume 453 (October 8, 1997), pages 2017–2041. The abstract of this article ends on this sour note, "Again no number of practical interest can be factorized." They did include the effects of error correction schemes available at the time but estimated their value to factoring large numbers to be negligible. We now believe that while difficult, things are not as bad as Raimond and Haroche, and Plenio and Knight, made them out to be. I should be careful to state that because nobody has yet built a quantum computer capable of factoring numbers large enough to be of interest to the NSA, perhaps these guys were right all along, but then it would not be for the reasons they put forth in 1996 and 1997.

13. A more rigorous proof is to assume there exists a "triplication" quantum gate TRIP that takes $|0\rangle + |1\rangle$ into $[|0\rangle + |1\rangle, |0\rangle + |1\rangle, |0\rangle + |1\rangle]$ and then show that this leads to a contradiction with the laws of quantum mechanics (particularly the law that such a gate is "linear"), hence proving that the assumed TRIP gate could not have existed in the first place—proof by contradiction.

14. A nice discussion of this idea for a quantum memory and quantum gates comes from the work of my friend and colleague, Swedish physicist Stefan Kröll at the University of Lund, where they host this nice web page, "Quantum Computation," Quantum Information Group (Division of Atomic Physics, Faculty of Engineering, Lund Tekniska Högskola, June 20, 2011), http://www.atom.fysik.lth.se/QI/research/quantum_computation.html. I recall fondly visiting Kröll's group in 2003 when I served as the "faculty opponent" on a student's PhD defense. In addition to the laboratory tours, Kröll took me on an outing to see the famous Viking standing stones or "Ale's Stones" (*Ales Stenar* in Swedish) in the province of Scania. The Viking stone-ship monument is marked by immense standing stones made from sandstone and arranged in the outline of a Viking ship. We had a nice lunch of dried, hard, smoked herring (*Strömming*) on the nearby wharf.

15. The first journal reference in print to the word "spintronics" that I can find is from "Will Spintronics Replace Conventional Electronics?" in *Research and Development Magazine*, Volume 41 (July 1999), pages 14–16. There is a an early reference in an

American Physical Society newsletter from 1998, "Quantum Computing, MEMs, Spintronics Mark 1998 L.A. March Meeting," *APS News* (American Physical Society, June 10, 2011), http://www.aps.org/publications/apsnews/199806/meeting.cfm. The first time I heard of this term was at the Defense Advanced Research Projects Agency (DARPA) Spins in Semiconductors (SPINS) workshop in Santa Barbara in 1999. I recollect (and my notes support) a close synergistic connection between the DARPA SPINS program and the DARPA Quantum Information Science and Technology (QuIST) program, which ran about the same time and funded research in both areas. An American Institute of Physics interview with American physicist and DARPA program manager Stuart Wolf supports my recollection. Wolf was the manager of both the QuIST and SPINS programs. See "Oral History Transcript—Dr. Stuart Wolf" in the *Niels Bohr Library and Archives* with the *Center for History of Physics* (American Institute of Physics, June 20, 2011), http://www.aip.org/history/ohilist/30668.html.

16. See "Measurement of Conditional Phase Shifts for Quantum Logic" by Quentin A. Turchette, Christina J. Hood, Wolfgang Lange, Hideo Mabuchi, and H. Jeffery Kimble in *Physical Review Letters*, Volume 75 (1995), pages 4710–4713: http://link.aps.org/doi/10.1103/PhysRevLett.75.4710. The article is available free in preprint format here: http://arxiv.org/abs/quant-ph/9511008. This "nonlinear polarizability" is the property of an atom or collection of atoms to interact and couple two independent photons together. Normally, the effect is very weak and has to be enhanced, in this case with the optical microcavity, which is made of two very good mirrors facing each other. The photons bounce back and forth millions of times between the mirrors interacting with the atom over and over and over again, building up enough interaction strength to get the desired ENT gate effect between the two photons.

17. See "Quantum Interfaces and Memory" in *Quantiki* (Quantum Information Wiki and Portal, July 5, 2011), http://www.quantiki.org/wiki/Quantum_interfaces_and_memory.

18. See the review "Optical Quantum Computing" by Jeremy O'Brien in *Science Magazine*, Volume 318 (December 2007), pages 1567–1570: http://www.sciencemag.org/content/318/5856/1567.full. A more detailed review can be found in "Linear Optical Quantum Computing with Photonic Qubits" by Pieter Kok, William J. Munro, Kae Nemoto, Timothy C. Ralph, Jonathan P. Dowling, and Gerard J. Milburn in *Reviews of Modern Physics*, Volume 79 (2007), pages 135–174: http://link.aps.org/doi/10.1103/RevModPhys.79.135. This can be found free in preprint format here: http://arxiv.org/abs/quant-ph/0512071.

19. This subtle but deep connection between "nonlinear optics" and "linear optics plus photon detectors" was suspected (mostly by me) as early as 2001 but not rigorously proved until 2003 in a paper by Canadian physicist John Sipe and me, as well as my postdoc Pieter Kok and Sipe's graduate student Geoff Lapaire. This development is an interesting lesson on the importance of serendipity in science. ("If we knew what we were doing we wouldn't call it 'research' now would we?"—attributed to Albert Einstein.) In the fall of 2002, Austrian Professor Anton Zeilinger had invited me to visit his experimental group at the University of Vienna for a few weeks. The visit fell over the US Thanksgiving holiday, which although not a holiday in Austria, nevertheless was celebrated by Prof. Zeilinger who had a Thanksgiving

dinner at his house with his family, his students and colleagues, and me. Perhaps the Austrian spin on it, instead of turkey, Zeilinger cooked a goose. During my first 2 days at the university, I was politely given tours of the Zeilinger laboratories and attended a 10:00 a.m. brunch meeting with his group where they served what looked like Weisswurst (veal and pork) sausages, which are a staple at breakfast in Bavaria. After that, I was left to my own devices and found John Sipe wandering around in the basement dungeon looking for another theorist to collaborate with. We broke into an unused and unheated classroom and worked out the details of this paper there over several days.

20. See "Quantum Computing Using Rotational Modes of Two Polar Molecules" by K. Mishima and K. Yamashita in *Chemical Physics*, Volume 361 (June 30, 2009), pages 106–117: http://www.sciencedirect.com/science/article/pii/S0301010409001724.

21. A nice review on optical quantum memories, including this EIT approach, is "Optical Quantum Memory" by Alexander I. Lvovsky, Barry C. Sanders, and Wolfgang Tittel in *Nature Photonics*, Volume 3 (December 2009), pages 706–714: http://www.nature.com/nphoton/journal/v3/n12/abs/nphoton.2009.231.html. The free preprint version is available here: http://arxiv.org/abs/1002.4659.

22. The more technical and historical theory reference is "Quantum Computation with Cold Trapped Ions" by J. Ignacio Cirac and Peter Zoller in *Physical Review Letters*, Volume 74 (May 15, 1995), pages 4091–4094: http://link.aps.org/doi/10.1103/PhysRevLett.74.4091. (This paper has over 1000 journal citations.) See also "Demonstration of a Fundamental Quantum Logic Gate" by Christopher Monroe, D.M. Meekhof, B.E. King, Wayne M. Itano, and David J. Wineland in *Physical Review Letters*, Volume 75 (December 18, 1995), http://link.aps.org/doi/10.1103/PhysRevLett.75.4714. (This paper has over 500 journal citations. Anything over 100 is considered to be a "classic" by the American Physical Society.) The 14-qubit entanglement is in "14-Qubit Entanglement: Creation and Coherence" by Thomas Monz, Philipp Schindler, Julio T. Barreiro, Michael Chwalla, Daniel Nigg, William A. Coish, Maximilian Harlander, Wolfgang Haensel, Markus Hennrich, and Rainer Blatt, in *Physical Review Letters*, Volume 106 (March 31, 2011), article number 130506: http://link.aps.org/doi/10.1103/PhysRevLett.106.130506, with the free preprint here: http://arxiv.org/abs/1009.6126.

23. Although making regular arrays of identical dots still poses a challenge, people have begun thinking about scaling up these and other schemes for quantum computation and considering the "architecture" of the future quantum computers. See, for example, "A Layered Architecture for Quantum Computing Using Quantum Dots" by N. Cody Jones, Rodney Van Meter, Austin G. Fowler, Peter L. McMahon, Jungsang Kim, Thaddeus D. Ladd, and Yoshihisa Yamamoto in Physical Review X, Volume 2 (July 31, 2012), article number: 031007: http://link.aps.org/doi/10.1103/PhysRevX.2.031007, with the free reprint here: http://arxiv.org/abs/1010.5022.

24. See, for example, "Quantum Hall Fabry–Pérot Interferometer: Logic Gate Responses" by S. Bellucci and P. Onorato in the *Journal of Applied Physics*, Volume 108 (2010), article number 033710: http://link.aip.org/link/doi/10.1063/1.3457357.

25. See "Making a Quantum Computer Using Neutral Atoms" by Tetsuya Mukai in the *NTT Technical Review* (July 10, 2011), https://www.ntt-review.jp/archive/ntttechnical.php?contents=ntr200801sp7.html.

26. See "Quantum Computer" in *NEC Innovative Engine* (July 12, 2011), http://www.nec.co.jp/rd/en/innovative/quantum/top.html.

27. See "Quantum Manipulations of Small Josephson Junctions" by Alexander Shnirman, Gerd Schön, and Ziv Hermon in *Physical Review Letters*, Volume 79 (1997), pages 2371–2374, http://link.aps.org/doi/10.1103/PhysRevLett.79.2371.

28. When looking at dust particles sparkling in sunlight coming in through your living room window, the smallest such particle a human eye can resolve is around 50 microns. Typical flux qubits are about 10 microns on a side.

29. See "Quantum Computing with Electrons Floating on Liquid Helium" by P.M. Platzman and M.I. Dykman in *Science Magazine*, Volume 284 (June 18, 1999), pages 1967–1969, http://www.sciencemag.org/content/284/5422/1967.abstract.

30. See "Solid State Quantum Computing" (IBM Research, August 2, 2011), http://www.research.ibm.com/ss_computing.

31. See "The Other ARPA" in *Aviation Week* (April 27, 2007), http://aviationweek.typepad.com/ares/2007/04/the_other_arpa.html.

32. See "ARDA Quantum Information Science and Technology Roadmapping Project," operated by the Los Alamos National Security LLC for the Department of Energy (July 19, 2011), http://qist.lanl.gov/.

33. Under the Clinton administration, the robotic exploration of the solar system was a big priority and JPL did very well as it has a history and charter for building and flying robotic spacecraft. Under the Bush administration, the goals changed to focus on the manned exploration of the Moon and Mars. Because by its charter (turf wars) JPL does not fly manned missions, the result was that much of the NASA funding for JPL shifted to other NASA centers that did do manned spaceflight. The joke at JPL in 2004, after massive Bush administration–induced budget cuts and the resultant 10% workforce layoffs (coincidently the year I left for Louisiana State University), was that JPL would be delighted to send a man to Mars, provided that man was George W. Bush in a one-way rocket.

34. See "Enrico Fermi and the Chain Reaction," an exhibition in the Department of Special Collections (The University of Chicago Library, July 31, 2011), http://guides.lib.uchicago.edu/fermi.

35. I realize that the acronym OXR does not really make much sense, as the three agencies do not even follow this pattern, but it is commonly used. See "What Does OXR Stand For" in *Acronyms and Abbreviations* (The Free Online Dictionary, July 31, 2011), http://acronyms.thefreedictionary.com/OXR.

36. In the United States, it is common for research active universities, such as my own (Louisiana State University), to fund only 9 months of a professor's full-time salary, coincident with the 9-month academic year September through April in most cases. Professors have three options for the remaining three summer months of salary. They can just take leave without pay (not vacation) and survive on their 9-month academic salary. Alternatively, they can volunteer to teach summer school classes (if there are any) or cover their summer salary from their research grants (if they have any). Only the 9-month academic year salary is guaranteed, and that salary is loosely allocated 50% for teaching and 50% for doing research or administration. Tenured professors at Louisiana State University do not get any paid vacation, which was a bit of a shock for me to learn, coming here from 15 years as a government employee or government contractor.

37. As far as I can tell, Senator Al Gore was not involved with the creation of ARPANET.

38. Sometimes, DARPA's goal of funding high-risk and high-payoff research can lead to problems. In the 2000s, DARPA spent tens of millions of dollars on the development of an "imaginary" weapon, the so-called hafnium bomb. This project is delightfully documented in the titillating book *Imaginary Weapons: A Journey Through the Pentagon's Scientific Underworld* by Sharon Weinberger (Nation Books, New York, 2006), http://www.worldcat.org/oclc/68109995. Much of the tens of millions of dollars went to a Texan physicist, Carl Collins, who was carrying out nuclear weapon research with the unstable radioactive isotope of hafnium using a used dental x-ray machine at the University of Texas at Dallas. The results were published in the flagship journal *Physical Review Letters* but never reproduced by any other research group. One difficulty in reproducing the results apparently came from the fact that Collins' wife owned the company that had acquired the entire American stockpile of radioactive hafnium and wasn't sharing any with his competitors.

39. I am pretty sure the proposal was from TRW Corporation, but I cannot remember for sure.

40. See "Quantum Mechanics Helps in Searching for a Needle in a Haystack" by Lov K. Grover in *Physical Review Letters*, Volume 79 (1997), pages 325–328, http://link.aps.org/doi/10.1103/PhysRevLett.79.325.

41. See "Quantum Search Without Entanglement" by Seth Lloyd, in *Physical Review A*, Volume 61 (2000), article number 010301, http://link.aps.org/doi/10.1103/PhysRevA.61.010301.

42. See "Bulk Spin-Resonance Quantum Computation" by Isaac L. Chuang and Niel A. Gershenfeld in *Science*, Volume 275 (January 17, 1997), pages 350–356, http://www.sciencemag.org/content/275/5298/350.short. See also "Ensemble Quantum Computing by NMR Spectroscopy" by David G. Corey, Amr F. Fahmy, and Timothy F. Havel in the *Proceedings of the National Academy of Sciences* (March 4, 1997), pages 1634–1639, http://www.pnas.org/content/94/5/1634.abstract.

43. During my graduate school days at the sometimes Ur-liberal University of Colorado at Boulder, the student government passed a resolution that the students should oppose "all things nuclear" and endorse green energy such as solar power. I responded in my column "A Ray in Hilbert Space," published in the local pseudo-student newspaper, *The Colorado Day*, that the Sun was powered by nuclear reactions and that therefore solar power was in the category of "all things nuclear" and should also be opposed, perhaps by bricking up the windows in the offices of the student government.

44. See "Toward a Table Top Quantum Computer" by Yael Maguire, Edward S Boyden, and Neil Gershenfeld, in *IBM Systems Journal*, Volume 39 (2000), pages 823–839, http://ieeexplore.ieee.org/xpl/freeabs_all.jsp?arnumber=5387036. Preprint is here: http://syntheticneurobiology.org/PDFs/00.11.maguire.pdf.

45. They did not factor 15-factorial, which would be the factoring of 1307674368000, but rather they just factored 15, which is 3 times 5, but everybody was all excited about it!

46. This graphic, taken from *Web of Science* on November 8, 2011, shows the distribution by year of the over 1000 citations for the original 1997 Chuang and Gershenfeld paper proposing NMR quantum computing (see note 49). The citation peak occurs

in 2000 and then begins dropping after that, indicating the decline in the field. The government funding for NMR quantum computing has followed a similar curve, which can be fit to a black-body spectrum curve. Like the black-body curve, there is a point in the past, 1997, when publications on and funding for NMR quantum computing were zero. There will never be a point in the future where both are zero ever again, but the curve will drop exponentially close to zero as time goes on.

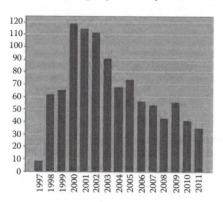

47. At the first conference I attended where Warren S. Warren was a speaker, the conference program listed, "Warren Warren (Duke University)" and I assumed it was a typo.

48. See "The Usefulness of NMR Quantum Computing" in *Science*, Volume 277 (September 12, 1997), pages 1688–1690, http://www.sciencemag.org/content/277/5332/1688.full.

49. See "12-Qubits Reached in Quantum Information Quest" in *Science Daily* (May 8, 2006), http://www.sciencedaily.com/releases/2006/05/060508164700.htm.

50. The Grover search was carried out on a four-item database, encoded in two spins (hydrogen and carbon nuclei) in a chloroform molecule. (Zzzzzzz?) See "Experimental Implementation of Fast Quantum Searching" by Isaac L. Chuang, Neil Gershenfeld, and Mark Kubinec in *Physical Review Letters*, Volume 80 (1998), pages 3408–3411, http://link.aps.org/doi/10.1103/PhysRevLett.80.3408. The Shor factoring of 15 into its two composite primes (5 and 3) was carried out on seven nuclear spins in a perfluorobutadienyl–iron complex molecule ("Borzhemoi! This I know from nothing!"—Tom Leher). See "Experimental Realization of Shor's Quantum Factoring Algorithm Using Nuclear Magnetic Resonance" by Lieven M.K. Vandersypen, Matthias Steffen, Gregory Breyta, Costantino S. Yannoni, Mark H. Sherwood, and Isaac L. Chuang in *Nature*, Volume 414 (2001), 883–887 (cited over 500 times), http://www.nature.com/nature/journal/v414/n6866/full/414883a.html.

51. See "Separability of Very Noisy Mixed States and Implications for NMR Quantum Computing" by Samuel L. Braunstein, Carlton M. Caves, Richard Jozsa, Noah Linden, Sandu Popescu, and Rudiger Schack in *Physical Review Letters*, Volume 83 (1999), pages 1054–1057, http://link.aps.org/doi/10.1103/PhysRevLett.83.1054.

52. Samuel Braunstein, private email communication (November 8, 2011).

53. I may have lost my ears to gangrenous frostbite but I still maintained my pride. See "Photos with Francis McDormand" in The Internet Movie Database (November 11, 2011), http://www.imdb.com/media/rm4265706496/tt0116282.

54. See "Local Realistic Model for the Dynamics of Bulk-Ensemble NMR Information Processing" by Nicolas C. Menicucci and Carlton M. Caves in *Physical Review Letters*, Volume 88 (2002), article number 167901, http://prl.aps.org/abstract/PRL/v88/i16/e167901.

55. See "Experimental Detection of NonClassical Correlations in Mixed State Quantum Computation" in *Physical Review A*, Volume 84, article number 044302 (2011), http://pra.aps.org/abstract/PRA/v84/i4/e044302. I cannot help but note that in its heyday, new experimental results on NMR machines, claiming to be universal quantum computation, appeared in such journals as *Nature* (impact factor 31.434), *Science* (impact factor 28.103), and *Physical Review Letters* (impact factor 7.621), and now they appear in *Physical Review A* (impact factor 2.866). Here, the bigger is the impact factor, the better is the journal. Dr. Passante surmised that she might be the last student to get her PhD in the Laflamme group in NMR quantum computation. I personally think the experiment should be converted into an undergraduate teaching laboratory, but perhaps that's a bit too harsh.

56. See "Entangled Systems Just Got Bigger" in *Quantum Factory* (WordPress, November 7, 2011), http://quantumfactory.wordpress.com/2011/04/05/entangled-systems-just-got-bigger/.

57. A byte is a classical computer register consisting of eight classical bits, and by analogy, a qubyte is a quantum register consisting of eight qubits. The classical term "bit" was first introduced by the grandfather of information theory, Claude E. Shannon, in his famous 1948 paper on classical information theory as short for "binary digit" (0 or 1). The term "bit" Shannon claimed was suggested by American statistician John W. Tukey. The word "byte" was introduced in the 1960s probably as a play on the words "bit" versus "bite" since bite is just a bit bigger than a bit (see "byte" in the *Oxford English Dictionary*). American theoretical physicists Benjamin Schumacher and William Wooters coined the word "qubit" in fun as a play on the biblical unit of length called the cubit. In his 1930 book, *Principles of Quantum Mechanics*, English theoretical physicist Paul Adrien Maurice Dirac (P.A.M.) introduced the hyphenated terms q-number and c-number for "quantum number" and "classical number" although such terminology is now somewhat outdated. However, this has led to the alternative (but rare) spellings of q-bit and q-byte in analogy. In 2007, physicist N. David Mermin (where N is large) particularly championed the alternative terms "q-bit" or "qbit" in an article in *Physics Today*, wherein Mermin declared the term qubit to be "orthographically preposterous." For a summary of Mermin's view and a rebuttal, see American physicist David "The Quantum Pontiff" Bacon's blog, http://scienceblogs.com/pontiff/2007/11/qubit_qbit_qbit_or_qbert_1.php. Mermin's protestations aside, qubit is by far the preferred and most used form.

58. Okay, so maybe I am exaggerating here a bit.

59. "Shetl" is Yiddish for "small town" and typically was any small European town with a large Jewish population. It also is the acronym for "Scottish Hydro-Electric Transmission Limited," but I think Aaronson has the first definition in mind. Aaronson's postings on D-Wave are compiled here: http://www.scottaaronson.com/blog/?s=d-wave, and his best is his 2009 post and tirade titled "Hopefully My

Last D-Wave Post Ever." (It was not.) Bacon's postings are compiled here: http://dabacon.org/pontiff/?s=d-wave, and the best is his 2009 post and rebuttal titled, "In Defense of D-Wave" (curiously not titled "In D-Fense of D-Wave").

60. See "The Vazmeister Enters the Fray" on Scott Aaronson's blog (February 17, 2007), http://www.scottaaronson.com/blog/?p=204.

61. See "Lockheed Martin Buys First D-Wave Quantum Computing System" in *Kurtzweil: Accelerating Intelligence* (May 26, 2011), http://www.kurzweilai.net/lockheed-martin-buys-first-d-wave-quantum-computing-system. See also "Google Collaborates with D-Wave on Possible Quantum Image Search" by Lisa Zyga in *PHYSORG .COM* (December 15, 2009) http://www.physorg.com/news180107947.html.

62. The Laboratory for Physical Sciences, http://www.lps.umd.edu/.

63. In the field of quantum electrodynamics, the Casimir effect of two perfect mirrors' attraction to each other in completely empty space is calculated by just such an artifice. See "The Mathematics of the Casimir Effect" by Jonathan P. Dowling in *Mathematics Magazine*, Volume 62 (1989), pages 324–331, http://phys.lsu.edu/~jdowling/PHYS4112/Dowling89d.pdf. Quantum electrodynamics is not unlike the curl with the curl from the old nursery rhyme: "There was a little girl, who had a little curl ($\nabla \times \mathbf{A}$), right in the middle of her forehead. When she was good, she was very, very good, but when she was bad, she was horrid."

64. See "Quantum Leap: The World's Biggest R&D Labs Are Racing to Build a Quantum Computer. Geordie Rose Thinks He Can Beat Them. Is He Just Blowing Smoke?" by Paul Kaihla in *CNN Money* (August 1, 2004), http://money.cnn.com/magazines/business2/business2_archive/2004/08/01/377387/index.htm.

65. See *Explorations in Quantum Computing* by Colin P. Williams and Scott H. Clearwater (TELOS, 1998), http://www.worldcat.org/oclc/36252825 for the first edition. A second edition has just been released in 2010 as well, *Explorations in Quantum Computing* by Colin P. Williams (Springer, 2010), http://books.google .com/books?id=QE8S--WjIFwC&lpg=PR3&dq=colin%20williams. While there are now perhaps better books out there, Williams' book in 1998 was the first such book and I credit it to widely introducing the field to technical audiences in the late 1990s.

66. Not to be confused with Without Loss of Generality (W.L.O.G.).

67. See "Adiabatic Quantum Computation Is Equivalent to Standard Quantum Computation" by Dorit Aharonov, Wim van Dam, Julia Kempe, Zeph Landau, Seth Lloyd, and Oded Regev in *SIAM Review*, Volume 50 (2008), pages 755–787, http://dx.doi.org/10.1137/080734479.

68. Mark Heiligman, who I have known for over 10 years, has given me permission to use his full name, which is why I do not refer to him only as Mark H.

Chapter 5

More Gadgets from the Quantum Spookhouse[1]

BLESSED ARE THE CODEMAKERS

In the last chapters, we led up to the idea of building a quantum computer using entangled particles as the underlying building blocks. If the goal is to run Grover's search algorithm, then a quantum search engine probably already exists and can be purchased from D-Wave for $10 million. If, on the other hand, the goal is to have a truly universal quantum computer capable of running Shor's factoring algorithm, in order to crack a 1024-bit public key, a quantum computer with all the bells and whistles and running full error correction would

---

* Photo: "Hamlet and His Father's Ghost," by Henry Fuseli (1780–1785). William Shakespeare, *Hamlet* 1.5.

be a tera-qubyte machine, that is, 10^{12} qubytes or approximately 10^{13} qubits. The current record in an ion-trap quantum computer is around 13 qubits not 10^{13} qubits—we are 12 orders of magnitude away from a machine that would be of use for factoring such large numbers. Nevertheless, work continues apace in the study and development of ever-larger machines. A recent Intelligence Advanced Research Projects Activity program, in which I participate, proposes to estimate the resources, both quantum and classical, that a future quantum computer would need to solve a pantheon of different algorithms, not just factoring. To quote the head of the program, American scientist Mark Heiligman, of the Office of the Director of National Intelligence, "National security decisions will now be made on the basis that there is no objective physical reality."[2] But as Nobel Laureate William Phillips is fond of saying, "The chances of building a quantum computer are 50–50: and by that I mean a 50% chance in 50 years." What are we then to do in the meantime?

Quantum computing is not the only game in Hilbert space. Once Pandora's Box of quantum spookiness—filled with such pests as unreality, uncertainty, and nonlocality—is opened, there is no putting any of it back. In this chapter, I will discuss a number of ideas for exploiting quantum weirdness for a practical technology, a quantum technology that does not directly involve computing. But I should be careful. The role of quantum computing in the field of quantum technology cannot be understated. Computer scientists, physicists, chemists, and engineers have all learned a common language in the goal of understanding quantum computer science, which is the language of quantum information theory with its qubits and entanglement and other strange beasties. That language is a Rosetta Stone that has allowed all these folks from these disparate fields to communicate in a common language, that of quantum information theory, and that common language has given voice to a number of new potential technologies that could be terribly useful for things other than computing. These quantum technologies are the focus of this chapter, and the premier near-term technology of quantum cryptography is the focus of this section.

Curiously, the first proposal for quantum cryptography was made in 1984, about the same time Feynman was ruminating on the uncomputability of certain physics problems—think thulium!—and proposed the idea of the quantum computer. Quantum cryptography is often pitched as the quantum fix to what the quantum computer has broken. If public-key encryption is no longer safe, bound to fall to the factoring prowess of a future quantum computer, then it should be replaced. But replaced with what? In this chapter, we discussed an unbreakable cryptographic system, one that does rely on the hardness of factoring, the one-time-pad cipher. Who can forget the woeful tale of our two Hawaiian star-crossed lovers, 'A'ala from Ahukini Landing on the island of Kauai and Pa'ahana from Puuohala Village on Maui, desperately communicating plans for their elopement on a one-time pad. Recall that the system works by, I will use their adopted English names here for standardization, Alice and Bob, sharing two copies of a pad that contains a matrix or array of random characters. The idea is for Alice to assign one character from the pad to each letter in her message and then for Bob to use the same pad and the same assignment to decode the message. If Alice and Bob never reuse any characters on the pads, and they are sure nobody (such as an evil eavesdropper named Eve) has a copy of the pad, then the system is utterly secure and uncrackable even by a quantum computer. Why then is this system not in wide use, and why then do we use the public key (factoring is hard) system instead. It's that somebody has to keep running the pads back and forth all around the Earth between all the various Alices and Bobs while ensuring none of the Eves are making any copies.

One-time pads are typically used in the diplomatic corps where long-term utter security is paramount. In the old days, the pads were actually pads of paper with arrays of random symbols on them.[3] The pads were typically secured in a diplomatic pouch and physically carried from Washington, DC, to US embassies in a locked briefcase handcuffed to the wrist of a trusted courier. There is no way, classically, to ensure beyond a shadow of a doubt that the pad is not copied in transit, and if Eve has a copy, she will know everything. These days, we have replaced the pads with CD-ROMs and flash drives and replaced the random symbols with random sequences of zeros and ones, but the principle is the same (see Box 5.1). Somebody can always borrow your flash drive out of your picked briefcase while you are dozing in a stupor in first class from a GHB-laced martini, copy it, and there is no way to tell. The situation is worse when we move to the modern Internet. The whole point of Internet encryption is to allow for encrypted data to be transmitted securely about the planet from computer to computer. If I had some mechanism to transmit secret cryptographic keys securely about the planet on the Internet, well I would not need to use those keys to send those messages secretly, I would just use that mechanism to send the messages themselves. It is a chicken-and-egg problem. I want to transmit the keys utterly securely so, consequently, I can use the keys to transmit data utterly securely. If I have the ability to do the former, I do not really need the latter. If I cannot do the former, then I cannot do the latter either. The idea is to break the symmetry. Use a mechanism or channel to transmit the key that is different from the one you use to transmit the message. In the classical case, a trusted courier transmits the key in a secure diplomatic pouch, and then once the pads are in place, the secret message is then transmitted over an insecure telegraph line. The method for sending the key is (hopefully) much more secure than sending the message. What if we could make the method of transmitting the key utterly secure? In a classical world, we cannot; classically, Eve can always copy the pad. The trusted courier may be more secure than the telegraph but it is never absolutely secure. We need to replace the druggable

BOX 5.1   NUMERICAL IMPLEMENTATION OF THE ONE-TIME-PAD

Let us see how the one time pad communication works with zeros and ones instead of an actual pad of alphanumeric characters. Alice first generates a string of 56 random zeros and ones for the pad.

```
11110100010001000011110101011100000101011010110011101011    (PAD)
```

Alice now securely transmits a copy of the pad to Bob and prays that Eve does not make a copy. Alice then takes the letters of "ALOHA" and converts them into a binary string of 56 zeros and ones.[4]

```
11110100010001000011110101011100000101011010110011101011    (ALOHA)
```

Now, the encryption process is clock arithmetic base two or arithmetic of a clock with only 2 hour marks on it instead of 12. In this arithmetic, we have a very simple rule when adding columns of numbers: $0 + 0 = 0$, $0 + 1 = 1 + 0 = 1$, $1 + 1 = 0$. The only new rule is that $1 + 1 = 0$ just like when the odometer on your car rolls over, $999,999,999 + 1 = 0$ (time to sell it) or on an ordinary clock $12 + 1 = 1$. Following the new rule, $1 + 1 = 0$, Alice adds PAD to ALOHA following the binary clock arithmetic to get the encrypted message MSG.

```
11110100010001000011110101011100000101011010110011101011+  (ALOHA)
01000001010011000100111101001000010000010000110100001010 =  (PAD)
10110101000010000111001000010100010101001010000111100001    (MSG)
```

It is easy to see, column by column, that the two-hour-hand clock arithmetic holds $0 + 0 = 0$, $0 + 1 = 1 + 0 = 1$, $1 + 1 = 0$. This gives us our encrypted message.

```
10110101000010000111001000010100010101001010000111100001    (MSG)
```

If the pad contains a true random sequence of zeros and ones, then this message is unbreakable, even on a quantum computer, for to break it a hacker would need to spot some pattern, but a string of truly random numbers has no pattern. That is the trick. If you add a message to a random string, the encrypted message is just as random looking as the pad. So long as there are only two pads (and you don't reuse the pads), you are safe. Alice transmits the message to Bob and then Bob just does binary clock addition one more time, adding the encrypted message to the random numbers on his copy of the pad. If the numbers on his pad are the same as on Alice's pad, he gets

```
10110101000010000111001000010100010101001010000111100001+  (MSG)
01000001010011000100111101001000010000010000110100001010 =  (PAD)
11110100010001000011110101011100000101011010110011101011    (ALOHA)
```

Comparing Alice's input message of zeros and ones for ALOHA to the output message of Bob for ALOHA, we see they are same. (The normal font–bold face coding is to help you see the pattern without having to check each zero and one.) This is the kind of operation a computer can do easily. Once again, it all comes down to the security of the pad. If I have an utterly secure way to transmit the pad over the Internet without anybody reading it or copying it, why do I not just use that same and utterly secure way to just send the message and forget about the fraking pad? This conundrum was solved by switching from one-time pads to public-key encryption where no secret key transmission is necessary but public-key transmission can now be hacked by a quantum computer, so back to the unbreakable pad. How can I break this vicious cycle? For the one-time-pad cryptography, I must solve the chicken-and-egg problem. Find a way to transmit a string of random zeros and ones utterly secretly, with insurance that nobody is copying them, and then use the shared random zeros and ones for my one-time pad for Alice and Bob to communicate in secret.

diplomat with Heisenberg's ghost—the ghost that cannot be bribed, drugged, or possibly even seduced.

Things are different in the quantum world. We have the elements of uncertainty, unreality, and nonlocality at our disposal. What can we do with them? In 1985, our old friend, IBM computer scientist Charles Bennett, as well as his colleague, Canadian computer scientist Gilles Brassard, invented quantum cryptography, known to the purists as quantum key distribution. It is typical in the cryptography community to name such protocols by the initials of the inventors followed by the year of the invention; hence, this protocol is called BB84. It makes use of quantum uncertainty and quantum unreality but not quantum nonlocality. Closely related to uncertainty and unreality is something called the quantum no-cloning theorem—it is impossible to make a perfect copy of an unknown quantum state. You may clone a marigold, a sheep, or a bounty hunter (named Jango Fett), but not a quantum state. This theorem is most easily explained by invoking Heisenberg's Uncertainty Principle.

Suppose Eve wishes to make a copy of one paper page of a classical document using an ordinary pocket-sized xerographic copying machine. And suppose that page contains a sequence of random zeros and ones that Alice and Bob are intending to use for their one-time-pad communication. (In the modern "pads," the plaintext message to be transmitted is converted to zeros and ones and the pad itself consists of random zeros and ones.) Eve drugs the courier and then just makes a copy of the pad, slips the original back in the stuporous attaché's attaché case, Alice and Bob start sending messages, and Eve uses her copy of the pad to read everything. The process of copying the pad hardly affects the pad at all. Maybe it bleaches the ink a bit, but if Eve uses a low-light copier, Alice and Bob will never know. Eve can copy classical data without affecting the data.

Back to the xerographic copying machine. So instead of sending a pad of paper with zeros and ones, let us suppose Alice and Bob decide to encode their zeros and ones in the spins of a bunch of ions, say 56 of them in an ion trap, with spin up a zero and spin down a one. Alice makes two traps, with identical spins in a random sequence of zeros and ones, carefully prepared with laser pulses, and then sends one of the traps to Bob in the attaché case of the untrustworthy attaché. Once again, Eve drugs him, slips the ion trap out of the attaché case, and attempts to copy it. But now, something goes horribly wrong. The photons from the copy machine bounce off the ions and completely scramble the spins from photon–ion collisions that collapse the wave function of the ions. Quantum uncertainty implies you cannot make a copy of these things without disturbing them. The photons that were meant to produce a copy will in fact produce only a picture of random noise in the copying machine.

In a panic, Eve slips the ion trap back into the attaché case of the stuporous attaché and hopes for the best. But now, we note two very important things in this quantum setup. First, Eve is unable to get a copy of the pad, and second, she has completely screwed up Bob's copy of the pad! This is exactly what we want. Classically, Eve gets a copy of the paper pad and Alice and Bob do not know— worst-case scenario—Alice and Bob communicate and Eve reads everything unbeknownst to them. Now, in the quantum setup with the ion pads, Eve gets nothing and Bob and Alice can tell they have been hacked! When the hacked ion-trap pad arrives at Bob's embassy in Burkina Faso, Bob can read out the 56 zeros and ones on the ion-trap pad (with laser pulses) and email the result to Alice as a test for hacking. If the zeros and ones are the same on both ion-trap pads, then they know Eve has not tried anything. If there is no relationship between Alice and Bob's strings of zeros and ones, they can conclude that Eve tried to make a copy and then not only do they not use those pads but they send an undercover air marshal on the next flight to watch out for Eve slipping a Mickey to the attaché.

Now, the astute reader may point out that instead of using a clumsy Xerox machine to copy the ion-trap pad, Eve should instead be more careful and use the same laser pulse system Bob has to read the spins, write down the zeros and ones she gets, then use the same laser pulse system Alice used to encode them in the first place to reprogram the ions, and then forward the trap on to Bob and then nobody will be the wiser. Eve can do this only if she knows in advance that Alice and Bob have agreed to encode zeros as spin up and ones as spin down in the ions. To thwart such a woman in the middle attack, Alice and Bob must resort to true quantum trickery and exploit not only quantum uncertainty but also quantum unreality and even quantum nonlocality.

The BB84 protocol exploits quantum uncertainty and unreality. To keep Eve from copying the data with laser beams, they need to use a CAT gate. In the original scheme where Alice and Bob always encode a zero as the quantum spin-up state $|\uparrow\rangle$ and a one as the quantum spin-down state $|\downarrow\rangle$, using the up–down spin encoding, if Eve knows this is the encoding, she can deterministically extract all the zeros and ones in the pad by deploying the same readout mechanism that Bob uses with his laser pulses. Eve gets the pad. Then, she deploys Alice's encoding scheme to write the zeros and ones from her now classical copy of the pad back into the ions with a duplicate of Alice's laser pulse system. Then, she slips the ion trap back into the attaché case and Bob gets an identical copy of what Alice prepared and extracts the pad data, and neither Alice nor Bob know that Eve has a copy. There is no improvement over the classical pad and the eavesdropping Xerox machine.

To fix it up, what Alice does during the encoding process is to randomly, with a 50–50 probability, decide to encode a zero as the quantum spin-up state $|\uparrow\rangle$ and a one as the quantum spin-down state $|\downarrow\rangle$, or to decide to encode a zero as the quantum spin-right state $|\rightarrow\rangle = |\uparrow\rangle + |\downarrow\rangle$ and a one as the quantum spin-left state $|\leftarrow\rangle = |\uparrow\rangle - |\downarrow\rangle$. She does this by flipping a coin and choosing the up–down (heads) versus the right–left encoding (tails) for each random zero and one in her pad. The right–left encoding is just the original up–down with a CAT gate applied. For decoding, Bob does the same thing. He also flips a coin for each of the 56 spins in the trap and randomly chooses to read out each spin in the up–down direction (heads) or the right–left direction (tails). See Table 5.1 to see a sample run using the first part of the pad from above.

What happens now is that Alice's pad will agree with Bob's pad only when they accidentally choose the same direction to prepare and to measure. That occurs when they both flips heads for a particular ion or both flip tails. The probability of both getting heads is $1/2 \times 1/2 = 1/4$ (25%) and that of both getting tails is also $1/2 \times 1/2 = 1/4$ (25%). Thus, the probability of both getting heads or both getting tails is 25% + 25% = 50%. Hence, 50% of the time they agree and then Alice's zero written in will be Bob's zero read out, and Alice's one written in will be Bob's one read out. For those cases, they share zeros and ones and can use that subset of the data for the one-time pad. What happens the other 50% of the time? If Alice writes in the right–left direction (tails) but Bob reads out in the up–down direction (tails), Bob will cause the spin to collapse randomly to either up or down. That is because what Bob is getting is either the cat state

TABLE 5.1　ALICE AND BOB'S QUANTUM ENCODING IN EIGHT IONS IN THE TRAP FOR THE FIRST EIGHT BITS OF THE CLASSICAL 56-BIT PAD USING THE BB84 PROTOCOL

Alice					Bob
0	H	$\lvert\uparrow\rangle$	H	$\updownarrow$	0
1	T	$\lvert\leftarrow\rangle$	H	$\updownarrow$	1
0	H	$\lvert\uparrow\rangle$	H	$\updownarrow$	0
0	T	$\lvert\rightarrow\rangle$	H	$\updownarrow$	1
0	T	$\lvert\rightarrow\rangle$	H	$\updownarrow$	1
0	T	$\lvert\rightarrow\rangle$	T	$\leftrightarrow$	0
0	H	$\lvert\uparrow\rangle$	T	$\leftrightarrow$	0
1	T	$\lvert\leftarrow\rangle$	T	$\leftrightarrow$	1
Alice's pad	Alice's coin	Write-in	Bob's coin	Readout	Bob's pad

Note: Alice flips her coin, and if she flips heads, she writes her zero or one in the ion spin up–down direction, and if she flips tails, she writes in the right–left direction. The ion trap is then transported to Bob where, for each ion in order, Bob flips his own coin, and if he flips heads, he reads out in the up–down direction, and if he flips tails, he reads out in the right–left direction. Only if Alice and Bob randomly both flip heads or both flip tails will the directions agree and will the zeros and ones agree (white background) and those shared zeros and ones can be used for the one-time pad. If Alice flips heads and Bob flips tails (or vice versa), then there is no correlation between Alice's zeros and one and they are completely uncorrelated from Bob's (gray background) and these data points are discarded and not used for the pad. (All random numbers generated using Mathematica.)

$\lvert\rightarrow\rangle = \lvert\uparrow\rangle + \lvert\downarrow\rangle$ or the cat state $\lvert\leftarrow\rangle = \lvert\uparrow\rangle - \lvert\downarrow\rangle$, which is an equal superposition of up and down. The point is that the quantum spin has no definite direction of spin until it is measured—quantum unreality—and if the right–left cat is measured in the up–down direction, it randomly collapses to either up or down—quantum uncertainty. How does all this help thwart Eve?

The final step in the BB84 protocol is for Alice to call Bob up on an insecure phone line and tell him (or email him) the encoding she used (the ordered list of heads and tails she got from flipping her coin). In our example from Table 5.1, it is *HTHTTTHT*. She does not tell him what she encoded, her pad list of zeros and ones, but just the choice of encoding direction, up–down versus right–left. The pad should never be revealed over an insecure phone line. Bob then compares his own list of readout directions, his list of heads and tails; from Table 5.1, it is *HHHHHTTT*[5] and he then tells Alice when they both agreed (both got heads or both got tails), set here in italics. They then throw away the zeros and ones when they both did not agree (one got heads and the other tails, or vice versa, set in bold face). They then renumber the list of the italicized agreed upon events, where they are sure they both share the same list of zeros and ones and then use this list for the new pad.

This is a very key-expensive process—on average, Alice and Bob will scramble and throw out around half of their shared (but random) zeros and ones for the one-time pad. In our example in Table 5.1, only four of the original eight random bits that Alice prepared survived. What do they gain from all this? They gain a foolproof test to see if Eve has copied the pad or not! Remember that the worst-case scenario is when Eve has a copy of the pad and Alice and Bob do not know and begin using it to encrypt and transmit secret messages. How can they tell? It is the noncloning theorem at work—quantum uncertainty and unreality. "National Security decisions will now be made on the basis that there is no objective physical reality."

Now, when Eve opens the attaché's attaché case and tries to extract the zeros and ones from the ion spins in the trap, she needs to know in advance the outcome of Alice's encoding, heads for up–down and tails for left–right. Eve needs to know for sure for each ion whether to measure in the up–down or right–left direction or else she has a 50–50 chance of picking the wrong direction and then Eve causes a collapse and scrambles the key even more! It turns out that Eve's best strategy is to do exactly what she did before, measure always in the up–down direction, extract the zeros and ones, and then prepare her measured list of zeros and ones back into the ions, put the ion trap back in the attaché case, and hope nobody notices. In Table 5.2, we illustrate what happens when Eve is in the mix. By doing this, Eve will get approximately 25% of the pad key bits, enough to read a good bit of the encrypted messages and then figure out the rest. Eve will get the measurement direction right (corresponding with Alice's write-in direction) about $1/2$ the time. Thus, the probability that Alice, Bob, and Eve all use the up–down direction is $1/2 \times 1 \times 1/2 = 1/4$ or 25%. The probability that they all use the right–left direction is similarly $1/2 \times 0 \times 1/2 = 0$ or 0%. The probability that all three use up–down or all three use right–left is the sum, $25\% + 0\% = 25\%$. Eve gets 25% of the pad key bits. That is bad! But now, Alice and Bob have a sure way to notice Eve has tampered with the pad. That is good!

In events where Alice and Bob both flip tails, both use right–left, they would expect for those results to have the same number, both have zero or both have one. But Eve measures in up–down and randomly collapses the ion into a zero or one that is not related to Alice's encoded zero or one. She retransmits this in the up–down direction but Bob has tails, measures in right–left, and again gets a number unrelated to Alice's input number. *But Alice and Bob's numbers should agree when they both flipped tails.* You can see one dark-shaded row in Table 5.2 where that does not happen. Alice sent zero and Bob got one. To reveal Eve, all

TABLE 5.2   ALICE AND BOB'S QUANTUM ENCODING IN EIGHT IONS IN THE
TRAP FOR THE FIRST EIGHT BITS OF THE CLASSICAL 56-BIT PAD USING
THE BB84 PROTOCOL WITH EVE NOW TRYING TO MEASURE THE IONS IN
THE MIDDLE

Alice			Eve	Eve			Bob
0	H	$|\uparrow\rangle$	0	0	H	$\updownarrow$	0
1	T	$|\leftarrow\rangle$	0	0	H	$\updownarrow$	1
0	H	$|\uparrow\rangle$	0	0	H	$\updownarrow$	0
0	T	$|\rightarrow\rangle$	0	0	H	$\updownarrow$	1
0	T	$|\rightarrow\rangle$	1	1	H	$\updownarrow$	1
0	T	$|\rightarrow\rangle$	1	1	T	$\leftrightarrow$	1
0	H	$|\uparrow\rangle$	0	0	T	$\leftrightarrow$	0
1	T	$|\leftarrow\rangle$	0	0	T	$\leftrightarrow$	1
Alice's pad	Alice's coin	Alice writes	Eve reads	Eve's result	Bob's coin	Bob reads	Bob's pad

Note: Alice flips her coin, and if she flips heads, she writes her zero or one in the ion spin
up–down direction, and if she flips tails, she writes in the right–left direction. The
ion trap is then transported to Bob where, for each ion in order, Bob flips his own
coin, and if he flips heads, he reads out in the up–down direction, and if he flips
tails, he reads out in the right–left direction. Only if Alice and Bob randomly both
flip heads or both flip tails will the directions agree and will the zeros and ones
agree (white background) and those shared zeros and ones can be used for the one-
time pad. If Alice flips heads and Bob flips tails (or vice versa), then there is no correlation between Alice's zeros and one and they are completely uncorrelated from
Bob's (light-gray-shaded rows) and these data points are discarded and not used
for the pad. However, because Eve always measures in the up–down direction,
independent of Alice and Bob's coin flips, sometimes Eve will use up–down when
Alice and Bob are in agreement about using right–left. Eve's measurement will then
randomly scramble what Alice and Bob could verify are good shared random bits.
As indicated in the dark-gray-shaded rows where Alice and Bob both got tails, measured in right–left, they should be either both zeros or both ones. You can see in one
of these rows that Alice sent one and Bob got zero. The only way that can happen is
if Eve tried to copy the ion state. They know now not to use the pad and also to send
an air marshal on the next flight to hunt down Eve.

Alice and Bob must do is, over the insecure line, sacrifice some more key bits,
in our example, say half of the remaining good key bits. They call each other up
again and they can choose a few rows at random to compare. If Eve did not tam-
per with the ion traps but is listening in on the telephone call, this act gives her
some information. Hence, Alice and Bob agree not to use these eavesdropped

check bits for the key but rather just to test for Eve. Bob can ask Alice, choosing a row at random for example, "In row five I flipped tails; what did you flip Alice?" Alice says, "I also flipped tails." "Great," replies Bob, "then our bits should agree. What did you send Alice?" Alice replies, "I sent a zero Bob." Bob exclaims, "I got a one, Alice!" They then know for sure that the traps were tampered with in transport. If they both flipped tails and Alice sent a zero, Bob should get a zero. The only way Bob could get a one is if Eve measured and retransmitted in the up–down basis where by chance Alice transmitted and Bob received in the right–left basis. Eve randomly scrambles such bits and the scrambling turns up in the check bit telephone call. Alice and Bob know two things. The key has been compromised and an eavesdropper should be looked for in transport.

However, if Alice and Bob check a number of bits, say half of the bits where they both flipped tails or both flipped heads, and they always both get zeros or both get ones, then they can be sure as they check more and more bits that the key is secure. This procedure reduces the probability that Eve even gets one key bit to an exponentially small number. The key is nearly perfectly secure and can be used now to transmit messages in an unbreakable fashion by using the two-hour clock arithmetic rules above. The no-cloning rule states that Eve cannot copy an unknown quantum state. If Eve does not know bit by bit if Alice encoded in up–down versus left–right, each quantum state is unknown to Eve and she cannot faithfully copy and retransmit it. In this way, Eve reveals herself to Alice and Bob.

Unlike quantum computing, where the goal of building the universal quantum computer seems many years off, the quantum cryptography schemes are here and now and can be purchased commercially. The trick is not to transport key bits via ions in traps, this would be slow and cumbersome and prone to error, but rather to send the equivalent of spin up–down and spin right–left states of photons—quantum particles of light. Like the ions, the photons have an intrinsic spin. This is usually written in the form of the polarization of the photons—the direction the photon's electric field seems to be wiggling as the photon approaches you. This polarization direction is related to the spin direction. As the photon approaches, you can measure if the polarization is vertical $|V\rangle = |\updownarrow\rangle$ or horizontal $|H\rangle = |\leftrightarrow\rangle$ versus $+45°$ $|+\rangle = |\nearrow\rangle$ or $-45°$ $|-\rangle = |\nwarrow\rangle$, where the angle is measured off the vertical (see Figure 5.1). Alice and Bob, like before with the ions, agree to assign a zero to $|V\rangle = |\updownarrow\rangle$ and $|+\rangle = |\nearrow\rangle$ versus a one to $|H\rangle = |\leftrightarrow\rangle$ and $|-\rangle = |\nwarrow\rangle$. Alice randomly switches her transmissions, by flipping a coin, from the $|V\rangle = |\updownarrow\rangle = |0\rangle$ and $|H\rangle = |\leftrightarrow\rangle = |1\rangle$ directions to the $|+\rangle = |\nearrow\rangle = |0\rangle$ and $|-\rangle = |\nwarrow\rangle = |1\rangle$. Bob randomly switches his detections between these directions by flipping his own coin. With this new notation, a typical run looks like those in Tables 5.1 and 5.2 and at the end with Alice and Bob getting the identical bit zero or one each time they both randomly choose H–V or $\pm45°$. Eve is caught because her best strategy, like before, is to always measure and retransmit in

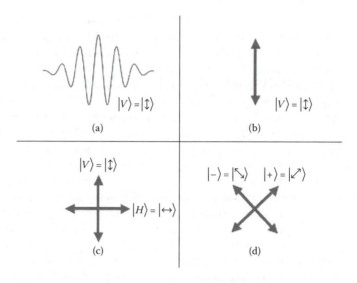

Figure 5.1 Polarized photos. In panel (a), we show a vertically polarized photon side view. The wiggles represent the oscillating electric field of the photon. In panel (b), we have the same vertically polarized photon front view as the photon moves toward you. The double arrow represents the direction of the electric field oscillations. In panel (c), we show both vertical and horizontal polarized photons, zero and one, front view. In panel (d), we show both +45° and −45° polarization photons, zero and one, respectively.

the H–V directions. In about a quarter of the events, Alice and Bob will both flip tails and be using the ±45° directions while Eve is using the H–V directions. In all those events, Bob's outcome on average will be completely uncorrelated with Alice's transmission and again Eve is caught, when Alice and Bob compare notes, and the copied key is discarded, and a secret agent is sent out to arrest Eve.

Whether it be photons or ions, the quantum unreality enters because, in the case of photons, the photon polarization state has no objective physical reality. This can be seen by writing the +45° photon as the cat state, $|\nearrow\rangle = |\updownarrow\rangle + |\leftrightarrow\rangle$. Is the +45° photon "really" a single +45° photon or the coherent superposition of a vertical and horizontal polarized photon? In classical theory, it is either +45° or vertical or horizontal. It is not, classically, all these three things at once. In quantum theory, a +45° photon is simultaneously a vertical and a horizontal photon in a cat state. It has no true direction until a measurement is made and then it collapses to either vertical or horizontal. If Alice sends in the ±45° directions and Bob receives in ±45° directions, then no collapse occurs and they share a zero (+45°) or a one (−45°) for the one-time-pad key. However, if Eve measures and retransmits in the vertical–horizontal directions, then the photon collapses randomly into either vertical or horizontal with a 50–50 probability. There is no connection to what Alice sent and what Bob got. Alice's random numbers will disagree with Bob's, on average, and by comparing a few of their numbers, they can tell there has been an eavesdropping attempt. The two Einstein-loathed properties of quantum mechanics come into play: unreality (the state has no direction until it is measured) and uncertainty (the photon collapses randomly if measured in the wrong direction).

The BB84 one-time-pad quantum key distribution protocol uses two of the three properties Einstein so hated about quantum mechanics, unreality and uncertainty, but what about the third, which bothered him most, nonlocality—spooky action at a distance. In 1991, Polish physicist Artur Ekert proposed a quantum cryptography system that exploits in addition just that aspect of quantum weirdness, spooky action at a distance. The Ekert protocol, called affectionately E91 from the name–date naming system for cryptographic protocols, uses shared remote quantum entanglement to generate a random shared key, shared by Alice and Bob, with a mechanism to see if the key has been copied and to eliminate the eavesdropper Eave. Because we have introduced the photon scheme for BB84, and because photon schemes have been implemented experimentally and are in common use (and available commercially), we go right to them.

The E91 protocol exploits the same setup as the Clauser and Aspect Bell-test experiments (see Chapter 2). A central source of entangled photons, of the form $|\updownarrow\rangle_A |\leftrightarrow\rangle_B + |\leftrightarrow\rangle_A |\updownarrow\rangle_B$, as discussed in Chapter 2, sends the photon pairs out either through free space or along optical communications fibers. Typically, the source is labeled "C" for "Charlie" and the photon flying to the left goes to Alice and the photon to the right goes to Bob.[6] If Alice and Bob both measure their photons in the vertical–horizontal directions, as in Chapter 2, then the quantum collapse of the wave function always gives $|\updownarrow\rangle_A |\leftrightarrow\rangle_B$ or $|\leftrightarrow\rangle_A |\updownarrow\rangle_B$. That is, when they both use H–V, when Alice gets a vertical photon, Bob gets a horizontal, or when Alice gets a horizontal, Bob gets a vertical. The results are called anti-correlated (opposite) provided they both measure in the H–V direction. In a slight change up from BB84, Alice and Bob agree that when Alice gets an H, she calls it a zero, but when Bob gets an H, he calls it a one, and vice versa. This is to account for the fact that the photons are anti-correlated (opposites) in the E91 scheme and are correlated (the same) in BB84. At the end of a run, Alice and Bob share an identical sequence of zeros and ones that they can use for the one-time pad to communicate in absolute secrecy.

What can go wrong? Well, an eavesdropper Eve could intercept both photons in transit and then measure their state H–V or V-H and then learn the one-time pad. She would just retransmit the states to Alice and Bob and they would not be the wiser. However, Eve cannot measure the two-photon state without irreversibly disrupting it. Therein lies the way to catch her! If Eve measures the entangled state $|\updownarrow\rangle_A |\leftrightarrow\rangle_B + |\leftrightarrow\rangle_A |\updownarrow\rangle_B$ in the H–V direction, she will randomly (with a 50–50 probability) get $|\updownarrow\rangle_A |\leftrightarrow\rangle_B$ or $|\leftrightarrow\rangle_A |\updownarrow\rangle_B$, just as Alice and Bob would have gotten. However, Eve has collapsed the entangled state and the outcome she gets, $|\updownarrow\rangle_A |\leftrightarrow\rangle_B$ or $|\leftrightarrow\rangle_A |\updownarrow\rangle_B$, is no longer entangled. The spooky, super strong, nonlocal correlations stored in $|\updownarrow\rangle_A |\leftrightarrow\rangle_B + |\leftrightarrow\rangle_A |\updownarrow\rangle_B$ are gone and only the paltry classical correlations stored in $|\updownarrow\rangle_A |\leftrightarrow\rangle_B$ or $|\leftrightarrow\rangle_A |\updownarrow\rangle_B$ are all that is left. As before, Eve's best strategy is to always measure in the H–V directions and then if she gets $|\updownarrow\rangle_A |\leftrightarrow\rangle_B$, she records Alice: zero; Bob: zero and then retransmits $|\updownarrow\rangle_A |\leftrightarrow\rangle_B$ to Alice and Bob. If Eve gets $|\leftrightarrow\rangle_A |\updownarrow\rangle_B$, she records Alice: one; Bob: one and then retransmits $|\leftrightarrow\rangle_A |\updownarrow\rangle_B$. At the end of many photon runs, Alice, Bob, and Eve will share the same list of random numbers and can all construct the same one-time pad and Eve can read all the encrypted communications and all is then lost.

The trick is for Alice and Bob, similar as in BB84, to switch randomly between the H–V direction measurement and the ±45° measurement. This works because there is no way to tell if Charlie sends a $|\updownarrow\rangle_A |\leftrightarrow\rangle_B + |\leftrightarrow\rangle_A |\updownarrow\rangle_B$ that it is not also a $|\nearrow\rangle_A |\nwarrow\rangle_B + |\nwarrow\rangle_A |\nearrow\rangle_B$. In fact, this is the infinite Library of Babel at work! Every possible anti-correlated two-photon state is stored in $|\updownarrow\rangle_A |\leftrightarrow\rangle_B + |\leftrightarrow\rangle_A |\updownarrow\rangle_B$. Hence, if Alice and Bob choose to measure in the ±45° directions, then the same entangled state collapses to $|\nearrow\rangle_A |\nwarrow\rangle_B$ or $|\nwarrow\rangle_A |\nearrow\rangle_B$. As before, the outcomes are always anti-correlated. If they get $|\nearrow\rangle_A |\nwarrow\rangle_B$, they agree to both write down a one. If they both get $|\nwarrow\rangle_A |\nearrow\rangle_B$, then they agree to both write down a zero. Charlie does not have to randomly change the orientations of the photons at the source, as Alice had to do in BB84; all possible opposite orientations are already built into the quantum entangled state $|\updownarrow\rangle_A |\leftrightarrow\rangle_B + |\leftrightarrow\rangle_A |\updownarrow\rangle_B$.

So how does this defeat Eve? Again, the trick is for Alice and Bob to switch randomly from the H–V measurement to the ±45° measurement. They do this by flipping coins again just before the photons arrive. If Alice and Bob get heads, they both measure in the H–V direction, and if Alice and Bob get tails, they both measure in the ±45°. As in BB84, all such measurements where they agree give useable key bits. The heads–heads gives zero–zero and tails–tails gives one–one. They will agree again about 50% of the time and so one in two transmitted entangled photons, on average, will give a usable key. As in BB84, they call each other up on their cell phones and go down the list, highlighting the events when they both flipped heads or both flipped tails and discarding the events where one flipped heads and the other flipped tails, or vice versa. The security relies on that they only share what the coin tossed showed but not what they measured (zero or one) for any bits that are used for pad key bits. See Table 5.3 for an illustration.

So what can Eve do to thwart this scheme? She can intercept the entangled photons that Charlie is sending, measure their states, and then resend them to Alice and Bob. But Eve cannot do this without irrevocably altering the

TABLE 5.3   ALICE AND BOB'S QUANTUM ENCODING AND DECODING IN THE FIRST EIGHT ENTANGLED PHOTONS OF THE CLASSICAL 56-BIT PAD USING THE E91 PROTOCOL

Alice						Bob
0	H	$\vert\updownarrow\rangle$	$\vert\leftrightarrow\rangle$	H		0
1	T	$\vert\diagdown\rangle$	$\vert\updownarrow\rangle$	H		1
0	H	$\vert\updownarrow\rangle$	$\vert\leftrightarrow\rangle$	H		0
0	T	$\vert\diagup\rangle$	$\vert\leftrightarrow\rangle$	H		1
0	T	$\vert\diagup\rangle$	$\vert\leftrightarrow\rangle$	H		1
0	T	$\vert\diagup\rangle$	$\vert\diagdown\rangle$	T		0
0	H	$\vert\updownarrow\rangle$	$\vert\leftrightarrow\rangle$	T		0
1	T	$\vert\diagup\rangle$	$\vert\diagdown\rangle$	T		1
Alice out	Alice's coin	Alice's photon	Bob's photon	Bob's coin		Bob out

Note: Charlie transmits entangled photon pairs to Alice and Bob. Alice flips her coin, and if she flips heads, she writes her zero or one if her photon is in the vertical or horizontal direction, and if she flips tails, she writes a zero or one corresponding to the +45° direction or the −45° direction. Bob does the reverse, because the results are anti-correlated. Bob flips his coin, and if he flips heads, he writes a one or zero if his photon is in the vertical or horizontal direction, and if he flips tails, he writes a zero or one corresponding to the −45° direction or the +45° direction, respectively. In the light-gray-shaded rows, they both either flip heads or both either flip tails and the zeros and ones are perfectly correlated and can be used for the one-time pad. If Alice flips heads and Bob flips tails, or vice versa, then there is no correlation between Alice's zeros and ones and Bob's zeros and ones (dark-gray-shaded rows) and these are discarded. (Occasionally, the dark-gray-shaded rows will agree but that agreement is random. It is clear in the fourth and fifth rows that Alice and Bob do not agree at least some times when they do not both flip heads or they do not both flip tails.)

entangled states, and then Alice and Bob can detect Eve's presence. Similar to the BB84 protocol, it turns out that Eve's best strategy is to pick one set of directions and always measure in those directions. In our example, Eve always chooses to measure in the H–V directions. Eve's measurement on the incoming entangled state $\vert\updownarrow\rangle_A \vert\leftrightarrow\rangle_B + \vert\leftrightarrow\rangle_A \vert\updownarrow\rangle_B$ will collapse it to either $\vert\updownarrow\rangle_A \vert\leftrightarrow\rangle_B$ or $\vert\leftrightarrow\rangle_A \vert\updownarrow\rangle_B$ with 50–50 probability. Eve then records a zero if she gets $\vert\updownarrow\rangle_A \vert\leftrightarrow\rangle_B$ or one if she gets $\vert\leftrightarrow\rangle_A \vert\updownarrow\rangle_B$ and then relays on to Alice and Bob exactly what she got, either $\vert\updownarrow\rangle_A \vert\leftrightarrow\rangle_B$ or $\vert\leftrightarrow\rangle_A \vert\updownarrow\rangle_B$. If Alice and Bob by chance both flip heads (25% of the time), then they get exactly what Eve sends. If they get $|\updownarrow\rangle_A |\leftrightarrow\rangle_B$, they write down a zero, and if they get a $|\leftrightarrow\rangle_A |\updownarrow\rangle_B$, then they write down a one, same as Eve, and in the end, Alice, Bob, and Eve share the same list of random numbers used in the one-time pad and Eve can read all of their future encrypted transmissions using her copy of the pad.

The trick is that Alice and Bob are no longer getting the original entangled states $|\updownarrow\rangle_A |\leftrightarrow\rangle_B + |\leftrightarrow\rangle_A |\updownarrow\rangle_B$ sent by Charlie but randomly they are getting either $|\updownarrow\rangle_A |\leftrightarrow\rangle_B$ or $|\leftrightarrow\rangle_A |\updownarrow\rangle_B$. This tampering is revealed on the measurements where Alice and Bob both flip tails. Eve is always transmitting in the H–V direction, but if Alice and Bob flip tails, they are measuring in the ±45° direction. There is no way Eve can know the outcome of Alice and Bob's coin flips as these flips can occur long after Eve measured and retransmitted.

As in BB84, there will be no correlation between Alice's zeros and ones and Bob's zeros and ones in the 25% of the runs where they both flipped tails. For example, if Eve sends $|\updownarrow\rangle_A |\leftrightarrow\rangle_B$, this can be written, in the language of cats, as $(|\nearrow\rangle_A + |\nwarrow\rangle_A)(|\nwarrow\rangle_B + |\nearrow\rangle_B)$ that can be simplified to $|\nearrow\rangle_A |\nwarrow\rangle_B + |\nearrow\rangle_A |\nearrow\rangle_B + |\nwarrow\rangle_A |\nwarrow\rangle_B + |\nwarrow\rangle_A |\nearrow\rangle_B$. This is unreality. The state $|\updownarrow\rangle_A |\leftrightarrow\rangle_B$ has no predetermined form until it is measured. Hence, when Alice and Bob do measure, they get, randomly, with 25% probability each, $|\nearrow\rangle_A |\nwarrow\rangle_B$, which they read off as zero–zero; $|\nearrow\rangle_A |\nearrow\rangle_B$, which they read off as zero–one; $|\nwarrow\rangle_A |\nwarrow\rangle_B$, which they read off as one–zero; or $|\nwarrow\rangle_A |\nearrow\rangle_B$, which they read off as one–one. About half the time on average they disagree! There is no correlation between Alice's zeros and Bob's ones and these cannot be used as shared random numbers for the one-time pad.

Let us compare this to what would have happened had Eve had not measured and retransmitted. Then, instead of Eve's doctored transmitted separable state $|\updownarrow\rangle_A |\leftrightarrow\rangle_B$, Alice and Bob would have gotten Charlie's unsullied entangled state, which for the ±45° measurement is best written as $|\nearrow\rangle_A |\nwarrow\rangle_B + |\nwarrow\rangle_A |\nearrow\rangle_B$. Alice and Bob's ±45° measurement produces the results, with a 50–50 chance, $|\nearrow\rangle_A |\nwarrow\rangle_B$, which they read off as zero–zero, or $|\nwarrow\rangle_A |\nearrow\rangle_B$, which they read off as one–one. With the entangled undisturbed state, they always get both zeros or both ones.

Hence, as in BB84, it is easy to check if Eve is doing this eavesdropping. Alice and Bob again call each other on the cell phone and compare at random about half the rows in Table 5.3, where they both flipped heads or both flipped tails, and now tell each other what they got, zero or one. Because they reveal both their measurement direction and their result on an unsecured phone line, these bits must not be used for secret keys. Instead, they are used to check for the presence or absence of Eve. In the absence of Eve, they should always both get the same thing, both zeros or both ones. In the presence of Eve on about half of the light-gray-shaded rows, they will disagree and can then send out a Viper attack squadron to hunt her down and eliminate her. At the very least, they know the key has been tampered with and they know not to use any of it to transmit secret messages. In the classical world, Eve can always make a copy of your key without you knowing. In the quantum world, Eve can never make a copy of your key without you knowing, if you are sufficiently tricky.

It gets even better. Eve could become devious and try all sorts of different schemes; herself randomly switching measurement directions or herself retransmitting entangled states instead of separable states. What Ekert proved in his 1991 paper was that whatever Eve did, in order for her to extract at least some of the key bits, her shenanigans—the correlations she encodes in the photons to fool Alice and Bob—can always be modeled by a classical, local, hidden variable theory. That is precisely the theory that Einstein so hoped for and that is the theory that Bell, Clauser, and Aspect showed cannot agree with quantum theory! To completely rule out any possible bag of evil tricks evil Eve might come up with, Alice and Bob need to randomly choose incoming photon pairs and instead of trying to extract a key from them, they instead run a test of the Bell formula. They crank up IBM-Charlie's pocket watch testing code (Chapter 2) and feed it in their data. Charlie's code spits out a single number, either 2.0 (classical hidden variable theory—Eve is present) or 2.83 (quantum theory—Eve is absent). They do not need to do more than that to eliminate the possibility of an eavesdropper. The final ingredient of quantum weirdness is now in the mix—nonlocality. Passing this Bell test requires, in addition to quantum uncertainty and unreality, also Einstein's bane, quantum nonlocality—*"Spukhafte Fernwirkung!"* (spooky action at a distance). The security of the key does indeed rest on the fact that there is no objective physical reality and that actions by Alice on Alpha Centauri influence measurements made by Bob on Beta Pictoris.

Unlike quantum computing, quantum cryptography is here and now. You can buy commercial off-the-shelf quantum key distribution systems from a number of companies for approximately $10,000. The most notable companies selling these are MagiQ Technologies[7] and NuCrypt[8] in the United States, QinetiQ in the United Kingdom, ID Quantique[9] in Switzerland, and QuintessenceLabs[10] in Australia. In fact, the 1994 UK demonstration (a prelude to the spin-off of QinetiQ) of quantum cryptography over 10 kilometers of fiber was one of my main motivations for going to the Army Research Office that year with the proposal to organize the 1995 workshop on Quantum Cryptography and Computing. That workshop kick-started the entire Department of Defense (DoD) program in these areas. Ten kilometers is enough to think of setting up a wide-area communications network in either Wall Street or Washington, DC. In 2004, a prototype quantum crypto system, developed by Zeilinger's group at the University of Vienna and that runs along optical fibers through hundred-year-old sewage pipes under the Danube River, was used to carry out the world's first quantum cryptography–based secure banking transaction. The Austrian system used a hybrid of the BB84 and E91 protocols with full entanglement with the photon source "Charlie" located in a central sewage pumping station on an island in the Danube.[11] In 2007, Swiss physicist Nicolas Gisin and his spin-off company, ID Quantique, used an E91 system that runs under Lake Geneva to transmit secret ballots for the elections that year in the Swiss Canton of Geneva. Single photons in the visible or near-infrared wavelength of light are quite robust things. The standard telecommunications fiber is most transparent for photons with a wavelength of 1.5 microns (six one-thousandths of an inch), and the core of the fiber is approximately 1.5 microns in diameter so such photons fit snugly. Such photons are in the infrared wavelength and not visible to the naked eye—just outside of the visible spectrum to the longer wavelength side of deepest red.[12] John Rarity and colleagues in the United Kingdom demonstrated quantum cryptography over 10 kilometers (6 miles) of optical fiber and Gisin's group in Switzerland has extended that limit to over 200 kilometers (120 miles) of fiber by using a special low-loss fiber developed by the Corning glass company.[13] (The fiber was wound up on a spool and they did not actually put out 200 kilometers of fiber under Lake Geneva, so the proof is still in the "putting.") Gisin's group claims they can extend this to approximately 300 kilometers (186 miles), which they call "inter-city" distances, provided you are in Switzerland where the cities are pretty close. The United States has its own distance scales to worry about.

There are three well-known quantum cryptography optical fiber–based local area networks (and one other less well known one) that use quantum cryptography to encrypt data transmission between several networked computers. Recall that quantum cryptography, more accurately quantum key distribution, uses the (sometimes special[14]) optical fibers and quantum states of photons to establish the one-time-pad key of shared random numbers. Once established, ordinary computers using the ordinary Internet can use that key to exchange unbreakable secret messages. The language is that the quantum key is established over the "quantum channel," and once the key is obtained, the communications take place over the "classical channel." Because the quantum channel is expensive in resources, such as time and money and special optical fibers, it is not used to send the encrypted messages but only to establish the cryptographic key. The three well-known networks are the Defense Advanced Projects Research Agency (DARPA) network in Boston (now mothballed) that was developed by BBN Technologies,[15] Harvard University, Boston University, and QinetiQ; the SECOQC network in Vienna; and the Tokyo QKD Network and SwissQuantum in Geneva.[16] There is an additional network (that I am at less liberty to discuss) that runs in a loop between various government facilities in the Washington, DC, area. It is anticipated that commercial and government use of quantum cryptography will begin moving out of the test-bed phase and into practical use over the next few years.

QUANTUM REPEATERS AND EARTH-TO-SPACE QUANTUM CRYPTOGRAPHY

There are two interesting technologies related to quantum cryptography to discuss now. First, what is a quantum repeater? Second, how does one send single photons from orbiting satellites to the Earth—and what good does that do? First, the repeater: In classical fiber optics communications, the classical bits, the zeros and ones that make up the classical internet, are not single photons traveling along optical fibers, but more robust little pulses of classical-like laser light (called coherent states of light) that travel down similar fibers. Just like the single photons, there is an upper limit to how far these classical pulses of light can go in a fiber, again on the order of tens of kilometers, before the pulse is lost or too degraded to tell if the pulse intensity represents a classical "zero" or a classical "one" bit of information. In the classical fiber links, a device called an optical repeater solves the distance problem. The early versions of the repeaters were optical–electronic relays or transponders that would measure the state of the incoming attenuated pulse (zero or one) and then fire off a new and reenergized pulse of the same bit state (zero or one). The optical–electronic devices have now mostly been replaced with all optical devices called fiber optical amplifiers, which are small laser amplifiers consisting of excited erbium atoms built right into the fiber that are triggered by the attenuated incoming pulse to add more energy (photons) to it and send it on down the fiber, all the while preserving the classical bits of data. These repeaters are typically placed every 20 kilometers or so (approximately 12 miles) along the fiber, even on long-distance undersea transmission fibers.

Immediately, we run into a problem trying to use these classical repeaters on single photons carrying quantum key bits. Whether we use the optical–electronic copy-and-forward transponder devices or the optical laser amplifier power-boosting devices, the classical optical repeater is essentially a copying machine for the laser pulses. It copies the bit state of the incoming weak pulse and transfers that state to a new outgoing strong pulse. Remember the no-cloning theorem—it is impossible to make a copy of an unknown quantum state! The no-cloning theorem has a sister in laser theory: the "no nonnoiseless amplification" theorem. One cannot build an amplifier that is noise free and that added noise (extra photons where you don't want them) destroys any quantum state trying to pass through. The same classical repeater technology cannot be used on the precious, delicate, single photons—to copy is to destroy—this is quantum uncertainty and unreality at work. The same feature that prevents Eve from copying the key bits is the bug that makes quantum cryptography useless in the face of classical repeaters. Put classical repeaters along the quantum channel and Alice and Bob's key bits will be randomly and hopelessly garbled and Alice's zeros and ones will have no correlation to Bob's zeros and ones. No shared and correlated and random zeros and ones are created and no one-time pad is generated. There is, however, a slick quantum fix to this problem of the repeater.

In the late 1990s, Austrian physicist Peter Zoller and collaborators proposed an idea for a *quantum* repeater.[17] Without going into too much detail, a quantum repeater is a somewhat complicated system that involves a source of entangled photon pairs (to be launched in opposite directions down the fiber), a quantum memory (something to store the state of the incoming key photons), and a small special-purpose quantum computer (something that does not exist yet). The DoD has spent millions on entangled photon sources and detectors, quantum memories, and few-qubit quantum computers, but nobody has (yet) assembled all these ingredients into a practical quantum repeater.

In parallel to quantum repeater development, there has been a consistent effort in the United States and Europe to specifically set up quantum cryptography links to satellites and more generally launch entangled pairs of photons into space to test quantum mechanics at large distances. The quantum theory says that Bell's inequality should be violated regardless of distance, Alpha Centauri to Beta Pictoris, but this should be tested over the largest distances we can muster, and Earth to space (medium Earth orbit) can be 1000 kilometers (600 miles). Most of the experimental tests have not actually involved satellites. They have instead involved sending single photons through the air from one ground station to another. The logic is that the Earth's atmosphere thins exponentially rapidly as you go up from the ground. Space is only a 4-hour drive away, if you could drive straight up. The International Space Station is in a 400-kilometer-high (250-mile-high) orbit, which you could reach in 4 hours driving at 100 kilometers per hour (approximately 60 miles per hour). But because of the thinning of the atmosphere, a photon traveling parallel to the ground over just a few tens of kilometers encounters the same amount of air (and probability of absorption) as a photon traveling hundreds of kilometers straight up (or straight down).

Most space-to-Earth quantum cryptography schemes, using the BB84 protocol, propose to put Alice and her single-photon machine gun on the satellite and Bob with a telescope and single-photon detector on the ground. This is the best configuration for two reasons. The first involves the turbulent fluctuating pockets of air in the atmosphere that cause a star to twinkle when you look up on a clear night. The star is not twinkling but pockets of air moving randomly in the atmosphere act like little magnifying glasses and cause the star to look like it is getting bigger and smaller and shifting around or "twinkling." This effect is most pronounced on the starlight when the observer is nearest to the ground where the air is the thickest. Hence, if you send single photons up from the ground, the twinkling effect will cause them to easily miss their target, a telescope on an orbiting satellite. An analogy would be putting in the game of golf. The golf ball is the photon. The putter is Alice. Bob with his telescope is the 18th hole. The fluctuating atmosphere corresponds to small bumps on an otherwise flat and well-manicured putting green. If Alice's golf ball hits such a small bump just after she putts, the deflection of the ball so early on its path will end up sending it way off course and miss the hole entirely. If the ball hits the bump just before getting to the hole, there is a much better chance it will not deflect much and go right in. Hence, the best configuration is to arrange for most of the bumps to be near the hole (Bob's telescope) and not near the source (Alice's putter). Because for free-space quantum cryptography the "bumpiest" air is on the ground, that is where Bob and his telescope should be.

A second consideration is that even in empty space, a single photon, once shot through space, tends to spread out or grow in size in right angles to its motion as it moves. This effect is called diffraction. You can see it yourself by shining a green laser pointer on a wall in your room and then on a distant building out your window hundreds of meters (hundreds of yards) away. The opening of the laser pointer is only a few millimeters wide (thousandths of an inch), the spot on the room wall will be maybe a centimeter wide (fraction of an inch), but the spot on the far building will be many centimeters wide (many inches). As the beam travels farther away from the pointer, it spreads out laterally. All the photons in the beam are each spreading individually as the beam travels. The precious little single photons become very chubby out at distances of thousands of kilometers. Shine your laser pointer straight up, and by the time it gets to space, the beam will be many meters (yards) wide. Ditto if you start in space and shine the pointer straight down. You will need a very large telescope to be sure to catch all of the photon. Large telescopes are very heavy. Heavy objects are very expensive to launch into orbit. Better put Bob and his heavy telescope on the ground where it will be much cheaper to install it.

Earth-to-Earth single-photon quantum cryptography has been carried out by a number of groups since the 1989 prototype experiment implemented by Charles Bennett and American physicist John Smolin at the IBM Thomas J. Watson Research Center. This first experiment was carried out in "free space" through the air on a laboratory bench where Alice's source and Bob's detector were located approximately 30 centimeters (1 foot) apart. The light source was a light-emitting diode that every once in a while would emit a yellowish green photon of wavelength 550 nanometers (approximately two hundred thousandths of an inch). The experiment was a tour de force not in the least because both Bennett and Smolin were theoretical physicists without much experience in the laboratory. To quote Smolin, "Neither Charlie nor I knew much about building anything, but we knew enough to be dangerous."[18] I was aware of both the theory and this experiment around 1990, when I heard Bennett give a talk about it, but I was not much impressed with an experiment that sent unbreakable cryptographic key over a distance of 30 centimeters. "Not very practical,"

I thought. I changed my tune only in 1994 when the distance was extended by Rarity and colleagues to 10 kilometers (approximately 6 miles). Also, the 1991 Ekert scheme using quantum entanglement and spooky nonlocal action at a distance met my minimum cuteness criterion.

The blizzard of DoD interest in all things related to quantum information processing, as well as the formal opening of the DoD wallets, as well as the interest in rekeying spy satellites on the fly, which led to a flurry of free-space quantum cryptography experiments in the mid-1990s, and the entire field rapidly began to snowball. American physicists James Franson and Brian Jacobs carried out the first free-space experiment (over a distance of more than just a foot) in 1996 with their results published as "Quantum Cryptography in Free Space," where the two intrepid experimentalists demonstrated the BB84 protocol using attenuated laser pulses with less than one photon on average per pulse.[19] Their first demo was in a fully lit hallway, at the Johns Hopkins University, Applied Physics Laboratory, over a distance of 150 meters (164 yards). I can imagine with each run Franson and Jacobs yelling "Fire in the hole!" as their Alice launched her single yellow photons down the hall.[20] Even more impressive, they were able to demonstrate the system outdoors over an inter-building distance of 75 meters (82 yards) in broad daylight. The reason this is impressive is that the billions of photons from the Sun flying about in full daylight very easily swamp the single photons being fired out of Alice's single-photon machine gun. The trick is to give Bob some very selective optics that let through only a small range of photons around the golden 633-nanometer-wavelength photons being sent by Alice. Few of the photons from the Sun are in this range (and moving in the right direction) and hence the vast majority of swamping solar photons are rejected utterly. This optical filter lets through just a very few of the solar photons in addition to mostly admitting Alice's precious photons carrying the bits and bytes of the secret key.

The initial groundbreaking experiment of Jacobs and Franson was followed up by a number of long-range free-space quantum cryptography demonstrations.[21] Of particular note was the 1999 realization of the same BB84 protocol by English–American physicist Richard Hughes, and collaborators at Los Alamos National Laboratories, over a distance of 1.6 kilometers (1 mile), in full daylight.[22] By 2001, the Los Alamos group had extended this daylight demonstration to a distance of 10 kilometers (6 miles) between two mountaintops near the laboratory, employing a modified version of Jacobs and Franson's scheme for filtering out the bad solar photons and letting through (mostly) only Alice's secret key photons into Bob's telescope.[23] The current record holder is a variant of the E91 protocol with entangled photons distributed a distance of 144 kilometer (90 miles) between two mountaintops located on two separate islands of the Canary Islands off the coast of Africa (Figure 5.2).[24] Austrian physicist Anton Zeilinger and his gang set up this system with Alice and the entangled photon source located on the Island of La Palma and Bob and his telescope on the Island of Tenerife. When I asked Zeilinger why he went all the way to the Canary Islands to carry out this experiment, he told me, solemnly, "There are four reasons: First, the atmosphere is often very clear over the Canary Islands. Second, the weather is often much warmer than in Vienna. Third, it is a nice change from crawling around in the Viennese sewers. Fourth, you have never had good Spanish wine?" This all added up! The system holds the record not only for free-space quantum cryptography but also for the longest distance over which quantum theory has been tested against classical local hidden variables with Bell's inequality.

The key point to remember is that the amount of air the photons encounter traveling 144 kilometers horizontally at 2500 meters (8000 feet) above sea level (the height of the mountains) from mountaintop to mountaintop is the same amount of air the photons would encounter traveling several thousands of kilometers going straight up (or straight down). That is because the density of air is constant (and large) at a height of 2500 meters, but the air thins exponentially quickly in density if you instead start at 2500 meters and then go straight up to space. This experiment is then a proof of principle that quantum cryptography

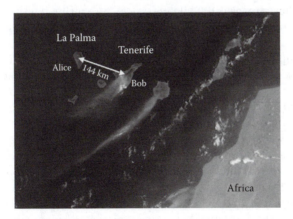

Figure 5.2   A satellite's-eye view of the quantum E91 cryptography system on the Canary Islands. Alice and the entangled photon source are located on a mountain on the Island of La Palma. Bob and his receiving telescope are located 144 kilometers away on a mountain on the Island of Tenerife. (This photo was taken July 30, 2007, when forest fires raged on two of the islands, Tenerife and Gran Canaria, the disk-shaped island to the right of Tenerife. You can see the gray streaks of smoke trailing the islands. This would not have been a good day for sending single photons between the islands because of the smoke.) (Photo is courtesy of NASA/JPL-Caltech. This photo is the property of the US government and is in the public domain, http://www.jpl.nasa.gov/imagepolicy/.)

Earth-to-space uplinks or space-to-Earth downlinks are indeed possible. A final proof-of-principle experiment clinched this case in 2008 when the Zeilinger team, in collaboration with Italian physicist Paolo Villoresi and his colleagues at the University of Padova, actually detected single photons reflected from a low-Earth-orbit Japanese satellite.[25] The team used the Matera Laser Ranging Observatory in Matera, Italy, for the demonstration. (Italian wine is as good as Spanish wine?) They launched a pulse of laser photons up toward the Japanese satellite, which conveniently has a mirror attached to it for just such ranging experiments, and the pulse had a carefully calibrated intensity so by the time it reached the mirror on the satellite, only one photon (on average) survived the passage to the satellite. Then, the game was to see if they could detect that one remaining photon on its way back to the ground with the telescope, which they could. That means if, instead of a mirror bouncing incoming photons, the satellite had aboard an Alice with a single-photon machine gun firing photons downward, it too would have been detected, demonstrating that a quantum cryptographic space-to-Earth downlink is feasible.

Laser ranging experiments are used to judge the distance to and fro by measuring the time of flight of photons that are sent to the target and then reflected back. (The word "ranging" means "distance finding.") Because distance equals speed multiplied by time (if you are driving 100 kilometers an hour and drive 1 hour, you have traveled 100 kilometers), by measuring the time it takes the photons to travel out and back, and then multiplying by the speed of light (300,000,000 meters per second or 700 million miles per hour), it gives you the round-trip distance. The particular mirror used on the satellite is called a "corner cube," which is a half of a hollow mirrored cube—a collection of such cubes is sometimes called a retroreflector or "cat's eye" reflector. This gizmo has the useful property that it reflects light incident on it exactly back along the direction the light came from, like a cat's eye, unlike a flat mirror that typically reflects light in another direction. This type of ranging or "distance measuring" experiment has been used for years to measure the distance between the Earth and the Moon to a very high degree of precision—we know that distance at any given moment is accurate to a few millimeters (a few hundredths of an inch). The US Apollo 11, 14, and 15 manned lunar missions, as well as the unmanned Soviet Lunokhod 1 and Lunokhod 2 rover missions, deliberately left arrays of retroreflectors on the Moon to facilitate such laser ranging experiments.

The Apollo 15 retroreflector array is the largest and most often used. A typical experiment involves launching a laser beam through a telescope aimed at the Apollo 15 array and then waiting for the 2 1/2 seconds round-trip time to collect any return photons. Typically, only a few return photons make it back to the telescope but they can be discriminated from stray light by their unique color and the approximate return time expected. These lunar ranging experiments have a precision equivalent to measuring the distance between

Los Angeles and New York with an accuracy of 0.025 centimeters (one one-hundredth of an inch). The laser beam pulse that is launched from Earth has a diameter of a few tens of centimeters and it contains 10^{17} photons. By the time the laser beam hits the Moon, it has spread out (owing to diffraction) to a spot size approximately 6.5 kilometers (4 miles) in diameter. Because the diameter of the Moon is 3500 kilometers, placing that spot square on the Apollo 15 landing site is akin to shooting a flying hummingbird with a rifle at a distance of 15 kilometers (9 miles), which is why the telescope is needed for aiming the outgoing laser as well as collecting the return photons. With 10^{17} photons out, you typically get only about one photon back (the light reflected from the retro-reflectors also spreads on the way back and makes another spot around 10 kilometers in diameter on the Earth, which is much larger than the telescope, so many photons are lost both coming and going).[26]

By combining this old technology of Earth-to-space laser ranging with sources and detectors of single photons, it is clear that Earth-to-space quantum cryptography links are just around the corner and likely will be demonstrated in just a few years (much sooner than the demonstration of a full-scale universal quantum computer). Well, what can we use this for? The Earth-to-space links do an end run around the need for quantum repeaters. The idea is to have quantum wide-area networks that span city-sized regions on the ground, using optical fibers, and then the city-to-city hookup is carried out via the satellites. For quantum cryptography, perhaps the simplest protocol would be to have Charlie on the satellite, positioned over New York City, fire pairs of entangled photons downward so that Alice in Washington, DC, gets one of the pair and Bob in Boston gets the other. Alice and Bob then execute the E91 protocol, and at the end of the run, they share a list of random zeros and ones that can be used for unbreakable one-time-pad communications. In this setup, even Charlie on the satellite has no information about the one-time-pad key that Alice and Bob distill from the photons, as he has no advanced knowledge of what polarization direction they will measure in. In this way, Charlie is no better off than Eve, who would try to intercept, measure, and resend the entangled photons, say while flying an invisible jet (that Eve stole from Wonder Woman) circling over the Empire State Building.[27] Given enough time (and money), every metropolitan area in the world could be equipped with a wide-area quantum cryptography network that is linked to all the others via an array of orbiting satellites. When the universal Internet hacking quantum computer comes online in 50 years, we'd be ready for it, for even a quantum computer cannot hack the uncrackable one-time-pad cryptography system. Eventually, say in 25 years, the quantum repeaters (which are just special-purpose quantum computers) would come online and then the expensive satellite links would be phased out. This worldwide quantum cryptography network would then also form the backbone for the future quantum Internet. Someday, there will be many quantum computers, distributed all over the world, and there must be a means to transmit quantum information reliably between them to form a quantum Internet. The quantum cryptography network provides this backbone. In the most exciting scenario, quantum information states would be teleported, using quantum teleportation discussed below, from quantum computer to quantum computer.

A more near-term goal of the Earth-to-space quantum cryptography link, at least a goal of the US government, is to rekey spy satellites on the fly. The US National Reconnaissance Office (NRO) is the US government agency responsible for the construction and operation of US spy satellites, formally called reconnaissance satellites. The NRO is a joint operation of the US DoD and the US Central Intelligence Agency (CIA). If you visit the NRO Headquarters in Chantilly, Virginia (by invitation only), they have a little museum in the visitor center of spy satellites through the ages, and by the ages, I mean since the 1960s. Even if you do not possess a security clearance, you can still have lunch in their nice cafeteria where a quick perusal immediately makes it clear that all the men at least are wearing either military uniforms or Armani suits, making the distinction between DoD and CIA personnel somewhat clear. You then may happily snack away in the well-lit room festooned with signs declaring "THIS IS A NONSECURE AREA—DISCUSSION OF CLASSIFIED OR SENSITIVE MATERIAL IS PROHIBITED!" When I once got beyond the cafeteria to a conference room in the inner sanctum, after having to stow all my electronic devices (including my watch) in a locker in the visitor center, even though I possessed a low-level security clearance, that was not enough to prevent the NRO escort from walking 10 feet in front of me down the hallway like the town crier and yelling out "UN-CLEARED PERSONNEL IN THE HALLWAY—UN-CLEARED!" Whereupon everybody leapt to their feet and slammed their office doors shut. I felt like a security-clearance leper.[28]

The NRO, like the ultra-secret National Security Agency (NSA), with whom they closely collaborate, is an agency whose very existence was a secret until the early 1990s. The existence of the NRO was declassified on September 18, 1992, and before that, even the acronym "NRO" was classified and could not be spoken or used in any type of communication.[29] Even after the declassification of the acronym NRO, I was hesitant to use even the letters in unsecured emails, and I would typically write "ONR" in quotes, which without the quotes would be the less secretive "Office of Naval Research," but my colleagues knew that when I added the quotes I was really talking about the NRO. A physicist colleague of mine, who will remain anonymous for his own protection (and because he can be a bit paranoid), tells a tale of visiting the "old" NRO headquarters near Washington, DC, in the early 1990s. He was given a set of directions that went something like this. Take Virginia State Route 267 from Dulles Airport and head IIII for II miles and take exit II and then turn IIII. Make a U-turn at the Dairy Queen and when you come to the cow pasture with the cows in it, make an immediate **IIII** and follow the unmarked paved road due **IIII**. Here is the tale my colleague told me.

*I'm driving along this unmarked, one-lane paved road thinking that I'm totally lost and I'm about to turn around when suddenly two jeeps, filled with four Marines each, pull out of the side of the road directly in front and behind of my rental car. Then, I think, well maybe I'm not lost after all! Each Marine is carrying an M16 assault rifle, which they are holding at an angle of 45° to the ground—not quite pointed at me but not quite not pointed at me. Two of them come up to the car, one to the passenger side and one to the driver side, and the driver side guy asks, may we help you sir? The other guy peers around in the back seat. So I state, I am Dr. **IIIIIII** and I have an appointment to see Mr. **IIII IIIIIIIIII**. Very good, Dr. **IIIIIII**, replies the Marine, may we see your ID? So I show him my ID and he radios back to somebody on his radio and gets a confirmation. He then motions to the other Marines and they lower their guns to point at the ground. Then he says, very good, Dr. **IIIIIII**, would you please follow us? Well I did not have much choice but to 'follow' them and with one jeep full of Marines behind me and one in front of me we went up this little hill on the paved road and then when we reached the crest of the hill the road led down into a little valley where there sat a huge windowless complex bristling with antennas. That was when I knew for sure I was in the right place....*

There are two basic spy satellites run by the NRO. The first is an eavesdropping type that listens in on radio and other communications. The second is an imaging satellite that takes pictures of things on the ground. The intercepted radio communications signals are then relayed to the ground and typically sent to the NSA for decryption and analysis. The NSA has banks of the best supercomputers in the world devoted to these tasks. The images are typically sent to the US National Geospatial-Intelligence Agency, formerly known as the US National Imaging and Mapping Agency (NIMA), again for analysis and interpretation. NIMA is responsible for providing accurate and up-to-date maps of the world to various government agencies.[30] Regardless if the satellites are taking pictures or listening in to radio communications, the resulting data are communicated back to Earth in encrypted form, as are all communications between the Earth control stations and such satellites. The satellites, as is common with most of the Intelligence community, do not use public key encryption for such communications. Quantum computers notwithstanding, there is no proof that public keys are not hackable by even classical computers, given that there is no proof that there is not an exponentially fast classical factoring algorithm. The plot from the movie *Sneakers* and the paper-and-pencil calculations of Indian mathematicians in an un-air-conditioned attic in Hyderabad weigh heavy on everyone's mind. They also do not use the one-time pad, at least not in its pure and unbreakable form. That is because to be unbreakable, there must be as many random bits of zeros and ones in the pads as in the data to be transmitted, and these satellites transmit a hell of a lot of data. There is not enough room to store all the random one-time-pad zeros and ones needed for all communications over the multiyear lifetime of the satellite. Instead, these satellites likely use a watered-down version of the one-time-pad system, one that still requires Alice and Bob to share large amounts of random key bits but not the key-intensive one bit of key per one bit of message that one-time pad requires.[31] This watered-down one-time pad, unlike the one-time pad itself, is not unbreakable, but its degree of breakability, given certain assumptions of computer power available to the eavesdropper, can be estimated, and a tolerable risk assessment can be made on the risk of hacking. The trouble with the factoring-based public-key encryption is that no such risk assessment can be made, because that would require an estimate of the probability of three Indian mathematicians cooking up an exponentially fast factoring algorithm tomorrow, which nobody can possibly estimate.

The problem is that the watered-down one-time-pad system still requires Alice on the satellite and Bob on the ground to share large amounts of random key bits. These key bits are installed in the satellite upon launch, and after it is all used up, then the spy satellite is useless. I don't know how these key bits are installed in the satellite, but given that NRO is a joint venture between the CIA and the DoD, in a moment of whimsy, I imagine a CIA agent and an NSA (DoD) cryptologist riding up together to the top of the satellite in an elevator, minutes before launch, and each inserting their own yottabyte flash drive (with half of the key on one drive and half of the key on the other drive so neither the CIA nor the NSA knows the entire key in advance) into two separate USB (universal serial bus) ports separated by a few meters on the satellite's nose-cone while the two agents point pearl-handled revolvers at each other to make sure each does his job.[32] Then, the satellite is launched and communications between the satellite and the ground continue apace until all that key is used up and then they blow up the billion-dollar spy satellite and steer the sizzling shards into a steep dive into the Earth's atmosphere where even those shards burn entirely up upon reentry, leaving nothing but ashes to float gently down to pepper the radioactive remains of the amoeba-shaped atoll of Mururoa in the Pacific Ocean.[33] So we see the problem. Sans carrying out a joint CIA–NSA space walk from the now defunct US Space Shuttle to install more keys in the orbiting satellite, which we cannot rule out because it has been done in the past, but we can rule out what will be done any time in the near future, it would be very useful (and much cheaper) for the NRO if they could instead rekey the spy satellites on the fly using single photons and quantum cryptography. This could greatly increase the life span of those spy satellites.

The idea of transmitting entangled photons between Earth and space has fundamental implications in addition to these communications applications. An ambitious proposal, again led by the team of Austrian physicist Anton Zeilinger, is called "Space QUEST: Quantum Entanglement in Space Experiments" (Figure 5.3). The proposal team consists of 43 scientists from Austria to Australia.[34] (It tells about the state of affairs of basic science research in the United States: that out of these 43 scientists, only 1 is from the United States.)

As stated in the figure caption, quantum communication and quantum cryptography are secondary goals of this proposal. The primary goal is to test the spooky action at a distance nature of quantum entanglement, quantum

Figure 5.3 Space QUEST: Quantum Entanglement in Space Experiments. This is a schematic of an ambitious proposal to the European Space Agency. Charlie (C) and his entangled photon source sit packaged into the European Columbus module at the International Space Station. The polarization-entangled photon pairs (arrows) are launched through simultaneous space to ground downlinks to two ground stations. Alice (A) and Bob (B) then can make measurements on their photons and exchange classical information (double arrow) about their measurement outcomes and test Bell's inequality, implement the E91 cryptography protocol, or carry out quantum teleportation. The primary goal is to test quantum theory, particularly the nonlocal nature of entangled particles, over distances of 1000 kilometers (600 miles), the distance between Alice and Bob. Secondary goals are to use the system as a prototype satellite-based quantum communications link for quantum cryptography and quantum communications.

nonlocality, over distances of 1000 kilometers (600 miles) between two ground stations. Charlie in the International Space Station's European Space Agency's Columbus module transmits entangled photon pairs to two ground stations, Alice and Bob, separated by 1000 kilometers. The first experiment is to see if the nonlocal nature of quantum entanglement, as encoded in the Bell test, still survives over such distances. (Previous tests have been carried out over 144 kilometers.) Quantum theory predicts that these spooky correlations should exist regardless of the distance, provided decoherence and other sources of noise are mitigated and accounted for. If the correlations do not survive, then quantum theory is wrong and must be revised. (I personally am betting on quantum theory.) Once the Bell correlations are established, then the system can be used for quantum cryptography and quantum communications, including the whacky implementation of quantum teleportation over 1000 kilometers, which we will discuss next.

BEAM ME UP, CHARLIE

We learned in the previous sections that one reason quantum cryptography is secure is because of the no-cloning theorem—according to the laws of quantum theory, it is impossible to perfectly copy an unknown quantum state. The modifier "unknown" is needed here because if the state is known with sufficient information, then one can build a quantum state source that duplicates it. However, if somebody hands you a photon in an unknown quantum state, you would have to know something about it to duplicate it. To know something about it, you have to measure or copy it. But then, Heisenberg's Uncertainty Principle kicks in and you destroy much of what you are trying to measure in the measurement process. In the end, you cannot get enough information to duplicate the unknown state, as you scramble the unknown state too badly in the measurement process to learn enough to duplicate it. Hence, with this quantum no-cloning theorem well understood, it came as a shock to many of us to learn in 1993 that it was possible for Alice (in the star system Alpha Centauri) to measure (and hence destroy) an unknown quantum state and then to arrange for the destroyed state to instantaneously appear (with some caveats) in Bob's laboratory (in the star system Beta Pictoris). This curious business is called "quantum teleportation" or more properly (and more boringly) "remote state preparation." None other than our infamous IBM-Charlie, Charles Bennett at IBM, and his collaborators (Canadian computer scientists Giles Brassard and Claude Crépeau, Australian computer scientist Richard Jozsa, the late Israeli physicist Asher Peres, and American physicist William Wootters) proposed the idea in a 1993 paper titled "Teleporting an Unknown Quantum State via Dual Classical and Einstein–Podolsky–Rosen Channels."

Three experimental groups demonstrated the idea with entangled photons in due course: Anton Zeilinger's group (then in Innsbruck, Austria) in 1997, H. "Jeff" Kimble's group at the California Institute of Technology (Caltech) in Pasadena in 1998, and Italian physicist Francesco DeMartini's group in Rome again in 1998.

When Texan physicist H. "Jeff" Kimble speaks to the popular press about his experiments in quantum teleportation, he always makes it a point to state that comparisons to the fictional transporter device of *Star Trek* fame are not useful and are in fact misleading. He will not be surprised, and will likely be dismayed, to find that I plan here to embrace the similarities and differences of quantum teleportation and the *Star Trek* transporter. One must keep in mind that the *Star Trek* transporter is a fictional device, based on unknown laws of physics, and that quantum teleportation experiments are reality, based on the known (if strange) laws of quantum theory. In addition, there is not one unified description of the *Star Trek* transporter; its properties and constraints are mostly unclear and those that are clear can change from episode to episode. However, consistently, the *Star Trek* transporter works as follows. The target to be transported, say Captain Kirk on the Starship *Enterprise*, is "dematerialized" and then the "energy pattern" of Captain Kirk is transported to the surface of, say, a nearby planet, where the energy pattern is used to "rematerialize" him. Because this is a fictional device, these words have no real physical meaning, so I am free to reinterpret them. "Dematerialized" means "destroyed," "energy pattern" means the "information" needed to reconstruct Captain Kirk, and "rematerialized" means "created." So with this interpretation, which is consistent with how the actors on *Star Trek* often view the process, Captain Kirk is destroyed on the *Enterprise*, the information needed to reconstitute him is beamed to the surface of the planet, and there that information is used to recreate a copy of him. I suggest that this is, Kimble's protestations to the contrary, a very good (classical) model of how the quantum teleporter works! The idea that the transported Captain Kirk is only a very good copy of the destroyed original, and not the original itself, becomes clear in the episode "Second Chances" from the series *Star Trek: The Next Generation*. In this episode, we learn that, owing to a transporter malfunction, a duplicate of Commander William Thomas Riker has been made in transport. The original, which should have been destroyed, is accidentally reconstituted on the planet he is transporting off of while the copy is also reconstituted on the Starship *Enterprise*. The duplicate Riker, who eventually takes the name Thomas (his middle name), is left marooned on the planet for some years, as the error goes undetected, whereupon he is rediscovered and rescued, but then he becomes somewhat bitter (as the other copy has advanced in his career while Thomas lived all alone in a cave) and Thomas Riker eventually ends up joining the Maquis rebellion. This somewhat convoluted tale confirms to us that the *Star Trek* transporter, at least in this incarnation, is a type of remote classical copying machine with a design feature that ordinarily prevents duplicate copies from cavorting about, unless there is a transporter malfunction required for some interesting plot twists.

Boiled down to its physical essence, the *Star Trek* transporter is like a fax machine with built-in paper shredder. You insert your document to be faxed and it is copied faithfully, converted to bits and bytes, and transmitted over the phone lines, while the original (instead of being returned to you) is dumped into the paper shredder and destroyed. If all goes well, there is never more than one copy. By design, the device prepares remote copies but prevents duplicate copies by eliminating the local one. But if the design fails, say the paper shredder jams, then it is possible to have your remote copy and your local copy, just like the two Commander Rikers. Thus, the *Star Trek* transporter (Heisenberg compensators notwithstanding) is a classical device. It cannot be a truly quantum device, as often claimed by the *Star Trek* scriptwriters, because of the quantum no-cloning theorem that is impossible to copy an unknown quantum state, and we allow that Commander Riker's physical state was unknown, at least at the moment of transport.[35] A truly quantum machine could never produce multiple copies of Commander Riker if the complete quantum mechanical state of Riker was required for transport and then shredding. Hence, a quantum transporter, called a quantum teleporter, cannot be the analog of a classical fax machine with a paper shredder attached to it. If a quantum copy is to be produced at a remote location, then it is mandatory by the laws of physics that the original must be destroyed in the process, not just an option via an add-on paper shredder. This is exactly how quantum teleportation works.

Because we are now (hopefully) comfortable with entangled photons flying about over long distances, I will explain quantum teleportation with these. The two main ingredients are a quantum channel, over which Alice and Bob share a pair of polarization-entangled photons (produced by Charlie), and a classical channel, such as a cell phone, by which Alice can communicate with Bob to complete the teleportation protocol. The quantum channel must be established first. To make things interesting, we shall put Alice in a spaceship in Alpha Centauri (a star system located 4 light-years from Earth) and Bob in one on Beta Pictoris (a star system located 63 light-years from Earth). We'll just place Charlie and his entangled photon source on Earth. From simple geometry, the farthest these two star systems can be from each other is approximately 67 light-years, but as they are not on opposite points on the sky from each other, they are closer, say a few tens of light-years[36] (see Figure 5.4). Because they are not equal distances from the Earth, Alice has to store her half of the entangled photon pair in a quantum memory for a few tens of years until the other half reaches Bob. Now, somebody, say Doug, hands Alice a photon in an unknown polarization state, at least not known to Alice, Bob, or Charlie. Such an unknown polarization state can be written as a coherent cat-like state that is a superposition of vertical and horizontal states of the photon, $a|\updownarrow\rangle + b|\leftrightarrow\rangle$, where a and b are related to the probability that when measured in H–V directions, the photon collapses to vertical or to horizontal. For example, if $a = 1$ and $b = 0$, then the photon is 100% vertical. If $a = 0$ and $b = 1$, then it is 100% horizontal. If $a = b$, then 50% of the time you get

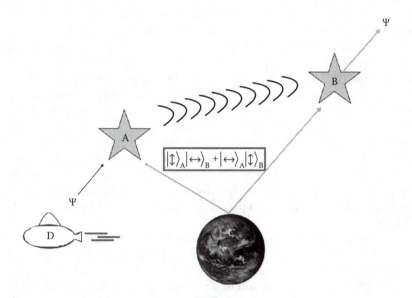

Figure 5.4   Quantum teleportation: Charlie on Earth (C) sends a pair of polarization-entangled photons (light-gray shaded arrows) to Alice on Alpha Centauri and Bob on Beta Pictoris (B) to establish the quantum communications channel. Then, Doug in the spaceship (D) provides a single photon (thick black arrow) to Alice in a polarization state Ψ that is unknown to Alice, Bob, or Charlie. Alice mixes Doug's unknown state with her half of the two-photon entangled state and makes a measurement that completely destroys Doug's unknown state Ψ and her half of the entangled photon pair. Only one photon now remains, and it is Bob's. Alice's measurement instantly transports the unknown state Ψ from her location into Bob's photon. To extract the unknown state Ψ (thick black arrow), Bob must apply one of four single-qubit gates to his photon, using information about the outcome of Alice's measurement. Alice transmits this information to Bob over the classical channel (radio waves in black). Once Alice's signal arrives, Bob applies the correct gate, and out pops the original unknown state Ψ. (Photograph of Earth courtesy of NASA/JPL-Caltech. This image is a production of the US Government and not subject to copyright, http://www.jpl.nasa.gov/imagepolicy/.)

vertical and 50% of the time you get horizontal. Other values of a and b give different percentages.[37]

The precise values of a and b are unknown to Alice, Bob, and Charlie. Alice cannot extract complete knowledge of a or b from a single measurement on a single photon. If she could, then she could make a copy, violating the no-cloning theorem. A single measurement will collapse the photon polarization completely in the vertical direction, giving some information about a but none about b, or completely in the horizontal direction, giving some information about b but none about a. No single measurement that Alice can make can give complete information about a and b simultaneously. This is the quantum uncertainty and unreality at work. The photon has no polarization direction until it is measured (unreality) and then it collapses irreversibly to one direction or the other (uncertainty) and thence the measurement process destroys much about what is to be measured. That is the essence of the no-cloning theorem. Any attempt to copy an unknown quantum state requires you measure it first to see what it is, but the act of measurement itself scrambles the state, preventing you from gathering complete information about it and certainly from gathering sufficient information to make a copy. For the quantum teleporter to work, the state cannot be copied, as in the *Star Trek* transporter; the state must be destroyed by the teleportation process and recreated remotely. No two copies can exist at once. That is what the quantum teleporter indeed does through the spooky mechanism of the nonlocal correlations between the entangled photons in the pair provided by Charlie.

As shown in Figure 5.4, first Charlie on Earth establishes the quantum channel by transmitting a pair of entangled photons to Alice on Alpha Centauri and Bob on Beta Pictoris. Then, Doug in the passing spaceship provides Alice with a single photon in a polarization state Ψ of the form $a|\updownarrow\rangle + b|\leftrightarrow\rangle$ where the state (and hence the values of a and b) is unknown to Alice, Bob, or Charlie. There are three photons in play now: the two in Charlie's entangled pair (now split between Alice and Bob) and the one from Doug that is now with Alice. Alice mixes Doug's unknown photon with her half of the entangled photon pair and makes a particular type of two-photon measurement. Alice's measurement completely destroys both Doug's photon and her photon from Charlie, but it also yields four possible outcomes from the measurement. After measurement, there is only one photon left, that of Bob. The unknown quantum state Ψ (a Greek letter pronounced "psi") is not lost upon Alice's measurement, but instead it is instantaneously teleported into Bob's remaining photon. This teleportation takes place because of the nonlocal nature of the quantum correlations between the two photons in Charlie's entangled pair—spooky action at a distance in this case really is spooky and at a great distance. The state is teleported in zero time, so infinitely faster than the speed of light, sending Einstein's ashes into a tailspin in the Delaware River.[38] However, there is a catch, an important catch, that saves the theory of relativity.

Bob cannot extract the unknown state Ψ without two bits of classical information from Alice, which are transmitted over the classical channel (radio waves in Figure 5.4). That is, the teleportation process itself performs a single-qubit rotation on the state Ψ in transport into one of four possible states, Ψ itself, or three possible other states closely related to (by rotation) but not equal to Ψ. To extract Ψ exactly as provided by Doug to Alice, Bob has to apply one of four different possible single-qubit gates to his photon, which correspond to rotating the polarization direction of his one remaining photon back into Ψ. However, and this is critical, which rotation he performs depends on the outcome of Alice's measurement. Alice gets four possible outcomes from her measurement, and she must relay this measurement outcome to Bob so he can apply the correct single-qubit rotation gate to guarantee the complete reconstruction of the unknown state Ψ. Alice and Bob can agree in advance on the following classical two bits to be transmitted via the classical channel: Alice sends 00 for measurement outcome number one, 01 for measurement outcome number two, 10 for measurement outcome number three, and 11 for measurement outcome number four. These are indeed just the numbers 1, 2, 3, and 4 written out in binary notation.

Without these two classical bits of information from Alice, Bob can only randomly guess which of the four rotations to apply to his photon. As we saw from the quantum cryptography schemes, random guessing is equivalent to completely scrambling the photon state. It is a bit beyond this book to prove this statement, but it is nevertheless true, without the two measurement bits from Alice, all that Bob can extract by random guessing are (on average) random quantum mechanical states with no relation at all to the original state Ψ. That is, upon measurement of his photon, he will produce a string of random numbers with no relation at all to the particular numbers a and b that make up Ψ. In *Star Trek* parlance, that is what you call a transporter malfunction and what Bob gets out his end is not very pretty.[39]

In Box 5.2 I summarize four important take-home points about quantum teleportation. Point number four is the one most misunderstood by novices trying to come to grips with quantum teleportation. When I worked for NASA (1998–2004), I must have received three different proposals to review that claimed quantum teleportation could be used to break the speed of light barrier in deep-space communications. This is a big deal to NASA program managers, particularly at the NASA Jet Propulsion Laboratory (JPL), which runs the Mars rover program so successfully. The general public often thinks these rovers on Mars are being driven about by drivers on Earth in real time, perhaps much the same way that US Air Force pilots, sitting at a desk in California, can remotely fly drone aircraft

**BOX 5.2   HERE ARE A FEW POINTS TO TAKE
HOME ABOUT QUANTUM TELEPORTATION**

1. Doug's original photon itself is not transported in quantum teleportation. Only its quantum mechanical state is transported.
   That is, nothing physical moves from Alice and Bob instantaneously but only the information needed to specify that state
   now embodied in Bob's photon instead of Doug's. This is then
   like the classical *Star Trek* transporter in that Captain Kirk
   himself is not transported physically down to the surface of the
   planet but rather only the complete information to reconstruct
   Captain Kirk is transported down. In the fax machine analogy,
   the information to reconstruct the document is transmitted
   over the phone line, not the paper and ink of the original document itself. Paper and ink must be provided on the far end to
   reconstruct the document there with the bits of information
   coming in through the telephone line.

2. In the classical *Star Trek* transporter, two copies of the thing to
   be transported ("William" Thomas Riker and William "Thomas"
   Riker) can exist simultaneously, like my classical fax machine with
   the optional paper shredder turned off, which is then just a fax
   machine. The classical fax machine produces a copy at the remote
   location but is not required to destroy the original. Such a machine
   cannot exist quantum mechanically, as copying an unknown
   quantum state would violate the quantum no-cloning theorem.
   In quantum teleportation, the destruction of the original is not an
   option—it is a critical part of the teleportation protocol itself.

3. At no time does Alice (or Charlie) know what the unknown quantum state Ψ actually is. In principle, the two complex numbers a
   and b required to reconstruct the unknown state could require
   an infinite number of classical bits to completely specify.[40] Alice
   has only access to two bits of information, 00, 01, 10, or 11, which
   specify one out of four of her measurement outcomes. These bits
   contain no information about the unknown state Ψ and so Alice
   could not provide Bob (on the classical channel) the information needed to reconstruct Ψ, even if she wanted to. It is all those
   potential infinite number of digits that are needed to specify a
   and b that are teleported upon Alice's mixing and measuring.
   Alice's four measurement bits merely tell Bob how to extract that
   information faithfully on his end, not what that information is.

4. Quantum teleportation does not violate Einstein's theory of relativity, which in simplest form states that information cannot be transmitted faster than the speed of light. While it is true that the necessary information to reconstruct Ψ is teleported instantaneously from Alice to Bob, the sufficient information required for the reconstruction is contained in Alice's two bits of measurement outcome information that are relayed on the classical channel (cell phone, radio signal, laser communication) that all move at or slower than the speed of light. Bob must wait for Alice's slower-than-light communiqué before he knows the correct single-qubit gate to apply to his photon to extract Ψ faithfully. Without those two slow-moving bits from Alice, his best strategy is to simply guess which of the four single-qubit gates to apply. It can be shown that, on average, this completely scrambles the state Ψ and all Bob gets out are random quantum mechanical states from which he cannot extract any information about Ψ at all! (In *Star Trek* lingo, this is called a "transporter malfunction" and you had better hope you are not the one being transported.) Random guessing, provided by measurement, will produce a (possibly infinite) string of random digits that are completely uncorrelated from those needed to specify a and b that specify Ψ. Those random digits are statistically just as random as if Bob made random rotations and measurements to his half of the Charlie-provided entangled photon pair while Doug and Alice did nothing at all. Another way to state this is that without the slow-moving classical bits from Alice, quantum teleportation can only instantaneously transmit random numbers faster than the speed of light, and those random numbers at Bob's location are completely uncorrelated to what is going on at Alice's location. Because a string of uncorrelated random numbers, by definition, contains no information—then Einstein's relativity theory is safe—I can transmit an uncorrelated string of random numbers as fast as I want, because I am not transmitting any information.

over Afghanistan in real time. The drone aircraft piloting can be done (nearly) in real time because the speed-of-light round-trip time between California and Afghanistan is about a tenth of a second. Not so the speed-of-light round-trip time between California and Mars, which is approximately 24 minutes (on average depending on the relative location of Earth and Mars in their orbits).

Imagine trying to drive down the freeway if your response time was 24 minutes! If the car in front of you suddenly brakes at 1:00 p.m., you only see the car braking at 1:12 p.m. and immediately you hit your brake pedal, then another 12 minutes later, at 1:24 p.m., your brakes are applied to your car wheels. Moving at 120 kilometers per hour (75 miles per hour), which is 2 kilometers per minute, you would have to allow a minimum spacing of 48 kilometers (30 miles) between you and the car ahead of you in order to keep a safe distance. Hence, the Martian rovers are driven less like the real-time Predator drones and more like a remote "point and click" video game. The JPL operator sees an interesting purple rock on the rover camera and points and clicks to move the rover to go over to inspect the rock. She then must wait 24 minutes before she can tell if the rover arrived safely at the rock or instead ran into the rock or instead fell into an invisible crevasse and landed upside down like a mechanical turtle on its back—wheels spinning uselessly—never to right itself ever again! (That would be a bad day for the JPL rover driver.)

That is why the JPL program managers were so enthralled with faster-than-light communication of information—instantaneously transmitting sequences of random numbers to the rovers is not what they had in mind—unless the goal was to make the poor little rover dance the Watusi. I had to carefully and repeatedly explain to them that quantum teleportation does not allow for faster-than-light communication and that nothing within the realm of known physics does.[41] The typical argument against the transmission of information faster than light is that such a communication system would allow you to send signals backward in time. With such a communication system, you could then send an anonymous "poison pen" letter back in time to your mother, urging her to call off her engagement to your father (who you cunningly and falsely posit is a scoundrel and notorious lothario), and if your mother complies and follows your advice to call off the marriage, then you were never conceived nor born in the first place, in order to grow up and send that poisonous message, in which case your mother does marry your father, in which case you are conceived and born and grow up to send the message, in which case. ... It is the existence of such paradoxes, associated with faster-than-light communication, which is often trotted out as metaphysical proof that it is physically impossible to communicate thusly. Perhaps a better answer is that we have no theory that allows for faster-than-light communication, and a very well tested theory that forbids it, and no experimental evidence to think it is possible, so we rule it out from the realm of plausibility (for now).[42]

However, the tension between quantum theory and Einstein's theory of relativity is palpable, and this tension was one of the reasons Einstein so disliked quantum theory in the first place. Quantum teleportation is clearly transmitting "something" faster than the speed of light, the necessary information for Bob to reconstruct the unknown state Ψ. But then, there is this caveat that he cannot do it at all without the sufficient two bits of slower-than-light information provided by Alice on her measurement outcomes. We sneak right up to this light barrier but then do not cross it. There is another curious and close connection between the no-cloning theorem and the faster-than-light business. Without going into detail, suppose Alice and Bob share an entangled pair of photons $|\updownarrow\rangle_A |\leftrightarrow\rangle_B + |\leftrightarrow\rangle_A |\updownarrow\rangle_B$, provided by Charlie. If Alice wants to transmit the classical bit 0, she measures her photon in the vertical direction ($\updownarrow$), and if she wants to transmit the classical bit 1, she does nothing. Then Bob—in violation of the no-cloning theorem—quickly makes many perfect copies of the state on his end. Then, he measures all of these copies in the horizontal direction ($\leftrightarrow$). Because of the entanglement, if Alice measures in the vertical direction, Bob's photon will collapse into the horizontal direction and all the measurements on all the copies will yield horizontal polarization for his photon ($\leftrightarrow$), and he will know that Alice is sending a zero. If instead Alice did nothing, you can show that in that case many repeated measurements of Bob's copies will just give the horizontal ($\leftrightarrow$) result or the vertical ($\updownarrow$) result, with a random 50–50 probability (like a coin toss), and in that case, he knows Alice is sending a one.

Even a single bit of information is important information if, for example, they have prearranged that if Bob gets a zero he should start a war with the Klingons and if he gets a one he should sign a peace treaty with them. If Alice and Bob have stored up many entangled photon pairs in advance of the faster-than-light communiqué, then the information content of the message can be more elaborate. Alice and Bob can store up the photons years before the message is transmitted, and even if they are light-years apart, they can then use these stored entangled photons to communicate instantaneously in violation of relativity theory, but only if Bob can make those pesky copies. It is the copying (or cloning) that gets around Einstein's speed limit. Without multiple copies, Bob has just one photon to deal with. If Alice measures in the vertical direction (sending a zero) and if Bob measures it in the vertical direction, he will get a vertical outcome ($\updownarrow$) with a 100% probability. However, if Alice does nothing (sending a one) and Bob measures in the vertical direction, he will get vertical ($\updownarrow$) with a 50% probability. Probabilities only apply to a large (and preferably infinite) number of events. Hence, he needs many copies to tell if he is getting vertical with a 100% probability or 50% probability, but the no-cloning theorem prohibits copying, in which case Bob with his one uncopyable photon gets nothing but a single random number of no use to anybody.

In summary then, the ability to clone unknown quantum states implies that it is possible (via entanglement) to perform instantaneous faster-than-light communication. Because the latter is widely thought to be impossible, this implies that the former must also be impossible, which then provides an independent (and relativity based) proof of the quantum no-cloning theorem. Why the no-cloning theorem, which is a pure by-product of quantum theory, has anything to do with the speed-of-light limit on communications, which is a pure by-product of Einstein's theory of relativity, is something of an unsolved mystery, to my mind. There is no single, universally agreed upon concrete physical theory that includes Einstein's (general) theory of relativity (with gravitation) as well as quantum theory. I would think that the connection between no-cloning and no faster-than-light communication would emerge naturally from such a theory, and sometimes—after I've had a few too many glasses of Pinot Grigio—the many wine interpretation of quantum mechanics—I wonder if this connection is some kind of hint for us to follow to find such a theory.[43]

As I mentioned above, there were three initial teleportation experiments using photons, carried out in 1997–1998, the first by the group of Zeilinger in Austria, the second by the group of DeMartini in Italy, and the third by the group of Kimble in the United States. Since then, in 2003, the Gisin group teleported photon states through 2 kilometers (1 1/4 mile) of telecommunications fiber[44]; in 2004, the Zeilinger group teleported quantum states of photons across Vienna (and under the Danube River through fibers in the sewer pipes)[45]; and the latest record (in 2010) is that by the group of Chinese physicist Jian-Wei Pan, which teleported the state of a photon a distance of 16 kilometers (10 miles) in the Beijing area.[46] There is no need to stick to photons, and indeed in 2009, the group of American physicist Christopher Monroe teleported the quantum state of the spin direction of one ion into another ion over a distance of 1 meter (1 yard).[47] There is no reason even that the two objects bearing the quantum mechanical state be the same. Quantum information, like classical information or money, is fungible. It does not depend on the physical system that is bearing it. Four quarters is a dollar just as sure as a single dollar bill is a dollar. Hence, it is possible to teleport a quantum state, sitting on a photon at Alice's location, into an ion sitting at Bob's location, although nobody has done that just yet.

The early days of the experimental demonstrations, particularly the first three experiments by Zeilinger, DeMartini, and Kimble, were not without controversy. This I learned firsthand in 1999 when I attended the Sixth International Conference on Squeezed States and Uncertainty Relations in Naples, Italy. As the three different experimental groups' publications had just appeared, the conference decided to have a panel discussion on these three different experiments. Curiously, the organizers asked me to chair the panel discussion. I was a bit puzzled by this request and carefully stated that while I was honored by the invitation, I was hesitant to accept, because I had never carried out research (either theoretical or experimental) on quantum teleportation and was certainly not an expert on the matter. The organizers looked about the room nervously and then in hushed tones took me aside and explained their rationale—apparently, there was a very contentious debate raging among the three experimental groups as to who had done the first "true" demonstration of quantum teleportation, and the organizers expected the panel discussion to be loud and chaotic and they feared the panelists might become unruly—quantum physicists run amok! Therefore, they explained to me nervously while averting their eyes from my gaze, that they needed a moderator who would be able to run the thing with a strong hand (and a loud voice) in order to keep order. "The organizers are unanimous in our decision that you, Dr. Dowling, are our only hope." I had been in a moment demoted from world's expert in quantum physics to the quantum mechanical equivalent of either Obi-Wan Kenobi or a barroom bouncer.

The panel was chaotic from the inception. As I expected, there was one panelist per experimental group, Austrian physicist Gregor Weihs from the Zeilinger group, DeMartini from his own group, and Australian physicist Samuel Braunstein from the Kimble group. (American physicist Marlan Scully once called me the "Bob Hope" of theoretical physics. If that is the case, then my good friend and colleague Sam Braunstein is surely the "Woody Allen" of theoretical physics.) Since the debate to be among these three competing groups, I was again puzzled when the organizers, at the last minute, added two additional Italian physicists to the panel, other than DeMartini. I was even more puzzled that these two last-minute additions seemed to have even less experience with quantum teleportation than me! I gently inquired as to why they should be on the panel—my role as bouncer was clear—but their roles were not. Again, more hushed tones and averted gazes and the organizers explained that they were both big-shot Italian professors who asked to be on the panel discussion on quantum teleportation, despite knowing absolutely nothing about quantum teleportation, but for the simple reason that they were big-shot Italian professors, and thought it would look prestigious to insert themselves onto the panel. I had lived in Italy for a year as a graduate student, and I knew a thing or two about Italian politics in the universities, and conceded to their admittance to the panel in the interest of keeping the peace. Peace, however, was not long kept.

The night before the panel discussion, I was in a bit of a panic myself in my hotel room as I wondered how to organize things. I decided that in the hour-long time slot I would give each of the six participants, including myself, 5 minutes to speak or present a few slides, and then reserve the second half hour for questions from the audience. Now, I knew there was this debate between the three experimental groups about whose teleportation experiment was the "best" experiment, where "best" is very subjective, but I had not followed this debate at all, and did not really have time to read through all three of the experimental papers and try to figure it all out. Hence, I had an experimentalist friend and colleague, German physicist Andreas Sizmann, give me a quick tutorial on the nature of the experimental debate, at the bar, and then went back and carefully reread the original theory proposal by Bennett and colleagues and, with the help of Sizmann, constructed a few overhead transparencies on how I thought a quantum teleporter should work and how, if handed such a device, one might be able to tell if it was working properly, and if handed three of them, how I might gauge which of the three was "best." It was then, for the first time, I devised the story of the mythical National Institute of Quantum Information Standards and Technology (NIQuIST) and the equally mythical quantum teleporter-testing machine NIQuIST had constructed to test the three claimed experimental implementations. In other words, I did not compare the three experiments at all; I figured the panelists could do that themselves, but instead I put up a series of tests that each teleporter should pass to get the NIQuIST seal of approval, or more accurately something like a Consumer Reports rating: Recommended, Best Buy, or Not Acceptable. The quantum teleporter-testing machine that I drew up by hand looked like that in Figure 5.5.

In Figure 5.5, we show a compactified version of the quantum teleporter, shown in Figure 5.4, being slowly lowered into the NIQuIST teleportation-testing machine. As before we have Doug, now a NIQuIST employee (D) who provides the teleporter with a photon whose quantum polarization state is unknown to Alice, Bob, or Charlie in the teleporter. Doug produces a large number of such unknown states using a machine called the "ensembler" that produces ensembles or collections of states Ψ all of the form $a|\updownarrow\rangle + b|\leftrightarrow\rangle$, where the ensembler can be programmed to choose the constants a and b at random, or in a preselected sequence. There are an infinite number of states of the form $a|\updownarrow\rangle + b|\leftrightarrow\rangle$, so the ensembler should produce a large number of different states so that the test of the teleporter will be statistically significant. That is, if the sample of states Ψ is random enough and samples enough of the possible infinite space in a way that the teleporter operators cannot anticipate, the more likely that it will be a fair test and that the teleporter operators cannot somehow cheat by using some inside knowledge of the states being transmitted.

The next part of the tester will be to see how good the teleporter is doing. If an unknown state Ψ is sent in by Doug to Alice, we want the state that emerges on Bob's side of the teleporter to be as close to Ψ as possible. This is quality control and one measure of the quality of such a state transport is called the quantum fidelity. The fidelity is 100% if the outputted teleported state is identical to the inputted stated to be teleported. The fidelity is 0% if the outputted teleported state is as far from the inputted state as possible, which would be hard to arrange without trying. If the teleporter simply completely scrambles the input state, then (on average) the fidelity of the output states, with respect to the input states, will be 50%. That is, if the teleporter sucks, on average the outputted state agrees with the inputted state only half the time. Hence, anything better than 50% is considered good; I would call it the bronze standard of teleportation. The gold standard would be 82% fidelity, which is the minimum required for the teleporter to be used to teleport one-half of an entangled photon pair and still violate Bell's inequality. Anything

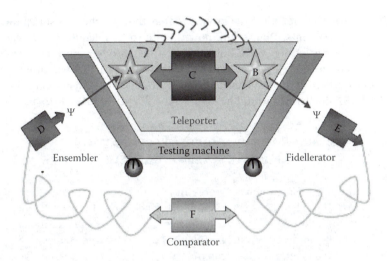

Figure 5.5 NIQuIST quantum teleporter-testing machine: The quantum teleporter to be tested (inverted trapezoid) is carefully lowered into the testing machine (bathtub shape with clawed-ball feet). As before, in the teleporter, Charlie (C) shares a pair of entangled photons with Alice (A) and Bob (B). Now, on the left side of the testing machine sits Doug (D), who works for NIQuIST and who has at his disposal a device called the "ensembler" that he can program to fire off a collection (an ensemble) of large numbers of photons in sets of different quantum polarization states Ψ that are unknown to Alice, Bob, or Charlie. On the right side of the testing machine sits Ellen (E), also an employee of NIQuIST, and she runs a device called a "fidellerator" that measures the teleported state outputted by the teleporter in order to gauge the output's fidelity or goodness compared to the state that was actually sent by Doug. All the data of the states Ψ Doug sent in from the ensembler and those states Ψ that Ellen collected at the fidellerator are forwarded to François (F) who runs a machine called the "comparator" that makes a record of all the states sent in and all the states teleported out and their relative quality with respect to each other.

less than 82% can be modeled with a classical local hidden variable and hence does not really test quantum mechanics. A new character in our pantheon, Ellen, on the right, extracts the fidelity; she runs a machine called the "fidellerator" that measures the state of the teleported photon and compares it with the state that was actually teleported.[48] Doug sends the complete information on the states he provided, and Ellen sends the complete information on the states she received, to François (bottom), who runs a machine called the comparator (which I do not put in quotes because "comparator" is a real word), which compares what Doug sent to what Ellen got and then ranks the teleporter in the test. In Box 5.3 I summarize six important take-home points about my quantum teleporter testing protocol.

The bottom line for all three groups is that they all cheated somehow. In all three experiments, Doug's "unknown" state to be teleported was known to everybody, including Alice, Bob, and Charlie. In one experiment, Doug's unknown quantum state was actually on the same photon that was provided to Alice by Charlie, a point that I don't really understand at all. In one experiment, the NIQuIST fidellerator would have rated the fidelity at just a little bit above

---

BOX 5.3   THE NIQUIST TELEPORTATION MACHINE TESTING PROTOCOL

1. The NIQuIST employees agree in advance how stringent their test should be. No experiment can measure anything to an infinite number of significant digits. To measure the width of a nickel, you would first use rulers of ever finer line spaces and then make many measurements with that same ruler to improve your accuracy, which improves (as in the presidential election poll) only quadratically with the number of measurements. NIQuIST must specify in advance to what precision they will test the teleporter, say accommodating a fidelity of 31.4159% but no better (six significant digits). That precision then tells NIQuIST how precise they need to specify the constants a and b for the state to be teleported and how many states with that same state-specifying constants, a and b, need to be teleported to ensure this six figure accuracy.

2. Doug opens the throttle on the ensembler, which then chooses random complex numbers a and b, with a precision specified in step 1, for the state to be teleported, and then begins firing off a large number of states with identical a and b and pitching them into the teleporter to Alice. The number of states provided by Doug depends on the number of significant digits specified in step 1. Because, in this example, the number of significant digits under test is six, corresponding to a precision of one part in 100,000, the number of states required to reach this precision is on the order of the square of that, so around 100,000 × 10,000, which is 100,000,000,000.

3. Alice, Bob, and Charlie carry out the teleportation protocol on all these copies of the single (unknown) state provided by Doug and the ensembler and then launch them out of the teleporter to Ellen who is ready to catch the many copies of the same teleported state in her fidellerator.

4. Doug and Ellen share (through a classical channel or in advance) what the teleported state was, and Ellen, with access to what was actually teleported from her measurements, slams her foot down on the fidellerator pedal and carries out the fidelity measurement, comparing the fidelity of that which was teleported to that which was intended to be teleported.
5. Doug and Ellen provide their data to François, who throws a large lever on the comparator, which then gauges the "goodness" of the teleporter for that particular state (choice of a and b).
6. Doug then hits the randomizer on the ensembler, which then chooses at random a different a and b, a different state Ψ to be teleported, and they do this all over again. They may need to test many hundreds of thousands of randomly different states to be sure that Alice, Bob, and Charlie are not cheating—guessing in advance what states Ψ will be used and optimizing their teleporter for this cheat.

50%, which the authors argued meant that they teleported a "nonclassical state" but it was well below the 82% limit and so their entire experiment could be explained with a local hidden variable theory and was therefore "classical" in that sense. However, these experiments were all experimental tour de force setups and the demonstration of quantum teleportation captured the imagination of the scientific and secular communities around the world with hundreds of press releases and television and radio shows on the topic.

So when we will be able to teleport a *person* to Mars? Well, even the *Star Trek* transporter had a limited range and typically could just teleport a person from orbit to the planet's surface or between closely located ships. Thus, even "Scotty" could not teleport a person from Earth to Mars. So far, experimentalists have teleported states of individual ions, electrons, or atomic nuclei over distances of nanometers to meters, and states of individual photons over tens of kilometers. Commander Riker is composed roughly of 10^{28} atoms. A very important question to ask is, do we need to teleport the full quantum mechanical state of the person to get a reliably good clone on the other end, or is teleporting the classical state of the person sufficient? If it is just the classical state of the person, then in addition to specifying the 10^{28} atoms needed to reconstruct Commander Riker, we'd need to specify which atoms (from the periodic table of elements), which would bring us up to 10^{30} bits of classical information, and then we'd need to specify how to hook the atoms together, and a classical calculation gives something like 10^{45} classical bits that would have to be transmitted from the Starship *Enterprise* to the planet Nervala IV.[49]

If teleporting the classical state of the person is sufficient, we need to teleport just the information about what atoms she is composed of and their arrangement, perhaps not much more than 10^{45} classical bits of information, and then the quantum teleporter becomes superfluous. Even though the *Star Trek* reference guides claim humans are teleported "at the quantum level," it sure does not seem like that or else we'd not have two copies of Riker running about in violation of the no-cloning theorem. If classical data suffice, then the future teleporter would be much like our classical fax machine with a built-in shredder. It would make a copy of Riker's classical state (while disintegrating the original in the shredder), transport that classical information to the planet's surface, and then using locally available atoms reconstruct him. Transporting Riker in 1 second would require the transference of a little more than 10^{45} classical bits in 1 second, which would be a very tall order. All the information traffic on the entire Internet has a bit rate transmission of approximately 100 exabytes a month, where an exabyte is 10^{18} bytes or around 10^{19} bits. Hence, the total transfer rate of all the information rate of all the data on the Internet is around 10^{14} bits per second. It would take 10^{31} Internets running in parallel to handle all the data needed to *classically* teleport Riker in 1 second. If we just use the Internet we have, then it would take 10^{31} seconds or 10^{23} years of total Internet traffic at today's transmission rates to transport Riker to the surface. The universe is only around 10^{10} years or 10 billion years old, so we have to wait 10^{13} times the total lifetime of the universe for one successful teleportation. That's *classical*.

Things get much, much worse if the essence of Riker requires transmission of quantum information and entanglement, as some have proposed that it is required for the working of the human brain. Then, we are talking about the transmission of exponentially more information to reconstruct Riker faithfully and, because of no cloning, any copy of him left behind would be left a blithering idiot, incapable of joining a rebellion unless it is at the Bedlam insane asylum. I will discuss these quantum versus classical origins of consciousness, and their ramifications for human teleportation, in Chapter 6, "Hilbert Space—The Final Frontier." While quantum teleportation may not be that useful for teleporting persons, it is very useful for teleporting quantum states. I briefly mentioned in quantum cryptography that one way to go distances longer than 100 kilometers (62 miles) is to extend the cryptography network using a quantum repeater. The quantum repeater is at its heart a quantum state teleporter. Along the quantum cryptography line, there are nodes or places that contain built-in quantum teleporters. The quantum state cannot be transmitted faithfully over more than approximately 100 kilometers (60 miles) without help. To go 200 kilometers (124 miles), we put in two quantum teleporters. At the 50-kilometer point, Charlie shares entangled photons with Alice at the 0-kilometer point and Bob, at the 100-kilometer point, then Claude at the 150-kilometer point shares another pair of entangled photons with Bob at the 100-kilometer point and Barthy, at the 200-kilometer point. Bob makes a certain type of destructive measurement and—hey presto!—Alice and Barthy share a pair of entangled photons, which then can be used to implement the E91 quantum cryptography protocol between Alice and Barthy. They could instead use the shared entanglement to teleport a random state from Alice to Barthy for the BB84 protocol, and if Alice and Barthy each have a quantum computer, they could then use this shared entanglement to teleport quantum information between their quantum computers, forming the first two nodes of the quantum Internet. All this has to be done by transmitting and manipulating photons![50]

Even if the goal is to move quantum information in short distances, quantum teleporters are very useful. In the all-optical approach to quantum computing, the quantum logical two-qubit ENT gates are made through a process of single-photon manipulation, detection, and—most critically—teleportation.[51] Without quantum teleportation, the ENT gates will not work at all. Then, there is the intra-computer communication problem. On any computer, classical or quantum, you have to move bits of information around the computer chip or around the computer chassis. Bits of information stored in memory have to be moved into the central processor for processing and back out again. On a classical computer, this is all accomplished with wires carrying the nuggets of current and voltages, which make up the zeros and ones of the classical computer, to and fro. On a quantum computer, you cannot just stick a quantum mechanical state into a room-temperature wire; the state will collapse into random nonsense instantly. You might use superconducting wires but that is tricky and they are not without their own loss and decoherence. Another approach is to convert the state from some electronic, ionic, or atomic state into a photon state and then run optical fibers all over and then convert back, but this too can be lossy. On the future quantum computer chip itself, there is no room for optical fibers. One approach to quantum information transport is the quantum mechanical equivalent of a fireman's bucket brigade. The quantum state stored on Alice's qubit is handed off to Bob's qubit using something called a SWAP gate, then Bob hands it off to Cedrick, who hands it off to Doris, who hands it off to,…, who hands it off to Yetta, who hands it off to Zenobia. This rather long-winded process is subject to error and decoherence at each step (spilling the water out of the bucket at each handoff) and you cannot go really far with it at all. A better approach is to establish two-party entanglement between Alice and Zenobia in advance of the moment the data transfer is to take place and then teleport the state instantly (using the free classical communication that can be carried out over the copper wires). It is widely believed that the future quantum computer chip will use massive amounts of quantum teleportation as the means to transport information quickly and efficiently (and with high fidelity) about the computer chip, around the computer chassis, and around the world on the quantum Internet.

THE TALE OF THE TRUE TIMEPIECE

Quantum entanglement can be used to improve the precision of atomic clocks. What is more, the atomic clock in which this has been demonstrated is none other than our old friend, the ion-trap quantum computer that has now been tweaked so that it is not much good for computing but it is very good for timekeeping.[52] To understand the connection between a quantum computer and a quantum atomic clock, I will have to introduce something I call the quantum Rosetta Stone. The original Rosetta Stone is an actual chunk of rock, now in the British Museum, that has writing carved into it that spells out a legal decree issued in Memphis, Egypt, in 196 BC by the Egyptian King Ptolemy V. On the stone, the King's decree is written out three times in three different languages, ancient Greek script at the bottom, an Egyptian form of writing called Middle (Ptolemaic) Demotic script (appropriately in the middle), and the ancient Egyptian hieroglyphics at the top of the stone. The stone was originally displayed in a temple in Memphis, Egypt, but eventually wound up used in building material and moved around a lot, finally coming to rest in Rosetta (Rashid), Egypt, where it was rediscovered in 1799 by one of Napoleon's soldiers during the French conquest of Egypt. Up until this time, despite many years of work, ancient Egyptian hieroglyphics had never been deciphered. The Rosetta Stone was the key to unlocking their secret, as the same text was written on the same stone in Demotic and ancient Greek, which were well known in 1799. It was as if all hieroglyphics consisted of a big Crypto-Quote puzzle and somebody just handed you the key (on a bit of rock) telling you what many of the enciphered letters were. Once you had some symbols, you (or a crafty linguist) could painstakingly work out the rest.

In our paper "A Quantum Rosetta Stone for Interferometry," we laid out a similar three-level decoder for understanding the connection between a quantum computer, a quantum atomic clock, and a quantum gyroscope.[53] The idea is that all three of them are a type of interferometer, a machine in physics that exploits coherence to measure something very accurately. The quantum gyroscope measures rotation, the quantum atomic clock measures time, and the quantum computer measures, in the case of Shor's factoring algorithm, the period of a function that will hand you the factor of a very large number. Hence, there is a mapping or relation between a simple quantum gyroscope, a simple quantum atomic clock, and a simple quantum computer circuit, shown in Figure 5.6.

We sent the "Quantum Rosetta Stone" paper for consideration in the *Journal of Modern Optics*, for an issue of compiled invited talks given at the *32nd Winter Symposium on the Physics of Quantum Electronics, January 6–10, 2002* (at the Snowbird Ski Resort). Papers in scientific journals undergo a peer review process whereby anonymous referees (your peers) peer inside your paper and they apply peer pressure to get you to make changes you do not want to make until you go absolutely mad and fling yourself off the edge of the Santa Monica wharf—whereupon a mysterious wind arises from the Pacific Ocean and blows you back onto the deck (pier pressure). For invited papers for special issues, they are typically looking for review papers and the refereeing is a little bit less harsh than it might be for a lone uninvited paper. In this case, the referee had only one comment, he or she did not like the title and demanded that we change it, as "…the term 'Rosetta Stone' should be reserved only for works that, like the original Rosetta Stone, are *important*…." This was a paper-thinly veiled insult and the referee clearly was stating that in his or her opinion, our work was not important.

In our rebuttal to the referee, we first pointed out that if the work was important, we would have not submitted it to the *Journal of Modern Optics* in the first place—bazinga!—and that, in common usage, the term "Rosetta Stone" is used to describe something that is the key that allows you to translate one thing into another, important or not, and to support our conclusion, we pointed out the existence of a foreign-language learning software (called Rosetta Stone), a video game (called Double Dragon 3: The Rosetta Stone), a guide to the UNIX computer operating system (called Rosetta Stone for UNIX), and a Scottish, rock, boy band from the 1970s (called Rosetta Stone). The referee completely capitulated and agreed that "…the term 'Rosetta Stone' is no longer reserved strictly for things that are important…," and we got that thing published with "Rosetta Stone" in the title!

In Figure 5.6, we show the Quantum Rosetta Stone. The origins of this came about when I was trying to explain simple quantum computing circuit diagrams, such as in Figure 5.6c, to scientists at the JPL who were not familiar with these diagrams, but who were familiar with spatial interferometers as in Figure 5.6a or temporal interferometers as in Figure 5.6b. The meeting room would light up and the interferometry experts would all say, "Ah, now we understand what that quantum computer circuit diagram is!" Hence, it really was a Rosetta stone, where experts in spatial interferometry (ancient Greek script) or temporal interferometry (ancient Demotic script) suddenly had the key to understand an equivalent device written out as a quantum computing circuit diagram (ancient hieroglyphics).[54]

Although initially an icon for interpreting quantum computer circuit diagrams, the Quantum Rosetta Stone eventually became to mean much more to us than that. In Figure 5.6, the researchers who built spatial interferometers (top) or temporal interferometers (middle) were interested in metrology, the science of measurement, not in computing. The spatial interferometers were exquisite at measuring the transit of distance and the temporal interferometers were superb

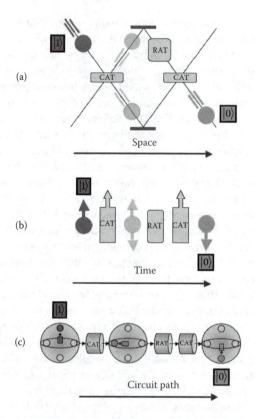

(a)

Space

(b)

Time

(c)

Circuit path

Figure 5.6   The Quantum Rosetta Stone: A photon or other particle in a spatial interferometer (a), an ion in an atomic clock temporal interferometer (b), and the equivalent quantum computer circuit with a RAT gate sandwiched between two CAT gates. In the spatial interferometer (a), two objects called beam splitters implement the CAT gate and a third object called a phase shifter implements the unknown RAT gate and then a second beam splitter carries out the final CAT gate. These gates are implemented on particles moving from left to right through the interferometer in space. In the ion-trap atomic clock temporal interferometer (b), the flow from left to right is in time and the stationary objects, typically ion spin states, are being manipulated by laser pulses that implement two CAT gates. Sandwiched between the two CAT gates is a RAT gate that corresponds to the passage of time between the two CAT gates. Finally, the bottom of the Quantum Rosetta Stone (c) contains a simple quantum circuit that can be implemented using either the spatial interferometer (a) or the temporal interferometer (b). The point is that panel (a) is a well-understood device to optical or matter wave interferometry experts, panel (b) is a well-understood device to atomic clock experts, and panel (c) is a well-understood circuit to quantum computer experts. The Rosetta Stone allows an expert in one of the three fields to immediately translate what the other two experts are doing, just like the real Rosetta Stone allowed writers of ancient Greek script or writers of middle Demotic script to decode Egyptian hieroglyphics.

at measuring the passage of time. Indeed, as shown in Figure 5.6b, the temporal interferometer is the basis for how an atomic clock works. An atom or ion in the upper spin state $|\uparrow\rangle$ is whacked with a CAT laser pulse (called a "pi-over-two" pulse in atomic clock jargon) to prepare a CAT state that is the coherent superposition of the up-and-down state, $|\uparrow\rangle + |\downarrow\rangle$. Then, the RAT gate is implemented by the simple passage of time itself. After a time, a second CAT laser pulse is applied to the thing that causes the spin to collapse either back into spin up $|\uparrow\rangle$ or instead into spin down $|\downarrow\rangle$, with probabilities that are a simple function of the time elapsed.

By repeating the experiment many times, or doing it once on many atoms, the experimenter can accumulate statistics, from the number of times she gets spin down versus spin up, on how much time has elapsed between CAT pulses. For a typical scenario where no ENT gates are used and no entanglement generated, in the absence of all other technical sources of noise, the accuracy in the time estimation improves like our old friend, the quadratic scaling law, discussed in Chapter 1. That is, the accuracy in our signal (how much time has elapsed) improves only quadratically slowly with increasing numbers of measurements. If 10 ions are used and the time is determined to within a margin of error of 10%, then the experimenter would have to increase the number of ions to 100 to knock the error in the time signal down to 1%. This quadratic scaling law for 50 years was thought to be the ultimate limit to the accuracy of atomic clocks. Then along came quantum entanglement.

To see how entanglement helps, let us look at the output of an atomic clock. In Figure 5.6b, we will start a collection of 100 ions all pointing up $|\uparrow\rangle$. (Only one ion is shown in the figure.) We apply the first CAT pulse, creating a collection of 100 cat states, $|\rightarrow\rangle = |\uparrow\rangle + |\downarrow\rangle$. The passage of time implements the RAT gate causing the cat state $|\rightarrow\rangle$ to spin around and around the qubit sphere, with the number of rotations proportional to the time. After 180°, it will rotate to $|\leftarrow\rangle$, and then after 360°, it will be back at $|\rightarrow\rangle$. After a certain amount of time, the second CAT pulse is applied, which collapses all the cats into either up $|\uparrow\rangle$ or down $|\downarrow\rangle$, with a probability $P_{up}(t)$ versus $P_{down}(t)$ that depends on the time between the two CAT pulses, and the difference between these two probabilities is shown in Figure 5.7a.

However, because of quantum uncertainty, there will be a spread or distribution of the ion probabilities. That is, suppose, in the absence of any other source of technical noise, the predicted probability of an individual ion pointing up after 1 second is 100%; that is, $P_{up}(1 \text{ second})$ is 100%. That does not mean that all the 100 ions measured will be pointing up but that most of them will be. Because of quantum uncertainty, a few will be pointing down, and the quadratic scaling law states that, on average, 90 will point up and about 10% or 10 will point down. These numbers will change randomly from run to run but the average will be 90 up and 10 down. This means that the data do not follow the nice smooth line of the probability curve but has some vertical uncertainty to the measurement probability, ΔP, indicated by the height of the box.

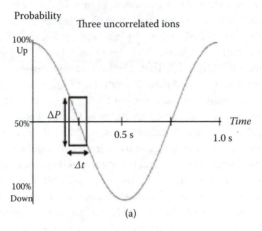

(a)

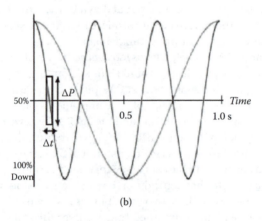

(b)

Figure 5.7   Atomic clock oscilloscope traces. In panel (a), we show the trace for three uncorrelated ions, and in panel (b), we show the trace for ions entangled in a three-ion Schrödinger cat state. The goal is to infer the passage of time (horizontal axis) from the measurement of the ratio of ions found in spin up to spin down (vertical axis). In both cases, the error in the vertical direction, ΔP, is the same. But you can see that the error in the time measurement Δt in the entangled trace (b) is smaller than that in panel (a). This is simply because the entangled (dashed line) ion trace in panel (b) oscillates three times as fast as in the unentangled (straight line) trace in panel (a) or (b). As the steepness of the sides of the oscillation increases, the timing error decreases.

But if there is some vertical uncertainty in the probability, there is a corresponding horizontal in the time, Δt, indicated by the width of the box. From the quadratic scaling law, for 100 ions, this uncertainty in probability ΔP is 10% and the corresponding uncertainty in the time Δt is one-tenth of a second. Hence, by monitoring the probability curve on the oscilloscope with 100 ions, the experimentalist can measure a time of 1 second with an accuracy of one-tenth of a second. If she wanted to knock this down to one-hundredth of a second accuracy, then she would need to use 10,000 ions, with a probability uncertainty of 100/10,000 or ΔP is 1%. This is the problem with the quadratic scaling law for errors. You need a whole lot more (quadratically more) data just to lower the error by a little bit. To drop the timing error an order of magnitude, from 10% to 1%, I have to increase the number of ions by two orders of magnitude, from 100 ions to 10,000 ions. What entanglement buys us is a change in the poor quadratic scaling law to a much better scaling law.

The insight is that, by employing ideas from quantum information processing, we can get a quadratic improvement over the quadratically bad signal-to-noise law. In this way, it is like the Grover search algorithm, where searching the phone book goes from hundreds of years to just a few hours. In the clock example, with the entanglement-induced improved scaling law, we can use 100 entangled ions to get a 10% timing error and then only increase the number of ions from 100 to 1000 (instead of 10,000) to knock the timing error down to 1%. This is a bit complicated to draw, but the point is that Figure 5.6b represents the clock gate sequence for a single ion. The idea is to draw a bunch of these gate sequences on top of each other, say 100 for 100 ions, and then put in 100 ENT gates entangling all the 100 spins with each other (right after the first CAT gate on ion one), then let the 100 independent RAT gates evolve in time, then apply 100 disentangling ENT gates between all the ions, and apply the final CAT gate again to ion one, and then measure the state of ion one to see how often it points up versus pointing down, and from that extract the time.

I illustrate the idea with 3 ions (instead of 100) in Figure 5.8.[55] In the figure, we start with three up ions and then apply a CAT gate to the top ion. We then apply an ENT gate between the top ion and the middle one, and then again between the top ion and the bottom one, which we label as just one big ENT. This creates a three-ion cat state (big blob) that acts like one big ion (composed of three little ions) with three times the spin magnitude of each individual ion. I have used small kittens to make a big cat! The big collective ion state is allowed to evolve via what is now one large RAT gate. The point is that the large RAT on the large cat state causes the thing to evolve three times faster than each individual ion cat state alone. Finally, a disentangling three-ion ENT gate is applied and the probability of the top ion pointing up or down is measured, as in Figure 5.7b. The uncertainty in the probability (vertical ΔP) in the measurement remains the same, but because the three-ion signal wiggles three times

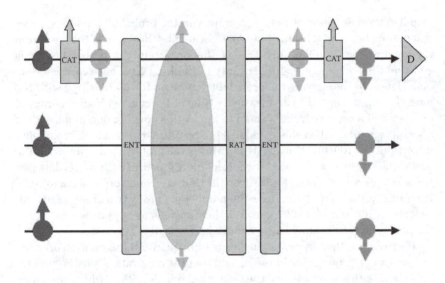

Figure 5.8 Using ideas from quantum computers to make better atomic clocks. Three ions are in an ion trap. The top ion is prepared in a CAT state. Then, all three ions are entangled in a three-ion ENT gate, which were allowed to evolve in time in the RAT gate. The ENT gate is reapplied to disentangle them and the state of the top ion is read out as a function of time in the detector "D." This procedure improves the timing accuracy of the atomic ion clock using three unentangled ions from approximately ±57% down to ±33% for this entangled case. The idea is to apply a quantum computer gate sequence to the ions and not to compute anything but to measure time more accurately.

faster, the associated error in timing (horizontal Δt) is much smaller than with three ions separately as shown in Figure 5.7b.

This method for improving atomic clocks has been championed particularly by American physicist Dave Wineland and his group at the National Institute of Standards and Technology (NIST), where they have had experiments beginning in the early 1990s through the present, exploiting entanglement and other quantum effects to improve atomic clock accuracy.[56] A future NIST atomic clock time standard may very well exploit quantum unreality, uncertainty, and nonlocality to make a better clock for some future global positioning system (GPS), which relies on accurate atomic clocks onboard each of the GPS satellites, to make accurate estimates of your location on the surface of the Earth.

How does the entanglement buy us this advantage? In the standard atomic clock, each ion is sampling the passage of time separately and independently of each other. The quadratic scaling law that applies to such a collection, going from 100 to 10,000 ions to drop the margin of error from 10% to 1%, is sometimes called the sampling theorem, and the derivation of the law assumes that the samples are *independent*. In the analogy of Galumph Polls, it is very important for their statistical analysis that each person they poll at random provides an answer that is his or her own, and independent of any other person. If the persons being polled—the pollees—colluded beforehand, somehow tipped off that the Galumph was doing an election year poll that day, the pollees could conspire to trip up Galumph by all claiming they are voting for Alice. Even if not all of the pollees collaborated thusly, if enough did so, it would still skew the poll results. Entanglement is precisely that collusion between pollees.

When the ions are entangled with each other, they are no longer considered independent ions sampling the passage of time. For the atomic clocks, if the ions are prepared in a massive 100-ion Schrödinger cat state, where the idea of separate ions breaks down, and you should then think of a single 100-ion quantum mechanical state with correlations built into it—correlations so strong it can skew the quantum mechanical error away from the poor quadratic scaling law into the more preferable linear law where improving the timing accuracy from 10% to 1% only requires increasing the number of ions from 100 to 1000 (linear) instead of 100 to 10,000 (quadratic). In fact, the 100 ENT gates that are deployed just after the first CAT gate produce a 100-ion cat state that acts as a single ion with one giant spin vector, which is 100 times bigger than the spin vector of any ion individually.

This idea is illustrated in Figure 5.8 with just three ions. In an ordinary atomic clock, the three independent ions would sample the passage of time independently and their results would be averaged to give the time with an error of ±57%. (Not very good, but you would in practice use far more ions.) Now, instead of taking the ions independently, we quantum correlate them into a type of three-ion Schrödinger cat state by applying a three-qubit ENT gate, which is just 2 two-qubit ENT gates, the first between the top ion and the middle ion and a second between the top ion and the bottom ion. Now, the ions are "talking" to each other as they sample the passage of time (the RAT gates) and they conspire to skew the sampling odds, so when the time is read out, the uncertainty drops from ±57% (for the three independent ions) quadratically down to ±33% (for the three entangled ions). This procedure scales up to an arbitrary number of ions, if one can control the noise from the environment; noise that tends to very rapidly collapse the big cat before it has much time to sense the passage of much time.

If I send in 100 independent ions, the error is ±10%, but if I entangle those same 100 ions, the margin of error drops by an order of magnitude to ±1%. To get this same improvement without entanglement, I would have to increase the number of independent ions from 100 to 10,000. I have beaten the classical quadratic sampling law! I get 10 times better results not by quadratically increasing the number of ions polled from 100 to 10,000 but by forcing the 100 ions to collude with each other before the poll is taken. Is this cheating? Yes. But it is the ghost of Werner Heisenberg cheating Mother Nature herself. The passage of time does not care if the clock ions measuring that passage are conspiring with each other or not. There is a connection between this improved scaling law and the Heisenberg Uncertainty Principle, and so the new linear scaling law is called the Heisenberg limit.

While 100 ions are beyond the limit of current technology, something similar has been demonstrated with over 10 ions; showing the proof of principle. But the important thing to note is that the Quantum Rosetta Stone is at work. The improved atomic clock (Figure 5.6b in the Rosetta Stone) is reinterpreted in terms of the equivalent quantum computer diagram (Figure 5.6c), and the laser pulses needed to manipulate the ions and read out the time are now coded in as quantum computer gate sequences. The end result is, however, not a mathematical calculation like searching a phone book with a quadratic improvement in speed (Grover's search algorithm) but instead an atomic clock with a quadratic improvement in timing accuracy. At its heart, the quantum computer is a quantum interferometer. But the atomic clock is also a temporal ion interferometer. There is a mapping, the Rosetta Stone, between them. Techniques you develop to make better atomic clocks can be used to make better quantum computers and techniques you develop to make better quantum computers can, via the Rosetta Stone, be almost immediately applied to making better atomic clocks. It is no accident that the same laboratories that perform atomic clock research with ions in ion traps are those that perform quantum-computing research with ions in ion traps. You adjust a few knobs and your atomic clock is a quantum computer. Adjust them back and then your quantum computer is an atomic clock. Advances in one field led to advances in the other almost immediately. In the next few sections, we will explore this connection further—exploiting ideas from quantum computing, not to compute things better (faster) than we can compute classically, but to measure, sense, and image things better than we can measure, sense, and image classically.

There is a pragmatic reason to think about these quantum sensors and imagers as well. A quantum computer capable of hacking the Internet will require millions or billions or trillions of qubits. Current-day quantum computers have about 10. It is hard to believe that the quantum military industrial complex will pay for 50 years' worth of research for a machine that is entirely useless for all those 50 years until you hit year 51. An atomic clock with only 100 ions is an order of magnitude more accurate than one with 10 ions. Everybody will pay for a clock or a sensor or an imager that is an order of magnitude better than the last generation. In this way, we can build much bigger quantum interferometers, from 10 to 100, then 100 to 1000, then 1000 to 10,000,..., all the way up to 1 billion—selling it at each stage as an atomic clock that is 10 times more accurate than the previous one. When we hit 1 billion ions, we turn a few knobs à la the Rosetta Stone and turn the billion-qubit quantum atomic clock into the billion-qubit quantum computer and then sell the same hardware to the NSA as the factoring engine that can hack the Internet. The atomic clock users pay for all the intermediate steps!

FROM QUANTUM COMPUTERS TO QUANTUM SENSORS

If you think of an atomic clock as a device that "senses" the passage of time, and a quantum entangled atomic clock as a device that "senses" the passage of time more accurately than its unentangled classical partner, then it is easy to think of other sensors that would see an improvement in the signal-to-noise ratio if you were to wave your quantum magic wand over the thing and entangle the particles doing the sensing. This takes us to the spatial interferometer of the Quantum Rosetta Stone in Figure 5.6a. The power of the Rosetta Stone is that ideas for improving computation in Figure 5.6c or improving atomic clocks in 5.6b can immediately be applied to improving the spatial interferometer in Figure 5.6a. If the math is the same, the physics must be the same! The trick is just translating the mathematics from the temporal atomic clock domain into the spatial domain.

The spatial interferometer in Figure 5.6a has a long history of making precise measurements, in which the particles or "stuff" passing through the interferometer are particles of light. Such a gizmo is called an optical interferometer and can be used to very precisely measure minute changes in spatial distance; in particular, if the length of the upper arm of the interferometer changes with respect to the lower arm, you can monitor the interference pattern of the beam at the output and measure very precisely these spatial changes. This is done in exactly the same way as the atomic clock, but now instead of stationary ions in traps, we have moving photons (particles of light) flying through the interferometer. In the classical setup, the beam of photons is just a beam of light from a lamp or a laser. The CAT gate is implemented on the beam of light with a partially reflecting device, like a one-way mirror, which is called a "beam splitter." With a 50–50 beam splitter (in the classical notion), half the light beam is routed through the upper arms of the interferometer and half the light is routed through the lower half. In the quantum interpretation, each photon is prepared in a cat state of the photon, simultaneously taking the upper branch and the lower branch.

This is the spatial analog of the ions in the trap: ion spin up is photon in the upper branch; ion spin down is photon in the lower branch. The entire optical interferometer may be described in the quantum computing language of CAT, RAT, and ENT gates, and these are the same gates used in the atomic clock interferometer with an appropriate mapping to the optical interferometer. The detectors now do not detect if the ions are pointing up or pointing down but whether the photon exits the upper arm of the interferometer or the lower arm. It may seem quite different from the clock, but the math and quantum circuit description are identical. Hence, anything we can do to improve the temporal interferometer—the atomic clock—we can use to improve the spatial optical interferometer—via the Quantum Rosetta Stone. We just have to take some care in translations.

What good are spatial interferometers? One of the first applications was an experiment that led to the support of Einstein's Special Theory of Relativity, the so-called Michelson–Morley experiment, carried out in 1887. In 1887, the classical theory of light postulated that the speed of light was different for different observers in motion. The classical theory postulated that the entire universe was filled with an invisible and nearly undetectable airy gelatinous goop, which permeated everything including our bodies, and which was called the "luminiferous aether," which is Latin for "light-bearing airy gelatinous goop." By the mid-1800s, it had been proved conclusively that light was a wave, after a hundred years of arguing back and forth whether it was a wave or a particle (until the 1900s when the argument started all over again in quantum theory where we now know light is simultaneously a wave *and* a particle). All waves previously encountered—sound waves, tidal waves, and earthquake waves—were waves *in* something that was doing the waving bit. Sound waves are waves in air. Tidal waves are waves in water. Earthquake waves are waves in dirt. What were light waves themselves waves *in*? Well, light travels not only through clear air but also through empty space between stars. Hence, the hypothesis was that there existed this invisible airy gelatinous goop filling all of empty space that was practically undetectable by any means and that was the stuff light waves were supposed to be wiggling about in. American scientists Albert Michelson and Edward Morley set up their 1887 experiment to directly detect the presence of this airy gelatinous goop.

The idea was simple. It was assumed (at least in 1887) that the goop was at rest with respect to the absolute spatial reference frame of the universe (a reference frame that was postulated by the famous English scientist Isaac Newton), and thus the Earth on its orbit around the Sun should then be plowing through the light-bearing airy gelatinous goop at the heady speed of 30 kilometers per second (67,000 miles per hour). The idea was then to measure the speed of light in a direction parallel to the Earth's motion and compare that to the speed of light in a direction perpendicular to the Earth's motion. The concept is that for Michelson and Morley, sitting in the laboratory on Earth, the laboratory and its spatial interferometer should feel a strong headwind from this moving gelatinous goop as the Earth plowed through it, in much the same way your Labradoodle feels a 130-kilometer-per-hour (80 mile per hour) wind when he sticks his head out your passenger side window when you are driving 130 kilometers per hour on an otherwise windless day.

Now, the light in the interferometer is supposedly also moving through this goop and should also feel this strong headwind, when it is moving in the direction the Earth is moving, and should not feel it at all when it is moving perpendicular or at right angles to the direction the Earth is moving. (If your Labradoodle spits in the forward direction, the spittle moves slower than his usual spit escape velocity, and if he spits in the backward direction, the spittle moves faster than his usual spit escape velocity.) Thus, the light moving against the headwind should move a bit slower than usual—the usual speed of light minus the headwind speed—and the light moving with the wind should move a bit faster than usual—the usual speed of light plus the tailwind. I see this effect every time I fly round-trip from Chicago to Beijing, where there is always a strong prevailing wind blowing from the west. I'm traveling this route in June and I can see that my flight from Chicago to Beijing (against the wind) is 13 hours and 25 minutes long (13.42 hours) but that the return trip from Beijing to Chicago (with the wind) is only 12 hours and 40 minutes (12.67 hours).

The plane flies the same exact route up near the North Pole coming and going. If I did not know about this wind, I would assume that the flight path from Chicago to Beijing was longer in *distance* than the flight path from Beijing to Chicago! Cruising velocity for passenger jets is approximately 800 kilometers per hour (500 miles per hour) and so, again not knowing about the wind, I would say that the flight path *distance* from Chicago to Beijing is 13.42 hours × 800 kilometers per hour = 10,736 kilometers and that the flight path distance from Beijing to Chicago is 10,136 kilometers—a 600-kilometer difference! In my jet plane with just my watch, I cannot tell the difference between two trips of different distances and two trips with the same distance, with and against the wind. Neither could the spatial interferometer sitting in Michelson and Morley's laboratory.

Hence, the setup is to have the upper arm of the interferometer pointing into the headwind/tailwind with the lower arm at right angles with neither headwind/tailwind. The net effect is that the light plowing through the headwind/tailwind moves a slightly longer effective distance than the light moving at right angles to the wind. The speed of light is 300,000 kilometers per second. The Earth is moving at 30 kilometers per second. The effect is very small, about a 0.01% change in the effective upper path length with respect to the lower path length. However, the Michelson–Morley interferometer was capable of measuring this as it was able to detect distances of a few hundred nanometers (a few millionths of an inch). The output of the interferometer is called an interferogram, similar to those shown in Figure 5.7 (interferograms in time rather than space). The interferogram is the difference of the light intensity of the light exiting the upper arm of the interferometer in Figure 5.6a and that exiting the lower arm. The idea is to monitor the movement of those wave peaks and wave troughs in the interferogram as the distances between the arms changed, which in this case was as the entire interferometer was rotated about. In the 1887 experiment, they were accurate enough to detect the movement of these peaks and troughs at the level of 0.01%. And, after all this work, they saw nothing. Nada. Nichts. Niente. No changes in the peaks, no evidence of a tailwind, no evidence of a headwind, no evidence of any wind at all. No wind? No airy gelatinous goop? No luminiferous aether? What was going on? What's the poop on the no-goop scoop? The Michelson–Morley experiment is sometimes called the most famous failed experiment of all time. They looked for a headwind of goop, that everybody (who was anybody) predicted should be there, but they found absolutely nothing instead. "O frabjous day! Callooh! Callay!" The theoretical physicists chortled in their joy. There was work to do.

The scoop on the airy gelatinous goop had to wait a few more years until Einstein's 1905 publication of his Special Theory of Relativity. In that theory, the primary postulate is that the speed of light is the same to all observers regardless of their speed. That is equivalent to stating that the speed of my passenger jet, relative to the ground, is independent of any tailwind or headwind. In that case, the presence or absence of the wind is irrelevant and the air that makes up the wind is irrelevant. No wind, no goop, and in one fell swoop Einstein rid the world of the notion of that notorious airy gelatinous goop once and for all.[57] Thus began the death spiral of the luminiferous aether—light waves (unlike sound waves that need air to wiggle in, or tidal waves that need water to wiggle in, or earthquake waves that need earth to wiggle in) need nothing to wiggle in at all. Einstein's postulate, that the speed of the light waves was a constant, independent of the speed of the observer, seemed to defy common sense. Obviously, the speed of my passenger jet depends on the speed of the headwind. But what is sensible for headwinds and tailwinds and a Boeing 777 is not sensible for light waves. In any case, the scientific method allows us to get around common sense in the laboratory.

The Michelson–Morley experiment looked for the headwind; Einstein predicted there should not be any, and the experiment did not see any. No matter how much this might defy common sense, if the data support the theory, you must accept them. Light waves are special and defy common sense. Common sense is our ability to predict the future on the basis of our past experiences. Prior to 1887, humans had no past experiences with experiments with things moving much faster than the speed of a galloping horse or a train or a steamship, where the relativistic effects are very small. You have to be moving very fast to see these effects, and the Earth's speed of 30 kilometers per second is just at the edge of what you would need to start running into problems with Newton's theory and Einstein's theory, at least in 1887. But the bottom line is that these spatial interferometers are very good at measuring distances very precisely. What else can we do with them other than measure the speed of light in a nonexistent light-bearing airy gelatinous goop?

One of the most sensitive measuring devices ever built is the great Laser Interferometer Gravitational-Wave Observatory (LIGO). This device (more properly "devices") also tests predictions of Einstein's theory of relativity as well as his general theory about gravitation. In particular, Einstein's theory predicts that a binary system of two co-orbiting neutron stars should emit gravity waves—ripples in the fabric of space and time. These ripples, when they arrive at Earth, would cause things in their path to stretch and shrink as the ripples go by—much like a large rubber lifeboat stretches and shrinks as ocean swells ebb and flow beneath it on the sea. Instead of a large rubber lifeboat, LIGO consists of two laser interferometers, of the same kind used by Michelson and Morley, with giant 4-kilometer-long (2 1/2-mile-long) arms. One interferometer sits just down the road from my office, in Livingston, Louisiana, and the other sits in Washington State at the nuclear weapons development laboratory at Hanford.

The idea is that as gravity waves from distant binary star systems pass through the Earth, they will stretch and shrink the perpendicular arms of the interferometers by just enough so the light traveling in the interferometer can detect their passage by shifts of the interferogram shown in Figure 5.7. One arm will be for a brief millisecond a bit longer than the other, as the gravity wave passes, and this bit of a stretch should show up in the data. The binary stars are far, and the gravity waves are weak; hence, the stretching and shrinking are minute. But LIGO is so sensitive that if the upper arm in Figure 5.6a stretches with respect to the lower arm by an amount around the size of a single proton (around 1 femtometer or 10^{-15} meters or 0.01 trillionths of an inch), then LIGO will catch it.

LIGO is the most sensitive length-measuring device around, but even at this sensitivity, it is still limited by the same classical quadratic scaling laws that limit the clocks and the pollsters. So long as the photons from the laser entering the interferometer are independent of each other, the improvement in sensitivity requires that I have to quadratically increase the number of photons (the laser power) to get a small increase in length-measuring sensitivity. Hence, if 100 photons (per second) gives me a 10% margin of error in the length measurement, then if the photons are uncorrelated, I need to quadratically boost the laser power to 100×100 or 10,000 photons to drop the margin of error from 10% to 1%. Once again, via the Rosetta Stone, I can instantly declare that if I can quantum correlate or entangle those photons, then I change the scaling law. If my photons are all entangled and 100 gives me a margin of error of 10%, I only have to increase the power to 1000 photons to get a margin of error of 1%. For LIGO, such a scaling is a big deal. They currently operate with around 10^{20} circulating photons (or approximately 10 kilowatts) of laser power. Already, they are at the point where the mirrors and beam splitters buckle, bend, and warp from being heated by the laser beam. The poor quadratic scaling law says that if they want to boost the sensitivity by one order of magnitude (say from a 10% margin of error to 1%), they would need to use $10^{20} \times 10^{20}$ or 10^{40} photons (per second), which translates to around a hundred megawatts of circulating laser power—about the same power generated by a good lightning strike. This would do far more damage than just bending some of the mirrors.[58] Laser heating induces serious disturbance to the warp and the woof of the immense optical latticework that comprises the great LIGO interferometers.

The idea of improving optical interferometers with quantum weirdness has been around since 1981 when American physicist Carlton Caves proposed using a nonclassical state of light call "squeezed" light to improve gravity wave interferometers such as LIGO. The squeezing in squeezed light is with respect to the Heisenberg Uncertainty Principle, which, in the case of LIGO, states that you cannot measure the lengths of the arms to infinite precision with a light field but that there is some trade-off. In the case of a spatial optical interferometer, the trade-off is in the fluctuations or uncertainty in the power of the laser beam. The particular Heisenberg Uncertainty Principle, in play here, states, more precisely, that you cannot measure the length of the arms to infinite precision and measure the power in the laser beam to infinite precision. Using standard laser light, this trade-off is equalized. The uncertainty in the interferometer arm length is balanced by the uncertainty in the power of the laser beam. For many years, it was thought that this balance (the origin of the quadratic scaling law) was the best you could do with laser light. What Caves showed is that you can upset this balance without violating Heisenberg's Uncertainty Principle. You can improve over the balanced situation by squeezing down the uncertainty in the arm length measurement at the cost of squeezing up the uncertainty in the optical field intensity. An analogy might be taking a perfectly round water balloon and squeezing it at the equator—it will have to pooch up at the north and south poles to compensate. If the diameter from the east pole to the west pole of the balloon represents uncertainty in spatial distance and that from the north pole to the south pole represents uncertainty in power, then by squeezing the balloon at the equator, you can reduce uncertainty in spatial distance (that you care about) at the expense of more uncertainty in power (of which you do not care a whit). Caves pointed out that as long as we are interested in the arm length and not the optical power intensity, we could do better measurements of arm length because we did not care much about the uncertainty in optical beam intensity. You cannot do this with normal laser light, but instead you must use "squeezed" light that comes from shining laser light into a particularly made crystal, something like a quartz crystal but special, which is designed to convert the laser light into squeezed light.[59]

If done correctly, typically by mixing ordinary light and squeezed light into the interferometer, you get the rollover from the quadratic error scaling law to the linear error scaling law. Hence, if ordinary laser light of 100 watts in LIGO gives me a 10% margin of error in the length measurement, then I need to quadratically boost the laser power to 100×100 or 10,000 watts to drop the margin of error from 10% to 1%. Instead, if I use squeezed light, then I change the scaling law to Heisenberg scaling. If the squeezed light of 100 watts gives me a margin of error of 10%, I only have to increase the power to 1000 watts to get a margin of error of 1%. Another way to look at this is that if my margin of error in length measurement is 10% using 100 watts of ordinary laser power, where I risk cooking the mirrors, I can get the same margin of error of 10% using only 10 watts of squeezed light, where the mirrors may be warm but not cooked. This squeezed-light approach has just recently been implemented in the GEO 6000 gravity wave interferometer near Hanover, Germany, which is part of the international LIGO gravity wave interferometer collaboration.[60]

Squeezed states of light can be used as qubits! Indeed, such squeezed-light qubits were the quantum information bearers in the Kimble team's quantum teleportation experiment and have been used in quantum cryptography and proposals for optical quantum computing. The most common idea is to assign a beam of light that is squeezed in the position direction to be the $|0\rangle$ state of the qubit and the intensity squeezed to be the $|1\rangle$ state, and then with mirrors and beam splitters and other optical devices, a quantum information machine may be constructed. There is the Quantum Rosetta Stone at work. States of light proposed in 1981 to make improved quantum spatial interferometers have found their way into quantum information processing interferometers. Because the amount of squeezing can in principle vary from 0% squeezing to 100% squeezing, this approach to quantum optical information processing has the flavor of analog quantum computing. In fact, ditching the zeros and ones of the digital notation entirely, a complete analog quantum optical computer has been proposed using such squeezed states.[61] To clarify, even in the classical world, we have analog and digital computers. Analog computers actually were in use long before their digital brethren and, loosely, a digital classical computer represents data in discrete bits. The bit can be in only one of two states, either a zero or a one. In an analog classical computer, the information is stored in something that can change continuously, such as electrical current or voltage in a circuit. Mechanical analog computers, powered by hand, steam, or electricity, were important in World War II where they were used for tide prediction on D-Day in support of the allied invasion of Normandy and also in ballistic calculations. The analog machines tended to be special purpose, one machine to compute tides and another machine to compute mortar trajectories,[62] and so when the digital electronic computers came online just after World War II, they quickly replaced their analog kin, as the digital machines were universal and could be programmed to do widely different tasks; one machine computed tides and mortar trajectories. The ENIAC (Electronic Numerical Integrator and Computer) was originally programmed to calculate ballistics tables for large guns and then quickly reprogrammed for nuclear weapons simulations. In the analog computer world, you would have to build one radically different machine for each task, which is why analog fell out of favor. But analog computers still find themselves used in niche applications where digital machines may have issues such as computational stability, such as in the modeling of classical chaos theory.[63]

So it did not take long to propose more digital-flavored quantum spatial interferometers using a discrete number of particles such as photons or atoms or even neutrons. (The squeezed-state interferometer uses optical squeezed states with an indeterminate continuous number of photons and is much more like an analog approach.) The first such idea I am aware of was that proposed in 1986 by American physicist Bernard Yurke, in the context of improving spatial neutron interferometers.[64] This came along just 5 years after Caves' squeezed-state proposal. Yurke proposed that an entangled state of the form $|N/2 + 1\rangle_{\text{Up}}$ $|N/2 - 1\rangle_{\text{Down}} + |N/2 - 1\rangle_{\text{Up}} |N/2 + 1\rangle_{\text{Down}}$ be inserted into the upper and lower input ports on the left-hand side of the spatial interferometer shown in Figure 5.6a of the Rosetta Stone. Here, N is the number of neutrons in the beam, and for N equals 10, we have a state of the form $|6\rangle_{\text{Up}} |4\rangle_{\text{Down}} + |4\rangle_{\text{Up}} + |6\rangle_{\text{Down}}$, where "up" means the upper input arm of the interferometer and "down" means the lower input arm.

This Yurke state is entangled in that you cannot say, even in principle, if there are 6 neutrons coming in the top and 4 coming in the bottom, or 4 coming in the top and 6 coming in the bottom. The state is something like a cat state. It can be shown that such a state has all the lovely whacky features we have all come to know and love in entangled states; it is unreal, uncertain, and nonlocal. If Alice in the up arm measures 6 then Bob will get 4 in the down arm, but if Alice gets 4, then Bob will get 6. The upper arm and the lower arm can be light-years apart and still this correlation holds, and the particle number 6 or 4 is indeterminate (unreal) until a measurement is made. What Yurke showed was that the same type of change in the quadratic behavior in the errors resulted in using this state, as opposed to a more typical unentangled state commonly used in neutron interferometry, $|N\rangle_{\text{Up}} |0\rangle_{\text{Down}}$, where in this example with N equals 10 we get $|10\rangle_{\text{Up}} |0\rangle_{\text{Down}}$, which means I send in all 10 neutrons into the upper port and none into the lower port, which is not an entangled state.

By exploiting entanglement, Yurke showed a change in the scaling law in the interferometer's margin of error for measuring path-length differences between the upper and lower arms. In this example, sending all 10 neutrons in the upper arm in the unentangled state $|10\rangle_{\text{Up}} |0\rangle_{\text{Down}}$ gives a margin of error of about 32%. Simply entangling the 10 neutrons into the Yurke state input drops this quadratic scaling to a more linear scaling and gives a margin of error of only 18%. If I increase the number of neutrons to 100, the unentangled approach gives a margin of error of 10%, but the entangled approach gives one of only 2%[65] (see Figure 5.9).

Everything about the Yurke proposal then hinges on how to make the entangled state like the Yurke state $|6\rangle_{\text{Up}} |4\rangle_{\text{Down}} + |4\rangle_{\text{Up}} |6\rangle_{\text{Down}}$ in the first place. The unentangled state $|10\rangle_{\text{Up}} |0\rangle_{\text{Down}}$ is easy to make. You open the door on your nuclear reactor that provides the neutrons, wait enough time until about 10 neutrons have shot out, close the door, and shepherd all 10 neutrons into the upper arm of the interferometer. The Yurke state would require placing something like an entangling ENT gate between the nuclear reactor and the input to the interferometer or having a magic nuclear reactor that spits out entangled states of this form in the first place. How did Yurke propose to make them? In perhaps the most understated infamous quote from his paper, he declares, in the conclusions, "Here I do not offer any means by which such states may be generated in the laboratory." The call to arms here was clear.

If you could make such entangled states, then you could build a better interferometer. What was unclear was what entangled states you could use and where the entanglement should be for the best effect. (The answers to these two questions are "a lot" and "inside the interferometer.") A popular one was the twin-number input state of the form $|N/2\rangle_{\text{Up}} |N/2\rangle_{\text{Down}}$, which in our 10 particles example becomes $|5\rangle_{\text{Up}} |5\rangle_{\text{Down}}$, which is not entangled at all! Slowly, at least it seemed slow to me, did the realization come that the entanglement should be inside the interferometer for best performance and the unentangled input state does indeed become a horrible entangled mess after passing through the first beam splitter in Figure 5.6a.[66]

My foray into this new field of digital particle number entanglement–enhanced spatial interferometry[67]—happened by happenstance in 1992 when I ran into my former postdoctoral adviser, the American physicist and quantum cowboy, Marlan Scully, at the International Quantum Electronics Conference in Vienna, Austria. As I tell my students, "Friends come and go, but bosses accumulate—at least at first." In 1992, I was working on projects for my then current postdoctoral adviser, American physicist Charles Bowden, and then for Marlan Scully, and then also for my PhD adviser, Turkish–Swiss–American physicist Asim Barut. Scully had buttonholed me at one of the conference coffee breaks and sat me down and began describing a calculation he had been working on. He had just come from another conference in Germany on atom spatial interferometry and he bemoaned to me that in the question-and-answer session, it became apparent that nobody really had a good idea what the sensitivity of an atom interferometer was.

Atom interferometry was a spin-off of neutron interferometry, where entire atoms (sodium atoms are popular) in a beam are launched into a spatial

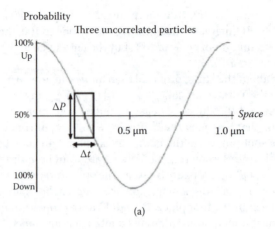

(a)

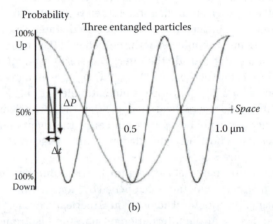

(b)

Figure 5.9 Spatial interferometer interferogram oscilloscope traces for the interferometer in Figure 5.6a. In panel (a), we show the trace for three uncorrelated particles, and in panel (b), we show the trace for particles entangled in a three-ion Schrödinger cat state. The goal is to infer the path-length difference between the upper and lower arms (horizontal axis) from the measurement of the ratio of particles that exit the top of the interferometer versus the bottom (vertical axis). In both cases, the error in the vertical direction, ΔP, is the same. But you can see that the error in the spatial measurement Δx in the entangled trace (b) is smaller than that in panel (a). This is simply because the entangled particle trace in panel (b) oscillates three times as fast as in the unentangled trace in panel (a) or (b). As the steepness of the sides of the oscillation increases, the spatial resolution error decreases.

interferometer, such as that shown in Figure 5.6a. Because entire atoms are much more massive than neutrons or photons used in spatial interferometers, atom interferometers are particularly sensitive to gravitational effects and also inertial effects, such as rotation, and have been developed as gravitational field sensors and rotation sensors (gyroscopes). Atom interference was first observed in the 1930s when sodium atoms were observed to diffract off of a salt crystal. Diffraction is a wave interference phenomenon, and from quantum theory, we expect all massive particles to exhibit wavelike effects. The trick is that the wave nature of a particle becomes less apparent the more massive it is. The bigger the cat, the less likely you will see it both dead and alive. Hence, it is easy to see waves of photons (zero mass), harder with electrons (small mass), even harder with neutrons (2000 times more massive than electrons), even harder with atoms, and hardest with molecules (the largest objects to be prepared in a cat-like superposition state to date). A renaissance in atom interferometry came about in the 1990s in part motivated by the great success in neutron interferometry in the 1980s. One of the most notable atom interferometry experts, Austrian physicist Anton Zeilinger, did his PhD work in neutron interferometry in the group of Austrian physicist Helmut Rauch at the Technical University of Vienna. Zeilinger then moved on to atom interferometers in the 1990s right about the time Scully became enthralled with calculating their sensitivity.

Scully sketched out his calculation on the atom interferometer sensitivity using a version of quantum theory called the "Schrödinger Picture" after the same Austrian physicist of cat fame, Erwin Schrödinger. I took his notes back to Alabama with me and rechecked his results by redoing the entire calculation in a different version of the theory called the "Heisenberg Picture" after the German physicist of uncertain fame, Werner Heisenberg. (It is well known in our field that certain calculations are much easier to do in the Heisenberg Picture.) I wrote up the manuscript with my short and sweet Heisenberg Picture calculations, added all the figures, and proudly faxed it to Scully for review … and he made me redo the whole thing back in the long-winded and confusing Schrödinger Picture. I tried to protest that my calculation was much simpler in the Heisenberg Picture but he vetoed that with the rejoinder that the calculations were much more physically intuitive in the Schrödinger Picture and made me rewrite the whole thing in that picture. Friends come and go, but bosses accumulate. The paper "Quantum Noise Limits to Matter-Wave Interferometry" appeared in the journal *Physical Review* in 1993. The result? If you send all the atoms into the upper port of the interferometer in Figure 5.6a, then you get back the usual quadratic scaling law! If 100 atoms buy you a margin of error of 10% in spatial measurement accuracy, then you need to quadruple it to 10,000 atoms to lower the margin of error to 1%.

As I mentioned before, a spatial interferometer can be used to sense rotation. Rotation sensors (called gyroscopes) are part of inertial navigation devices found on everything from intercontinental ballistic missiles to nuclear submarines. The measurement of rotations, combined with that of acceleration and local gravitational fields, for example, will allow a submarine to navigate with great precision and entirely circumnavigate the Earth while completely submerged underwater where no signal from the GPS can reach. The US military is also adverse to the use of the GPS for critical systems, as in a wartime situation the GPS satellites can be shot down or the signal from the satellites can be easily jammed. Inertial navigation can be performed without any need for an external positioning system and hence is robust against such threats.

To measure rotation, we imagine the spatial interferometer in Figure 5.6a is rotating clockwise about its center. As it rotates, the particles in the upper arm have to travel a bit farther to catch up as the arms rotate away from them and the particles in the lower arm have to travel a bit less as the arms rotate toward them. The net effect is that there arises an effective path-length difference between the two arms. The upper arm is effectively shorter and the lower arm is effectively longer. (Think headwinds and tailwinds.) This effective path-length difference between the two arms can be measured in as a shift in oscillations of the spatial interferogram in Figure 5.9. The accuracy, to which such a shift in the oscillations can be measured, depends on the steepness or slope of the curve of the interferogram. As we illustrate in Figure 5.9, the sides of the wiggles get steeper if you switch from unentangled particles (Figure 5.9a) to particles entangled inside the interferometer in a cat state, with all particles simultaneously in the upper arm and in the lower arm in a coherent superposition. Running the numbers, the end result is a quadratic improvement in signal-to-noise ratio with entanglement. So how to produce the entanglement in the first place? My colleagues and I have spent the past 15 years trying to figure that out. My first attempt was the 1998 gyro paper, where I was working with cold atoms (instead of neutrons).

The idea, as illustrated in Figure 5.10, is to make something American physicist Marlan Scully has popularized as the quantum eraser. We have two buckets labeled A and B with six identical particles in each bucket. The state of the two buckets is the unentangled state $|6\rangle_A |6\rangle_B$ with the total particle number 12. Next, we run pipes from the buckets to a beam splitter and then pipes from the beam splitter out to two detectors. The beam splitter for a single particle acts as a CAT gate and also is the quantum eraser—it erases which path the particle took from the buckets. We wait for one of the detectors to fire (lightbulb lights up) and then we know for sure one particle is missing out of one of the buckets. But because the beam splitter erases the which-path information, we cannot know from which bucket the particle is missing, A or B. That lack of knowledge collapses the unentangled state into the new state $|6\rangle_A |5\rangle_B + |5\rangle_A |6\rangle_B$ with the

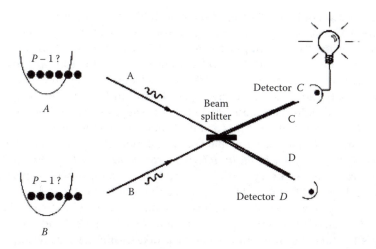

Figure 5.10 How to make an entangled state with a quantum eraser. We start with two buckets containing the same number of indistinguishable particles P in each, say we take P is 6 particles. Then, the initial state of the two buckets is the unentangled state $|6\rangle_A |6\rangle_B$. We now hook the buckets up to the front half of a spatial interferometer and attach pipes from the buckets to the beam splitter. Then, we run pipes from the beam splitter to the two detectors C and D and wait for one detector to fire (e.g., half the time the lightbulb lights up on detector C) so that a single particle is missing from either bucket A or bucket B. The beam splitter acts as a quantum eraser, erasing which path the particle took, so we cannot be sure if the particle detected at detector C came from bucket A or bucket B. Thus, we know that there is a particle missing but from which bucket we cannot say. In addition, because the particles are indistinguishable, we cannot say which particle in the bucket is missing even if we knew which bucket it was missing from. The measurement at C projects the initial unentangled state of the particles in the bucket into the entangled state $|6\rangle_A |5\rangle_B + |5\rangle_A |6\rangle_B$. That is, we know one particle is gone, but we don't know wherefrom. This state is almost exactly the Yurke entangled state! The detectors are now removed and the rest of the spatial interferometer in Figure 6.6a is inserted in their place and we allow the entangled state in to get a quadratic improvement in signal-to-noise ratio.

total particle number 11. This is not quite the Yurke state but it is close enough. The plus sign indicates that we have a state where there are six particles in bucket A with five in bucket B and simultaneously five in A and six in B. It is the spoon in the coffee cup being in my office and the bathroom sink at the same time all over again. The detectors are now removed and the buckets and beam splitter are hooked up to the rest of the spatial interferometer in Figure 5.6a. Then, we let her rip and we get a gyroscope with a quadratic improvement in signal-to-noise ratio. Without entanglement, 100 atoms in the gyroscope give a margin of error of 10%, but with entanglement, 100 atoms in the gyroscope give a margin of error of only 1% in sensing the rotation rate. Usually, engineers are happy with just 1% or 2% improvement but 10%? Well that is just great.

What is the general trick for making such entangled states for spatial interferometry and which entangled states are best? This is a long story, but many of the details can be found in my paper, "Quantum Optical Metrology—The Lowdown on High-N00N States."[68] As the title suggests, the optimal state for a certain strategy that gives this quadratic improvement is something we call a N00N state. The name is an artifact of the notation, the optimal state has the form $|N\rangle_A |0\rangle_B + |0\rangle_A |N\rangle_B$ *inside the interferometer between the two beam splitters.* Here, N is a positive integer, and so for our previous examples with 10 particles, we would want $|10\rangle_A |0\rangle_B + |0\rangle_A |10\rangle_B$ inside the interferometer in Figure 5.6a. This is a type of entangled Schrödinger cat state where all 10 atoms take the upper branch of the interferometer (and none take the lower) and simultaneously all 10 take the lower branch (and none take the upper). If a measurement is made, you find all 10 atoms up (but none down) or all 10 atoms down (but none up). Reading from left to right, $|N\rangle_A |0\rangle_B + |0\rangle_A |N\rangle_B$ spells out "N00N."

This term "N00N" was invented in our Quantum Computing Technologies group at the NASA JPL around the year 2000 when we were working on quantum spatial interferometry. It was just an accident that the letter "N" was chosen and they equally could have well been the less euphonious "P00P" states but thankfully N00N instead stuck. At one point the file name for a draft of the paper was "noon.ps" back in the day when "postscript" (.ps) was more popular than the Adobe PDF (.pdf). It was German–American scientist Chris Adami, then in our group, who suggested we call N00N states with N larger than two "high-N00N" states, as this would allow for jokes based on the cowboy movie *High Noon* starring Gary Cooper. Indeed, such jokes were deployed and hilarity ensued and I recall peppering a talk I gave on the topic at Caltech with such references when American physicist John Preskill began, in the middle of my lecture, reciting dialog verbatim from the movie. I mentioned later to his postdoctoral researcher, American physicist Dave Bacon, that Preskill must have really liked the movie *High Noon* because he knew all the dialog. Bacon looked at me seriously and whispered, "Preskill memorizes all the dialog from every movie he's ever seen, even if he's just seen it once!" I hear Preskill is also a whiz at memorizing vast databanks of baseball statistics.[69]

Now, when you try to introduce a novel and cute term like "high-N00N states" into the pages of a venerable physics journal like *Physical Review Letters*,[70] it is often rejected outright as "technical jargon" and replaced with the more accurate but less cute "maximally path-entangled states of large particle number." This makes me wonder how cute high energy physics words like "charmed quark" or "strange quark" or just plain "quark" ever made it into the physics literature at all. Our own protocol for rolling out "high-N00N" began with a footnote added in proof to the Quantum Rosetta Stone paper, a footnote that read, in a slightly different notation than used here, "We call the state of the form $|N, \phi\rangle + |\phi, N\rangle$ the NOON state, and the High NOON state a large N." I'm not sure what went wrong but I suspect we added the footnote by hand and put slashes through the zeros, to be sure they were typeset as the number zero and not the letter "O" (as is common to do in computer programming) but instead they were misread as the Greek letter phi (ϕ). Hence, the first appearance of the now somewhat infamous term "NOON state" appeared as a typographical error in a footnote in 2002 in our paper in the *Journal of Modern Optics*. Typographical errors and footnotes and corrections added in proof are a long-standing and infamous tradition in physics.[71] Eventually, the term NOON state worked its way up from the footnotes to the appendix to the body to the abstract and now can even be found in the title of an article published in 2010 in the most prestigious journal, *Science*.[72] But it was a long haul.

I spend a little time here on NOON states because they appear again in the next section where they are used in something called "quantum lithography." I also wanted to point out the Quantum Rosetta Stone in action. Our group at the NASA JPL came up with the high-NOON state working directly with the Rosetta Stone and by reading articles on atomic clocks or temporal interferometers. In the study of ion atom clocks in particular, a paper appeared in *Physical Review Letters* in 1997 titled "Improvement of Frequency Standards with Quantum Entanglement," where here "frequency standards" means "atomic clocks." In this paper, they argued that the optimal entangled state for a temporal interferometer (Figure 5.6b) was a maximally entangled cat state of the form $|\uparrow\uparrow\uparrow...\uparrow\rangle + |\downarrow\downarrow\downarrow...\downarrow\rangle$ between the two light pulses, which can be seen to be a quantum superposition of all the ions pointing up ($\uparrow$) and simultaneously all the ions pointing down ($\downarrow$). With one look at Figure 5.6 and a little thought, it becomes clear that such a state of ions between the two pulses in the temporal interferometer in Figure 5.6b becomes a NOON state between the two beam splitters in the spatial interferometer in Figure 5.6a. If the math is the same, then the physics must be the same! Hence, it is easy then to prove, using the same math as for the ions, that if the maximally entangled cat state is optimal for the ions, then the NOON state is optimal for the path-entangled particles and gives the same quadratic improvement in signal-to-noise ratio.

Now, making the NOON state between the two beam splitters in a spatial interferometer is tricky. For low-NOON states, an ordinary 50–50 beam splitter will do. For example, in Figure 5.6a, inject the unentangled state $|1\rangle_A |0\rangle_B$ into the first beam splitter on the left (one particle up and none down) and out pops inside the interferometer the entangled low-NOON state $|1\rangle_A |0\rangle_B + |0\rangle_A |1\rangle_B$ between the two beam splitters (one particle up with none down and simultaneously one particle down with none up). Alternatively, if one starts with the unentangled state where two identical particles, photons in particular, are inputted in the form $|1\rangle_A |1\rangle_B$ (one up and one down), out pops the low-NOON state $|2\rangle_A |0\rangle_B +$ $|0\rangle_A |2\rangle_B$ inside the interferometer (two up with none down and two down with none up). To go further, you cannot use an ordinary 50–50 beam splitter. For example, if you send in the twin-number state $|2\rangle_A |2\rangle_B$ into an ordinary beam splitter, then what comes out is not the four-particle NOON state $|4\rangle_A |0\rangle_B +$ $|0\rangle_A |4\rangle_B$ but something much more complicated called a "bat" state, which is a NOON state mixed with a little bit of "poop."[73] Because inside the interferometer the bat state is transformed into almost a NOON state, it does perform well, but not as well as a pure NOON state between the beam splitters. Indeed, in the Rosetta Stone paper, we "pulled a Yurke" when it came to how to produce high-NOON states with more than two particles. That is, we said we had no simple idea how to do it, but if you could, then you could build an optimal spatial interferometer with a quadratically improved signal-to-noise ratio. We did have a complicated idea, and that would be to go to Figure 5.6c of the Rosetta Stone and use the same quantum computing circuit proposed to make entangled cat states in the ion temporal interferometer to make the entangled NOON states in the spatial interferometer. That would mean that you would have to replace the first beam splitter, which is just normally a slab of specially prepared glass, with a small special-purpose quantum computer, a device American physicist Dave Kielpinski calls a "magic" beam splitter or magic BS.

In the end, we know of at least one way to build such a magic beam splitter. This can be done again via the Quantum Rosetta Stone by mapping the entangling ENT gate in Figure 5.8 from a temporal interferometer into a spatial interferometer. Using some ideas from optical quantum computing, our group at the JPL first proposed such an ENT gate or magic beam splitter for photons that beat the low-NOON state limit, thus making a $|4\rangle_A |0\rangle_B + |0\rangle_A |4\rangle_B$ high-NOON state with four photons. Our ENT gate was readily lifted from the toolbox of optical quantum computing and then applied to the new field of quantum optical sensing with spatial interferometers, and the scheme was recently demonstrated in the laboratory of Australian physicist Jeremy O'Brien.[74] Now to be fair, to be of any use in something like the LIGO interferometer, we'd need many, many, many more photons than four, but the ideas leading up to the prediction, design, and production of such nonclassical states, at the interface between quantum interferometry and quantum information processing, have been very important. There is also the matter of flux or number of photons passing through per second. In LIGO, there are around 10^{20} circulating photons emanating from a laser that puts out approximately 10 watts of laser power. While it would be difficult to imagine a source that puts all 10^{20} photons into a single giant NOON state, you could imagine a source that is something like a laser that puts out 10 watts of $|4\rangle_A |0\rangle_B + |0\rangle_A |4\rangle_B$ NOON states, which would give an improvement of a factor of two in signal-to-noise ratio, which is not too shabby. Even a small high-NOON state might come in handy for imaging objects better than can be imaged classically, particularly objects such as bacteria and viruses that might be killed if you tried to image them with say ultraviolet light. We'll discuss these imaging applications next.

LIGHT BLIPS SHRINK CHIPS

In April of 1999, Welsh–American scientist, Colin Williams, then the supervisor and principal scientist of our Quantum Computing Technologies group at the NASA JPL, arranged for a meeting with an Italian entrepreneur named Giovanni Della Rossa. Della Rossa is an industrialist interested in commercializing quantum technologies, and that was the topic of the meeting, which took place in a small boardroom on the third floor of building 126 on the JPL campus. The meeting consisted of just the three of us, and we were impressed to learn Della Rossa was president of a small company called Quantumatics (and had been since 1998). We began a free-ranging discussion of various projects we were working on in the group at that time, from quantum computing to quantum gyroscopes. I was gleefully telling the tale of our work on using entangled particles in spatial interferometers to make more sensitive optical gyroscopes when Della Rossa suddenly interrupted and asked, "Does this have any application to lithography?" Williams, perhaps a bit frustrated that we were already running late, turned to him and said curtly but politely, "No, no we are talking about gyroscopes not lithography. This has nothing to do with lithography." Upon hearing this word "lithography," I pushed myself backward into my plush boardroom chair and immediately went into a daze. My eyes glazed over, and I began just staring out the slatted boardroom window into a patch of scrub oak on the nearby hillside. Williams snapped his fingers a few times and called softly, "Jon? Are you still with us?" I snapped out of it and generated the idea for something now called quantum lithography, which did help us land several hundreds of thousands of dollars in grant funding over the years and resulted in what is still to this day Williams and my most highly cited paper, "Quantum Optical Interferometric Lithography: Exploiting Entanglement to Beat the Diffraction Limit." I did not sadly keep my notes but the diagram looked like that in Figure 5.11.

The gestalt moment that caused my momentary coma was that the spatial interferometer interferogram with the increased number of wiggles (Figure 5.9b) could be translated from wiggles on the screen of an oscilloscope to wiggles on a multiphoton absorbing substrate. This idea emerged only because I was thinking about wiggles and gyroscopes and interferograms and—for a different project—multiphoton absorption. All it took was the question from Della Rossa to tie it all together, "Does this have any application to lithography?" The insight was to realize that the mathematics of the multiphoton detection

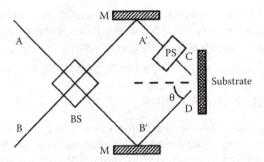

Figure 5.11 Quantum lithography. Incoming photons (A and B) are prepared in a N00N state by the magic beam splitter (BS) and then routed to an *N*-photon absorbing substrate. The photons write a pattern of parallel lines on the substrate similar to the spatial interferogram of Figure 5.9b. The lines are spaced closer together (by a factor of *N*) than what is usually allowed by the Rayleigh criterion of classical optics. Such a technique might someday be used to manufacture computer chips with more transistors and hence more processing speed and memory.

scheme used in the gyroscope was identical to the mathematics of the multiphoton absorption used in lithography—if the math is the same, then the physics must be the same!

So let's go back now and figure out what this is all about and why it is (or could someday) be important. In the computer industry, the classical computer industry, the computer chips that Intel and other companies make that power your iPad, iPhone, iPod, iMac, or Dell are made through a process called optical lithography. Very simply, the design for the computer chip is prepared as a complicated black and clear silhouette on a type of glass slide (called a mask) and then laser light is shone through the slide and focused down onto a light-sensitive substrate called a photon absorbing resist. The image on the glass slide, that of a complicated circuit, is then shrunk to microscopic size with some lenses set like a backward microscope (making big things smaller) and transferred to the substrate. Then, the substrate is processed chemically until the image of the computer chip circuit is transformed into the computer chip itself. This process has been used in making computer chips for decades now. There is a limit on the size of the features the demagnification process can write on the substrate, called the Rayleigh criterion, or Rayleigh diffraction limit (named after the English physicist John Strutt, the Third Baron Rayleigh), which says you cannot write features smaller than the wavelength of the light you are using to do the writing process.

Now, Intel and its competitors would like to write the smallest circuits possible on the chip, so they can pack as many circuits on there as close together as possible, and so the semiconductor chip industry for decades has invested billions of dollars on optical imaging systems that use ever smaller and smaller wavelengths of light, typically in the ultraviolet part of the spectrum. Systems in use from the 1960s through the 1980s used ultraviolet light of wavelengths around 400 nanometers (around 10 millionths of an inch) and more modern systems use ultraviolet lasers around 200 nanometers. With each decrease in wavelength size, Intel must invest in a totally new imaging system that is optimized to the new wavelength and this is where the billions and billions of dollars come in. The promise of quantum lithography is that when you entangle the photons into a NOON state, you get an effectively shorter wavelength without going through all the trouble of radically changing up your optical imaging system. This is the Quantum Rosetta Stone in action. The increased frequency in the wiggles of the interferogram in Figure 5.9b directly translates to increased (and narrower) wiggles on the substrate that will then be turned into your transistors and other chip circuitry.[75]

For example, if you look at the spatial interferogram in Figure 5.9b, you can see there are *three* times as many wiggles using three-entangled particles in a $|3\rangle_A |0\rangle_B + |0\rangle_A |3\rangle_B$ high-NOON state than there are in Figure 5.9a when there is no entanglement. For an optical interferometer, the size of the wiggles in Figure 5.9a—the peak-to-peak distance—is about the wavelength of the light when translated to the minimum feature size on the substrate. That is the Rayleigh diffraction limit. If we are using unentangled photons of 400-nanometer wavelength, this means the wires and circuits that we put on the chip cannot be less than approximately 400 nanometers in size and cannot be spaced closer than approximately 400 nanometers. But suppose I wave my quantum magic wand and replace the classical photons with a stream of photons prepared instead in a $|10\rangle_A |0\rangle_B + |0\rangle_A |10\rangle_B$ high-NOON state? Well, even though each individual photon has a wavelength of 400 nanometers, the combined entangled 10-photon entangled NOON state has an effective wavelength that is 10 times smaller, in this case, 40 nanometers. That is because there are 10 times as many wiggles such as in Figure 5.9b that are now packed into the same space as before, making them 10 times more closely spaced. That means I can put 10 times more circuits on the chip lengthwise and another 10 times as many widthwise with the net result that I can package 10 × 10 or 100 times as many circuits on the same-sized chip. To do this classically, I would have to replace the entire optical imaging system running at 400 nanometers (ultraviolet light) with a new, much more expensive optical system running at 40 nanometers (almost x-ray light). To do this quantum mechanically, I just keep the optical imaging system that is optimized for a 400-nanometer light but draw features as if I had a 40-nanometer light.

That is the hype and the promise of quantum lithography. The computer chip manufacturers could invest in entangled light sources, instead of moving to ever-shorter wavelengths, to continue onward with Moore's law, which is a rule of thumb that states that the number of transistors that can be put on a computer chip doubles every 2 years. They double the number of transistors by halving their size. The modern computer revolution is closely tied to Moore's law as it is an exponential scaling law. The more transistors you put on a chip, the faster the processing speed and the more the computer memory. This law implies that the computer chip speed and memory double every 2 years and has held approximately true since about 1970. In high school and college, I programmed on one of the first supercomputers, a CDC-6600 scientific computer that was the size of a large room and had 600,000 transistors in its central processing unit (CPU). Today, an iPhone 4 has a CPU with 177 million transistors, and the whole thing fits in my pocket. Compare this with the ENIAC, which had only 17,468 vacuum tubes (that functioned like transistors) and filled an entire warehouse. Your single pocket-sized iPhone has 10,000 times the processing power of the entire warehouse-sized ENIAC. The ENIAC required a team of personnel to program and operate it. A single 7-year-old can operate an iPhone. That is what Moore's law means to you.

For years now, various experts have been predicting the end of Moore's law. The issue is that as you make the transistors smaller and smaller, you can fit more and more of them on a chip, and you will eventually hit a roadblock that keeps you from going even smaller. Commercial transistors today are around 40 nanometers in size, in keeping with the wavelength of the ultraviolet light to write them on the chip. A single silicon atom, the dominant material that makes up most of the chip, is approximately 0.2 nanometers in diameter. Thus, a square transistor that is 40 nanometers on a side has approximately 200 atoms on a side or 200 times 200 total atoms or 40,000 atoms total, which is proportional to the area of the transistor. A single-atom classical transistor was recently demonstrated in a research laboratory at the University of New South Wales in Australia—a "spin-off" of their spintronic quantum computing research!—but the single atom transistor has not yet found its way into a commercial computer chip. Most experts agree that when the transistors hit the size of a single atom, you cannot go any smaller and Moore's law will come to an end. The total number of transistors you can fit on a chip is limited by the area of the transistor. Hence, following Moore's law, in 2 years, the area of the transistor will be half and have 20,000 atoms on it, then in two more years, it will have 10,000 atoms, and so on, and then following this logic, commercial computer transistors will consist of a single atom in 15 years, somewhere around 2027, and the computer improvements will come to an end. People liken this to the improvement in cars. At first, there were great improvements in speed and power and so forth but then they hit a wall imposed by basic physics and the power and maximum speed have not improved much in many years and so each new-year model is sold with cosmetic improvements or safety features or reliability or gas mileage. We no longer expect the maximum speed of a car to double every couple of years as we might have in 1900. We'll hit this same wall with the computer chips and then the iPhone 20 will be no more powerful than the iPhone 19 but might come in a nicer set of colors or have better software or be more reliable or have other tangential features. The computer age will come then to an end. In the early heady days of quantum computing, circa 1994, people predicted that the quantum computers would take over, with their exponential scaling in Hilbert space, when Moore's law came to an end. But the exponential scaling is there only for a few very special problems like factoring. The quantum computer will not give a general exponential speedup in processing power on all problems, and I doubt that switching from a classical computer chip to a quantum computer chip in 2027 will give you any improved performance in watching movies or playing video games.[76]

Where does quantum lithography fit in to all this? Well, the hope was that it could help take us down to the single-atom transistor for commercial fabrication where lots of transistors have to be made on the chip and all wired together. To do this with ordinary light, make transistors the size of silicon atoms that are 0.2 nanometers across, you would have to use light of a wavelength on the order of 0.2 nanometers from the Rayleigh criterion, and we're now talking light that is made up of high-energy x-rays. Making optical imaging systems for x-rays is notoriously difficult, owing to the fact that x-rays like to go straight through things and not bounce off them. In addition, as we know from Einstein's theory of the photoelectric effect, the shorter the wavelength of the photon, the bigger the wallop it packs in terms of energy and momentum kicks. Instead of drawing features on the substrate, high-energy x-rays have a tendency to just blow holes in it instead. But if entangled photons could be used in a N00N state well, then a $|100\rangle_A |0\rangle_B + |0\rangle_A |100\rangle_B$ state of 20-nanometer ultraviolet light could be used to draw transistors a hundred times smaller, 0.2 nanometers, without having to resort to x-rays. That, at least, is how we pitched the idea.

We worked on the project throughout the summer and fall of 1999, we applied for several patents through a somewhat tedious and opaque process with the JPL patent office. The JPL patent office was actually a sub-branch of the main Caltech patent office, because Caltech runs JPL for NASA. We wrote up and submitted a formal invention disclosure to a kindly but slightly delusional woman at the JPL patent office, whose job it was, apparently, to make sure that nobody at JPL ever patented anything or at least to make sure the number of patent files was kept to a minimum. (Because it costs money and involves a lot of work to file a patent, I suppose she figured that she was saving herself time and JPL money.) Dealing with the layers of bureaucracy at JPL was often nightmarish. This kind lady kindly rejected our invention disclosure for a potential patent filing because we had not actually produced a working device, only some theory. I kindly explained to the kind lady, supposedly an expert on the US patent code, that it was not necessary to produce a working device, so long as the description and theory were clear enough for someone to do so. To back up my assertion, I provided her copies of three of my previous patents that were based only on theory but for which we never produced a working device. The kind lady kindly then told me, "Oh the US Patent Code has changed since those patents of yours were granted. You now have to produce a working device." I did not believe it and went online and downloaded and printed out the US patent code, all 500 something pages of it, went carefully through it and highlighted the sections where it explicitly stated that a working device was not required, and then came and dumped it on her desk with yellow sticky notes festooning the stack, marking each place where I was right and she was wrong. It was then she confessed, "Dr. Dowling, we are under a lot of pressure to keep costs down in this office and so I tell all the engineers they have to produce a working device to cut down on the number of patent applications." That was it. (She called me an engineer!) I went over her head to her boss on the main Caltech campus, told him that quantum lithography was the greatest invention since somebody used a mousetrap to slice bread, and convinced them to file not one but three separate and interlocking patents on quantum lithography. This whole process took a while—time well spent improving the theory and the design.

Particularly good progress was made in the summer of 1999 with the collaboration of Agedi Boto, who was my Summer Undergraduate Research Fellow that year (from Caltech), when we extended the theory to arbitrarily high-N00N states (and hence arbitrarily small transistor manufacturing). Agedi, who last I heard is pursuing his MD at Johns Hopkins Medical School, was an undergraduate freshman in 1999, and for his efforts, he was anointed first author of the manuscript that we eventually submitted for publication. (In keeping with the tradition of our community, the person who does all the work is the first author and the person who had the idea is the last author.) I set him to the task of carrying out the calculations on the computer using the algebraic symbolic manipulation package Mathematica, while I carried out the calculations by hand, and we then could check each other's work. One particular pleasing moment was when the Mathematica code produced, on a Friday afternoon, 76 pages of mathematical output. I told Agedi to just run the "Simplify" command over the weekend and see what the 76 pages of mathematical output reduced to. On Monday, the 76 pages of mathematical expressions had simplified to a single, one-line expression, $1 + \cos(2N\varphi)$, the correct result. We both stared at that for a bit and then I said, "Well maybe we can get this by hand without the computer in less than 76 pages and 48 hours?" We did.[77]

It is not a good idea to publish the work at least until the patent is filed, as you have only a year under US rules to file after public disclosure of the idea, and so we kept polishing the calculations and extending the result to lithography with arbitrarily high-N00N states. The original idea was just with two-photon N00N states. So other than leaking a preprint to the Army Research

Office, in a preemptive attempt to secure funding for the work, we kept the paper quiet as the patent application wound its way through the Byzantine Caltech patent office process. We did not discuss the work at any conferences, nor did we post the preprint on the Los Alamos quantum physics preprint server, as either would have counted at public disclosure and hurt our chances for the patent.[78]

I learned much later that the Army Research Office immediately sent the draft of our paper to two experimental physicist colleagues of mine (one a Nobel laureate) and asked them to check our math and report back on the correctness and the importance of the result. As Welsh physicist Daniel James conveyed to me years later, the two experimentalists not only could not reproduce our calculation, they also could not even get their answers to agree with each other. As the American physicist Robert Boyd is always fond of saying, "The only thing worse than experimentalists trying to do theory is theorists trying to do experiments!" In desperation, they asked Daniel James for help and he graciously reproduced the results of our paper up to correcting a pesky factor of two that had crept into the document as a typo and relayed their correctness to the two experimentalists, who then relayed this result and their assessment to the Army Research Office. What I learned from this experience is that when you give a draft of a proprietary paper to the Army Research Office, they immediately forward it to all your competitors asking for their opinion about it. This realization will be important later.

By the end of the fall of 1999, the provisional patent had been filed, and we were free to announce our results to the world. We posted a preprint on the Los Alamos ArXiv preprint server on December 11, 1999, and simultaneously submitted it for consideration in the premier journal *Physical Review Letters*.[79] I first presented the work in a public lecture entitled "Quantum Lithography" at the Workshop on Quantum Electronics, held at the ski resort of Snowbird, Utah, during the week of January 10–12, 2000.

The paper languished for over 6 months at the editorial offices of the journal *Physical Review Letters*.[80] Submitted manuscripts are sent out for review to anonymous referees, and despite the fact that we thought it was a great idea, the referees, as usual, decided to be nitpicky. One referee really liked it but another, a person who clearly was an expert in classical lithography, had some issues with it. This person did not like how we defined the Rayleigh diffraction criterion in terms of the wavelength of the light and insisted in the classical lithography that it was always half the wavelength of the light and forced us to insert factors of $1/2$ all over the place in our equations and text. Considering that one of our primary results showed an improvement by a factor of two, with available quantum light sources, this insertion somewhat muddled things. (And introduced typos, because we missed a few places where the pesky factors of two should have been inserted.) Second, this referee did not like that we only showed how to write parallel lines better than was possible classically and insisted that we needed to show how to write arbitrary two-dimensional patterns to make the case that the work was important. The two primary criteria for publication in *Physical Review Letters* are "broad interest" and "importance." If our results panned out, broad interest was assured, but without arbitrary two-dimensional patterns, importance was less so.

Fortunately my friend and collaborator, Australian physicist (and Woody Allen impersonator) Samuel Braunstein had a PhD student, Dutch physicist Pieter Kok, working on the two-dimensional pattern problem, immediately after our preprint appeared on the ArXiv. About the time we got the referee reports back, Kok and Braunstein sent us their preliminary results on this, and so we were delighted to embrace them and their calculations and folded the result into a revised manuscript, with Kok and Braunstein now added as coauthors, and retorted to the referee, "See now we know how to make two dimensional patterns!" I figured that would be enough but *Physical Review Letters*, which is supposed to be a venue for rapid publication of important results of broad interest, continued to process the paperwork at a snail's pace. They took forever to send our revised manuscript back to the referees, it took forever for the referees to respond with even more comments, and so it went dragging on for months until July 2000. It was then that I had the pleasure to meet in person, for the first time, George Basbas, the editor of *Physical Review Letters* in charge of shepherding our paper through the review process.

Basbas and I both attended the Workshop on Quantum Optics, held in Jackson Hole Wyoming from July 30 to August 2, 2000. Basbas, a large, jovial, and gregarious fellow, gave (and still gives) a hilarious talk about his life as editor at *Physical Review Letters*. In the talk, he presents, with names removed, actual snippets of referee reports and responses to the referees that he has collected over the years. After his amusing talk of "PRL bloopers," I went up and introduced myself and told him that for years we all had thought that he was a fictitious person, since we could never get him on the phone. He laughed at that and then intoned, jokingly, "Well Jonathan, as you can see I am a real person, and if you ever need help with a manuscript don't be afraid to call me!" I replied, "Well, George, since you never answer the phone, and now that I have got you here in person, I do have a paper for which you are the editor and which has been sitting in PRL hell for six months." I described the manuscript and how quantum lithography would revolutionize the semiconductor industry and mouse traps and sliced bread, and he nodded thoughtfully, "Well that sounds very interesting, I will have to take a look at it!"

I figured it would be another month at least before anything was done, but after dinner, I went back to my hotel room to check my email and lo, there in my inbox was an email from *Physical Review Letters* that stated, "Your paper has been accepted for publication." Basbas must have rung up his staff and had them send him all the correspondence and made the decision right there on the spot! When a paper is accepted, there is some paperwork to fill out and there is a check box next to "Check here if you think this paper merits an American Physical Society press release." I gleefully checked that box, and upon doing so, my life, according to the Many Worlds Interpretation of Quantum Mechanics, collapsed into a completely different life that I might have otherwise had. Someone at the American Physical Society contacted me for information about the manuscript, in order to construct an American Institute of Physics press release, timed to be released simultaneously with the publication of the paper on September 25, 2000. This was the first time any paper of mine had an accompanying press release and so I decided to cover all bases and contacted the NASA JPL Public Affairs office, told them what was coming down the pipeline, and they too decided to emit a press release. The American Institute of Physics news release appeared on September 21, 2000, and had the titillating title "'SPOOKY PHOTONS' MAY BREAK MINIATURIZATION BARRIER FOR COMPUTERS—'Entangled' Light Can Potentially Create Smaller Circuit Patterns." The JPL press release soon followed on September 25, 2000, with the somewhat more subdued title "ENTANGLED PHOTONS COULD PROMISE LIGHTNING-SPEED COMPUTERS," and then, well then that did it—my desk phone did not stop ringing for months.[81] The *New York Times*, CNN, the *Chicago Tribune*, *The Christian Science Monitor*, *Science Magazine*, *Nature Magazine*, *Scientific American*, and a bunch of other places picked up the story.[82] I was not really ready for such a response to a four-page theory paper with two figures and whose single offset equation (equation number one) contained only the rather innocent-looking $1 + \cos(2N\varphi)$. However, we concluded the paper with the more astounding claim, "Entanglement turns out to be a useful resource, which can be employed in a technology such as lithography to overcome seemingly unbreakable constraints such as the diffraction limit."

Clearly, when spooky quantum entanglement meets the computer chip industry, things take off… or perhaps it was just a slow news week. With apologies to Thomas Edison: "Genius is one percent inspiration and ninety nine percent public relations." (Here I am quoting myself.) The publicity led to two, not one, but two Hollywood consulting gigs and a $50,000 grant from NASA Headquarters to investigate whether or not entangled photons could be used to propel light-sailed ships from Earth to Alpha Centauri and eventually to an even larger grant from the Defense Advanced Research Projects Agency to study whether or not entangled photons could be used in remote laser sensing. Oh those were the heady days! The idea of using entangled or other nonclassical states of light and matter for improved imaging and remote sensing still percolates about. However, while a lot of work has been invested in these applications, no real practical technology has emerged—yet! Even with quantum lithography, as American physicist Robert Boyd is fond of saying, "Quantum lithography is a really great idea that is really hard to implement." The quantum states of light are hard to produce, fragile, and so weak in intensity that the needed, efficient, N-photon absorbing resist still remains elusive. There has been work on applying nonclassical states of light to other forms of imaging, such as microscopy, but, while intriguing, nothing of commercial importance has panned out yet. But I am not too disappointed. Quantum lithography is still trotted out as an example of a noninformation processing (computing or cryptography) application of quantum entanglement. In addition, I'd like to think that the idea caught the people's imagination—maybe entanglement can be used to improve all sorts of things, which is one of the precepts of the field of quantum technology: "We are currently in the midst of a second quantum revolution. The first quantum revolution gave us new rules that govern physical reality. The second quantum revolution will take these rules and use them to develop new technologies."[83] Certainly, there have been a great many papers published since the year 2000 reexamining the classical field of optical lithography with a new quantum mechanical eye.

THE GREAT CLOCK SYNCHRONIZATION SAGA

One of my goals for this book is to illustrate, using tales from my own personal experience, how scientists and government program managers interact in unpredictable fashions in a rapidly evolving field such as quantum technologies. The tale of the great clock synchronization saga illustrates this curious interaction to a degree that would be difficult for a layperson to believe. Just coming off the heady experience of quantum lithography with press releases ricocheting all over the Earth and reporters buzzing about the laboratory, our Quantum Computing Technologies group at the NASA JPL was laying low and trying to make some progress on what seemed like an innocent enough research project, using quantum entanglement to synchronize clocks. Clock synchronization has a long history in physics, particularly from the legacy of Albert Einstein, whose discussions on the ability (or inability) to synchronize clocks over large distances played a great role in the development of his special theory of relativity. It has been conjectured that, during his stint as a patent examiner at the Swiss Patent Office, Einstein's obsession for synchronizing clocks was in part caused by the deluge of patent applications by punctual Swiss inventors who were themselves obsessed with synchronizing the clocks between Swiss train stations by using electrical signals that ran back and forth on the telegraph lines connecting those stations at the time.[84] Telegraph signals move nearly at the speed of light and so Einstein extrapolated these ideas to propose a protocol, now called Einstein clock synchronization, by which Alice and Bob synchronize their respective clocks via the exchange of carefully timed pulses of light.[85]

A variant of Einstein synchronization is used to synchronize atomic clocks on the Earth to the atomic clocks located in every GPS satellite. Each satellite then rebroadcasts a radio signal that contains information about its location and the time on its onboard atomic clock. The GPS receiver in your car then uses this information to triangulate (more properly quadrangulate) its own position on the ground. If four independent satellite signals are readily visible, with an unobstructed view of them in the sky, then your GPS receiver knows fairly precisely its latitude, longitude, altitude, and the correct local time. It finds out this by simultaneously solving four algebraic equations for these four unknown variables. Interestingly, the GPS requires corrections from both Einstein's special and general theories of relativity in order to work at all! This is because, from relativity, clocks that are in motion and that are at different altitudes in the Earth's gravitational field with respect to each other will run at slightly different rates of speed. Once synchronized, a ground-based clock and a satellite-based clock will have times that drift away from each other, because of relativity, at a rate of tens of thousands of nanoseconds a day. Now, a nanosecond is a billionth of a second and does not seem like that much, but a fleet light signal travels approximately 30 centimeters (about 1 foot) in 1 nanosecond. Hence, doing the math, if the clocks disagree by 10,000 nanoseconds, after 1 day, your GPS receiver would be off by approximately 300 meters (328 yards)—more than enough for you to miss your exit on the freeway. The satellite clocks are purposely loaded with a relativistic correction offset before launch, and then tweaked with further tweakings from the theory of Einstein after launch, to compensate for these relativistic effects. This is the only practical application of Einstein's theory of relativity in modern everyday life I can think of![86] Without Einstein's theory of relativity, the GPS unit in your car would not work at all.

Initially, our quantum clock synchronization work did not have lofty goals of improving the GPS in mind, but we lofted that goal out from conversations with Richard Jozsa, who was visiting the JPL group in January of 2000. Colin Williams brought up the idea that given that distributed quantum information networks, the quantum Internet, would have distributed quantum entangled states at the nodes of the network, and given that communications and computations on even classical communications networks, the classical Internet, had to be carefully synchronized, perhaps there would be some quantum advantage to using entangled states to synchronized clocks on the network. Working with our new group member, American physicist Daniel Abrams, after several days of discussion, Jozsa, Abrams, Williams, and I cooked up a protocol, which eventually resulted in a paper, "Quantum Clock Synchronization Based on Shared Prior Entanglement," which we posted on the ArXiv preprint server on April 27, 2000, submitted to *Physical Review Letters* on June 15, 2000, and which finally appeared in print there on August 28, 2000. Notice that the time between submittal and publication was only about 3 months! There is a reason for that, which will soon become clear.

While the original motivation of synchronizing clocks in a distributed quantum communications network was innocuous enough, when I was an army scientist in 1995, I had organized a workshop on the sources of noise and error in the GPS. Particularly, I had become well versed in the relativistic corrections and their implementations. In addition, I learned, from discussions at this workshop, that the primary source of error in positioning accuracy in the GPS was due to the effect of turbulent fluctuations of the Earth's atmosphere on the synchronization clocks, the same fluctuations that cause a star to appear to twinkle in the night sky. When Einstein's relativity and other sources of noise were corrected for, this twinkling-star error was the one remaining dominant error. It cannot be modeled away, it cannot be subtracted away, and the effect was to make the time of arrival of the radio signals between clocks on the ground and clocks on the satellites uncertain to approximately 10 nanoseconds or 300 centimeters or 3 meters (approximately 3 yards) accuracy in positioning.[87] Not enough to miss your exit on the freeway but enough so you would not try to land an airplane by GPS signal alone—misgauging the height of the runway by 3 meters would make for a very rough landing. I had convinced myself that our quantum clock synchronization protocol offered a way around this 3-meter error in the GPS!

Thus, let us review briefly how our quantum clock synchronization protocol is to work.[88] The setup was much like that in quantum teleportation or the Ekert quantum cryptography protocol (see Figure 5.4) where Charlie transmits to Alice and Bob a large number of pairs of entangled qubits of the particular form $|\uparrow\rangle_A |\downarrow\rangle_B - |\downarrow\rangle_A |\uparrow\rangle_B$, let us say 100 of them, where the arrows are the spin directions of ions in two ion-trap atomic clocks. The idea is that while, say when exposed to a magnetic field, the individual ion states $|\uparrow\rangle_A, |\downarrow\rangle_B, |\downarrow\rangle_A,$ and $|\uparrow\rangle_B$ evolve in time and can thence be used to measure time, all the time evolution in the entangled state $|\uparrow\rangle_A |\downarrow\rangle_B - |\downarrow\rangle_A |\uparrow\rangle_B$ cancels out, in part because of that minus sign, and we say that this state is "idling." Now, Alice performs a measurement of the spin direction of one of her ions, say ion number one, along the vertical axis, and with a 50–50 probability, her ion collapses to either $|\uparrow\rangle_A$ or $|\downarrow\rangle_A$, which does indeed evolve in time. This starts Alice's clock, and she can then extract which way it collapsed, either up or down. The measurement causes Bob's ion one state to collapse the opposite direction spin to either $|\downarrow\rangle_B$ or $|\uparrow\rangle_B$, which starts Bob's clock. To clarify if Alice's ion collapses to up $|\uparrow\rangle_A$, then Bob's collapses to down $|\downarrow\rangle_B$ (or vice versa); hence, we say the measurements are anti-correlated.

To complete the protocol, Alice makes a simultaneous measurement of all 100 of her ions at once and marks the time on a second atomic clock when the measurement was made. She notes in her laboratory notebook the time on her second (classical) atomic clock when the measurement was made, say 12:00

noon on her clock, and records the spin state (up or down) for each of the 100 ions. Then, she relays this information to Bob as a classical message over a telegraph wire, cell phone relay, or pony express (assuming you have ponies that can achieve low-Earth orbit). Bob's clocks are already running by the time Alice's message arrives; he just does not know what time (on Alice's primary atomic clock) Alice's measurement was made. However, all he has to do is collect say all the ions in his trap that corresponded to Alice's measurement outcome of spin down, which he knows collapsed to spin up at Alice's time zero, and then he can use laser pulses to synchronize these time-evolving ions in his trap to his primary clock, which will be running in sync with Alice's primary atomic clock, which is both agreeing on what 12:00 noon actually is. That is, Alice and Bob now have states that are synchronized.

Now, we have to handle the disruption of the entangled states in transport. The state $|\uparrow\rangle_A |\downarrow\rangle_B - |\downarrow\rangle_A |\uparrow\rangle_B$ must somehow be transmitted from Charlie to Alice and Bob. If the state is in two ion traps, the transport of the traps from Charlie's location to Alice and Bob's locations could be error prone. Acceleration or buffeting of the transport ships could cause the state $|\uparrow\rangle_A |\downarrow\rangle_B - |\downarrow\rangle_A |\uparrow\rangle_B$ to drift into another state that is not idling properly and mess up the synchronization. You could imagine transmitting entangled pairs of photons (instead of entangled pairs of ions) in the polarization-entangled state $|\updownarrow\rangle_A |\leftrightarrow\rangle_B - |\leftrightarrow\rangle_A |\updownarrow\rangle_B$ from Charlie to Alice and Bob and then swapping out the entanglement from the photons to the ions at Alice and Bob's locations, but the twinkling star effect of the atmosphere could mess with the photons in transit, corrupting the entangled state we are trying to transmit, taking us back to square one. But there is a save for fixing this error.

There is a protocol called "entanglement purification" that allows Charlie to send a large number of crappy entangled pairs with noise on them and then Alice and Bob perform a multi-round communications protocol to "distill" a smaller number of good entangled pairs from the large number of crappy ones. Entanglement purification would get rid of all the errors accumulated in the entangled pair transport phase of the clock synchronization process, we hoped, and then we could carry out the synchronization as described above. Very proud of ourselves, Jozsa, Abrams, Williams, and I began completing our calculations, buttressing our arguments, and slowly preparing our manuscript for submission to the quantum physics preprint ArXiv followed by submission to *Physical Review Letters*.

While we were working away on the project, I happened to have a discussion with Henry Everitt at the Army Research Office, in early February of 2000, and I happened to mention that we were working on a new quantum clock synchronization protocol, and I believed I used words to the effect, "Our protocol will allow you to teleport timing information between the Earth and a satellite— past the turbulent atmosphere!" Little did I know that the DoD and other more secretive departments in the government were *very* interested in precisely synchronizing ground clocks with satellite clocks for a number of applications that had nothing to do with the GPS and everything to do with spy satellites. Everitt immediately asked if they could have a copy of the preprint of the article, but in January, we were just beginning work on the calculations and did not have a preprint to give him. As it turned out, we submitted the paper to the ArXiv on the afternoon of Thursday, April 27, 2000.

We immediately got a flurry of comments from a number of quantum physicists around the world on the paper, pointing out some inconsistencies, and so forth. This is not unusual. If you post a new paper to the ArXiv you can expect a lot of comments from almost everybody who would be a potential referee of the paper, once it is submitted for publication in a refereed archival journal, and if you are lucky, the referees for the journal article have already made all their comments on the ArXiv version and the referring process goes much smoother. However, in this case, there were numerous comments, and some took longer to answer properly than others, so due diligence we did do and worked out responses to the comments and revised the manuscript, uploaded version two to the ArXiv on June 8, 2000, and then submitted the work to *Physical Review Letters* on June 15th.

Then, a curious thing happened. Robert Garisto, the editor in charge of the paper at *Physical Review Letters*, called me up the next day and told me that they had just accepted a paper by American physicist Isaac Chuang on quantum clock synchronization and wanted to know how our paper compared to it. We were aware of Chuang's work, as it had appeared on the ArXiv about 3 weeks after our own, on May 22, 2000. I explained that our paper was a different and entirely new protocol, using prior shared quantum entanglement as a channel to do the synchronization, and that Chuang's was instead a quantum improved version of the Einstein synchronization protocol, using quantum entanglement to exponentially reduce the number of bits of information Alice and Bob needed to exchange (via light pulses) to synchronize their clocks to a given accuracy. I also pointed out that our paper had been on the ArXiv about 3 weeks *before* Chuang's had appeared there. I suggested the two papers were both important and complementary and might have more impact if they appeared conjointly in *Physical Review Letters*. Garisto said he would see what he could do and then, much to my surprise, I received an email from him 48 hours later stating that our paper had been accepted for publication. (Recall that the time between submission and acceptance of the quantum lithography paper was more like 6 months.)

I have never had a paper submitted to *Physical Review Letters* accepted in 48 hours before or since. Years later, I asked Garisto how that happened and he told me they sent it to a member of the editorial board to referee, given the apparent urgency. The two papers appeared back to back (Chuang's first and ours second) in the August 28, 2000, issue of *Physical Review Letters*. Once again, there was a JPL and American Institute of Physics press release, but the post-publication buzz was nothing like with quantum lithography. I imagine that the arcane art of synchronizing clocks just did not have an obvious appeal to the general public, but I didn't imagine that the arcane art of making computer chips with light beams did either.

Once these two papers on quantum clock synchronization appeared, and the rumor of government funding got around, there was a sudden flurry of activity in the area. Looking at the Science Citation Index, as I type this, our paper has been cited 80 times and Chuang's paper has been cited 49 times. (Compare this to our quantum lithography paper that has been cited 488 times.) As Turkish–American physicist Ulvi Yurtsever in our group at JPL used to joke, "Chuang's protocol was useless and our protocol was wrong." Well, his is not quite useless and ours is not quite wrong, and our first author Richard Jozsa bristles visibly at any such suggestion. However, there was a subtle point that we missed, which was pointed out to us by John Preskill from Caltech. That is, that entanglement purification, which we were relying on to get rid of errors in transporting the entangled qubits, if applied to qubits that evolved in time, required Alice and Bob to synchronize their operations in order to carry out the purification steps. Well, if Alice and Bob already had synchronized clocks, they did not need our clock synchronization protocol! Hence, in some sense, our argument was a tad circular, which explains why we did not get as many citations as with lithography. However, we still hold out hope that some purification protocol could be found that would not require Alice and Bob to synchronize their operations in advance and then our protocol would work just fine. At least nobody has proved that such a purification protocol cannot exist. Hope springs eternal.

NOTES

1. Title taken from "Gadgets from the Quantum Spookhouse" by Peter Weiss in *Science News*, Volume 160 (December 8, 2001), pages 364–366, http://www.sciencenews. org/pictures/yawn/112010/spookhouse.pdf.
2. Mark Heiligman, private communication (June 29, 2011), permission to quote him granted.
3. See "One-Time-Pad (Vernam's Cipher) Frequently Asked Questions" by Marcus J. Ranum (November 26, 2011), which also has a cool photograph of a Russian one-time pad captured by the MI5 British spy agency, http://www.ranum.com/security/computer_security/papers/otp-faq/.
4. See "Binary Conversion" in *My Play Ground* (August 29, 2012), http://www.roubaixinteractive.com/PlayGround/Binary_Conversion/Binary_To_Text.asp.
5. For Dilbert's amusing take on a random number generator, click on the following URL: http://dilbert.com/strips/comic/2001-10-25/.

6. Here, Charlie is the source "S" in Figure 2.4 in Chapter 2.

7. See *MagiQ* (August 29, 2012), http://www.magiqtech.com/MagiQ/Home.html.

8. NuCrypt was founded by my old friend and collaborator, Prem Kumar, who I affectionately now call "The NuCrypt Keeper,"

9. See IDQuantique (August 29, 2012), http://www.idquantique.com/.

10. See Quintessence Labs (August 29, 2012), http://www.quintessencelabs.com/.

11. See "Prologue: Under the Danube" in *Dance of the Photons: From Einstein to Quantum Teleportation* by Anton Zeilinger (Macmillan, 2010), http://books .google.com/books?id=Oykwl_269KsC (search for the word "sewage"). See also "Quantum Secured Bank Transfer" (Institute for Quantum Information, University of Vienna, December 15, 2011), http://www.quantum.at/research/ quantum-cryptography/quantum-secured-bank-transfer/.

12. The late American physicist and Nobel Laureate, Willis Lamb, in his later years gave lectures and published a paper entitled "Anti-Photon," in which he complained about the popular notion of a photon as a fuzzy ball of energy and so he provided instead a rigorous mathematical definition of the photon, which almost nobody ever used except Willis Lamb. In one such lecture, an audience member, American physicist and Nobel Laureate Roy Glauber, retorted, "I can't define a photon, but I know one when I see one," as a spoof on US Supreme Court Justice Potter Stewart's similar and famous definition of pornography. To quote from Lamb's paper, speaking of himself in the third person, "It should be apparent from the title of this article that the author does not like the use of the word 'photon,' which dates from 1926." In his view, there is no such thing as a photon. See "Anti-Photon" by Willis E. Lamb in *Applied Physics B: Lasers and Optics*, Volume 60 (1995), pages 77–84.

13. See "Researchers Set New Record for Quantum Key Distribution" by Lisa Zyga in *PHYSORG* (July 21, 2009), http://www.physorg.com/news167390366.html.

14. Also note that many of the prototype quantum cryptography systems do not use infrared photons—yet!—since the photon detectors that are used to generate the key, with currently available technology, work better at shorter wavelengths, photons somewhere in the blue green part of the spectrum. Hence, some prototype systems use special fibers optimized for these shorter wavelengths. But because optical fibers are most transparent in the infrared, there is a great push to move the whole technology down to these longer wavelengths, and for the sake of simplicity, I am discussing all the systems as if they used infrared photons of 1.5 microns wavelength.

15. For a schematic of the DARPA network in Boston, see "Quantum Key Distribution" by Jennifer Ouellette in *The Industrial Physicist* (American Institute of Physics, December 24, 2004), http://www.aip.org/tip/INPHFA/vol-10/iss-6/p22.html.

16. See "Long-Term Performance of the SwissQuantum Quantum Key Distribution Network in a Field Environment" by D. Stuki,..., P. Monbaron, in New Journal of Physics, Volume 13, Article No. 123001, http://iopscience.iop.org/1367-2630/13/12/ 123001/.

17. See "Quantum Repeaters for the Novice" by Matthieu Legré (QuRep, April 21, 2010), http://quantumrepeaters.eu/index.php/qcomm/quantum-repeaters.

18. See "The Early Days of Experimental Quantum Cryptography" by John A. Smolin in the *IBM Journal of Research and Development*, Volume 48 (January 2004), pages 47–52, http://ieeexplore.ieee.org/xpl/freeabs_all.jsp?arnumber=5388927.

19. See "Quantum Cryptography in Free Space" by Bryan C. Jacobs and James D. Franson, *Optics Letters*, Volume 21 (November 16, 1996), pages 1854–1856, http://www.opticsinfobase.org/abstract.cfm?URI=ol-21-22-1854.

20. It is very unlikely that a single photon of visible wavelength would cause any damage. The Franson–Jacob experiment shot about a thousand photons per second down the hall, which, if converted into power, would be about 14 orders of magnitude weaker (10–14) than the power you would get standing next to a 100-watt lightbulb. Charlie Bennett once asked me, as a puzzle, if there was any wavelength of photon so energetic that a single photon might cause physical damage to a human. I consulted with Bradley Schaefer, an astronomer and colleague at Louisiana State University, and we decided that a single photon with a wavelength 10^{16} times shorter than the visible photon would work. Such a photon would carry around 1000 watts of power—more than enough to cook your brain and kill you if it was aimed at your head. However, the most probable thing such a photon would do would be to go clean through your head without depositing any of its energy.

21. The term "free space" means in mostly empty space, like air or outer space, unlike in an optical fiber. It is used for optical communications systems in actual outer space or in the atmosphere.

22. See "Practical Free-Space Quantum Key Distribution Over One Kilometer" by W.T. Buttler, Richard J. Hughes, Paul G. Kwiat, et al. in *Physical Review Letters*, Volume 81 (1998), pages 3283–3286, http://link.aps.org/doi/10.1103/PhysRevLett.81.3283.

23. See "Practical Free-Space Quantum Key Distribution over 10 Kilometers in Daylight and at Night" by Richard J. Hughes, Jane E. Nordholt, D. Derkacs, et al. in *New Journal of Physics*, Volume 4 (July 12, 2002), article number 43, http://iopscience.iop.org/1367-2630/4/1/343.

24. See "Quantum Spookiness Spans the Canary Islands" by J.R. Minkel in *Scientific American* (March 9, 2007), http://www.scientificamerican.com/article.cfm?id=entangled-photons-quantum-spookiness.

25. See "Secure Communication via Space" in *PHYS.ORG* (April 22, 2008), http://www.physorg.com/news128096976.html.

26. These experiments were championed by American physicist Carroll Alley during the Apollo years; see, for example, "What Neil and Buzz Left on the Moon" in *Science News* (NASA Science, December 29, 2011), http://science.nasa.gov/science-news/science-at-nasa/2004/21jul_llr/. These experiments form a simple rebuttal to the so-called Moon Landing Conspiracy Theorists (one of whom once called Buzz Aldrin, the second man to walk on the Moon, a liar to his face and in response the then 74-year-old Aldrin punched the Lunar Conspiracy Theorist in his face). These conspiracy nuts typically claim the Moon landing never occurred and that NASA faked the films of the landings in conjunction with Hollywood. My retort to these guys is that US President Bill Clinton could not keep the fact that he had a "girlfriend" in the White House a secret for 1 year. A faked series of lunar landings would require a vast net of continuous secrecy over 40 years and over the terms of eight US presidents, Nixon through Obama. But there is a simple test. Take a powerful laser, rent some telescope time, and point the laser at the Moon and launch a laser pulse with 10^{17} photons in it at the Moon. If you carefully aim your laser at the Apollo 11, 14, or 15 landing sites, then you'll always get one photon on average back. If you aim your laser anywhere else on the Moon, you'll never get any photons back, ever. The only rational explanation

is that the Apollo astronauts placed retroreflectors on the Moon when they landed there. Of course, the lunar landing conspiracy nuts are anything but rational. To be fair to Buzz, the conspiracy theorist not only called Aldrin a liar, he also called him a coward. (Aldrin flew 66 combat missions in the Korean War.) The conspiracy theorist that Aldrin punched is a "filmmaker." (I refuse to mention the conspiracy theorist by name as his 15 minutes of fame are long over.)

27. I would imagine a far more common problem for Wonder Woman than locking her keys in her invisible jet plane would be remembering where she parked it in the first place! "I'm just sure I parked my invisible jet around here somewhere...."

28. "As for the leper who has the infection, his clothes shall be torn, and the hair of his head shall be uncovered, and he shall cover his mustache and cry, 'Unclean! Unclean!'" See Leviticus 13:45 in the *New American Standard Bible* (The Lockman Foundation, 1995), http://bible.cc/leviticus/13-45.htm.

29. It is amusing to note that even though the agency did not "exist" until 1992, in their lobby there is a wall of photos of the dynasty of NRO directors going back to 1960.

30. The 1999 NATO bombing of the Chinese Embassy in Belgrade, during the Yugoslav Wars, was originally blamed on an incorrect map provided to the North Atlantic Treaty Organization by the National Imagery and Mapping Agency (NIMA). The original assessment was that the map was out of date and did not account for the fact that the Chinese Embassy had moved 3 years earlier from another location in Belgrade to a location that was marked for bombing as a warehouse for storing arms. Follow-up assessments by the CIA do support that the bombing was accidental but do not particularly blame any single mapping error by NIMA.

31. I must state clearly that I have no idea what encryption protocol the NRO uses to communicate with its satellites, nor do I want to know, nor do I have any means of knowing, but it is reasonable to assume it is a modified version of the Digital Encryption Standard developed in 1976 by the National Bureau of Standards, now the National Institute of Standards and Technology, in consultation with the National Security Agency, as a US Federal Information Processing Standard.

32. A yottabyte is 1 septillion bytes and should not be confused with a Yoda bite. This protocol of requiring at least two people to carry out some protocol that requires protection against accidental or malicious execution is known as the "Two-Man Rule." For example, two persons were required to launch a Minuteman nuclear missile. In the 1983 film *WarGames*, during a simulated launch, one soldier pulls a gun on the other when he refuses to turn his launch key.

33. I don't know that they blow up the spy satellite when it runs out of key so I am just surmising. Mururoa is the atoll in the Pacific where the French once carried out aboveground nuclear weapons testing.

34. See "Space-QUEST: Quantum Entanglement Experiments in Space" by Rupert Ursin, ..., Anton Zeilinger in *Europhysics News*, Volume 4 (2009), pages 26–29, http://dx.doi.org/10.1051/epn/2009503. This file is licensed under the Creative Commons Attribution-Share Alike 3.0 Unported license.

35. The *Star Trek* series' writers introduced the Heisenberg compensator in order to address their understanding of the Heisenberg Uncertainty Principle and their statement that the transporter produced copied and transported matter at the quantum level. Apparently, some physicists complained that the transporter violated the Heisenberg Uncertainty Principle, so the writers just fixed it by adding

the Heisenberg compensator, whatever the hell that is. This of course is just a non-physical gimmick. The Heisenberg Uncertainty Principle, and its close relative, the no-cloning theorem, cannot be compensated for, at least within the known laws of physics. However, much of *Star Trek* technology (such as the warp drive) are outside the known laws of physics. When asked "How does the Heisenberg compensator work?" by *Time* magazine, *Star Trek* technical adviser Michael Okuda responded: "It works very well, thank you." See "Reconfigure the Modulators" in *Time* magazine (November 28, 1994), http://www.time.com/time/magazine/article/0,9171,981892,00.html.

36. A "light-year" despite the name is a unit of *distance* and not of *time*. It is the distance light travels in a year moving at the speed of light—299,792,458 meters per second (186,282 miles per second). A light-year is about 10^{13} kilometers (6×10^{12} miles). Alpha Centauri is the closest star system to our own solar system at about 4 light-years' distance. It takes the light emitted from the surface of our Sun about 8 minutes to reach the Earth. Hence, when you look up at the Sun (not for too long as you'll hurt your eyes), you do not actually see the Sun where it *is* but where it *was* 8 minutes ago, due to the Earth's rotation. Since the light-year is a unit of distance, it is a pet peeve of mine to see "light-year" misused as a unit of time, typically by misinformed marketing writers looking for a fancy-sounding technical word (that they do not understand), and we end up with such mangled and knuckleheaded grammatical atrocities as, "The design for the solar-powered car is light-years ahead of its time." The design is simply *years* ahead of its time. Light-years ahead of its time makes no sense. The design could be light-years ahead of its competitors only if the designers put the actual design on a rocket ship and launched it off at the speed of light some years, say toward Alpha Centauri, but this is clearly not what his writer means. The Starship *Enterprise* can be light-years ahead of a cubical Borg spacecraft, but only if they are having a race (and the Borg are losing). See "Light Years Ahead" in *Idiom Quest* (January 5, 2012), http://www.idiomquest.com/learn/idiom/light-years-ahead/. For a spectacularly bad example, in a newspaper that should know better, see "A Device Light Years Ahead of its Time" by Dan Vergano in *USA Today* (November 30, 2006), http://www.usatoday.com/tech/science/discoveries/2006-11-30-antikythera-mechanism_x.htm.

37. In general, a and b are complex numbers that satisfy $|a|^2 + |b|^2 = 1$. Hence, if $a = b$, one possible choice is $a = b = 1/\sqrt{2}$ because $\left|1/\sqrt{2}\right|^2 + \left|1/\sqrt{2}\right|^2 = 1/2 + 1/2 = 1$. That is, in quantum theory, the absolute value square of a gives the probability of a vertical and the absolute value square of b gives the probability of a horizontal, 50–50 in this example.

38. Einstein's ashes were scattered in an unknown river near Princeton, New Jersey. Throughout this book, for concreteness, I guess various rivers near Princeton as the scattering spot.

39. To quote from the film *Star Trek II: The Wrath of Khan*, "*Enterprise*, what we got back didn't live long, fortunately."

40. For example, a typical photon polarization cat state is written as $a|\updownarrow\rangle + b|\leftrightarrow\rangle$ with a and b both equal to $1/\sqrt{2}$, which is an irrational number specified (in base 10) by an infinite sequence of repeating decimal digits, namely, 0.70710678118654752440084 36210484903928483593768847403658833986899536623923105351942519376716 38207864....

41. Faster-than-light travel has been proposed using a cosmological space–time construction called an Einstein–Rosen bridge, more colloquially known as a "wormhole" in space. Sadly, there are no such wormholes between Earth and Mars (nor anywhere else) that we know of.

42. There is an amusing history of several such "faster than light" or "superluminal" communication schemes told in the recent book *How the Hippies Saved Physics* by David Kaiser (W.W. Norton, New York, 2011), http://www.worldcat.org/oclc/668194856. The first was a 1978 patent disclosure by theoretical physicist and parapsychologist Jack Sarfatti entitled "Faster-Than-Light Quantum Communication System," which deployed entangled particles shared by Alice and Bob. Sarfatti missed the point that no useful information could be extracted from the system until Alice and Bob correlated their measurements over the slowly moving classical channel. The second was the widely circulated 1981 preprint by Nick Herbert entitled "FLASH—A Superluminal Communicator Based upon a New Kind of Quantum Measurement," which invokes the (noiseless) amplification of laser signals. Kaiser's book credits these two interesting but wrong papers for stimulating research into the development of the quantum no-cloning theorem, the laser no-nonnoiseless amplification theorem, and the resultant nonsuperluminal signaling theorems. A laser amplifier is like a copying machine; it takes a weak laser signal and makes a strong one. This cannot be done without adding noise to the signal and thence is closely related to the no-cloning theorem. You cannot perfectly copy a quantum state (no cloning) and you cannot noiselessly amplify a laser signal (no nonnoiseless amplification). If you could do either of these, then you could indeed send signals faster than the speed of light, and sometimes the inability to do the latter (via Einstein) is taken as proof that you cannot do the former. But there are proofs of the no-cloning theorem and no-nonnoiseless amplification theorems that depend only on the assumed linearity of the theory of quantum mechanics that are independent of the speed-of-light argument. In fact, if quantum mechanics was nonlinear—the input state was not simply proportional to the output state—for certain types of nonlinearities, quantum theory would indeed allow superluminal communication. I even got involved in an imbroglio over this issue of superluminal communications, a battle among the *Journal of Parapsychology*, ESP-wielding martial arts students, and a friend of Sarfatti named Henry Stapp, which played out in the letters-to-the-editor section of a physics trade journal and got a reference in Kaiser's book. See "Parapsychological Review A?" by Jonathan P. Dowling in *Physics Today* (Letter to the Editor), Volume 48 (July 1995), page 78, which is reference 68 in Kaiser's book, http://www.deepdyve.com/lp/american-institute-of-physics/parapsychological-review-a-0qD7VUn1Py. This is an epic story that I will tell in the next chapter in full here but some of it appears in Kaiser's book, pages 257 to 258. My version is a bit longer than that of Kaiser and I know the true identity of the anonymous referee—not me!

43. Lest some of my colleagues think I have gone off the deep end with such ruminations on the relation between strange quantum effects and relativity, at least I am in good company, as noted Swiss physicist Nicolas Gisin and his colleagues have carried out several experiments on entangled pairs of photons, within the context of a particular model, to test the speed at which a measurement by Alice on her photon induces the collapse of the state of Bob's photon 18 kilometers away. Their conclusion was that, "…the speed of the influence would have to exceed that of light

by at least [a factor of 10,000]." That is, this speed of influence is consistent with instantaneous. See "Testing the Speed of 'Spooky Action at a Distance'" by Daniel Salart, Augustin Baas, Cyril Branciard, Nicolas Gisin, and Hugo Zbinden in *Nature*, Volume 454 (August 14, 2008), pages 861–864, http://www.nature.com/nature/journal/v454/n7206/abs/nature07121.html. See also the commentary *News and Views* article, "Quantum Mechanics: The Speed of Instantly," by Terence G. Rudolph in *Nature*, Volume 454 (August 14, 2008), pages 831–864, http://www.nature.com/nature/journal/v454/n7206/full/454831a.html.

44. See "Long-Distance Teleportation of Qubits at Telecommunications Wavelengths" by I. Marcikic, H. de Riedmatten, Wolfgang Tittel, Hugo Zbinden, and Nicolas Gisin in *Nature* (January 30, 2003), pages 509–513, http://www.nature.com/nature/journal/v421/n6922/abs/nature01376.html.

45. See "Quantum Teleportation Across the Danube Demonstrated" by Sarah Graham in *Scientific American* (August 19, 2004), http://www.scientificamerican.com/article.cfm?id=quantum-teleportation-acr.

46. See "Quantum Teleportation over 16 Kilometers" by Lin Edwards in *PHYS.ORG* (May 20, 2010), http://www.physorg.com/news193551675.html. More recently, this same group has reported a free-space quantum teleportation experiment over 97 kilometers (60 miles) through the air over Qinghai Lake, the largest saltwater lake in China. See "Teleporting Independent Qubits through a 97 km Free-Space Channel" by Juan Yin, ..., and Jian-Wei Pan in the *Quantum Physics ArXiv* (May 9, 2012), http://arxiv.org/abs/1205.2024.

47. See "Quantum Teleportation between Distant Matter Qubits" by S. Olmschenk, D.N. Matsukevich, P. Maunz, D. Hayes, Lu-Ming Duan, and Christopher Monroe in *Science*, Volume 323 (January 23, 2009), pages 486–489, http://www.sciencemag.org/content/323/5913/486.short.

48. To be precise, Doug should first set the constants a and b on the ensembler and send an entire collection of states Ψ with these exact a and b through the teleporter. Only in this way can repeated measurements on the corresponding output collection by Ellen with her fidellerator extract the teleported a and b and compare it with a and b that should have been teleported. In this way, through repeated measurements of the ensemble, can Ellen compute the fidelity for that particular identical set of states.

49. I am assuming that each of around 100 possible atoms from the periodic table needs to be labeled, giving around 10^{30} bits. The classical information content can be estimated from something called the Bekenstein Bound on the amount of classical information that can fit in a region of space with a fixed total amount of energy.

50. See "Quantum Repeaters for the Novice" by Matthieu Legré in Quantum Repeaters for Long Distance Fiber-Based Quantum Communication (April 21, 2010), http://quantumrepeaters.eu/index.php/qcomm/quantum-repeaters.

51. See "Linear Optics Quantum Computation" in *QuanTiki* (January 29, 2012), http://www.quan tiki.org/wiki/Linear_optics_quantum_computation.

52. See "Entangled States of Trapped Atomic Ions" by Rainer Blatt and David Wineland in *Nature*, Volume 453 (June 19, 2008), pages 1008–1015, http://tf.nist.gov/general/pdf/2284.pdf.

53. See "A Quantum Rosetta Stone for Interferometry" by Hwang Lee, Pieter Kok, and Jonathan P. Dowling in *Journal of Modern Optics*, Volume 49 (2002), pages 2325–2338, http://dx.doi.org/10.1080/0950034021000011536.

54. This business of understanding the quantum computing circuit diagrams may seem trivial, but to the uninitiated, they can be very daunting. For example, a simple quantum circuit diagram for the Shor quantum error correction code looks like this:

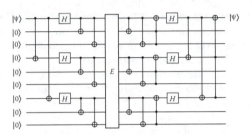

55. This discussion and Figure 5.8 are adapted from that of "Improvement of Frequency Standards with Quantum Entanglement" by Susana F. Huelga, Chiara Macchiavello, Martin B. Plenio, Thomas Pellizzari, Artur K. Ekert, and Ignacio Cirac in *Physical Review Letters*, Volume 79 (November 17, 1997), pages 3865–3868, http://link.aps.org/doi/10.1103/PhysRevLett.79.3865.

56. See "Atomic Clocks Use Quantum Timekeeping" by David J. Wineland, James C. Bergquist, John J. Bollinger, Robert E. Drullinger, and Wayne M. Itano in the *Proceedings of the 6th Symposium on Frequency Standards and Metrology*, edited by Patrick Gill (World Scientific, 2002), pages 361–368, http://books.google.com/books?id=MgQW6Tg10uAC&lpg=PA361. A preprint of this single article can be obtained here: http://tf.nist.gov/general/pdf/1581.pdf.

57. Einstein was apparently not aware of the Michelson–Morley experiment and based his postulate of the constancy of the speed of light on his analysis of the mathematical properties of the Scottish physicist James Maxwell's equations that accurately describe how light should propagate in empty space. However, soon after Einstein's theory appeared, people rushed to point out that the Michelson–Morley experiment provided experimental evidence to the postulate and this connection led to the rather quick acceptance of the counterintuitive Special Theory of Relativity of Einstein.

58. See "Quantum Optical Metrology—The Low-Down on High N00N States" by Jonathan P. Dowling in *Contemporary Physics*, Volume 49 (2008), pages 125–143, http://www.tandfonline.com/doi/abs/10.1080/00107510802091298. A preprint version may be found here: http://arxiv.org/abs/0904.0163.

59. For many years, in the late 1980s and early 1990s, the best (and cheapest) of these so-called nonlinear optical crystals were fabricated in China. As Chinese–American physicist Yanhua Shih relates, this is because during the Chinese Cultural Revolution (1966–1976) the government rounded up many intellectuals and academics such as physicists and chemists and banned them from working at their universities and instead sent them to work on farms. I myself have Chinese colleagues who were forced out of university as students and sent to work on pig farms, for example. But the government found this group of Chinese crystal growers not in an office or in a clean laboratory but out in a shed stirring a giant steaming smelly pot of goo that they were growing the optical crystals in. The government officials noted that, whatever they were doing, it looked like hard work. So

the scientists were told, "Okay, you just keep doing that." So when their colleagues were slopping the pigs, the Chinese crystal growers perfected their art in that hot shed for 10 years, and by 1977, they were producing the highest-quality and most inexpensive nonlinear optical crystals in the world. The Cultural Revolution over, they re-embraced capitalism and started a very successful company to start selling and marketing these crystals.

60. See "A Gravitational Wave Observatory Operating Beyond the Quantum Shot-Noise Limit" by the LIGO Scientific Collaboration in Nature Physics, Volume 7 (September 11, 2011), pages 962–965, http://www.nature.com/nphys/journal/v7/n12/full/nphys2083.html. See also "Gravity Waves, Scientists Wave Back: Squeezing Light Beyond Quantum Limit" by David Blair in *The Conversation* (September 13, 2011), http://theconversation.edu.au/gravity-waves-scientists-wave-back-squeezing-light-beyond-quantum-limit-3342.

61. See "Analogue Logic for Quantum Computing" in *PHYS.ORG* (February 21, 2008), http://www.physorg.com/news122827796.html.

62. An analog computer is a device that computes using continuous variables rather than the discrete variables (zeros and ones) of a digital computer. Probably the most familiar analog computer to anybody over the age of 50 is the slide rule where the slide moves continuously on the rule.

63. See "Chaos" by Jim Yorke and Timothy D. Sauer in *Scholarpedia* (June 20, 2012), http://www.scholarpedia.org/article/Chaos.

64. See "Input States for Enhancement of Fermion Interferometry Sensitivity" by Bernard Yurke in *Physical Review Letters*, Volume 56 (1986), pages 1515–1517, http://link.aps.org/doi/10.1103/PhysRevLett.56.1515. In a curious coincidence, Bernie was a housemate of mine in Austin, Texas, when I was an undergraduate in physics and he was in graduate school. He still looks the same. I remember when he graduated he left a note on the bulletin board that said something like, "Goodbye to all. Off to build A-Bombs!—Bernie"

65. For this particular setup, the margin of error as a percentage for the unentangled case is given by $100/\sqrt{N}$ and that for the entangled case is $200/(N+1)$, which gives 32% and 18%, respectively, when N is 10 and 10% and 2% when N is 100. The square root symbol indicates the quadratic (poor) scaling and the absence of it indicates the linear (good) scaling.

66. The twin-number state $|5\rangle_{Up} |5\rangle_{Down}$ becomes, after the first beam splitter, something now called a "bat" state (as a plot of the photon number distribution by arm looks like the two ears of a bat) and is approximately of the form $|10\rangle_{Up} |0\rangle_{Down} + |0\rangle_{Up} |10\rangle_{Down}$, which is now popularly called a N00N state, because it has the form $|N\rangle_{Up} |0\rangle_{Down} + |0\rangle_{Up} |N\rangle_{Down}$, a type of cat state where all 10 photons are up and none are down and all 10 photons are down and none are up.

67. Or *"Digitalteilchenanzahlräumlichverschränkungsverstärkteinterferometrie"* in the original German.

68. See "Quantum Optical Metrology—The Lowdown On High-N00N States" in *Contemporary Physics*, Volume 49 (2008), pages 125–143, http://www.tandfonline.com/doi/abs/10.1080/00107510802091298. The free version may be found here: http://arxiv.org/abs/0904.0163.

69. When the English physicist Stephen Hawking conceded the bet (over an information paradox involving astronomical black holes), Hawking paid off the bet by presenting Preskill with a book of baseball statistics. Most of us puzzled over this

payoff as Preskill had long ago memorized all these statistics and certainly did not need such a book.

70. See "Anagrams for the Electronic Age" by Jonathan P. Dowling in *APS News*, Volume 4 (1995), http://www.aps.org/publications/apsnews/199512/anagrams.cfm.

71. When German physicist Max Born introduced the probability interpretation of Schrödinger's wave function ψ, he wrote in the text that ψ should be interpreted as the probability of finding the electron. That statement is wrong. But in a note added in proof, he wrote that on second thought perhaps it should be $|\psi|^2$, and for that footnote added in page proof, Born won the Nobel Prize in physics. See "Max Born's Statistical Interpretation of Quantum Mechanics" by Abraham Pais in *Science*, Volume 218 (December 17, 1982), pages 1193–1198, http://www.science mag.org/content/218/4578/1193.abstract. A similar but perhaps less innocent story occurred in the arena of high-temperature superconducting physics where in 1987 the group of Chinese–American scientist Paul Chu submitted two papers to *Physical Review Letters* on a new class of high-temperature superconductors called "yttrium barium cuprate" abbreviated "$YBa_2Cu_3O_7$." The controversy arose in that in the original submission, the wrong abbreviation—"$YbBa_2Cu_3O_7$"—was used, which is a different compound ("ytterbium barium cuprate") and is not supercon-ducting at high temperatures. The error was repeated two dozen times and only corrected in the page proof stage when the papers were accepted for publication. The innocent version of this typo, held by Paul Chu himself, is that it was a simple mistake. There are others who claim that Chu's group deliberately submitted the wrong abbreviation so that word would not leak out before publication, either from the editorial staff or the anonymous referees, on what the correct compound was, so that the group would not be scooped by competitors. In the end, word did leak out and a number of laboratories immediately began experiments on the wrong compound $YbBa_2Cu_3O_7$ and some even gleefully trumpeted their null results before the typo was caught. Some physicists point at that competition in Chu's defense that he (if he did) was right to disguise the compound since clearly the information was leaked in review. Others complain that it is unethical to submit something you have deliberately falsified for publication. I, personally, lay the blame where it clearly and solely belongs, on the town of Ytterby, Sweden, for which an inordinate number of similar sounding chemical elements are named: yttrium, erbium, terbium, and ytterbium (as well as holmium—thulium!—and gadolinium), which were all discovered in that little quarry near Ytterby. See *The Breakthrough: The Race For The Superconductor* by Robert M. Hazen (Summit Books, 1988), page 62.

72. See "High-NOON States by Mixing Quantum and Classical Light" by Itai Afek, Oron Ambar, and Yaron Silberberg in *Science*, Volume 328 (May 14, 2010), pages 879–881, http://www.science mag.org/content/328/5980/879.short.

73. It is called a "bat" state because if you plot the statistical distribution of the particle numbers distributed over the two arms of the interferometer, the plot looks some-thing like a pair of bat ears or perhaps the top of batman's mask.

74. See our paper, "Linear Optics and Projective Measurements Alone Suffice to Create Large-Photon-Number Path Entanglement," by Hwang Lee, Pieter Kok, Nicolas J. Cerf, and Jonathan P. Dowling in *Physical Review A*, Volume 65 (March 1, 2002), arti-cle number 030101(R), http://link.aps.org/doi/10.1103/PhysRevA.65.030101. For

the experiment that exactly exploits the theory, see "Heralding Two-Photon and Four-Photon Path Entanglement on a Chip" by Jonathan C.F. Matthews, Alberto Politi, Damien Bonneau, and Jeremy L. O'Brien in *Physical Review Letters*, Volume 107 (October 11, 2011), article number 163602, http://link.aps.org/doi/10.1103/PhysRevLett.107.163602. The first experiments with photons to beat the two-photon N00N-state limit, using a slightly different approach, were carried out in 2004 by the groups of American physicist Aephraim Steinberg and Austrian physicist Anton Zeilinger. See "High Noon for Photons" by Di(r)k Boumeester in *Nature*, Volume 429 (May 13, 2004), pages 139–141, http://www.nature.com/nature/journal/v429/n6988/full/429139a.html.

75. See "Quantum Lithography" by Pieter Kok, Samuel L. Braunstein, and Jonathan P. Dowling in *Optics and Photonics News* (September 2002), pages 24–27, http://www.osa-opn.org/Content/ViewFile.aspx?id=1618. (To access the paper, you must sign up for an account but the account is free of charge to sign up for.)

76. See "Single Atom Transistor Is End of Moore's Law; May Be Beginning of Quantum Computing" in *Science Daily* (February 19, 2012), http://www.sciencedaily.com/releases/2012/02/120219191244.htm.

77. As usual, I circulated a preprint of the publication to various funding agencies, including Henry Everitt at the Army Research Office. According to Welsh physicist Daniel James, Everitt, sensing the potential importance of the work, sent the paper to two American physicists, Paul Kwiat and William "Bill" Phillips (Nobel Laureate, 1997), to check our calculations. Why Everitt sent our theory paper to two experimentalists I will never know. As James tells the story, neither Kwiat nor Phillips could reproduce our results, nor could they reproduce each other's results. (If you are not careful, there are some pesky factors of two floating around.) In desperation, Kwiat sent the paper to James, who found the errors in the Kwiat and Phillips calculation, reproduced our calculation, and they then all three reported back to the army that Dowling and his undergraduate could indeed do basic math and get the right answer and that the result claimed in the paper was correct.

78. The preprint ArXiv was developed in 1991 at the Los Alamos National Laboratory by American physicist Paul Ginsparg to replace an email distribution service for distributing scientific preprints that had been in place since 1989. The initial database was housed in a 486-processor PC in a basement, and it was assigned the domain name xxx.lanl.gov in order to fool the primitive search engines of the time into thinking it was a pornography site so as to keep these engines from constantly downloading the preprints and hogging all the bandwidth to the PC so that the scientists could use it. Later, when the ArXiv moved from Los Alamos to Cornell University, they changed the domain to www.arxiv.org. The ArXiv revolutionized the distribution of preprints and other unpublished work, opening up scientific publications, particularly institutions at Third World countries that could not afford the journal subscriptions for the reprints. The preprints were, and still are, free.

79. See "Quantum Interferometric Optical Lithography: Exploiting Entanglement to Beat The Diffraction Limit" by Agedi N. Boto, Pieter Kok, Daniel S. Abrams, Samuel L. Braunstein, Colin P. Williams, and Jonathan P. Dowling in the Cornell University Quantum Physics ArXiv (December 11, 1999), http://arxiv.org/abs/quant-ph/9912052

and also in *Physical Review Letters*, Volume 85 (received January 4, 2000, and published September 25, 2000), pages 2733–2736, http://link.aps.org/doi/10.1103/PhysRevLett.85.2733.

80. For a related and amusing list of anagrams formed from the words, "Physical Review Letters," see "Anagrams for the Electronic Age" by Jonathan P. Dowling in the *American Physical Society News*, Volume 4 (December 1, 1995), http://www.aps.org/publications/apsnews/199512/ana grams.cfm.

81. See " 'Spooky Photons' May Break Miniaturization Barrier for Computers— 'Entangled' Light Can Potentially Create Smaller Circuit Patterns" by Ben Stein in News Release from *Inside Science News Service* (American Institute of Physics, September 21, 2000), http://www.eurekalert.org/pub_releases/2000-09/AIoP-Spmb-2009100.php. See also "Entangled Photons Could Promise Lightning-Speed Computers" by Gia Scafidi from the Media Relations Office (NASA Jet Propulsion Laboratory, September 25, 2000), http://www.jpl.nasa.gov/releases/2000/quantum.html.

82. See "Quantum Leap May Transform Chips" by Ian Austen in the *New York Times* (October 2000), http://www.nytimes.com/2000/10/26/technology/what-s-next-quantum-leap-may-transform-chips.html?pagewanted=all&src=pm; "Quantum d iscovery could mean faster computer chips" in *Tech: Quantum Mechanics* (CNN, September 27, 2000), http://articles.cnn.com/2000-09-27/tech/photon.chips_1_entangled-photons-quantum-mechanics-computer-chips?_s=PM:TECH; "Chips May Take Quantum Leap" by Ian Austen in the *Chicago Tribune* (November 13, 2000), http://articles.chicagotribune.com/2000-11-13/business/0011130138_1_quantum-leap-quantum-physics-quantum-theory; "Atomic Superhighways … on a Silicon Chip" by Alexander Colhoun in the *Christian Science Monitor* (November 2, 2000), http://www.csmonitor.com/2000/1102/p14s1.html; "Yoked Photons Break a Light Barrier" by Charles Seife in *Science Now* (*Science Magazine*, October 3, 2000), http://news.sciencemag.org/sciencenow/2000/10/03-03.html; "Fine Lines" by Phillip Ball in *Nature* (September 27, 2000), http://www.nature.com/news/2000/000927/full/news000928-7.html; "Getting Past Point One" by Kristin Leutwyler in *Scientific American* (September 26, 2000), http://www.scientific american.com/article.cfm?id=getting-past-point-one.

83. See "Quantum Technology: The Second Quantum Revolution" in the *Philosophical Transactions of The Royal Society of London, Series A*, Volume 361 (August 15, 2003), pages 1655–1674, http://www.jstor.org/stable/3559215.

84. See "Einstein: His Life and Universe" by Walter Isaacson (Simon and Schuster, 2007), page 126, http://www.worldcat.org/oclc/76961150.

85. See "Einstein Synchronization" in *Wikipedia* (Wikimedia Foundation, May 16, 2012), http://en.wikipedia.org/wiki/Einstein_synchronisation. A less commonly used method of clock synchronization, attributed to the English astronomer Arthur Eddington, is called "slow clock transport," whereupon Alice synchronizes two clocks and then ships one of them very slowly to Bob. If the shipment is slow enough, then the two clocks, after transport, will be synchronized to any required degree of accuracy. When we at JPL were discussing synchronizing atomic clocks on the Earth to those on satellites, Colin Williams in a moment of whimsy suggested that we float the clocks up to the satellites in slow-moving

weather balloons. See "Conventionality of Simultaneity: Transport of Clocks" in the *Stanford Encyclopedia of Philosophy* (July 15, 2010), http://plato.stanford.edu/entries/spacetime-convensimul/#3.

86. See "Einstein's Relativity and Everyday Life" by Clifford M. Will (Physics Central, May 16, 2012), http://www.physicscentral.com/explore/writers/will.cfm. I knew a great deal about relativity and the GPS, since 1995, when I was an army scientist and had to organize a DoD workshop on the sources of error in the GPS, especially those from relativity. There were claims in 1995 that relativistic corrections in the GPS were not done right, but fortunately, these claims proved to be false.

87. Stars appear to twinkle as the Earth's atmosphere churns turbulently because of wind and convection of airflow. This churning moves pockets of air of random size and randomly higher or lower average air density between your eye and the star. The index of refraction of air shifts with the shifting density and, not unlike looking at the pebbles on a stream bed through flowing water, causes the apparent positions of the pebbles to move about randomly. The speed of light (or radio signals in this case) in air is also proportional to the air density. As the air density fluctuates randomly, so does the speed of light, and so the arrival time of the clock signals also fluctuates—sometimes arriving a bit early and sometimes a bit late. The accumulated effect of all the fluctuations of the signals traversing 200 kilometers (124 miles) of the Earth's atmosphere is to cause an error in timing of around 10 nanoseconds.

88. I should state that this is not meant to be a review of all possible quantum clock synchronization protocols, a number of which were independently proposed by different groups in the 5-year period between 1998 and 2003. See, for example, "Quantum-Enhanced Measurements: Beating the Standard Quantum Limit" by V. Giovannetti, Seth Lloyd, and Lorenzo Maccone in *Science Magazine*, Volume 306 (November 19, 2004) pages 1330–1336, http://www.sciencemag.org/content/306/5700/1330.abstract.

Chapter 6

Hilbert Space—The Final Frontier

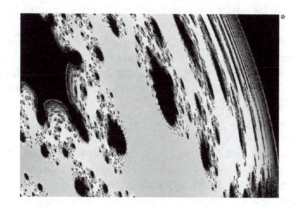

QUANTUM TECHNOLOGY IN A FLASH

Australian physicist Gerard Milburn and I coined the phrase "The Second Quantum Revolution" in an invited paper, entitled "Quantum Technology: The Second Quantum Revolution," which was published in 2003 in *The Philosophical Transactions of the Royal Society of London*, a journal founded in 1662 that claims to be "the world's first science journal." (A number of Isaac Newton's papers appeared in this journal in the 1600s and 1700s, and then came ours in 2003. So much for quality control.) This paper began its life as a National Aeronautics and Space Administration (NASA) Technical Brief that I routed to various NASA program managers in the early 2000s as a tutorial on quantum technologies. I had stolen the term "Quantum Technology" from a book Milburn published in Australia in 1996 by that same name.[1] In the fall of 2002, I mentioned to Milburn at a workshop that I was working on this paper and he informed me that he had been asked to write a review article for *The Philosophical Transactions of the Royal Society of London*, on exactly that

---

* Photo: a Mandelbrot fractal produced by the Wikipedia user "Esse."

topic, and so we joined forces, and this publication was the result. It is likely the first (and probably the last) time a NASA Technical Brief made it into *The Philosophical Transactions of the Royal Society of London*. That paper has been cited only a paltry 47 times but I often see it quoted in talks and public lectures, particularly the first sentence of the abstract: "We are currently in the midst of a *second quantum revolution*. The first quantum revolution gave us new rules that govern physical reality. The second quantum revolution will take these rules and use them to develop new technologies." Without reproducing this entire paper, which is easy enough for you to read yourself, let me blunder on a bit about what ideas we were trying to get across.[2]

First, I need to lay out for you the *first* quantum revolution. As we sketched in Chapter 1, quantum mechanics had its beginnings around 1900 when German physicist Max Planck first realized that by introducing a new fundamental constant of nature, Planck's constant, he could explain some anomalous experiments on the spectrum or energy content of heat radiation emitted from a small hole poked in the side of a hot oven. While this may not seem like a spectacular experiment, the classical theory for such a holey oven was very simple to work out and write down—but then that simple classical theory turned out to be spectacularly wrong.

The classical theory stated very clearly and conclusively that, if you went to your local potter, and asked her to close up her ceramic kiln oven, crank up the temperature to 1305°C (2381°F), and poke a small pin-sized hole in the side of the oven, then an *infinite* amount of radiant heat energy would come blazing out of the hole—vaporizing the entire universe. Classical theory typically does not do so badly in its predictions. The theory was checked and rechecked, but as the theory was very simple, with very simple ingredients, there could be no mistake. Classical theory predicted universal vaporization by pottery kiln. But this is not what we observe. The classical theory of heat radiation must be therefore wrong. Indeed, in the late 1880s, experiments with just such ovens were made, the temperature was cranked up, the holes were poked through, and the universe did not vaporize; the experimenters were able to measure the power coming out of the hole as a function of the heat radiation's optical wavelength and—far from being vaporized—they measured a nice smooth hump-shaped data curve with finite and not infinite total energy. The classical theory predicted that, in particular, the short (or ultraviolet) wavelengths would correspond to where the heat energy went to infinity, as the wavelengths got shorter. The experiments showed instead that the heat energy of the ultraviolet wavelengths went very quickly to zero, as the wavelengths got shorter. Theory predicted infinity. Experiment demonstrated zero. Rarely has a theory done so poorly. The fact that the theoretical energy-versus-wavelength curves "blew up" at ultraviolet wavelengths was called "the ultraviolet catastrophe" and nobody in the late 1800s knew how to fix it. Nobody at least until Planck came along in 1900.

By 1900, there were two wrong theories in play both predicting infinite heat energy coming out of the hole, in radical disagreement with the humped data curve that went smoothly to zero at short (ultraviolet) wavelengths and also smoothly to zero at long (infrared) wavelengths. The first theory agreed well with the data at the infrared end but blew up at the ultraviolet end of the spectrum. The second theory (Wien's theory) agreed well with the data at the ultraviolet end but disagreed with experiment at the infrared end. Planck's brilliant idea was to take the two theories and mash them together in a shotgun marriage. He had a fit. And by that, I mean he was able to fit the mashed-up theory to the data, using a fitting parameter that he called the fundamental constant h, whose numerical value he was able to estimate from the data. But Planck did not have a theory; Planck had a fit. He then bent over backward trying to construct a new theory that explained his fit. He finally, kicking and screaming, cooked one up. But he did not like it. Planck did not like it one bit. He was forced to add a new assumption to the classical theory. That was the little atoms in the walls of the oven, which absorbed and emitted the heat radiation, did so not continuously but in energetic little chunks of energy that he called "quanta" (which is Latin for "chunks"). From his fit, he had to take the quanta of energy to be in chunks whose size was determined by his new fundamental constant h, Planck's constant. Planck did not like this theory at all as it was at odds with classical theory. He even tried to take h to zero at the end of the calculation to fix it up, and he did recover classical theory, but he also recovered the prediction that the oven would vaporize the universe in an infinite blast of heat energy. Not good. For the safety of the universe, h must not be zero, its value must be taken from the data, and the new theory of Planck could not be classical theory. It had to be something new. It was quantum theory.[3]

By 1905, Einstein got into the act. Planck was ready to admit that the atoms in the walls of the kiln emitted and absorbed the heat radiation in chunks, but he held out hope that this chunkiness was a property of the atoms and not the heat radiation itself. Einstein smashed that hope and proved the heat energy itself was moving about in chunks or quanta of light, now commonly called photons. This was the beginning of the new theory of quantum mechanics. People, mostly Germans (as well as a pale, thin, short, squeaky-voiced Englishman named Paul Dirac), worked out the details of the theory, and by 1930, the theory was in final shape. Now that we had the new theory, we could use it to calculate things, primarily things that had never been understood classically. Why were the elements in the periodic table periodic? That is, why did the chemical properties of atoms seem to repeat in a regular way? Classical theory had nothing to say about this. Quantum theory explained it all. Why are some solid materials electric conductors (like copper where electricity moves freely) and others insulators (like glass where electricity does not move at all)? Classical theory tells us nothing about this. Quantum theory explains it all. Why do some metals, when cooled down to near absolute zero, become superconductors and conduct electricity with no resistance at whatsoever? Classical theory? Zip. Quantum theory? You bet!

And so it went throughout the 1900s that the quantum theory was wielded like a Bohr's magic wand to explain all this unexplainable stuff that physicists and chemists had pondered about for years and for which classical theory was of no help. The primary theme of the 1900s was to use quantum theory to explain things lying about in our laboratories and on our shelves filled with chemicals. This is what I mean by the first quantum revolution: using the base quantum theory to explain the world around us. Now, to be sure, new technologies did emerge from the first quantum revolution. Making the first transistor revolutionized the world, and to make it, we needed to understand the theory of quantum mechanics as applied to semiconductors—materials that are sometimes a conductor and sometimes an insulator. But not the full theory of quantum mechanics was used in such technologies, with all its entanglement and spooky action at a distance and Alices and Eves and Bobs, but the less spooky base theory of Schrödinger's simple wave theory of quantum mechanics or Heisenberg's matrix theory of mechanics was used—a theory that seldom had need to invoke such far-out concepts as nonlocal entanglement.

In the 1930s, Einstein and Schrödinger did indeed point out that quantum theory contained such spooky notions as entanglement and action at a distance, but they pointed this out not to embrace quantum theory (and its technological implications); they instead pointed this out to attack quantum theory. In the most extreme case, Einstein, Rosen, and Podolsky hoped to dethrone quantum theory and replace it with a more palatable classical, statistical mechanics–like hidden variable theory. These discussions on entanglement were mothballed for 30 years until Bell came along in the 1960s with his famous theorem, a proof that quantum theory and classical hidden variable theory were different, and that difference could be definitively decided one way or another in the laboratory. Behold the scientific method at work! Enter Frey and Clauser and Aspect and the gang, in the 1970s and 1980s with their clanking photonic contraptions. Much to the surprise, of particularly Bell and Clauser, the experiments agreed with quantum theory and not the classical hidden variable theory.[4] As we have seen, these experiments led to the begrudging acceptance of the weirder predictions of quantum theory, and it is the technologies that are emerging from these weirder predictions that, primarily, Milburn and I have christened quantum technologies—technologies that require quantum entanglement. Certainly, technologies like quantum teleportation, universal quantum computing, and the E91 quantum cryptography protocol fall into the fold, as entanglement is requisite for their operation. Here, then, the quantum nonlocality of entanglement is at the fore in a new technology. But also, too, are the lesser weirdnesses, quantum uncertainty and quantum unreality, at play in such technologies as the BB84 quantum cryptography protocol and the D-Wave "quantum" computer, important technologies in their own right. The hippies, who supposedly saved physics, tried their hands at spooky technologies almost immediately, in the mid-1970s, as soon as they became aware of Clauser's experiment [See, *How the Hippies Saved Physics*, by David Kaiser (W.W. Norton & Co., 2011).] Alas—their minds, perhaps muddled with too much lysergic acid diethylamide (LSD), too many "office hours" ruminating naked in hot spring hot tubs, too much Eastern mysticism, and too much talk of paranormal oddities—such as extrasensory perception (ESP), telepathy, telekinesis, remote viewing, and spoon bending—did not produce any working technologies at all. But they did produce a whole lot of nonworking ideas for technologies on the basis of their misunderstanding of quantum theory.

In the 1970s, the hippies focused on the mangling of ideas from quantum theory with those of the paranormal, human consciousness, and Eastern mysticism.[5] That was their entangled path to perdition, perhaps because by the 1980s and 1990s, ever more exacting quantum experiments were ruling the weirdness of quantum theory in, while ever more exacting experiments in parapsychology were ruling all this paranormal stuff just plain *out*. The 1998 experiment on Bell's inequality, by the Zeilinger group, rules in quantum theory at the level of 30 standard deviations and rules out quantum hidden variable theory at that same level (minus the detection loophole). This is an incredible level of surety. What does that level of surety mean? It means the probability that quantum mechanics is right (and therefore local hidden variable theory is wrong) is 99.999999999999999999 9999999999999999999999999999999999999999999999999999999999999999 9999999999999999999999999999999999999999999999999999999999999999 99999999999999999999999999999999999999999999999%. It does not get any more certain than that. What about experiments hoping to provide proof of the paranormal? Well, their percentages went in the other direction, ruling paranormal explanations of the real world *out* to more and more significant digits. The early parapsychology experiments that early on claimed to show scant evidence for ESP and such things were never independently replicated and their statistical analyses repudiated. More careful and more modern experiments showed no effects of the paranormal at all.

In the 1970s, the hippies hoped that the weirdness of quantum mechanics could explain paranormal phenomena such as telepathy, which they posited would involve nonlocal communications mediated by pairs of quantum particles, lodged in human noggins, and interacting nonlocally and instantaneously over vast distances. Never mind that, at body temperature, any quantum entanglement would be destroyed in a septillionth of a second by all the thermal fluctuations percolating around a living brain. But first, to tie quantum theory to parapsychological effects such as telepathy, you would have to prove that such parapsychological effects existed in the first place. In the 1980s and 1990s, parapsychologists set out to do just that with ever better and more controlled experiments. And as the parapsychological experiments got better and more rigorous, all the earlier reported paranormal effects simply disappeared. Remember the scientific method? Stuff you make up in your head isn't really real until you verify that the stuff exists in the laboratory. If you find it does not exist in the laboratory, you might be wise to then purge it from your head.[6]

My favorite parapsychologist-turned-skeptic is the British psychologist Susan Blackmore, who got her PhD in parapsychology—or as I like to call it—"perhaps psychology." She launched her career by devising exacting and rigorous tests of the paranormal, and as her experiments improved, her data supporting the existence of what they call "psi" phenomena vanished while the data supporting the absence of any such effects accumulated. At the many sigma or percentage level, she systematically ruled out the existence of any such thing as telepathy or the like in her laboratory. Puzzled by this, as other laboratories were claiming positive results supporting the existence of such things as telepathy, Blackmore went to visit these laboratories. She found that all their positive results could be explained by faulty experimental protocols (nondouble-blind experiments), faulty data analysis, wishful thinking, and, in some cases, outright fraud. The scientific method had spoken, at least to Blackmore; there is no experimental support for the existence of telepathy. Therefore, it does not exist.[7] When Blackmore's ever-more careful experiments showed null results for the existence of psi, while other parapsychologists were getting positive results, her "perhaps psychologist" colleagues told her that her own negative thoughts were causing her mind to emit a negative psionic dampening field that ruined all her own experiments but not anybody else's. Try adding "negative psionic dampening field" to your list of experimental systematic errors. One by one, all the other paranormal phenomena fell to the ever-improving and exacting parapsychological experiments. Paranormal phenomena are a pipedream, a religious belief perhaps, but not science. There existence is now almost completely ruled out by over 60 years of experiments, no matter what you hear on *The History Channel*. There is now certainly no need to explain paranormal phenomena with quantum theory any more than there is a need to explain the flight patterns of wood fairies with quantum theory. Neither of these two phenomena exists. As they say in New Jersey—fugget it!

Hence, some of the hippies turned away from the paranormal and toward technology. And the technology they were most enthralled with was superluminal or "faster-than-light" communication. Remember, if you can send messages faster than light, you immediately run into Einstein causality and *Star Trek*–like paradoxes such as sending messages backward in time to order your grandfather to divorce your grandmother so that you are never even born to send the message in the first place. Most physicists trot out such paradoxes as proof that superluminal communication cannot exist. Not the hippies. American physicist-hippies Nick Herbert and Jack Sarfatti nearly simultaneously produced (wrong) ideas for such superluminal technologies, claiming that quantum entanglement could be used to build machines to send signals faster than the speed of light—in fact, instantaneously at infinite speeds. Sadly, this does not work. (But even I have to review proposals on this topic 30 years later, as it is so alluring to the incognoscenti.) This tale of the hippies is told in the amusing book *How the Hippies Saved Physics* that regales us with tales of counterculture types in the 1970s, stirring into a metaphysical jambalaya pot, Bell's theorem, Clauser's experiment, and a healthy dose of Eastern mystical mumbo jumbo, to cook up a mélange of quantum mechanical explanations for everything from telepathy to human consciousness. In my opinion, the hippies did not save physics, but instead they nearly ruined it, at least for much of the interested lay audience.

Take quantum mechanics and human consciousness. As far as I can tell, the hippies' argument goes like this: "I don't understand quantum theory and I don't understand human consciousness—therefore quantum theory must be used to explain human consciousness." (Yes it is that bad.) In this type of muddled thinking, ideas about quantum theory, waves, crystal vibrations, consciousness, ESP, spoon bending, superluminal communication, and telekinesis are loaded up into multiple, multicolored, metaphorical tubes of gloss-enamel paint and then squirted willy-nilly, Jackson-Pollock style onto a prostrate canvas to form an impasto pastiche of utter crapola. This type of marled, maniacal, quantum-mechanical paint job gives me the creeps. Quantum theory is so interesting by itself without having to lard all this other junk into it, and quantum theory has the advantage of being provably correct in the laboratory where things like ESP—after years and years of trying—are not. The hippies ruined physics for the lay people interested in quantum theory, most of whom have become convinced, by reading all the whacko books out there on the connection between quantum theory and the paranormal and Eastern mysticism and consciousness, that quantum theory and all this New Age mumbo jumbo are provable and definitively linked and that this link has been experimentally vindicated. This is the only explanation that I have, when sitting on the middle seat of a Boeing 777 jet on a transpacific flight, as I reveal to my seatmate that I am a quantum physicist, that I am immediately accosted with such bubbling retorts as, "A quantum physicist? Oh then you must have loved the movie, *What the (Bleep) Do We (K)now!?*"[8] I then have to explain that, in my professional opinion, the film was a steaming pile of (bleeping bleep). (At least then I'm left alone in peace for the rest of the flight so I can nap.) I am a skeptic. Scientific method, remember? Stuff you make up in your head needs to be checked in the laboratory before you can claim it is an element of reality. That's physics. Stuff that you make up in your own head and never check in the laboratory but just keep in your own head (or tell to the heads of your own friends), that's metaphysics not reality.

Connections between quantum theory and Eastern mysticism: what the heck is there, there, to check? Quantum theory as the basis for human consciousness? Wild-ass conjectures with no experimental proof whatsoever.

Quantum theory to explain telepathy? Telepathy does not exist; there is nothing to explain. For the skeptics' rejoinder to the film *What the (Bleep) Do We (K)now*, see the amusing blog posting "What the (Bleep) Were They Thinking?"[9] Yes, the hippie physicists nearly ruined physics for the layperson and we nonhippie physicists are left to try and repair all that damage. Quantum mechanics is wonderfully strange and weird and, provably so in the laboratory, there is no need to add extra unprovable mumbo jumbo to the pot, if you are only after the wonderful weirdness of physical reality. If you want an unearthly jolt of hold-your-head-and-wince noncommonsensical weirdness, then learn quantum mechanics! Leave the mysticism to the mystics. Okay, now back to the hippies.... There is some truth in the notion, posited in *How the Hippies Saved Physics*, that Nick Herbert and Jack Sarfatti's erroneous papers on faster-than-light communication was the motivation for correct quantum technologies to come.[10] But this motivation came not from the hippies but from the nonhippie-physicist community, as that community had to prove over and over again that the hippies' proposals for superluminal telegraphs were just plain flat wrong. All the hippie superluminal communication proposals had one thing in common: they proposed to exploit quantum nonlocality and entanglement to send signals of information between Alice and Bob faster than the speed of light. Not just faster but instantly at infinite speeds. The hippies were not the first to cook up such schemes. Shortly before the Einstein, Rosen, and Podolsky paper appeared in 1935, Swiss philosopher Karl Popper published an EPR-like analysis of nonlocal correlations of his own in 1934. In this paper, in modern terms, Charlie sends pairs of entangled particles, one pair to Alice on Alpha Centauri and another pair to Bob on Beta Pictoris. I will change the setup from Popper's original scheme to polarized photons, with which we are familiar.

Popper in 1934, Sarfatti in 1978, and Herbert in 1979 thought up thought experiments for what turned out to be the underlying same entanglement-based superluminal communications scheme. However, from their schemes, they reached radically different conclusions. In 1934, the nonlocal nature of entanglement was 30 years away from being put to the experimental test by Fry and Clauser and Aspect and Zeilinger and Shih and.... However, Einstein's relativity, and its ultimate speed limit, that of the speed of light, was well tested in 1934. Popper declared that Einstein's relativity theory trumped quantum theory and declared that his superluminal thought experiment showed that quantum theory was wrong. What actually was wrong was Popper's understanding of quantum theory. By 1978, the Clauser experiment had ruled quantum theory and its nonlocal EPR correlations in, so Sarfatti and Herbert took the position that the relativity theory must be wrong, asserting that their superluminal communicators worked as claimed. Sarfatti submitted a patent for a "Faster than Light Communication System" and circulated widely a preprint on the idea, particularly to the Department of Defense and various intelligence agencies. To quote from *How the Hippies Saved Physics*, "Presumably the image of CIA agents doped up on LSD, communicating instantly with operatives half a world away via correlated brain impulses, seemed no more far-fetched than the parapsychological effects in which Sarfatti had been immersed for years." Enough said. The Sarfatti scheme was very closely related to the Popper scheme, although I don't know if Sarfatti was aware of it, and the Herbert scheme was most similar to my own simplified presentation in Box 6.1.

The hippie agenda was clear; ESP must require superluminal communication and EPR must provide it. To be fair to the hippies, in 1978, Clauser's experiment for EPR was not all that conclusive for the EPR effect, and the parapsychologists' experiments for ESP were not all that conclusive against the ESP effect. But now, such experiments are. There is no need to explain ESP with EPR, as ESP does not exist. We now know that ESP is but a frivolous flight of fancy in a hippie's LSD-saturated cerebrum. To quote Sarfatti, "…superluminal precognitions … *exist as facts in abundance in my own laboratory of the mind*. Am I to ignore facts simply because old men are afraid to experience them?" (see note 10) (Italics mine. As paraphrased from "Seeds of Superluminal Quantum Physics" by Jack Sarfatti, unpublished [October 6, 1979].) Unfortunately for Sarfatti, who is now an old man himself, the scientific method requires us to check and establish facts by carrying out experiments in laboratories in the real world and not laboratories in our own minds. The Greek philosopher Aristotle tried carrying out laboratory experiments in his own mind, and thereby concluded (incorrectly) that heavier objects fell faster than lighter objects in the Earth's gravitational field. Aristotle never bothered to check his "own laboratory of the mind" with that of the real world, say by tossing a large Grecian urn and a small drachma coin off his balcony, to see if

---

BOX 6.1   NOT-SO-SUPERLUMINAL COMMUNICATIONS SCHEME

So let us say that Charlie sends out entangled states of the form $|\updownarrow\rangle_A|\leftrightarrow\rangle_B + |\leftrightarrow\rangle_A|\updownarrow\rangle_B$. This is just the state used in tests of Bell's inequality, teleportation, and the E91 cryptography scheme where the photons are anti-correlated in the vertical–horizontal directions (see Figure 6.1). If Alice chooses to measure in the vertical–horizontal direction and gets $|\updownarrow\rangle_A$, then Bob instantly gets $|\leftrightarrow\rangle_B$, and vice versa. This is code for the infinite Library of Babel; the state contains an infinite number of anti-correlated polarized states, and in particular, it can be written as $|\nearrow\rangle_A|\nwarrow\rangle_B + |\nwarrow\rangle_A|\nearrow\rangle_B$, the ±45° directions. If now instead Alice chooses to measure along these ±45° directions and gets $|\nearrow\rangle_A$, then Bob is sure to get $|\nwarrow\rangle_B$, and vice versa. Thus, to

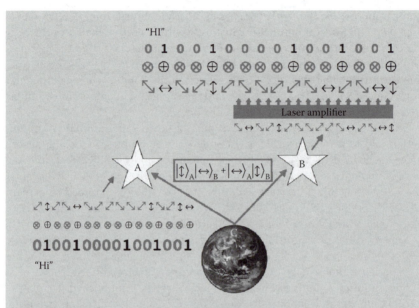

Figure 6.1   Superluminal communications in a "FLASH." This is Herbert's modification to the scheme in Box 6.1. Charlie sends 16 pairs of polarization-entangled photons in the state $|\updownarrow\rangle_A|\leftrightarrow\rangle_B + |\leftrightarrow\rangle_A|\updownarrow\rangle_B$ to Alice on Alpha Centauri and Bob on Beta Pictoris, 60 light-years apart (drawing not to scale). Alice then converts her message "HI" into a binary string of zeros and ones 0100100001001001. She sets her measurement direction to horizontal–vertical ($\oplus$) for the ones and to the ±45° direction measurement ($\otimes$) for the zeros. This causes her photons to collapse randomly into horizontal, vertical, +45°, or −45° photons, depending on what direction she picks. A sample run would look like, ⤢↕⤢↖↔↘⤢⤡↘↘⤢↕↘⤢↕↔, where now the binary message is encoded in the pattern of red and black arrows. For this run, Bob's 16 photons instantly collapse to the anti-correlated states of ↘↔↘⤢↕⤢↘↘⤢⤢↘↔⤢↘↕. Again, the binary message is encoded in the pattern of H-V versus ±45°. If Bob could read off the single-photon polarizations at his end perfectly, he would know Alice's choice of directions, the message sequence, and he could then instantly extract the binary message, which would imply faster-than-light or superluminal communication. However, quantum uncertainty and unreality prevent Bob from measuring the states of the single photons without error and all he gets out is random noise. In Herbert's revised scheme that he called "FLASH," Bob places a perfect laser amplifier, "a perfect xerox machine," in front of the incoming single photons and amplifies the 16 single photons into 16 bright (classical) laser beams. Bob then measures the polarization of those bright beams instead and then extracts the message. FLASH does not work either, as a perfect laser amplifier cannot exist, and also because the no-cloning theorem prohibits the copying of the single-photon states into many.

make a superluminal communications scheme, we deploy what is called "delayed choice" whereby Alice, photon pair by photon pair, waits until the last minute to make her measurement direction choice. Alice then takes her message she wishes to send to Bob, says "HI," and converts it to a binary string of 16 zeros and ones, as before, to get **0100100001001001**, where the bold font is just used to make the zeros stand out so you can see the pattern of zeros and ones more easily. What Alice then does next is to assign a "one" to the choice of a horizontal–vertical measurement ($\oplus$) and a "zero" to the choice of a ±45° measurement ($\otimes$). As the first 16 entangled polarized photons arrive from Charlie, Alice measures in the following way by choosing directions that match the pattern of zeros and ones in the message: $\otimes\oplus\otimes\otimes\oplus\otimes\otimes\otimes\otimes\oplus\otimes\otimes\oplus\otimes\otimes\oplus$, which you can see is the same pattern as that of the zeros and ones that encode the message "HI." These measurements collapse Alice's photons randomly into measurement outcomes that look like, for one particular run, ↗↕↗↘↔↘↗↗↘↘↗↕↘↗↕↔. Note that the direction outcomes are random, Alice gets a 50–50 chance of a ↕ or a ↔ if she measures in the vertical–horizontal directions and also a 50–50 chance of a ↗ or a ↘ if she measures in the ±45° directions, but her measurement outcome, unlike E91 quantum cryptography, is not the bit being transmitted, but rather the choice of which direction to measure in, the pattern of the plain and bold sequence, is.

Now, according to quantum mechanics, Bob's half of all the 16 entangled photon pairs also collapse *instantaneously* into the same sequence of blacks and reds but with measurement outcomes anti-correlated to Alice's: ↘↔↘↗↕↗↘↘↗↗↘↔↗↘↕. Now, again, the measurement outcomes are not what Bob is after; he is after the pattern of plain and bold, which contains Alice's choice of measurement direction and her message in binary code.

If Bob had a device that could with a 100% fidelity, at the single-photon level, distinguish between photons that were polarized in the vertical–horizontal direction, ↕ or ↔, versus those polarized in the ±45° ↗ direction, or, ↘ then he'd be done. He could then read out the sequence pattern of plains and bolds, convert it to zeros and ones, and reconstruct Alice's message instantaneously, even though he was 60 light-years away from her on Beta Pictoris. This is the entanglement-based superluminal communications scheme in a nutshell. The problem? Such a measurement device cannot exist. If it did, it would violate the laws of quantum theory as now understood in the quantum no-cloning theorem.

the urn actually fell at a faster rate than the coin. It took a thousand years until the Italian scientist Galileo bothered to check this notion out in the real world by flinging a golf ball and a bowling ball off the top of the Leaning Tower of Pisa. Guess what? The result from Aristotle's laboratory of his own mind was wrong. Heavy objects and light objects fall at exactly the same rate and hit the ground at the very same time. That's reality, folks. The laboratory of the mind is a good place to cook up ideas but not a good place to verify them. It is amusing to think that there exist people who think that the ideas they cook up in their own head are "facts" simply because they are ideas that they cooked up in their own head. That's why we call people who claim as facts the results of experiments carried only in their own mind, instead of in their own laboratory, "philosophers" or "lunatics" but not physicists. Philosophers and lunatics seldom invent useful new technologies only in their laboratories of their own mind.

There is still the claimed superluminal signaling to explain, but alas that's all too good to be true. Recall from Box 6.1 that Bob, to read out Alice's message, must be able to, without error, distinguish between *single* photons that are polarized in the vertical–horizontal directions ($\updownarrow$ or $\leftrightarrow$) and photons that are polarized in the ±45° directions ($\nearrow$ or $\searrow$). Therein lies the bug. How can Bob do this without error? Well he can't. Any attempt he makes to measure the polarization direction, *without knowing the answer in advance*, irreversibly destroys the state of the photon, giving him, on average, no information at all. Quantum unreality and uncertainty bite. It can be shown that Bob's best strategy is to randomly, with 50–50 probability, switch back and forth between attempts to measure in the vertical–horizontal directions ($\oplus$) and attempts to measure in the ±45° directions ($\otimes$). But that optimal scheme does not produce the correct sequence of bolds and plains (zeros and ones) that contains Alice's message; it produces instead *an absolutely random* string of zeros and ones that contain no information whatsoever about anything, much less Alice's message. This no-go theorem for Bell-enabled superluminal communication was pointed out, also in 1978, by one of Sarfatti's colleagues,[11] the French physicist Philippe Eberhard.[12] This point was made by a number of other authors soon after. According to the laws of quantum theory, you cannot use quantum entanglement to carry out superluminal signaling. (Remember that next time *before* you send me a research proposal claiming to do just this.) In order to extract the message "HI," Bob must call up Alice on her cell phone and then Alice must tell him all of her measurement settings (horizontal–vertical vs. ±45°) *and* all of her measurement outcomes ($\updownarrow$ or $\leftrightarrow$ vs. $\nearrow$ or $\searrow$). But that ruins the whole scheme. Instead of going through that long-winded and elaborate ritual, Alice, as soon as she answers the phone, could simply just say, "Oh Bob, by the way, the message was 'HI.'" The cell phone signal, or classical channel, travels at (or less than) the speed of light. Bob would have to wait many tens of years for the cell phone call to be completed in order to extract the message and thus any superluminal advantage is lost. Both Sarfatti's and Herbert's schemes suffered the same sad fate. They don't work.

Now, their superluminal signaling setup, while not a superluminal signaling setup after all, is not completely useless. If Bob calls up Alice and shares with her only the measurement settings, *but not the measurement outcomes*, then you have Artur Ekert's E91 quantum cryptography protocol! Alice and Bob can use EPR correlations to set up a shared list of *absolutely random numbers* that can be used later, over a subluminal (slower-than-light) classical channel, to send unbreakable messages. But sending a random sequence of zeros and ones is not sending a message—by classical information theorist Claude Shannon's own criteria—a random sequence of zeros and ones contains no information whatsoever, much less a message. Sarfatti and Herbert missed this cryptography point and we had to wait another 23 years for Ekert to propose it as his E91 quantum cryptography scheme. The direct superluminal telegraph scheme itself falls before quantum uncertainty and quantum unreality. Einstein's theory of relativity is safe from the spookiness of quantum theory! Or is it? Herbert called his initial superluminal telegraph, replete with polarized photons and with now known to be manifestly nonsuperluminal behavior, "QUICK," "...an acronym so clever even he can't remember what it stood for anymore" (see footnote 10). While Sarfatti spent the next few years trying to convince everybody that his idea really would work, to no avail, Herbert conceded Eberhard's point about measurement destroying the single-photon states, and so Herbert cooked up a new scheme around 1980, which he called "FLASH," for "First Laser-Amplified Superluminal Hookup."[13] The name was catchy, and it took some time for the physicists, hippie and nonhippie alike, to figure out what was wrong with it. Herbert's original QUICK scheme is nearly identical to that in Box 6.1. Herbert now realized, from Eberhard's critique, that if Bob only had one photon at a time arriving from Charlie, any measurement in the wrong measurement direction would destroy that one photon's state, scrambling Alice's message irretrievably. The only way Bob could be sure of the measurement direction is if he communicated with Alice, slower than the speed of light, spoiling the superluminality. His "QUICK" fix was to stick a laser amplifier in front of the incoming photons arriving from Charlie and give them a power boost.

A laser amplifier is a device made of atoms in their excited state that takes an incoming stream of photons (a few photons per second) and converts them into a high-intensity stream (many photons per second). This is how a laser works. Inside each laser is an amplifier. The excited atoms in the laser amplifier at first emit a few stray photons. Then, some of these strays begin bouncing back and forth between a pair of mirrors making many round-trips through the amplifier that is jammed between the mirrors. The few stray photons are upgraded again and again, more and more new photons added on each passing, into what finally emerges as an intense laser beam that you can then use to do surgery on your own retina (well maybe not your own) or shoot down an intercontinental ballistic missile. Herbert reasoned that if the laser amplifier was absolutely perfect, noiseless in every way, then his single incoming photons, with the binary encoding ↖↔↘↗↕↗↘↘↗↗↗↖↔↗↖↔↕, would be up-converted into bright beams of laser light with exactly the same polarization encoding—one separate beam of polarized output laser light per incoming photon. Now, reasoned Herbert, you have converted each fragile incoming quantum photon into a robust classical laser beam, with a gazillion photons each, which gets around quantum unreality and uncertainty. Heisenberg has nothing to say about such classical beams and the polarization directions could be measured on such bright beams without error. Herbert even called it a "...perfect photon xeroxing provided by the laser effect" (see Figure 6.1).

The phrase "perfect photon xeroxing" should be ringing alarm bells in your head, particularly if you did not skip over the section on the no-cloning theorem in Chapter 4. Within the laws of quantum theory, one cannot build a "perfect photon xeroxing" machine. This is the ruling of the no-cloning theorem, which states that you cannot perfectly copy an unknown quantum state, and the Alice-collapsed states, arriving from Charlie, are quite unknown to Bob. The no-cloning theorem has a less euphonious cousin, the "no-nonnoiseless-laser-amplification theorem," which states that you cannot perfectly or noiselessly amplify the unknown polarization state of a single photon.[14] Herbert's proposed noiseless laser amplifier cannot exist as its very existence would violate the foundations of quantum theory itself. His presumed perfect laser amplifier violated both the no-cloning theorem and the no-nonnoiseless-amplification theorem. It was a double whammy of no-go theorems for FLASH. We have discussed the no-cloning theorem. What about no-nonnoiseless laser amplification? What happens when you stick the best laser amplifier you can possibly make, according to the laws of quantum mechanics, in front of the incoming stream of polarized photons? Two things. First, along with the incoming stream of polarized photons sent by Charlie, there are also so-called "vacuum" photons coming in along with them. The quantum theory of light predicts that there are such vacuum photons everywhere, seething and foaming and jetting about in the bubbling froth that is the vacuum of empty space. These vacuum photons get amplified just as perfectly as Charlie's incoming real photons and the vacuum photons are upgraded to a stream of a gazillion real photons intermixed with the photons amplified from Charlie's message-bearing sequence. However, the vacuum photons have random polarizations going into the amplifier and, hence, random polarizations for the gazillion amplified photons coming out, and that randomness completely swamps the polarization encoded in the amplified versions of Charlie's photons. Even worse, second, the bazillion atoms that make up the laser amplifier must, and I mean must according to the laws of the quantum theory of light (quantum optics), emit photons of their own randomly into the stream, all with totally random polarizations, even further scrambling the mix. These gazillions of amplified vacuum photons, as well as the gazillions of amplifier photons, with random polarizations, completely swamp the amplified photons from Charlie's stream—there is no way (even in principle) to separate the amplified photons of Charlie from the amplified photons of the vacuum and the amplifier. All Bob gets out, once again, is pure random noise. And, as we have argued time and time again, pure random noise (by definition) contains no message. Alas then in a flash, Herbert's "FLASH" was unceremoniously demoted to "FLUSH" ("Faster-than-Light Ultimately Shown to be Hogwash").

Over the years, the connection between the no-cloning theorem and the no-superluminal-signaling hypothesis has been made more concrete. You can prove that if you can perfectly clone an unknown quantum state (or perfectly laser amplify one), then you can carry out faster-than-light communication. This statement is now taken as additional support for the no-cloning theorem—if you could clone, you could communicate superluminally. Einstein tells us you cannot communicate superluminally; therefore, you cannot clone. But one must be careful. These so-called no-superluminal-signaling theorems, in the context of shared entanglement, state that you cannot send messages *instantaneously*. There is no speed of light, at least in the version of quantum theory we are using here. This no-signaling result is consistent with Einstein's theory of special relativity, which states that you cannot send messages faster than the speed of light. But there is a gap between the speed of light and infinite speed. And puzzling, to me at least, is the question: Why are theorems from ordinary nonrelativistic quantum theory consistent with Einstein's theory of relativity? What is it in the ingredients of nonrelativistic quantum theory that forces it to be consistent with Einstein's theory of relativity? There are upgrades to quantum theory that incorporate Einstein's special theory of relativity but not the general theory of relativity. Superstrings notwithstanding, nobody has produced a theory that consistently handles both quantum theory and Einstein's general theory, the theory of gravity. I sometimes wonder if the no-superluminal-signaling theorem provides a hint on how to unify these two things.

So did the hippies save physics? As far as I can tell, the hippies' failed attempts to design an entanglement-enabled superluminal telegraph did lead directly to the discovery of the no-cloning theorem. It is true that the no-cloning theorem is at the heart of all quantum cryptography schemes, but the hippies did not invent any such cryptography schemes; their work simply motivated the early discovery of an important piece, the no-cloning theorem. Would quantum cryptography and the no-cloning theorem have ever been discovered without the hippies? I say certainly yes. First of all, many of the ideas of quantum cryptography were contained in a 1970 unpublished (but widely circulated) paper called "Quantum Money" by American physicist Stephen Wiesner. Note that 1970 is 10 years before Herbert invented his superluminal telegraph scheme FLASH. All the basic ideas of BB84 are in that paper of Wiesner's. This paper is widely viewed as the first paper ever in quantum information theory.[15] Wiesner's idea was to use quantum mechanics to make noncounterfeitable currency by using, say, quantum states of ions stored in the bill as a kind of a quantum serial number. The serial number is converted into a series of zeros and ones by the bank. The bank then chooses randomly, binary digit by binary digit, which measurement direction to encode the zeros and ones in the ions, randomly switching from the up–down direction ($\updownarrow$) to the left–right direction ($\leftrightarrow$). Sounds familiar? It is exactly the scheme used in BB84 quantum cryptography; this is where Bennett, a collaborator of Wiesner, got the idea. Then, the zeros and ones are encoded as spin up and spin down ($|\uparrow\rangle$ and $|\downarrow\rangle$) for the up–down measurement direction and as spin right and spin left ($|\rightarrow\rangle$ and $|\leftarrow\rangle$) for the right–left measurement direction. In order to copy the bill and the serial number, a counterfeiter, named Eve, would have to know the measurement direction for each bit used by the bank. But only the bank knows this and keeps it secret. Eve's best strategy would be to always measure in the up–down direction but on average she would scramble approximately 25% of the bits of the serial number, half of those encoded in the left–right direction, both in the original and in the copy, so that the serial number in the copy would no longer exactly match that at the bank. When the copy is taken back to the bank, the bank would measure the bits in the correct basis and see that 25% of them had been scrambled, and so Eve would be detected and the counterfeiting would be revealed. This is nearly the exact setup as BB84 quantum cryptography. The only catch was that Wiesner did not *prove* the no-cloning theorem; he *assumed* it. Wiesner and Clauser were both colleagues of American computer scientist Charles "IBM Charlie" Bennett, coauthor on the BB84 protocol, and Bennett was already working on quantum cryptography in the early 1970s, well before FLASH, using ideas from Wiesner's (unpublished) quantum money manuscript.

Four or five different groups proved the no-cloning theorem nearly simultaneously around 1980 in response to circulating preprints of Herbert's FLASH paper.[16] My position is that quantum cryptography was already in the works before the FLASH and the no-cloning papers appeared. The hippies did not motivate the invention of quantum cryptography, Wiesner did. All the hippies motivated was a strict proof of the no-cloning theorem, which did indeed then allow proofs that quantum cryptography was certainly unbreakable. But quantum cryptography would have been invented eventually, hippies or no hippies. Sarfatti and Herbert had their hearts in the right place though, suppressing the talk of the connection between entanglement and the paranormal, and focused on exploiting the nonlocal correlations for a technology. But, by becoming obsessed with the quixotic quest for that holiest of grails, the quantum superluminal telegraph, they missed the low-hanging fruit, the holey goblet, the E91 protocol of Ekert, to which they were tantalizingly close. Sarfatti even touted to the Department of Defense that his superluminal telegraph would allow for instantaneous and untappable communications. Superluminal? No. Untappable? Yes. The no-cloning theorem also immediately shows up when you try to design the circuitry for a quantum computer. Deutsch's 1989 paper on quantum computer circuit design quickly followed up Feynman's 1982 paper on quantum computing.[17] One thing you do when trying to design a quantum computer circuit is to try and mimic as closely as possible what is done in a digital classical computer circuit. You see what classical computer gates may translate into quantum gates. Some gates, like NOT and SWAP, go through just fine. By then, you immediately hit the problem of copying. In classical computers, there are COPY and FANOUT commands that take a single unknown input bit; in the case of COPY, one copy is made, and in the case of FANOUT, many copies are made. It is almost immediately obvious, by the simple no-cloning argument in Box 6.2, that these copying commands will not work in a quantum computer. Had the hippies not motivated the discovery of the no-cloning theorem in 1980, it would have certainly been discovered in 1989 by Deutsch. Deutsch is a big fan of the many-worlds interpretation of quantum mechanics, and I would happily live in an alternative universe where

---

BOX 6.2   A SIMPLE PROOF OF THE NO-CLONING THEOREM

Don't you love farce?
My fault, I fear
I thought that you'd want what I want
Sorry, my dear.
But, where are the clones?
Send in the clones.
Don't bother they're here.[18]

In Chapter 4, I gave a heuristic argument in favor of the no-cloning theorem; here, I will give a simple proof. Suppose I have a perfect copying machine. We insist that the machine only obey the laws of quantum mechanics. One of the principles of quantum theory is "linearity" in that the output of a machine must be proportional to its input and cat-like superpositions must be preserved, if quantum mechanics is to be obeyed. Now, Alice has an ion that is in the spin-right state $|\rightarrow\rangle_A$ that she wants to make a copy of for Bob. She feeds it into the copying machine and gets $|\rightarrow\rangle_A \Rightarrow |\rightarrow\rangle_A|\rightarrow\rangle_B$. So far so good, Alice keeps her original and Bob gets his copy. Now, quantum unreality rears its ugly head. The state $|\rightarrow\rangle_A$ actually is shorthand for an infinite number of possible linear superpositions of states, the infinite Library of Babel at work again, and it can, for example, be rewritten as the cat state $|\rightarrow\rangle_A = |\uparrow\rangle_A + |\downarrow\rangle_A$. That is, it is a *linear* superposition of alive (spin up) and dead (spin down). These two things are the same state, just written two different ways.

Written this way, as a cat, if we feed it into the perfect copying machine, and if the machine obeys quantum mechanics, particularly it preserves a linear relationship from input to output, it does the following: $|\uparrow\rangle_A + |\downarrow\rangle_A \Rightarrow |\uparrow\rangle_A|\uparrow\rangle_B + |\downarrow\rangle_A|\downarrow\rangle_B$. Fair enough, but now we can rewrite this output, in the language of cats, back in terms of spin right and spin left as using $|\rightarrow\rangle = |\uparrow\rangle + |\downarrow\rangle$ and $|\leftarrow\rangle = |\uparrow\rangle - |\downarrow\rangle$, where here the minus sign is needed in the latter expression to ensure everything is linear. In math, this is called a linear basis change. Using those two rules, then we can take the copy machine output of the cat state, $|\uparrow\rangle_A|\uparrow\rangle_B + |\downarrow\rangle_A|\downarrow\rangle_B$, and rewrite it as $(|\rightarrow\rangle_A + |\leftarrow\rangle_A)(|\rightarrow\rangle_B + |\leftarrow\rangle_B) + (|\rightarrow\rangle_A - |\leftarrow\rangle_A)(|\rightarrow\rangle_B - |\leftarrow\rangle_B)$, which then can be simplified (with some algebra) to $|\rightarrow\rangle_A|\rightarrow\rangle_B + |\leftarrow\rangle_A|\leftarrow\rangle_B$.[19]

Following the chain of reasoning, we have a copying machine that *simultaneously* carries out, on a single state, both $|\rightarrow\rangle_A \Rightarrow |\rightarrow\rangle_A|\rightarrow\rangle_B$ *and* $|\rightarrow\rangle_A \Rightarrow |\rightarrow\rangle_A|\rightarrow\rangle_B + |\leftarrow\rangle_A|\leftarrow\rangle_B$, a result from which we are forced to then conclude that $|\rightarrow\rangle_A|\rightarrow\rangle_B = |\rightarrow\rangle_A|\rightarrow\rangle_B + |\leftarrow\rangle_A|\leftarrow\rangle_B$. Do those two things look equal? Well, they do not look equal and they definitely are not equal. We have arrived at the quantum theoretical equivalent of proving that one is equal to two, a nonsensical result. We must have made a mistake! What was the mistake? It was the assumption, Herbert's assumption, that a perfect copying machine for quantum states exists in the first place!

In mathematics, this type of proof is called "proof by contradiction" and is the last refuge of a scoundrel (or at least a scoundrelly mathematician). We make an assumption, arrive at a contradiction, and if we have made no mistake in our logic, we must then conclude that our assumption was wrong in the first place. From the linearity of quantum mechanics alone, we have proven that there cannot exist a copying machine that can perfectly copy an unknown quantum state, because by assuming such a thing exists, we immediately arrive at a contradiction.

That, then, is the no-cloning theorem in a nutshell.

the no-cloning theorem was invented in 1989, quantum cryptography was invented a few years later, and where I never would have to sit on an airplane and listen to the person in the seat next to me yammer on about the film *What the (Bleep) Do We (K)now!?*

The proof of the no-cloning theorem in Box 6.2 exploits a very important property of quantum theory called "linearity." Linearity is a mathematical property of quantum theory that allows us to write the state of a spin-up ion in terms of a cat state as $|\rightarrow\rangle = |\uparrow\rangle + |\downarrow\rangle$, where we say $|\rightarrow\rangle$ is in a cat-like *linear* superposition of $|\uparrow\rangle$ and $|\downarrow\rangle$ (alive and dead). The linearity of quantum theory is well established in the laboratory, but it is possible that there could be a new version of quantum theory, which would be different from the old version of quantum theory. Let us call this new nonlinear version "quantum-schmantum" theory. That is, you could add extra terms to the equations of regular run-of-the-mill quantum theory to ruin linear superpositions so that you would end up with extra bits floating around in your equations, which are not there in ordinary quantum theory. These extra bits, if they were very large bits, would give radically different predictions than quantum theory never before seen in the laboratory, so they must be very small bits just to agree with current-day experiments. American physicist and Nobel Laureate Steven Weinberg proposed just such a nonlinear quantum-schmantum theory in 1989, careful to keep the nonlinear bits small in order for quantum-schmantum theory to agree with predictions of ordinary quantum theory in experiments to that date.[20] Weinberg's idea is that you might want to test in the laboratory if quantum theory was linear or nonlinear, quantum theory versus quantum-schmantum theory, and so he worked out some of the predictions of quantum-schmantum theory so that the experimenters would know what to test for in their laboratories so as to look for such small effects.

Almost immediately, well at least in 1991, American physicist Joseph Polchinski proved that Weinberg's quantum-schmantum theory allowed for superluminal signaling, no matter how small those extra bits were, unless those extra bits were precisely zero, in which case quantum-schmantum theory becomes just plain old quantum theory.[21] In quantum-schmantum theory, the noncloning proof breaks down and superluminal "flashing" is again possible. As superluminal signaling contradicts Einstein's theory of relativity, as well as allowing you to send backward in time (advising your grandfather against marrying your grandmother so that you would not ever be born to send that message), everybody in the physics community embraced Polchinski's result as a call to reject quantum-schmantum theory. But not the hippies! A colleague of Sarfatti and Hebert, American physicist Henry Stapp, embraced quantum-schmantum theory in order to resurrect the superluminal telegraph, which he needed to send messages backward in time in order to explain some experiments, and this is where in 1994 I myself first became aware of the hippies who were trying to save physics, or at least one of them. As my battle with that hippie is very briefly mentioned in the book *How the Hippies Saved Physics*, I figured it would be enlightening to expand on that Waterloo here from my point of view. It was July of 1994 when the battle began, innocently enough, the very month I was hired for my first permanent position as research physicist in the quantum optics group at the US Army Missile Command (now the US Army Aviation and Missile Command), in Huntsville, Alabama (after postdoctoring there since 1990). I was doing something that physicists, young or old, seldom do anymore, I was sitting in the library flipping through the *printed* copy of the July issue of the premier physics journal *Physical Review A*. (These days, all the flipping is done online, don't you know?) It was in this flippant moment that an article with a curious title caught my eye, "Theoretical Model of a Purported Empirical Violation of the Predictions of Quantum Theory," by Henry P. Stapp. In 1994, I had never heard of American physicist Henry Stapp or of the hippies trying to save physics, but the title almost screamed at me: empirical violations of the predictions of quantum theory? That translates to an experimenter who had found laboratory evidence showing that quantum theory was wrong! Why had I not heard anything about these experiments? Who did them? Pipkin? Frey? Clauser? Aspect? Zeilinger? What was up?

Henry Stapp was a close associate of Sarfatti, Herbert, and even Clauser and was part of the same San Francisco "Fundamental Fysiks Group," whose antics are so gleefully detailed in *How the Hippies Saved Physics*.[22] This is the same group that was enamored with quantum theories of the paranormal, Bell's inequalities and nonlocal correlations, the Clauser and Aspect experiments, quantum theories of consciousness, quantum connections to eastern mysticism, and of course the superluminal telegraph.[23] I knew none of this at the time, and thence I did not know what I was stepping into as I turned the pages and carefully read Stapp's paper in that July 1994 issue of *Physical Review A*. The point of Stapp's paper is that Weinberg's superluminal quantum-schmantum theory, once resurrected from the dead, could be used to explain some recent experiments that measured superluminal signaling in a quantum mechanical system. *What* experiments? I pored through Stapp's paper until I came to the sentence "According to the report in Ref. [8], it would appear that in certain experimental situations willfull (sic) human acts, selected by pseudo-random numbers generated at one time, can shift, relative to the randomness predicted by normal quantum theory, the timings of radioactive decays that were detected and recorded months earlier on floppy discs, but that were not observed at that time by any human observer." *What* the *hell*? I paused over this sentence for a full minute, the hair on the nape of my neck standing straight up, and then I turned the pages furiously to the end of the paper to look up reference eight, and when I read it I audibly gasped—so loud that the librarian told me to shush. Reference eight was to "H. Schmidt, J. Parapsychol. **57**, 351 (1993)."[24] The experiments were published in the *Journal of Parapsychology*, the flagship journal of all things paranormal. Now I never had heard of Henry Stapp but I was very familiar with the experimental tests of the paranormal by the German-born parapsychologist Helmut Schmidt! At this point, my readers, and particularly my good colleagues who purchased this book by accident, and who are now reading this tirade against their own good judgment, may be asking each other, "Why the heck is Dowling so familiar with experiments in parapsychology in the first place?" When I moved to Boulder, Colorado, in the 1980s to pursue my graduate career in mathematical physics, I found myself immediately immersed in a maelstrom of New Age malarkey that was Boulder in the 1980s. The city had shops selling crystals infused with "quantum energy," had night schools that offered classes in psychic healing, and was home to the breatharians—a spin-off sect from the fruitarians that was itself a spin-off sect from the vegans that was in turn a spin-off from the vegetarians that was indeed a spin-off from people who eat food.[25]

Seeking a life raft to preserve me from being sucked into the eye of this very Boulderian paranormal vortex, I joined a local group called *The Rocky Mountain Skeptics* (now defunct), which was an organization of *skeptics* of the paranormal—the anti-hippies if you like. The Boulder group was loosely affiliated with a national organization that was then called the *Committee for the Scientific Investigation for the Paranormal* ("CSICOP," which is pronounced, as a pun, as "psi-cop"), which is now called the *Committee for Skeptical Inquiry* or CSI. From their web page, "The mission of *The Committee for Skeptical Inquiry* is to promote scientific inquiry, critical investigation, and the use of reason in examining controversial and extraordinary claims."[26] This organization, founded in the late 1970s to combat the ever-rising tide of pseudoscience in the popular media, was my vortex-eye survival life raft.[27] CSICOP even has a journal, more of a magazine, *The Skeptical Inquirer*, which I subscribed to for many years, and it was in the pages of this magazine that I had read about parapsychologist Helmut Schmidt's experiments testing psychic phenomena. The comments in *The Skeptical Inquirer* were critiques of his experimental techniques, his data analyses, and the irreproducibility of his experiments. Yes, I knew indeed who Helmut Schmidt was.[28] I stalked indignantly over to the circulation desk at this army library and asked for a copy of reference eight, "Observation of a Psychokinetic Effect Under Highly Controlled Conditions," by Helmut Schmidt, in the *Journal of Parapsychology*. The librarians by now were used to my odd requests—but this?—this threw them for a loop. Never before (and never again they hoped) would an army scientist ask for a copy of an article from the *Journal of Parapsychology*. (In fact, the army and the Central Intelligence Agency had been in cahoots with the hippies in the 1970s, trying to test all sorts of paranormal phenomena, such as "remote viewing," for use in warfare and espionage.) I was told I would have to wait for a few days for it to arrive by interlibrary loan, but I could not wait to fire off a letter to the editor of *Physical Review A*, Bernd Crasemann, with a kindly but firm missive, informing him that he blew it.

I assumed that Stapp's paper had flown in under the radar, and that the referee missed Stapp's reference eight to Schmidt's article in the *Journal of Parapsychology*, and I assumed wrong. While I was waiting from a response from *Physical Review A*, Schmidt's article from the *Journal of Parapsychology* arrived, but it was not very revealing. "Observation of a Psychokinetic Effect Under Highly Controlled Conditions" was not a report of any one experiment, but rather a statistical "meta-analysis" of five experiments, carried out by Schmidt and his team and published in five more articles in the *Journal of Parapsychology*. Back to the library I went and requested the other five articles, where I hoped eventually to find the description of the experiments. Before I got those five articles, there came the response from *Physical Review A*. I was told that that they did not miss anything, that Stapp's paper was very carefully refereed, and that to prove it, *Physical Review A* provided me with an expurgated version of the entire editorial correspondence, including Stapp's originally submitted manuscript, the revised version that appeared in print, and all the referee reports. It was much worse than I feared. The original submission by Stapp contained detailed descriptions of all the five experiments, carried out by Schmidt, which I will get to in a minute. The referees had insisted that Stapp *remove* all or most of the references to the experiments and *remove* the description of the experiments in his manuscript. Why? Because the experiments were just plain too bizarre. It looked to me as if the editor, the referees, and the author conspired to disguise the true nature of Stapp's paper before letting it appear in *Physical Review A*. I was furious.

When I submit a theory paper to *Physical Review A*, I am often encouraged to *add* references to experiments but not *delete* them. I responded to *Physical Review A* much more sternly, made this point about the conspiracy theory, and then I (unwisely) used the "F" word. (And by that, I mean that I stated that the reviewing process bordered on scientific *fraud*.) No more correspondence on that topic from *Physical Review A*! But I did get a nice phone call from Benjamin Bedersen, the editor-in-chief of all the *Physical Review* journals (then A through E and now A through X), who supposed I was going to raise a fuss. (He supposed correctly.) He tried to nip me in the bud and to calm me by explaining that the refereeing process had run its due course, they had turned the crank on their editorial review algorithm, and the refereeing machinery had produced the end result—publish!—so there was nothing more to be said or to be done. Bedersen even told me that he had a nice conversation with Stapp, who implied to Bedersen that he was merely an innocent bystander with only a passing interest in these experiments, and who thought it would be nice to cook up a superluminal theory of quantum-schmantum mechanics to explain them. But *what* were these *damn* experiments? The interlibrary loan produced, finally, the remaining five publications of Schmidt, all from the *Journal of Parapsychology*, and now it was even *würst* than I thought. In addition to being quite peculiar, one of the experimental papers had a curious coauthor, none other than *Henry P. Stapp*. Recall that Stapp had implied to the editor-in-chief of *Physical Review*, Bedersen, that he, Stapp, was merely a curious but passive observer of these experiments. This paper with both Schmidt and Stapp as coauthors provided evidence to the contrary; far from a disinterested observer, Stapp directly participated in the experiments. I faxed this Stapp–Schmidt paper off to Bedersen and, I hear, he was *not* very happy to find out about it.

In fact, Stapp, far from being a disinterested observer, had long been a cheerleader for the superluminal crowd. Indeed, he wrote a 1977 paper on this topic— "Are Superluminal Connections Necessary?"—in which he concludes, "The theorem of this paper supports this view of Nature by showing that superluminal transfer of information is necessary, barring certain alternatives that have been described [in this paper] and that seem less reasonable."[29] (By "Nature" he does not mean the journal of that name but I suppose Mother Nature herself.) For Stapp, the only reasonable conclusion was that quantum correlations were required for faster-than-light signaling. But the no-cloning theorem ruled out such signaling in ordinary linear quantum theory. What to do? Embrace new and improved nonlinear quantum-schmantum theory. The experiments are too involved to describe in detail here, but this is the gist.[30] I will describe one of the five experiments, a description that gives the flavor of all of them: a teenage karate student, located in Syracuse, New York, stares at a video of a swinging pendulum on a computer screen and focuses his psychic powers to alter the decay rate of a radioactive atom located in San Antonio, Texas, and succeeds— in altering the decay rate months *in the past*, long before the karate student ever sat down in front of the computer in the first place. The purported mechanism for this experiment is "retroactive telekinesis." Telekinesis is the hypothetical paranormal effect whereby a human can, by some unknown mechanism, move things about in the real world with the power of her mind—bending spoons was a popular sport. Retroactive telekinesis is the new and improved version, where she can use her mind to move things about in the past, perhaps before she was even born. Let me illustrate this experiment by an analogy to popular culture.

In American writer Stephen King's 1974 novel, *Carrie* (and 1976 film of the same name), the heroine-villainess, Carrie White, is blessed–cursed with the power of telekinesis. At the climax of the film, she stands in a sports gymnasium upon a dais at her senior prom and, while in a trance, incinerates the building (and most of its occupants) in revenge for a cruel prank. When she snaps out of the trance, the whole place has gone up in a telekinetically powered inferno that she stalks through and out of the gym unharmed. Let us now rewrite the script so that Carrie has *retroactive* telekinesis, the presumed workings of the Schmidt experiment. When retroactive-telekinetic Carrie emerges from *her* trance, instead of the inferno, she finds a quiet, windswept, ash-strewn lot. She has used her *retroactive* telekinesis to burn the gym to the ground 6 months in the past, long before her senior prom ever took place. (A plot likely too implausible for even Stephen King.) And in the spirit of the Rube Goldberg-esque nature of the Schmidt experiment, she does not use her brainwaves to burn the place down in the past directly. Instead, operating backward in time, she uses her brainwaves to alter the hitherto random decay of a radioactive atom in Syracuse so that it decays in a nonrandom fashion such that the times of arrival of the emitted decay particles, converted to zeros and ones, contain a secret binary message—"BURN DOWN THE GYM"—that is dutifully recorded (unseen) on a Geiger counter by an experimenter named Larry "The Flame" Inflämmibilé. Larry transfers the bits of the secret message to a personal computer, then to a floppy disk, and then to a printing station. He then prints the message out on a piece of cardboard and, while blindfolded, covers up the zeros and ones that make up the message with bits of masking tape.

Larry sends the cardboard, with its hidden secret message, by ordinary US mail to San Antonio, addressed to a second experimenter, Max "The Match" Incendiarió, who peels off the tape and decodes the numbers (using a secret protocol that he constructs from numerical temperatures he reads in the weather report in that day's local newspaper, the *San Antonio Gazette*). Max thereby collapses the massively entangled multiparty state across time and space (reaching all the way back to the future into Carrie's brain), decodes the secret message, "BURN DOWN THE GYM," and then heads out in his Mercedes-Benz 600 Pullman with a canister of gasoline to torch the place—9 months before Carrie ever set foot in it. *This* then is a close analogy of the experiment that Stapp and Schmidt would have us believe requires the modification of the very foundations of quantum theory in order to explain. *This* then is why I gasped audibly in the army library. To quote from my own (unpublished) comment on the Stapp paper, "There is much to be said about the fundamental misunderstanding of the quantum mechanical notion of state reduction inherent in this hypothesis. In particular, the far-fetched implicit assumption is simply preposterous, namely, that one has a coherent entangled wave function for the complete system of: radioactive sample, Geiger counters, monitoring PC, floppy disk drive, floppy disk, printer, print-out, United States Postal Service, plus anything else that has been interacting with any one of the components in the time between the clicks of the Geiger counter and the actual [telekinesis] experiment. Consider—even in the best EPR-type experiments it is extremely difficult to get microscopic, entangled quantum states of matter that extend just across a laboratory bench and last for a few microseconds" (see note 30).

As implied above, together with German physicists Berthold Englert and Axel Schenzle and American psychologists Ray Hyman and James Alcock, we wrote up a comment to Stapp's paper and submitted it to *Physical Review A* for publication. I enrolled Englert and Schenzle for the project at the hotel bar while we were attending the *Third Crested Butte Workshop on Quantum Coherence and Interference,* held August 8–11, 1994, in Crested Butte, Colorado. After I told them this story, and after only three glasses of scotch whiskey (each), they were onboard. I enrolled Hyman and Alcock, psychologists whose papers regularly appeared in the *Skeptical Inquirer,* because they were experts on the statistical analysis of experiments in parapsychology and particularly experts on the paranormal experiments of Schmidt. In our manuscript, "Comment on 'Theoretical Model of a Purported Empirical Violation of the Predictions of Quantum Theory,'" we three physicists focused on a critique of the quantum-schmantum theory deployed by Stapp, while the psychologists focused on the statistical analyses and experimental protocols set out by Schmidt. Our joint conclusion was that, "In the absence of overwhelming experimental support that [these superluminal] effects indeed occur, such [modifications to quantum theory] do not constitute new physical theories but rather mathematical exercises of little value. Should a theory that abandons Einstein causality be true, it would shatter the very foundations of physics as we know it. Such a modification to our basic physical laws should be made only if a vast body of repeatable experimental evidence requires us to do so. We shall argue ... that the experiment of Schmidt does not provide this evidence." (Brackets mine.) In a nutshell, our argument was that Schmidt's experiments were flawed in their setup and statistical analysis, and never ever repeated by any other experimenters, and, thus, that retroactive telekinesis did not exist. No new-fangled quantum-schmantum theory was needed to explain the experimental results, because there were not any reproducible experimental results in need of explaining.

There is always a problem when submitting a comment to *Physical Review* on a paper published by *Physical Review.* Things are weighted in favor of the author of the original paper being commented upon, as the editors first send it to him or her to referee. Authors do not like negative comments about their papers appearing in print and tend to reject them, and editors fearing a public kerfuffle tend to uphold the rejection. We could have appealed, but once the alcohol wore off, my coauthors' stamina waned, and so in the end I decided to take matters into my own hands. In addition to the long comment, eventually rejected by *Physical Review A,* I instead sent a letter to the premier US physics trade journal, *Physics Today,* as sole author, which appeared in July of 1995, a year to the day after Stapp's original paper. My letter and Stapp's reply appeared together with the catchy caption, "Parapsychological Review A?"[31] My letter began with the opening salvo, "It appears to me that there is a small but dedicated group of scientists—some with quite respectable reputations—who nevertheless dabble in things that most of us would not call science. (The terms "pseudoscience" and "pathological science" come to mind.)" Stapp replied, "A scientist does not become 'dedicated' to pseudoscience by accepting a challenge to examine purely physical facts created under highly controlled conditions. Indeed, to refuse to look at such physical evidence on ideological grounds would be pseudoscience." He went on to describe, in detail, his participation in the one of the five Schmidt experiments on which he appeared in print as coauthor, "I received by mail (in batches) a set of thick cardboard sheets. Each sheet had a set of rows, with each row consisting of a pair of short strips of black tape. After receiving a batch I waited for at least a week and on a day prescribed by the fixed protocol, I extracted from the weather table of *The New York Times*, by a fixed recorded procedure known only to me, a pair of 'random numbers' that I then used as seed numbers in a computer program devised by myself and divulged to no one (until after the experiment was completed) ... the procedure that I myself carried out was purely a 'physics experiment.'" The brains of the karate students in Syracuse (in the future) were apparently entangled with a weather report in the *New York Times* (in the past).

I thought I had made my case (or that Stapp had made my case for me), but then came the hate letters attacking *me*. Oh boy it was fun. Paranormally inclined websites for years maintained a reprint of my letter "Psychological Review A?" with captions such as, "An attack from a US Army physicist on Stapp and his theory of [telekinesis] just [*sic*] presented. Stapp, like Harvard psychiatrist, John Mack, who studies UFO abductions, is under serious peer pressure (e.g., head of LBL [*sic*] where Stapp works) to supress [*sic*] his fringe research. This is a new threat to academic freedom in our elite universities." [Square brackets mine.][32] The underlying theme here is that there is some plot by the US Army to supress research into the paranormal. Ah, conspiracy theories, how I love them. In my defense, I did state clearly in my letter that I was not trying to censor Stapp's work, but that a more appropriate venue for its publication, other than *Physical Review A*, might be instead in the *Journal of Parapsychology*, where the original experiments were carried out. Nobody seemed to get that point. In parallel to the comment and the letter, I decided to go around the editors at *Physical Review* and approach the editorial advisory board for *Physical Review A* directly. I put together a large printed file, with Stapp's original publication, my original letter to *Physical Review A*, the reply from *Physical Review A*, my response to the reply, a copy of the infamous reference eight (Schmidt's meta-analysis), and copies of the five actual experimental papers that reference eight summarized (including the one with Stapp as coauthor), and then I iced that paper–wedding cake of a stack with our still unpublished but detailed comment. The stack of paper was 500 sheets high, exactly a ream, and fit nicely into the ream-sized cardboard boxes our flimsy army printer paper came in. Cloning allowed, I xeroxed off 20 copies of the stack and packed each copy in a separate printer paper box and mailed them all out (fourth class printed matter rate—your tax dollars at work) to each and every one of the members of the editorial board—and then waited for the reaction.

I understand that the editorial board meeting was very lively that year and they recommended tightening up the acceptance criteria for purely conjectural theory papers on the foundations of quantum theory with little or no experimental evidence. Of course, that policy led to even more accusations, leveled against me in part, of academic censorship. I was in a curious position, as a dabbler in the foundations of quantum theory myself, that I had just made it harder to get my own papers into *Physical Review A*.[33] But I never submit my papers on the foundations of quantum theory to *Physical Review A*—I submit them to the *Foundations of Physics*, where they belong. Herbert's original FLASH paper was published there.

A new round of attack letters appeared in *Physics Today* in April of 1996, under the even catchier caption, "More Spirited Debate on Physics, Parapsychology and Paradigms."[34] Alexander Berezin wrote, "...we need more, not fewer, independent experimental verifications (or refutations) of the effects claimed by Helmut Schmidt, Robert Jahn and others." I replied, "...Berezin argues that we need even more experiments on paranormal phenomena. For the most part, the results of relevant experiments undertaken over the last four decades have not proved to be reproducible by independent experimenters. Accordingly, I really do not see the value of conducting yet more experiments." Shimon Malin wrote, "[Stapp] studied the Schmidt effect both experimentally and theoretically. I believe that he should be commended for his efforts, and his results should be publishable in the most prestigious journals." I replied again, "Malin believes we are on the verge of a paradigm shift having to do with the role of consciousness in relation to the physical world, and that Stapp is a brave pioneer whose results 'should be publishable in the most prestigious journals.' Again, I disagree—and I don't think I am alone in taking this position." (I did, however, get to check off "paradigm shift" that day on my official US Army buzzword-bingo scorecard... [see note 3].)

The hippies could not quite disconnect their quixotical quest for a superluminal telegraph and their understanding of entanglement from the goal to hook this stuff all up to consciousness, Eastern mysticism, and the paranormal. What I think is most interesting is, that despite the hippies, work continues at a rapid pace in the new field of quantum communications, a brand new quantum technology. Quantum communication theory explores, in a rigorous fashion testable to many decimal places in experiments that are ever repeatable, the role of quantum theory in communications. Quantum communications theory is more than just cryptography, there is not only teleportation but also quantum super-dense coding, quantum secret sharing, all sorts of interesting technologies in the pipeline.[35] These quantum communications technologies are readily and rapidly tested in the laboratory in repeatable experiments that do not require faster-than-light signaling, messages that travel backward in time or that require entangling the minds of karate students with the pages of the *New York Times*. And that development of well-grounded theory of quantum communications development is a good thing.

So I have tried to address five of the hippie hypotheses. Let me summarize them now. First, that quantum theory is needed to explain experiments in parapsychology. Such experiments, after decades of trying, have produced no reliable or reproducible results. There is nothing there that quantum theory needs to explain. Second, that quantum theory is needed to explain superluminal communications. No experiments have ever seen such communications. Even the Schmidt effect disappeared as more data were taken and independent researchers tried to reproduce his results. There are no superluminal signaling experiments—superluminal neutrinos be damned!—that require quantum theory to explain.[36] There are plenty of reasons (Einstein causality, no-cloning, no-nonnoiseless laser amplification) to think that there will never be any such experiments. Nothing again there for quantum theory to explain. Third, that quantum theory is needed to explain human consciousness. I will save this discussion until a bit later. Fourth, that heavy hallucinogenic drug use is required to understand quantum theory. Who knows? All the experimenters in this field of research have subsequently lost their minds! Fifth, that Eastern mysticism is needed to understand quantum theory. I will briefly address this fifth hypothesis here.

The idea that Eastern mysticism is needed to understand quantum theory was promoted initially in two wildly popular books, written by two members of the Fundamental Fysiks Group, that is, two of the hippies. The first was the 1975 publication of *The Tao of Physics* by Fritjof Capra, and the second was the 1979 publication of *The Dancing Wu Li Masters* by Gary Zukov. Both of these books played neatly into the western 1970s peak of interest in all things Eastern mystical, a fad that was likely the direct result of the philosophy's popularization by the rock band *The Beatles*. In the preface to the Capra book, he states that the work was the result of a vision that he had while doing psychedelic drugs. (It certainly reads like that but then so do some of my own writings....) I have nothing against Eastern mysticism as either religion or a philosophy, but rather I address the specific hypothesis, promoted in these two books, that a deep knowledge of Eastern mysticism is a prerequisite for a full understanding of quantum theory. (Or that at least it makes quantum physicists feel better about themselves.) This hypothesis cannot be tested in the laboratory, but it can be tested in a fiery crucible fueled by the incandescent flames of anecdotal evidence.

The inventors of quantum theory were, for the most part, German, Danish, Swiss, Austrian, and English. These are the four *least* Eastern-mystical nationalities that I can possibly think of. I can assure you that German physicist Wolfgang Pauli did not first commit the Bhagavad Gita to memory before plunging into his development of the full quantum mechanical theory of atomic spectra. Dour Danish physicist Niels Bohr did add an Eastern yin-yang symbol to his family coat of arms (in 1947) but also had a lucky horseshoe nailed up over his barn door. When asked whether he believed in such a superstition as a lucky horseshoe, Bohr replied, "Of course not, but I am told it works even if you don't believe in it."[37] Capra and Zukov simply draw artificial parallels between words and phrases in Eastern mystical texts and similar words and phrases in quantum theory texts and then claim that's understanding.

I could just as well spend a week on a farm dropping acid each morning, draw fanciful correlations between the rooster's morning crow and the daily rise of the sun, and then write a wildly popular book that claims a deep knowledge of the biomechanics of rooster crowing is required to understand the astronomical phenomenon of the sunrise. *Postdoc ergo proper doc!* I must sheepishly confess I know almost nothing about Eastern mysticism, but I can proudly proclaim that, despite that deficit in my education, I understand quantum theory quite well, thank you very much. That, then, is my singular and personal anecdotal data point. No knowledge of Eastern mysticism at all is required to understand quantum theory at all well.

Quantum communication, for example, is a fascinating and evolving new quantum technology that exploits the weirdness of quantum theory without having to invoke the paranormal or the role of human consciousness in quantum effects. Quantum communications will be the framework for many if not all communications technologies in our new millennium. And I conjecture that today's state of the art in quantum communications technologies owes very little to the hippies who tried to save physics. Physics, as usual, did a pretty good job of saving itself.[38] To paraphrase the words of the British evolutionary biologist Richard Dawkins, I encourage you my readers to keep an open mind ... but not so open that your brain falls out.[39]

QUBITS, FOUR BITS, SIX BITS, A DOLLAR

When will the quantum computer be built? As American physicist and Nobel Laureate William Phillips is fond of saying: The probability of building a quantum computer is 50–50—and by that I mean there is a 50% chance of building one in 50 years. Well Bill what about the D-Wave "quantum" computer? You can buy one now for $10,000. Phillips I believe meant to qualify the probability of building a universal quantum computer capable of cracking a public-key encrypted message of interest to the National Security Agency (NSA). But the public-key scheme was only invented in the 1970s, 40 years ago. The beginnings of the Internet are from only 50 years ago, when the Defense Advanced

Research Projects Agency (DARPA, then known as Advanced Research Projects Agency) invented the ARPANET to hook supercomputers together to preserve the data in a distributed grid. It was the first remote backup and emailing system. In 50 years, who knows what the Internet will become? Perhaps we'll all have Borg implants with laser beams shooting out of our eyeballs and communicate brain to brain using quantum communications while flying about in a giant glowing green cube. In any case, we'll certainly not be encrypting our communications using the current public-key system based on the hardness of factoring. Even now, mathematicians in secret NSA lairs are working out new public-key systems based on mathematical problems so difficult that even a quantum computer cannot hack them. In 50 years, if we are not using uncrackable quantum cryptography, we'll be using an uncrackable version of public-key encryption and the primary goal for building a tera-qubit quantum computer will have evaporated. What then? Will the quantum computer already be obsolete in 50 years?

American computer scientist Scott Aaronson is fond of saying that asking "When will the quantum computer be built?" is a bit like asking British inventor Charles Babbage, in 1850, when the Internet will be built? In the 1820s, Babbage designed a mechanical steam-powered difference engine that might be called the first digital computer. The machine, like the ENIAC (Electronic Numerical Integrator and Computer), would have weighed many tons and, perhaps for this reason, was not built until the 1980s, using only technology available in Babbage's day, where it was shown to be able to carry out mathematical calculations accurate to 30 digits. Clearly, Babbage, plagued by organ grinders (and their monkeys) that ground away outside his home at all hours of the day and night, could not have predicted the rise of the Internet in 1850.[40] Why could he not?

In 2008, I was invited by the US Army to attend the annual 3-day *Future Technology Seminar*, which is more popularly known as *The Mad Scientist Conference*, held that year from August 19 to 21, in Portsmouth, Virginia. To this day, I proudly display on my office note board my name tag from the event that declares, "Jonathan P. Dowling (Mad Scientist)."[41] This is one of the most curious events I have ever had the pleasure to attend in my life. The attendees consisted of military rank army officers, civilian rank army analysts, scientists (mad and otherwise), science fiction writers, and "futurists." On the morning of day 1, we mad scientists were read our marching orders; our job was to come up with creative ways to destroy the world, preferably to destroy it in a few days or less. Why was I invited? One of the organizers had complained that quantum technologies had not been represented in the previous years and so I was hauled in and then trotted out as their token quantum technologist.

One scientist, perhaps not so mad, inquired meekly of the army administrators running the show: wouldn't it be better for the assemblage of great thinkers to think up ways to keep the world from being destroyed? The army colonel responded gruffly, "No. We have plenty of people already working on that. You are our red team!" (In war games, the "red team" consists of your allies pretending to be the bad guys so that the good guys, on the "blue team," have something real to shoot at.) Thus instructed, we mad scientists, like Montgomery Burns from the animated television series, *The Simpsons*, twiddled our fingers together and chortled in glee, "Excellent!" Then we were off—hopping mad as March hares. What was my favorite part of the conference? Like most conferences, it was the evenings spent at the bar where the real work gets done. Each evening, the army would host a happy hour at a cash bar (your tax dollars not at work), where I mingled with inebriated attendees such as army colonels (dressed in military fatigues), army civilian analysts (dressed in suits and ties), scientists (dressed in business casual khakis and button-down shirts), science fiction writers (dressed in mismatched Hawaiian shirts and Bermuda shorts), and futurists (dressed in long black capes while wielding tall oaken staffs). Yes, at the bar we barking mad scientists would become positively unhinged, prattling on about the end of the world and cackling away long into the night.

I did inquire, quietly, of the army organizers why were there so many science fiction writers in attendance? Perhaps they're mad but certainly they're mostly not scientists. I was immediately scolded, "Science fiction writers have a much better track record of predicting future scientific and technical breakthroughs than you scientists do." Unfettered by the laws of physics (or much anything else), the SciFi writers are free to extrapolate much more wildly than the scientists. This is why poor organ grinder monkey–plagued Babbage could not have predicted the future Internet back in 1850. To do so, he would have to predict a number of technological inventions that did not yet exist: the triode vacuum tube (1907), Turing's theory of computation (1936), the vacuum-tubed ENIAC (1946), the transistor (1947), the first integrated circuit (1958), Moore's law (1965), ARPANET (1969), optical communications fiber (1970), the very large-scale integrated circuit (1970s and 1980s), the invention of the fiber amplifier-repeater (1986), and finally the development of the World Wide Web (1990). To quote Hungarian–British electrical engineer and physicist Denis Gabor (the inventor of optical holography), "The future cannot be predicted, but futures can be invented."[42]

Why do we scientists and engineers do so badly, compared to science fiction writers, in predicting the future? That is, why do we scientists and engineers always come up way *short*? It is because we are used to making *linear* extrapolations forward into time based only on what we currently know now. What we fail to do is anticipate what we don't yet know, integrated over long periods. Babbage would have been able, on the basis of 1850 technologies, to predict the dystopian and Orwellian future of the alternative history thriller, *The Difference Engine*.[43] In this novel, there is a parallel universe that splits away from our own in the 1880s, in which Babbage (despite the monkeys) gets his difference engine working, mass produces them, and invents a steam-powered Internet. This pneumo-mechanical Internet is replete with information-bearing punch cards incased in black gutta-percha cylinders, which shuttle back and forth between the massive, warehouse-sized difference engines, which are themselves distributed all over Great Britain. The whole thing is hooked up to an immense network of compressed air–powered mail tubes—the Internet really *is* a series of tubes![44] But as we know now, that is not what happened.

Recall from Chapter 4, "You're in the Army Now," my bet in 1999 with "Keith" from the NSA. I bet him a pizza and a beer that the quantum computer would be ready in 10 years, not 50. I was trying to think more like a science fiction writer and less like a physicist. I lost that bet, but only after we agreed in 2009 to rule out the D-Wave machine as evidence that I had won. The bet explicitly specified that it would be "a quantum computer capable of cracking a 512-bit public crypto-key provided by the NSA." The D-Wave machine is not that, but it is a start. When I paid Keith up in 2009, he asked me again in the pizzeria, "When do you *now* think the quantum computer will be ready?" and I replied again, "In 10 years." If Keith asks me today, I will still say 10 years. If Keith asks me in 10 years (God willing and the creek don't rise), I will *still* say 10 years. But if we count D-Wave, perhaps I have already won but we just don't know it. There is a logic, followed by the fusion energy community for over 60 years, to always respond with 10 years, when asked when the fusion reactor will be ready. If you say 5 years, and well the typical time a government program manager is in a job slot is 5 years, you might have to actually produce something. If you follow Bill Phillips' example and say 50 years, well in 50 years the program manager will have retired and won't get any credit for starting the program. That's too far out to keep the funding flowing. Ten years is just about right. Close enough to be tantalizing but far enough away to allow you some wiggle room. This is the fusion energy model and their funding has not dried up (yet).

Current estimates for the resources for a quantum computer capable of cracking a 512-bit public key are that it would require trillions of entangled qubits, or a terabyte's worth. Most experiments are following a bottom–up approach. Every year or so, the ion trappers add another qubit to their trap. Then, they carefully fine-tune and characterize and polish that new qubit and its associated quantum gate fidelities. The current limit to this approach was the 2011 demonstration of 14 qubits in the ion trap of German physicist Rainer Blatt. At this rate, it will take a trillion years for the quantum computer to be ready. There is even a more immediate pressing problem. The few-qubit experimenters characterize their qubit states and gates by taking lots and lots of data and then feeding it into a (classical) computer algorithm that executes either quantum state tomography (to characterize the qubit states) or quantum process tomography (to characterize the quantum gates). The word "tomography" is derived from the Greek words *tomē* (to cut) and *graphein* (to write or to draw).

In medical imaging, x-ray tomography reconstructs three-dimensional images from cuts or slices through the body. In quantum tomography, the slices are through Hilbert space to reconstruct the quantum state (or gate).

To completely characterize a 14-qubit state, you must remember that the number of possible states scales exponentially with the number of qubits, so the number of possible states is $2 \times 2 \times 2 \times 2 \times 2 \times 2 \times 2 \times 2 \times 2 \times 2 \times 2 \times 2 \times 2 \times 2$ or 16,384 states. For each of these states, the experimenter must take hundreds or thousands of pieces of data for good tomographic (slices through Hilbert space) characterization—but there our old friend the quadratic scaling law rules. To reduce the margin of error (per state) from 10% to 1%, one must go from 100 measurements per state to 10,000 measurements per state. So let's say we can tolerate a 1% margin of measurement error per state characterization, then that is 10,000 measurements per state multiplied by 16,384 states or 163,840,000 measurements needed to characterize the entire 14-qubit machine. The exponential scaling of the Hilbert space, which is a blessing when trying to build a machine that factors large numbers, is a curse when trying to delineate the darn thing's performance. At one measurement per second, it would take over 5 years to see if your 14-qubit quantum computer was working correctly or not. Because graduate students in a PhD program are expected to graduate in 5 years, this then is the hard upper limit for building and characterizing a few-qubit quantum computer. They could add more qubits, but then they would not be able to tell you if the machine was working properly or not, as they could not characterize its performance.

This is what happens when you use classical data collection and feed it into the classical computer to characterize the quantum computer. We know that approach will ultimately fail because you can prove that a classical computer cannot efficiently simulate the performance of a quantum one; that's the whole point, and that is precisely what the experimenters are trying to do here. The Hilbert space grows exponentially with the number of qubits but the classical computational power grows only linearly with the number of bits in the classical computer chip. The classical computer cannot possibly keep up, which is the whole reason why quantum computers are more powerful than classical ones. What to do? When Intel wants to characterize the performance of their new "Unobtainium" chip, they do not try to do it using a Chinese abacus (*suàn-pán*), which is analogous to trying to characterize a quantum machine with a classical one. Intel instead wires together hundreds of their current generation Itanium chips into a supercomputer and then using something called a Monte Carlo approach makes many random runs of characterization of the new Unobtainium chip with the bank of old Itanium chips. (The term "Monte Carlo" just means "random" in the case, like the chances of winning at the casino of that same name.) Like playing roulette at the Monte Carlo, Intel is taking a bet. It bets that if it makes enough random characterizations of the new Unobtainium chip using the banks of old Itanium ones, then it is unlikely that a defect will slip through the Monte Carlo net uncaught. Sometimes they lose the bet but mostly they win it.[45]

In the bottom–up approach to quantum computing, the only hope to getting way beyond 14 qubits is to follow this lead of Intel. We should not characterize the 16-qubit quantum machine with a classical computer. Following the same logic as above, a 16-qubit machine has an exponentially large Hilbert space containing 65,536 states that, to characterize it (classically) to the 1% margin of error, would take 655,360,000 data measurements, which at 1 second per measurement would take nearly 30 years—you had better hope your graduate students don't take 30 years to finish their PhDs. Instead of this classical characterization, what needs to be done is to hook up a bank of well-characterized 14-qubit quantum machines and entangle those old 14-qubit machines with the new 16-qubit one. Just like Intel, some future manufacturer of quantum computer chips, call it "QuIntel," would have to use lots of the previous-generation quantum chips (running a kind of quantum Monte Carlo algorithm) to characterize the next generation of quantum chips. We call that approach "bootstrapping," and to save the bottom–up approach, bootstrapping is our only hope. Even bootstrapping is a tall order. Blatt typically has only a few ion traps in his laboratory at one time, certainly only one 14-qubit trap at a time, and the idea that he would have banks of tens or hundreds of 14-qubit machines lying around to characterize the 16-qubit machine seems far-fetched. But the Internet would have seemed far-fetched to Babbage in 1850. We await the development of future technology, such as the very large integrated arrays of ion trap quantum computer circuits. I predict that we'll have them in ... 10 years.

But aside from the bottom–up, there is another top–down approach. Throw a huge number of ion qubits (gazpacho-like) into a very chilled pot, season with gleaming laser pulses, and see what you cook up (or cool down). Maybe all these qubits are not all entangled. Maybe they are not all individually controllable so as to make lots of quantum gates. Maybe you cannot characterize their performance at the 1% level. Probably you cannot run Shor's algorithm and crack a 512-bit public key with them. But likely you can do simulations of complex quantum systems with them, such as simulating the energy levels of the intractable atom of thulium, or simulating the properties of exotic magnetic "spintronic" materials. Throw hundreds or thousands of qubits in the pot and keep cooling and sifting and straining until you get a reduction of thousands of qubits all linked up in a way that might be useful for something.

Australian physicist Robert Clark first championed this top–down approach to quantum computing in the early 2000s for the design and fabrication of a solid-state semiconductor-based quantum computer where many phosphorus ion qubits are embedded into a silicon semiconductor host. At first, the qubits may not be pure, and even one-qubit (and much less two-qubit) gates are hard to come by, but the team started by putting down hundreds of them and then seeing how they could improve the performance.[46] (They have the one- and two-qubit gates working now. This program has recently demonstrated a single-atom transistor—a "spin"-off application.) The top–down approach was then picked up by D-Wave, which currently manufactures a quantum computer with a 128-qubit processor. The qubits are not perfect. They are probably coherent but probably not entangled and probably will never be able to crack a 512-bit public-key encryption. Who cares? (Well other than the NSA....)

In 2011, D-Wave sold two of these quantum computers, at $10,000,000 a pop: one to Google and one to a collaboration between Lockheed–Martin Corporation and the University of Southern California (USC). Google, I hear, is interested in running a variant of the Grover search algorithm, and Lockheed–Martin and USC are after the quantum simulations—thulium my friends, thulium. The most recent salvo in the top–down approach was launched in 2012 by the group of American physicist John Bollinger at the National Institute of Standards and Technology (NIST) that recently demonstrated a crystal of hundreds of ion qubits in a large two-dimensional array in a large electromagnetic ion trap. The goal? Not to make a factoring engine but instead to make a quantum simulator. If you give up the idea of addressing and carefully characterizing the individual qubits in a quantum computing circuit (the bottom–up approach) but just dump them all in and let Heisenberg's ghost sort them out, then you might, just might, get a super-duper quantum simulator capable of simulating new types of magnetic materials that may someday make up the classical computer memory of a future terabyte hard drive.[47] I predict that, in 50 years, the Shor factoring algorithm will be irrelevant, nobody will be using that type of public-key encryption anymore, but that in 2062, the thumb-sized yottabyte hard drive will be quite the Christmas stocking stuffer.[48]

Which is better, the bottom–up or the top–down approach? Both have their merits but until now almost all experiments in quantum computing have focused on the bottom–up. Every new qubit is precious and must be completely characterized and the fidelity optimized as it is carefully placed near its old neighbors in the ion trap. The advantage of bottom–up is that you have a great deal of knowledge about what is going on in the few-qubit machine. But remember, without quantum bootstrapping, it is impossible using today's classical computing resources to carefully characterize what is going on for 16 or more entangled qubits. In the new NIST experiment, they have trapped 300 ions in a two-dimensional crystal lattice and, again with laser pulses and such, they can precisely control the interactions between the qubits (see Figure 6.2). Not so much control that they can make a universal quantum computer out of the thing, but enough that they can make a special-purpose quantum simulator.

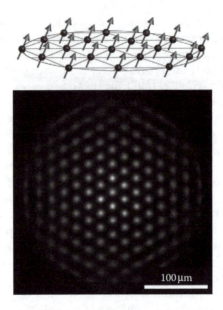

Figure 6.2 The NIST 300-qubit quantum simulator. At the top is a schematic showing the qubit spins (balls with arrows sticking through them) in the two-dimensional ion trap. The lattice of lines represent the nearest-neighbor qubit–qubit interactions that will (someday soon we hope) generate the massively entangled 300-qubit entangled state. At the bottom is a top–down photograph of this top–down scheme showing about 100 of the actual beryllium ions glowing in the trap that make up the simulator. The trapped ions, when illuminated with a laser beam, begin to glow so brightly that you can see the individual ions (individual atoms with an outer electron missing) with a digital camera, which is how the bottom photo was taken. (The image shows only a small fraction of the 300 ions, namely, those located near the center of the trap.) (Both images are courtesy of Joseph W. Britton and NIST. For the top image, see "Quantum Simulator Illustration" by Joseph W. Britton at the National Institute of Standards and Technology [2012], http:// patapsco.nist.gov/imagegallery/details.cfm?imageid=1044, and for the bottom image, see "Quantum Simulator Crystal" by Joseph W. Britton at the National Institute of Standards and Technology [2012], http://patapsco.nist.gov/imagegallery/details.cfm?imageid=1045. Both images are contributions of the US National Institute of Standards and Technology and are not subject to US copyright.)

In this case, what NIST is hoping to simulate are the electron spins in solid materials such as metals that give rise to a type of magnetism we observe in some particular materials. Each electron spin, treated as a tiny bar magnet, might tend to line up (point in the same direction) or anti-line up (point in the opposite direction) with its neighboring electron. Immediately, you have a cat state that is simultaneously lined up and anti-lined up (called a frustrated state) and pretty soon this entanglement of frustration spreads to all the electron spins in the material's lattice. This frustrated entangled state of many spins is called anti-ferromagnetism and leads to a magnetic behavior called giant magnetoresistance, an effect that has led to new technologies in the field of spintronics (electronics but with spins).

Research in giant magnetoresistance led directly to the development of very sensitive magnetic field sensors that were then placed in the readout heads of magnetic computer hard drives. These heads could read out tiny magnetic field changes encoding the zeros and ones on the hard drive, allowing for a great increase in hard drive storage capacity. When you apply a magnetic field to this entangled state, you get a very complicated behavior of the collection of spins, behavior that is impossible to predict or simulate on a classical computer because there are so many entangled electron spins on the read head. Typically, to simulate the behavior of such a magnetic material, you would like to be able to simulate the forces that couple one electron spin to its neighbor and also simulate the temperature, as the entanglement tends to be destroyed at high temperatures. This is the connection between the NIST quantum simulator and a yottabyte hard drive. If you could simulate the properties of such engineered magnetic materials, you could use your simulations to improve the magnetic response and thus could make more sensitive magnetic field sensors and store more data on magnetic disks and hard drives.

So what to make of the NIST contraption with 300 spins? Well, when I started out in the quantum computer game in 1994, the most powerful classical computers of the time could simulate the performance of approximately 20 entangled qubits. This was one of the selling points in the 1997 proposal for NMR quantum computing, sold as able to do calculations out of reach by any other machine, classical or quantum mechanical. (In 1997, the ion trappers had demonstrated only two entangled qubits.) We now know that the NMR machine, without any entanglement in it, was not a quantum computer at all, and it did not matter if it had 2 or 20 qubits, they were not entangled qubits, and so there was no exponentially large Hilbert space to execute our quantum algorithms in. (The performance of a 20-qubit NMR machine could easily be simulated on a 1997 Power Macintosh 6500 desktop computer.) Meanwhile, the ion trappers plodded along, adding 1 qubit per year until they hit the current record of 14 in the tiny trap that Blatt built. (He has 14 well-characterized qubits that can do any universal but small quantum calculation.) But Moore's law for classical computing power is also an exponential law and the classical computers continued to become more powerful too.

Today, according to my colleague here at the Beijing Computational Research Center (where I am currently on a 3-week visit typing away on this book), Argentine physicist Alejandro Muramatsu, a classical supercomputer can today in 2012 simulate a 30-qubit universal quantum computer.[49] Muramatsu was chiding me at dinner that computational solid-state physicists, such as he, could already simulate 30-qubit systems on their classical supercomputers, while we quantum physicists had only built a 14-qubit device. I emailed him the new NIST 300-qubit publication with the subject line, "300 qubits? The race is on!" He then walked into my office with the reprint and pointed out that, at least in the current experiment, the NIST researchers had simulated the 300-spin system only in a regime where there was no entanglement, and that it too could be simulated on his supercomputer or even on a 1997 Power Macintosh. Is it NMR quantum computing all over again? Not quite. The NIST researchers plan, in future work, to move into a regime where the 300 qubits are all entangled and where Muramatsu's classical supercomputers will be of no avail. Such an option was never available to the old NMR experimenters. And unlike in NMR, such a highly entangled regime is sure to exist. We eagerly await the results of these intrepid NIST experimenters.

The drawback to this particular top–down or 300-qubit NIST design is that it is not good for much more than simulating these types of special magnetic materials. It is not a universal quantum computer, which is a quantum computer capable of running any algorithm any quantum computer scientist might want to run. Particularly, this NIST gadget, once they get the entanglement up and running, still cannot run Shor's factoring algorithm. There are no scalable arrays of controllable CAT, RAT, and ENT gates. Even with 300 qubits, it still cannot crack a 512-bit public crypto key. But perhaps it can be used to design a yottabyte magnetic hard drive. I repeat that in 50 years, nobody will much care about Shor's algorithm (except as a historical curiosity). In 2062, public-key encryption, if it is used at all, will certainly not use the (unproved) hardness of factoring as its engine but rather some other hard mathematical problem uncrackable even on a quantum computer. But I bet in 2062, the yottabyte hard drive will in great vogue.

When I first cooked up the title of this book, I wondered what I meant for *Schrödinger's Killer App* to be (other than catchy). I suggested in the early chapters that it might be Shor's factoring algorithm. But in this end chapter, I will reveal a dirty little secret. Among ourselves, out of the earshot of the funding agencies, particularly those of the intelligence agencies—shhh!—we quantum physicists speak of the true killer app—Feynman's dream—the universal quantum simulator. We tell each other that the greatest contribution of quantum computers to science and technology, particularly quantum technology, will be to the field of physics simulations of massive quantum entangled systems. It is from that field that new quantum technologies will rise apace. Quantum simulators will be particularly important to the field of quantum materials science, such as these entangled magnetic systems, where new breakthroughs in simulations can lead to new technologies such as hard drives and new computer chips. We joke that the NSA, if it wants to only use the quantum computer to hack the Internet, will buy only one—and then work hard to make sure nobody else's national security agency has one.[50] That would hardly correspond to the saliva-inducing business plan sought by a venture capitalist. But to develop a quantum simulator that can simulate entirely new states of matter for the development of an entire ensemble of new, moneymaking, solid-state quantum technologies? I'd buy *that* for a dollar.

It was Shor's factoring algorithm and the promise of hacking the Internet that kicked off the world's race to build the first quantum computer. But the real application will be to quantum simulations of quantum materials for new quantum technologies. In this prediction, the history of quantum computing repeats itself. Recall that the ENIAC was designed and built during World War II in order to compute US Army artillery firing tables. However, it never carried out that task for which it was built—World War II was over in 1946—and instead the ENIAC was co-opted for simulating hydrogen bomb explosions. (World War III had not begun.) The ENIAC cost $6 million (in 2010 dollars adjusted for inflation) on par with the $10 million D-Wave is charging for the D-Wave One computer. As of today (July 2, 2012), they have sold precisely two: one to Lockheed–Martin and the other to Google. D-Wave claims that it is a special-purpose machine good for quantum simulations as is the new NIST machine. Like the history of the ENIAC, the original motivation for building a quantum computer, factoring, may already be eclipsed by the next new best quantum application in town, physics simulations. Are quantum simulations the real killer app? If so, then Feynman would be proud; it was for physics simulations that he predicted the quantum computer would be good in the first place.

When will the quantum computer be built? The answer depends on your definitions of "quantum computer" and "built." If you think the D-Wave One is a quantum computer, then it is ready now. If you want to crack a 512-bit public-key encryption, then wait … 10 years. Scientific revolutions do not occur overnight but typically over many years or even a generation. We are in the middle of one right now, the second quantum revolution, but scientists typically do not see the revolution that is going on all about them and only note that it happened after the fact. In this context, I quote the noted philosopher of science, Thomas Kuhn:

> I suggest that there are excellent reasons why revolutions have proved to be so nearly invisible. Both scientists and laymen take much of their image of creative scientific activity from an authoritative source that systematically disguises—partly for important functional reasons—the existence and significance of scientific revolutions. Only when the nature of that authority is recognized and analyzed can one hope to make historical example fully effective. Furthermore … the analysis now required will begin to indicate one of the aspects of scientific work that most clearly distinguishes it from every other creative pursuit except perhaps theology.[51]

Scientific revolutions are invisible to the very revolutionaries themselves. One hundred years from now, scientists will point back to the 30-year period between 1984 and 2014 and call it the beginning of the second quantum revolution. Today, scientists are sitting in the middle of it but do not realize it. Asking when the quantum computer will be built is a bit like asking when quantum mechanics was invented. Unlike, say, the discovery of x-rays, the discovery of quantum mechanics took place over 30 years. (In comparison, German physicist Wilhelm Röntgen's discovery of x-rays took place over *3 weeks*, but we scientists are seldom so lucky.)

Was quantum mechanics invented in 1905 when the proper German, Planck, discovered the correct theory of holey radiant ovens or when the pacifist Swiss, Einstein, discovered the photoelectric effect? Was it invented in 1913 when the dour Dane, Bohr, discovered his model for the hydrogen atom? Was it instead invented in June of 1925 when the wheezing German, Heisenberg, escaping from the toxic clouds of pollen that fueled his horrible hay fever, fled to the desolate island of Helgoland in the German North Sea and there invented matrix quantum mechanics in a fortnight? Was it invented in 1925 over the Christmas holiday when the lusty Austrian, Schrödinger, invented wave quantum mechanics? Or was it finally invented at last in 1927 when the frail Englishman, Dirac, unified Heisenberg and Schrödinger's theories with his own transformation theory of quantum mechanics? Who can say? In 1900, we can surely state that quantum mechanics had not yet been invented, but by 1930, we can surely say that it had. There is no precise date. But if we can't even answer the question "When was quantum mechanics invented?" by studying well-documented historical records, how on earth are we to answer such a question about the future, "When will the quantum computer be built?"

A closer analogy to the past would be to ask, "When was the *classical* computer built?" There again it depends on what you mean by "classical," "computer," and "built." The Babbage difference engine was designed as a digital, universal, and programmable computer in the mid-1880s, and although Babbage built a nonworking prototype of part of it, a working version of the whole thing was not built until the 1990s. Had the difference engine been built in 1850, it would have been in some ways superior to the 1942 Atanasoff–Berry Computer (ABC), which was not universal, not programmable, and had an unreliable punched paper card memory. The ABC was the direct precursor to the ENIAC but got much less press coverage. (Atanasoff did however wrest the patent rights for the first digital computer back for himself from the inventors of the ENIAC.)

The great, analog, mechanical, tide-predicting machines were first built in the 1870s and perfected into massive machines that helped "turn the tide" of World War II. The last tide-predicting machines were built in East Germany in the mid-1950s, when the advent of digital machines like the ENIAC made them suddenly obsolete. The tide-predicting machines were certainly classical computers. But they were not universal classical computers and they were not digital computers; they could not compute anything a classical computer scientist might wish to compute. They were special-purpose devices designed to do just one thing and do it well—predict the tides. Like the D-Wave One computer and the NIST quantum simulator, the tide-predicting machines were special purpose but far from useless.

But perhaps you insist on "universal" as part of your definition of computer. The 1946 ENIAC was universal but had no computer memory. By 1952, the Hungarian physicist John Von Neumann and his team at Princeton had demonstrated the Institute for Advanced Study digital universal computer that did have a memory—5.1 kilobytes worth. Confusing matters even more, the Colossus Mark I was built in 1943 and the Colossus Mark II digital computer became operational in Britain in June of 1944, just in time to start breaking Nazi U-boat secret messages for the D-Day invasion of Normandy. The Colossus, like the ENIAC, was also universal and also had no memory, but unlike the ENIAC, its very existence remained classified until the 1970s, right about the time when the patent dispute between the inventors of the ENIAC and the ABC was finally resolved.

When was the *classical* computer built? Even with 20–20 hindsight, we cannot answer that question definitively. Most American computer scientists say it was the ENIAC in 1946, but more and more say it was the less publicized underdog ABC in 1942, but now the Brits claim it was the Colossus in 1943. There are still a few holdouts (mostly organ grinders) for Babbage in 1871. And what about the banks of human computers (mostly women) who in Paris calculated astronomical tables for Le Verrier in the 1800s and at Los Alamos computed A-bomb detonation dynamics for Feynman? These rows of women with calculators were in fact called "computers." The first phase of the classical computer revolution took place over a period of nearly 100 years. In 1850, we can say that the classical computer had certainly not been built, but by 1950, we can safely say that it had been built. There is not one date when "the classical computer" was built. With precise questions about our own recent technological past so muddled, how can we even hope to make precise predictions about our future? We can't. When will the quantum computer be built? There is no particular date when this will happen anymore than there was a particular date quantum mechanics was invented or the classical computer was built. In 100 years, the historians of science will only be able to say that in the year 2000, the quantum computer had certainly not been built, but by the year 2100, it certainly had. With this margin of error of ±50 years, the "50–50" prediction of Phillips seems right on target.

It is a feature of scientific revolutions, and the second quantum revolution is no exception, that they are generated or embraced by either young scientists new to the field or older scientists who have changed fields. The old scientists who have no desire to change fields typically ignore the revolution or fight rabidly against it. Think of Einstein battling quantum theory for 50 years, a theory that turned out to be correct, while searching in vain for a unified theory of gravity and electromagnetism that turned out not to exist (not at least as a classical theory). These gray-haired ones have spent most of their lives building up toolboxes filled with theoretical and experimental apparatuses to do what Kuhn calls "normal science," which is to investigate all the nooks and crannies of the currently accepted paradigm or dogma. When the revolution starts, the paradigm shifts, and to stay abreast of the insurgency, your karma must run over your dogma. These old tools either become useless or have to be used in new and unexpected ways. Scientists nearing retirement typically do not like to change gears at that stage of their careers and instead take solace in beating the dying horse.

Beginning in the 1980s, a few physicists and a few classical information scientists began to realize that information science, in the end, was based on physics. You cannot ask questions like Turing did, about what is or is not computable, independent of the hardware you plan to compute with. Even the Turing machine, although a thought experiment, assumed a particular classical type of hardware: readers, writers, and erasers. If your hardware is classical mechanical, you get one answer, and if your hardware is quantum mechanical, you get a different one. By the 1990s, most of the physicists and a few of the information scientists realized that all the proofs that classical computer scientists had cooked up over the past 50 years were either flat wrong or at least had to be reinvestigated. Before 1990, all these proofs assumed that the computer was classical. After 1990, we knew that was not correct and we had to assume that the computer was quantum mechanical. What was the reaction of computer science departments across the globe? Dead silence. Most people working on quantum information science today are either in physics or in engineering departments. (Seth Lloyd at Massachusetts Institute of Technology is in the mechanical engineering department, making him truly a quantum mechanic.) I can only think of a handful of quantum information scientists who are in computer science. Anecdotal thinkology you say? Let's do an experiment!

We will compare two high-impact journals, *Physical Review Letters* (impact factor around 7) and *Institute of Electrical and Electronics Engineers* (*IEEE*): *Transactions on Information Theory* (impact factor around 4).[52] Both claim to encourage papers on quantum information. *Physical Review Letters* is where physicists send their best stuff. The *IEEE* is where traditional information theorists send their best stuff. Using the *Science Citation Index*, I will search for articles that have the exact term "quantum information" in their title. In the *Transactions on Information Theory*, I get precisely six hits; the first, a paper published in 1983, and the last, a paper published in 2010. In *Physical Review Letters*, I get 44 hits, with the first published in 1991 and the last published in 2012. That's 50 papers between them with 12% published in the *IEEE* journal and 88% published in the physics journal. (The fact that the *Transactions on Information Theory* has a lower impact factor than *Physical Review Letters* would typically mean you would expect to see more papers published there and not less.)

You may try many variations on this theme, but in the end, you will conclude that papers on quantum information theory are not published in journals that are the traditional venues of classical information theory. To be sure, journals of quantum information theory have sprung up, but few of the old classical information theoretical guards are publishing there. Classical information theorists did not run out and embrace quantum information theory but rather the physicists did. The classical information theorists, with a few notable exceptions, continue to plod along as if quantum information theory does not exist. This is typical when you are in the middle of a scientific or any revolution. The old guard has too much investment in the old system and either ignores the revolution or tries to quash it. Very few information scientists who have spent their careers steeped in the classical arts of information have much incentive to go out and relearn the basics, starting with quantum theory. As Kuhn points out, the old guard eventually retires or dies off to be replaced with the new guard. There were many physicists who never accepted Einstein's theory of relativity and many more (including Einstein) who never accepted quantum theory. They are mostly all gone now. In the year 2100, there will be very few purely classical information theorists left and the term "information theory" will apply only to quantum information theory with classical information theory relegated to an approximation of the quantum theory that remains mostly of historical significance. That's how the paradigm shifts, folks. To quote Kuhn quoting Planck, "And Max Planck, surveying his own career in his Scientific Autobiography, sadly remarked that 'a new scientific truth does not triumph by convincing its opponents and making them see the light, but rather because its opponents eventually die, and a new generation grows up that is familiar with it.'"[53]

I attended the conference on *Algorithms in Quantum Information Processing*, held January 18–22, 1999, on a cold and rainy week in Chicago, Illinois, where I had the pleasure to hear a lecture by American computational complexity theorist Lance Fortnow. Computational complexity theory is the study of algorithms and their relative "hardness" or the scaling of their use of resources such as time or space. We know multiplication is easy and (we think) factoring is hard to do on a classical computer, and it is upon this rock that we built our public-key encryption scheme. But then along comes the quantum information science revolution and we now know that factoring is easy if you switch to a quantum computer. Complexity theorists classify different algorithms as easy or hard or super-duper hard. These classifications all had to be redone when quantum computation came on the scene, as in the case of factoring; sometimes, what was classically computationally hard was now quantum computationally easy.[54] These classifications are typically depicted in PowerPoint lectures as Venn diagrams, colored oval blobs depicting classes of problems that are hard or easy. Typically, the easy problems are depicted in a small oval and the hard problem is a big oval that either overlaps the small oval or engulfs it completely. I seldom understand much from these lectures. (I remember at one such lecture the speaker drew a small pink oval completely surrounded by a large oval blue oval but then confessed that he was not sure if that was really the case and it was possible the blue oval was small and completely enclosed instead by the pink one.)

Fortnow, trained as a classical information scientist, was giving such a talk, replete with colored, oval, overlapping Venn diagrams comparing quantum algorithms and classical algorithms, and he confessed that he did not really understand what quantum entanglement was, but that he also did not understand how his car worked, but nevertheless he could still drive it. I raised my hand and pointed out that he at least knew enough to put gas into it once in a while. (This got a few laughs.) My story illustrates how the classical information scientists really keep the quantum in quantum information theory at arm's length. Younger information scientists, such as American computer scientist Scott Aaronson or Israeli computer scientist Dorit Aharonov, handle quantum complexity theory and classical complexity with equal ease. While I would call Aaronson a quantum computer complexity theorist, I rather doubt he calls himself this, and in some years, this distinction will be lost and he will just be a computer complexity theorist. The idea of working only on problems in the old paradigm of classical computer science separate from problems in the new paradigm of quantum computer science will then seem odd. It is clear that Fortnow, trained in the old classical paradigm, was struggling (at least in 1999) with the new quantum paradigm and such whacky notions as quantum entanglement. Aaronson and Aharonov have no such struggles and they embrace entanglement with ease. In 100 years or likely much sooner, it will be just complexity theorists with no distinction between quantum and classical. Then, the quantum information revolution will be over and the period of what Kuhn calls "normal" science will have commenced.

Again, from Kuhn, we understand that it is critical for a new scientific paradigm to do two distinct things to be accepted. It must first explain and predict new things that are unexplainable under the old paradigm and it must, second, in some approximation, include the old paradigm. Unlike in art where the works of the expressionists exclude the works of the realists, in science, comfort with (and acceptance of) the new paradigm flows from the fact that it, in addition to producing new results, also reproduces the results of the old paradigm. A famous example is that of Einstein's new paradigm, the theory of general relativity and gravity, which in the approximation of weak gravitational fields can be shown to contain the older Newton equations for gravity. Similarly, the equations of quantum mechanics can be shown to reproduce the equations of Newtonian mechanics when averages are taken and Planck's constant is taken to be zero or quantum numbers are taken to be large. Such limiting cases give comfort to the old guard of the old paradigm, that their life's work was not for naught, and also gives credence to the reliability of the new paradigm.

Quantum information science is no exception to such approximate limiting cases. For example, a quantum computer can always efficiently simulate a classical one. More precisely, you can always take your quantum computer and run it in a classical computing mode by deliberately turning off quantum unreality, uncertainty, and nonlocality. Entanglement vanishes and cats collapse to dead or alive and the qubits become just plain bits. Thence, classical information science (small pink oval) in this sense is entirely contained in quantum information science (large blue oval), but (provably in this case) not the reverse. There are problems that are easy on a quantum computer and hard on a classical computer and this is provably so. Thus, a problem that is easy to solve on a classical machine is always just as easy to solve on a quantum device, but the reverse is certainly not true. There are some computer algorithms for which quantum computers give an exponential speedup that provably cannot be had on the classical computer. This observation satisfies the second of Kuhn's requirements for the acceptability of a new paradigm, that the new paradigm in some sense contains the old as an approximation. Kuhn's first requirement for the acceptance of the new paradigm is that there must exist problems only solvable in the new and not the old. That is, that quantum information science must solve problems that cannot be solved within the classical information science paradigm. An example of such a problem is one-time-pad cryptography. Unbreakable cryptography based on one-time pads cannot be had from classical theory owing to the lack of a classical no-cloning theorem. Classically, Eve can always copy the pad. This is not true in quantum information science, with its quantum uncertainty, unreality, and no-cloning rules, and where pads cannot be copied. Quantum information science brings to the table solutions to problems that are not solvable at all within the old classical informatics paradigm.

The second quantum revolution has numerous fronts, with the quantum information science revolution being one of these, and the most progressed at that. There are other fronts. The greatest contribution of quantum information science to physics may be its own version of a Quantum Rosetta Stone, that of a unifying common language. A common language that spreads from subfield to subfield is a hallmark of a rapidly spreading new paradigm. Before 1994, researchers in disparate fields of physics—quantum optics, electronics, atomic and molecular physics, nuclear physics, and superconductor physics—all worked with rudimentary quantum cat states in their own technical jargon or *lingua franca*. The atomic, optical, and molecular folks called these "two-level atoms." The nuclear and electronic chaps called them "two-level spin systems." The superconducting practitioners called them "two-level cat states." The protocols for creating and manipulating such two-level systems were remarkably similar across fields, but there was not much interaction across fields. Quantum opticians went to their own workshops (or their own parallel sessions in large conferences) and published in their own journals, as did the electronic, atomic and molecular, nuclear, and superconducting physicists. Before 1994, I doubt I had ever attended a lecture on two-level systems in superconductors, perhaps because before 1994, the superconductor physicists had not succeeded in making such a system, but more likely that I would avoid sessions and workshops on superconductivity like a dead cat.

Even the labeling could not be agreed upon. In the summer of 1989, I was having a heated discussion at the Max Planck Institute of Quantum Optics with American physicists (and quantum opticians) Joseph Eberly and Marlan Scully, along with my PhD advisor Asim Barut, over Barut's alternative theory of quantum electrodynamics as applied to a two-level atom. We were discussing Barut's theory as applied to the hydrogen atom, with an infinite number of levels, when Eberly declared, "I won't be able to understand anything unless you apply this to a two-level atom." Each distinguished professor would take turns going to the whiteboard and draw two, horizontal, closely spaced lines—one line for each of the two levels—but the three could not agree on how to label these two lines. Eberly would label the lower level "a" and the upper level "b" (the so-called northeast labeling). Scully would stalk up and erase Eberly's letters and relabel the lower level "b" and the upper level "a" (the southwest labeling). Barut would then go up and calmly erase the letters and relabel the lower level "1" and the upper level "2" (the mid-west labeling). And these three guys were in the same field of quantum optics![55] Nowadays, all we physicists, if we know what is good for us, call this "two-level atom" by its quantum computational term "qubit" and we now all label the bottom level $|1\rangle$ and the top level $|0\rangle$. (Although there are still a few holdouts for labeling the bottom level $|0\rangle$ and the top level $|1\rangle$.)

I have made this point earlier in the book but it bears repeating. The most important contribution the field of quantum information science has made to that of physics is this new language of qubits and all that. The second most important contribution has been the focus of funding in quantum information science regardless of what subfield of physics was working in it. The third most important contribution was a physical unification of these fields in such a way that disparate researchers publish, and more importantly read, the same papers in the same journals as well as attend the same meetings. The iconic example of the fruits of such inter-subfield cross-pollination is the tremendous recent theoretical and experimental advances in superconducting physics, particularly the coupling of superconducting qubits to each other via microwave photons. This particular field, an area of my own research since my PhD, is called cavity quantum electrodynamics. The "cavity" in this case is composed of a pair of highly reflective mirrors that bounce single or few photons back and forth millions or billions of times between the mirrors before they become absorbed or otherwise lost. Quantum electrodynamics is the theory of matter, particularly atoms, interacting with photons. The field of cavity quantum electrodynamics typically involves a single atom sitting in the cavity (between the mirrors) so that the atom can interact with a single photon millions or billions of times. The photon might excite the atom from the bottom level to the top level $|0\rangle$, or if the atom is already excited, the interaction with the cavity can alter the rate that the atom decays from the top to the bottom emitting a photon. The emission rate can be enhanced or suppressed dependent on the wavelength of the photon and the spacing between the mirrors. The best enhancement of the decay rate comes when the mirror-to-mirror spacing is about half the wavelength of the emitted photon. In such a configuration, the single photon is absorbed and reemitted by the single atom millions or billions of times.

There are at least two atomic quantum-computing schemes that exploit cavity quantum electrodynamics. In the first scheme, two photons in the cavity are the qubits and the atom acts like an intermediary that allows the state of one photon to flip or not flip depending on the state of the second photon. In this way, the photon is the qubit and this flip or not flip is the fundamental two-photon ENT gate. In a second protocol, two atoms are the qubits, with the excited atom the $|0\rangle$ state and the unexcited atom the $|1\rangle$ state. Here, we put two atoms in the cavity and then one photon is the intermediary and we can flip the state of one atom dependent on the state of the second atom via the interaction with the photon and we have a two-atom ENT gate. These two approaches to quantum computing have been pursued by a number of groups over the past 15 years, and in 2012 French physicist Serge Haroche was awarded the Nobel Prize for his work in this field.[56]

The group of Texan physicist H. Jeff Kimble at Caltech demonstrated the first photon–photon ENT gate in 1995.[57] The atom–atom ENT gate was much harder to come by. American physicist Michael Chapman has championed the atom–atom approach but the issue is that in all these experiments, it is darn hard to get two atoms into the cavity at the same time to interact with the photon. In Kimble's experiment, he would drop the atoms through the cavity one at a time and then when one entered the cavity at just the right position they could then implement the photon–photon ENT gate in the brief time the atom passed through the cavity. Getting two atoms to arrive in the cavity at the right place and the right time by dropping them has been technically impossible. Chapman has built an optical conveyor belt that routes two chains of atoms in and out of the cavity two at a time to hold them there long enough to make the atom–atom ENT gate. Chapman's group has got the conveyor belt working but to this date, after 10 years of trying, has not demonstrated the atom–atom ENT gate. Many technical hurdles still need to be overcome.[58]

For the past 15 years, the US government holds yearly summer review meetings of all the groups working on quantum computers. The great thing about this meeting has been that the photon–photon folks, the atom–atom folks, the superconductor folks, the ion trap folks, the quantum dot folks, the NMR folks, and even the electrons floating on liquid helium folks, all were at the same meeting discussing their latest results. Before 1994, the people from these subfields rarely attended each other's talks but indeed after 1994 (and the publication of Shor's algorithm), they were *paid* to listen to each other. In addition to these yearly review meetings, workshops and conferences and sessions at conferences sprang up with the focus on quantum computing (and not the specific hardware to build one) and again the scientists pursuing the superconducting approach were forced to listen to those pursuing the photon–photon or atom–atom approach. Even better, before 1994, they could not even understand much of each other's talks, because each subfield spoke in its own impenetrable technical jargon.

After 1994, we all had learned the secret language of qubits and quantum gates and circuits and suddenly all the talks became accessible to us all. Now we had overlap in space, time, and—most importantly—language. The cross-fertilization began. Then, the field of cavity quantum electrodynamics was wrested away from the atom and laser jocks and very successfully deployed by the superconducting mafia. This must have caused some chagrin on the part of the laser jocks; remember that in 1996 H. Jeff Kimble told me at coffee that the superconducting guys would never get a working CAT gate in 10 years. Now, not only do they have a CAT gate and an ENT gate, they are doing this all with cavity quantum electrodynamics, Kimble's own area of expertise, now translated into the superconducting domain. I conjecture that this transference occurred precisely because the superconducting folks had to listen to Kimble's talks for 10 years and then suddenly realized that they could do cavity quantum electrodynamics as well and even a bit better.

American physicist Robert Schoelkopf and his group at Yale University were the first to demonstrate strong coupling of superconducting qubit "atom" with a microwave photon in a type of engineered superconducting cavity.[59] The primary advantage of switching from the atomic system to the superconducting system is that, unlike the atoms that are flying all about and hard to keep still, the superconducting qubits are fabricated on a metal chip in place exactly where you want them. You don't drop them through the cavity and pray, instead you build them in the cavity and play. A second supercool feature is that not only is the superconducting qubit an artificial atom—it is a very big artificial atom. That is, a real atom is only about a few tens of nanometers in diameter but the superconducting "atom" is around a few tens of microns in diameter—a thousand times bigger. The bigger the atom, the more likely it will interact with the photon and so the atom–photon interaction is much stronger in the superconducting cavity quantum electrodynamic system than it is in the atom system. Two-atom superconducting ENT gates are much easier to make for these two reasons and have already been demonstrated whereas the two-atom atomic ENT gates are still being struggled with.

Primarily because of these ideas lifted from the atomic cavity quantum electrodynamic field, superconducting cavity quantum electrodynamic systems are now considered a frontrunner for a scalable quantum computing architecture. But quantum computing is not the only game in town, quantum metrology and sensing with such systems is in sight. In 2011, the group of American physicist John Martinis used a superconducting cavity quantum electrodynamic system to generate my own favorite NOON states of the microwave photons in the cavity through the strong interactions between the photons and the superconducting qubits in the cavity. As we recall, NOON states can be used to sense and image things better than is possible classically.[60] I even dabbled in this field myself in 2006 with my former postdoc Alexandre Guillaume from the Jet Propulsion Laboratory. I knew a lot about cavity quantum electrodynamics but almost nothing about superconducting qubit theory. For Guillaume, it was the reverse. Our collaboration consisted of Guillaume writing down one set of equations after another of what could be made in the laboratory with superconducting qubits interacting with microwave photons in a cavity. I would look at each equation in turn and say, "Nope. Don't like that one." Finally, about halfway down the list, we hit one I liked, and I yelled, "That's exactly the Jaynes–Cummings Hamiltonian! I know how to solve that equation!" The result was a scheme to use the photons to make a superconducting qubit–qubit ENT gate and make something like a NOON state of the superconductors in order to sense magnetic fields better than possible classically. Superconductors are some of the best magnetic field sensors around, but if you entangle them, you can even make them better.[61]

The point of this section's object lesson is that quantum technology and the second quantum revolution are progressing much more rapidly because of the language and funding provided in the race to build the world's first quantum computer. This stimulus of the second revolution is likely to lead to an immense array of new technologies independently of the construction of the quantum computer itself. Another example of such technological advance in an unexpected area is that of classical optical computing, which I have mentioned before, but I will recap again in this context. In the 1970s, there appeared a number of publications claiming that Moore's law for the exponential growth of classical computer chips was coming to an end. The argument was that, precisely because techniques such as optical lithography were being improved, soon it would be very possible to put very many transistors and wires very close to each other on the computer chip. Making transistors and wires smaller and smaller, so that more could be packed onto computer chips, would make the chips more powerful in processing power and memory.

However, the argument went, there was a physical barrier that emerged precisely because the widgets were getting too close to each other. When two wires carrying electric current are close to each other, there is a type of cross talk between the wires called mutual inductance and mutual capacitance. For mutual inductance, the current in one wire produces a magnetic field that will alter the current in a second wire. For mutual capacitance, the electric charge on one transistor produces an electric field that in turn will affect the charge on a second transistor. Naively, the arguments went, this cross talk gets worse the closer the wires and transistors are together on the chip—as the wires get infinitesimally close, the electronic cross talk gets infinitely bad, or so the argument went. All these publications predicted that we were reaching the point where the transistors and wires were so close together that this electromagnetic cross talk between them would eventually completely ruin their performance and they would not function at all as intended. These research papers predicted that there was a minimum distance that could be tolerated between wires and transistors; a distance of tens to hundreds of microns that was just being reached in chip fabrication technology. Once this limit was hit, the transistors and wires could be placed no closer and the yearly increase in computer processing power predicted by Moore's law would come to an ignoble end. (When my colleagues tell me their own pet scheme for a quantum computer is scalable, because they can put many qubits close together on a chip, I point out to them that, in the 1970s, it was for that very reason engineers predicted that *classical* computers were *not* scalable.)

These haughty "no-go" predictions for the end of large classical integrated electronic circuits on a chip panicked a great deal of industrial concerns and governmental agencies that relied on the continuance of the classical computer revolution and Moore's law for commercial and military gains in computational power. Then, in pranced the laser jocks whirling about their bright red beams of light with a bold new proposal—make all the wires and transistors using light beams instead of electrical currents. Called "optical classical computing," the idea was that, unlike electrical signals, optical signals had no cross talk. Light beams routed about hither and yon on the optical equivalent of electrical wires (called optical waveguides) did not couple to each other at all. Evil electronic cross talk would be eliminated and then Moore's law would continue and the military industrial complex would rejoice. The problem was that while it was reasonably easy to make optical wires, it was much harder to make optical transistors. An electronic transistor is a device in which the flow of one electrical current is controlled by a second electrical current. This control feature of the current in the transistor is used to make the elementary set of universal classical computing logic gates, such as the NOR or the NAND gate, with which any classical computation may be carried out. The feature of optical computers, the lack of optical beam coupling and the lack of cross talk, was also its flaw. If you could not couple the light beams, you could not easily make an all-optical transistor. For that to happen, one optical beam had to control a second, and for that, they had to talk.

The US government, in particular in the late 1970s and the early 1980s, dumped millions of dollars into the research and development of classical optical computers, which were to replace the electronic computers and keep Moore's dying law alive. Much of the focus of this research was to construct integrated circuits of light beams with all the widgets on them that you have at your disposal in integrated circuits of electrical currents. Such widgets are optical switches, transistors, diodes, and other things commonly found in an electronic circuit. But at the peak of the government funding for classical optical computing, American computer scientists Carver Mead and Lynn Conway launched a second electronic computing revolution with the discovery of design rules for the very-large-scale integration (VLSI) of electronic circuits.[62] This VLSI revolution came about when Mead and Conway and others showed that, if you designed your integrated electronic circuit carefully and obeyed certain rules (called design rules), then the electronic cross talk that so worried the chip builders of yore could be made to *vanish* (as opposed to becoming overwhelmingly large) as the spacing between the wires got smaller and smaller. That is, if you obeyed the Mead–Conway design rules, things got *much better* as the electronic circuits got smaller and closer together, not worse. There was then no need for the optical classical computer—and thus Moore's law for the electronic circuit continues apace to this very day.[63]

However, what was good news for the electronic circuit designers was bad news for the optical circuit designers. The funding for optical classical computing collapsed utterly and to even say that you were working on it at all became the kiss of death in the research community. A big waste of taxpayer money it was? No it was not. The application was just wrong. All the optical gizmos such as switches and diodes and transistors, developed with millions of your tax dollars, never did find their place in a large-scale integrated optical circuit for optical classical computing, but all those widgets were invented just in time to find their way in to small-scale integrated optical circuits that ended up in the optical routers and switches that today run the entire Internet. The money was not wasted and I would argue that the successful development of the Internet was a far more important outcome of the government funding than the failed development of a scalable optical classical computer. Just like the ENIAC, never to produce the ballistic gun trajectory tables it was built to calculate, the optical integrated circuits found their home not in computing but rather in linking up the fiber optical communications grid that forms the backbone of the

Internet, and they arrived just when the Internet needed them the most. The US government, particularly the military and the intelligence agencies, seldom brook the funding of basic research for its own sake. But often the particular application for which they fund a project is not the application that turns out to be the true killer app.

Even now, my colleagues in the field of quantum computing are beginning to whine that funding for quantum computing in the United States in particular is drying up and that the field is at risk of going the way of optical classical computing—the kiss of death blows coldly on the napes of their necks. I hear rumors of similar mood shifts in Europe. But the vast government funding for quantum computing, even though it did not produce a Shor factoring engine capable of hacking the Internet, the original killer app, produced something much more valuable. It produced a generation of researchers versed in the arcana of quantum information science and the dark arts of quantum technologies. I predict that the immediate application of optical quantum computing, following the trajectory of optical classical computing, will be to the quantum Internet. Communications is the realm of optics. If you want to send quantum information over intercontinental distances, then you must learn the lost language of photons, and then you must speak with me.

THE CHURCH OF THE LARGER HILBERT SPACE

As in most scientific endeavors, there are experimental things we can test in the laboratory and theoretical constructs, scaffolding of the theory, that are not really testable within the framework of the theory. For example, in Newton's theories of mechanics and gravity, the motion of objects took place with respect to theoretical constructs known as "absolute space" and "absolute time." Absolute space was something like a giant three-dimensional stage and absolute time was something like a giant master clock that ticked out the same time for all the actors and dancers on this stage, synchronizing their movements. The orbits of planets about the sun and their moons about themselves took place on this stage like an elaborately choreographed waltz with many dancers dancing in time to the unerring beat of a giant invisible clock, something like the robotic animated figurines that come out of the Munich town hall clock every evening exactly at 7:00 p.m. to put on an exquisitely timed show for the hordes of gaping tourists. Many natural philosophers of Newton's time, the late 1600s, objected to these axiomatic concepts of absolute space and time, particularly because they were axioms that seemed untestable in the laboratory and, generally, anticipating Einstein's theory of relativity by more than 200 years, because notions of relative times on different clocks and relative distances on different stages had for them more philosophical appeal. But Newton's theory, despite these suspect and originally contentious philosophical constructs, was very, very, very accurate in predicting everything from the fall of an apple to the orbital trajectory of the yet to be discovered planet Neptune. It was the accuracy of Newton's theory that allowed Le Verrier in the 1800s to find Neptune, the first new planet discovered since antiquity. After that, no one dared doubt Newtonian mechanics, suspicious philosophical constructs notwithstanding. With such great predictive and calculational successes, the scientists forgot their original objections to the notions of absolute space and absolute time. As we say here in Louisiana, "If it ain't broke don't fix it."

If Le Verrier's great success in analyzing the motions of the known planets and using perturbations in their expected orbits to find Neptune was the peak of Newton's success, the inability of Le Verrier to find the planet Vulcan using the same approach applied to the weirdness of the orbit of Mercury heralded the downfall of Newton's absolute space and time. Mercury's mercurial orbit could not be explained by Newton's theory and attempts to explain it by introducing a yet undiscovered planet, a scheme so joyously successful with Neptune, was utterly ruinous when it came to predicting the existence of Vulcan. That nonexistent planet, carved prematurely with jubilation into Le Verrier's statue at the Royal Paris Academy, had at last to be plastered over in ignominy. Planet Vulcan did not exist. It was Einstein's theory of gravity, the general theory of relativity, which explicitly rejected Newton's notions of absolute space and absolute time and replaced them with relative space and relative time, which finally explained Mercury's errant orbit. After 200 years of embracing the absolute over the relative, the scientists had a test, and experiment, to objectively measure in the night sky Mercury's motion. Newton was wrong and Einstein was right and that was the end of absolute space and time.

So too in quantum theory there are such axiomatic beliefs that cannot be readily tested within the confines of the theory. Until a competing theory comes along that relegates the axioms to testable hypotheses, quantum theory is in its Le Verrier era, making astoundingly accurate predictions that agree with the data to fantastic precision and with no experiments contradicting quantum theory at all. To be sure, we expect quantum theory to fail at very small distances around 10^{-34} meters where the yet undiscovered quantum theory of gravity is expected to prevail. This is the domain of such exotic theories as superstring theory or loop quantum gravity. Despite what you may have heard watching the television show *Nova*, superstring theory is not really a theory at all but rather a collection of ideas that theorists hope will someday lead to the long dreamed theory of everything (TOE)—one set of equations that will describe all the subatomic particles found and to be found in nature and the rules by which they interact. However, dancing about with a theory of everything glittering before you in tantalizing relief makes it much easier to stub your "toe."

The scientifically inclined lay public, the intended audience of this very book, may think from all the popular press that string theory is (as American physicist Brian Greene calls it) "the only game in town" and we are agonizingly close to the end of physics where we will have one set of equations from which we can derive the properties of everything in the universe—all the particles and their interactions. At that time, basic research in physics will come to an end and all the rest to come afterward will be clean up and engineering. That is what the superstring theorists would have us believe, but I'm afraid nothing could be further from the truth. When string theory emerged 30 years ago, its first initial success was that it appeared to be a quantum theory that unified the theory of all known subatomic particles (and the forces that guide them) and gravity. After 30 years of research, the field has produced vast amounts of beautiful (and bewildering) mathematics but precious few testable physical predictions. As I said, it is not even a theory but a hope for a theory. In fact, it is a vast collection of theories, none of which can conclusively be ruled in or ruled out in the laboratory, because most of the few predictions they make are out of range of any laboratory experiment and can only be tested by listening to the whispers of the cosmos. Not a single Nobel Prize in physics has been awarded to a single string theorist for his or her work in string theory, because no single prediction of the theory has ever been observed. (They do hand out the Fields Medal to string theorists like candy, but the Fields Medal is the equivalent of the Nobel Prize in mathematics, not physics.)[64] Just because a theory is mathematically beautiful does not make it right and the history of physics is littered with the carcasses of theories that were "…too mathematically beautiful to be false."[65] Loop quantum gravity is a more respectable attempt to find just a quantum theory of gravity (without the clutter or all the other particles and forces of nature) but it too suffers from the same difficulty as strings in that the predictions it makes are so far untestable in any laboratory. As you know by now, I am big fan of the scientific method. A physics theory that cannot be tested in the laboratory is only metaphysics or philosophy until such tests can be constructed and carried out. Then, when the tests come in, the theory is due a promotion.

With string theory, I fear that promotion will never come. When invented 30 years back, it seemed to offer slots in its machinery to explain all known particles and forces including gravity. But in the past 30 years, we have discovered evidence of new particles that interact by new forces that we completely do not understand and for which it is unclear and I think unlikely that string theory can manage ever to explain, once we sort it all out. As we have learned in the past 30 years, 95% of our universe is composed of mysterious stuffs that we have not a clue as to what the stuffs are. We now know that every galaxy in the universe is embedded in a spherical blob of something called dark matter that makes up approximately 23% of all the stuff in the universe. All we know for sure about dark matter is that it is dark, you can't see it, and that it only interacts with itself and other common visible matter (like you, me, and a tree) by gravitational attraction. The astronomical evidence supports that this stuff is at this very moment gushing through every cell in your body at super high speeds, as our solar system plows through it in its journey around our Milky Way galaxy's massive black hole of a galactic core, and we don't even feel it at all in its overwhelming violation of our personal inner space. For all we know, dark matter beings with dark matter telescopes are sitting right on top of us peering at their own dark matter planets and stars, which they note are gravitationally perturbed in their motion by some invisible stuff they call "invisible" matter that really is just us and all our planets and stars and galaxies. To these beings, *we're* the dark matter!

To be sure, the string theorists conjecture that dark matter is made up of particles predicted by curiosities in the standard model and hoped to be consistent with string theory, such things as the axion particle (or its super-symmetric partner the "axino") or a pure superstring-predicted but never seen neutralino. Thus, the superstring experimenters construct experiments to look for dark matter particles flowing through their apparatus in the hopes that their passage will appear as a phantom blip on a detector screen.[66] The neutralino is a pure superstring construct and is a candidate for dark matter and is predicted to have a mass around that of the now infamous Higgs boson, found just this year in the Large Hadron Collider, but the "dark" secret is that there is no sign in the collider of the neutralino. The Higgs boson and the axion can be considered predictions of the standard model alone, but the neutralino is a string theory prediction beyond the standard model. No neutralino, no string theory? But there is a Catch-22 here. Particle theorists construct their experimental search for the neutralino on the basis of the properties the neutralino is predicted to have based on the constructs of string theory. What if dark matter is not an axion or any particle in the panoply of particles the theorists predict are made up of infinitesimal vibrating strings? Kuhn comes to our rescue (or to our defeat)—we can only look where the current theory tells us we can look. If dark matter requires a theory radically different from axions or strings to explain its existence, then dark matter is not made up of axions, and then string theorists have us looking in all the wrong places and we'll never find out what it is by following their lead. Often when we found new particles in physics, say the neutrino, there was no theory around to predict it before the discovery. There's more.

In addition to dark matter, which really we don't have a clue what it is, the entire universe is filled with something else, which really we don't have a clue what it is, called "dark energy" that makes up another 72% of all the stuff in the universe. (Well-known and visible things made of ordinary atoms like you, a tree, visible stars, gas, and me make up only about a paltry 5% of the stuff in the universe. Approximately 95% of the universe is made of invisible stuff and we don't know what that stuff is.) Thirty years ago, astronomers and cosmologists predicted that our expanding universe was slowing down in its rate of expansion. Like throwing a baseball up in the air where it eventually slows down and returns to Earth, if you can throw a pitch at Earth's escape velocity of 40,320 kilometers per hour (approximately 25,054 miles per hour), the ball never turns around and falls back down but instead climbs slower and slower against the Earth's gravitational pull, finally coming to rest at infinity. Even if you throw it faster than that, it still slows on its entire trip and ends up with enough energy to meander through space at infinity. But what it never does, no matter how fast you throw it, it is never sped up after leaving your hand.

In our theory of the universe, the big bang 14.6 billion years ago threw the universe up. All the galaxies are racing away from each other at high speeds, but like that baseball, we expect those speeds to slow over time as the mutual gravitational attraction of each galaxy pulls back on every other. We would not, for example, expect the galaxies to suddenly start speeding up! That would be equivalent to the baseball at the peak of its trajectory, just before it fell back to Earth, experiencing an invisible ghostly throwing hand shoving it all the way to Alpha Centauri ever faster and faster never to return and never to slow down ever again. Approximately 72% of our universe is composed of and permeated with precisely that invisible ghostly galaxy throwing hand—all the galaxies are not slowing, they are speeding up and racing away from each other faster and faster. What is doing this? We have no fraking idea. So we call the giant, universal, ghostly, invisible throwing hand of God the "dark energy" and keep an eye on it. Like dark matter, dark energy was not around to be explained when string theory was invented in the 1970s and 1980s. It's new. What is dark energy made of? We don't know. Could string theory explain it? As I have said above, string theory is a collection of ideas and at least five different theories that could be tweaked to explain just about anything, and so it therefore explains just about nothing. As every engineer and statistician knows, given enough adjustable mathematical parameters, you can fit any possible data curve. But then you don't have a theory, you just have a curve that fits your data. You might as just as well draw a curvy line through your data points and claim the line is your theory. (It is not.)

Then there is the "inflaton." The properties of the observed cosmos, particularly its "flatness" (in the language of Einstein's curved space and time), can be best understood if we invoke a hypothetical particle called the "inflaton" that caused the entire universe, moments after the big bang, to inflate at warp speed like a toy balloon affixed to a gasoline station's car tire air pump opened to full throttle. The inflationary period halted just in time to keep the universe from bursting and then ... the inflaton disappeared forever, never to be seen again. What the heck was the inflaton? We don't know. Could string theory explain it? Who knows? My point is that there are now a number of new barely known and little understood forces and particles that string theory was never designed to explain, and now string theory (or string theories) has grown into an unwieldy and parameter-strewn imbroglio of beautiful mathematics—too beautiful to be wrong!—that can simultaneously explain everything and hence explain nothing at all. The hope of string theorists is that all this new stuff being discovered will neatly fit into string theory.[67] My fear is that it will not and we'll have to go back to the drawing board.

The current state of string theory reminds me of the jury-rigged astronomy theory of yore, the Ptolemaic theory of the ancients that had the Earth (instead of the Sun) at the center of the solar system. Ptolemaic theory, because it was wrong to begin with, could only explain all the apparent loopty-loop motions of the planets in the night sky ("epicycles" they called them) by adding ever more and more adjustable parameters to fit the ever more precise astronomical data. Eventually, Ptolemaic theory had so many adjustable parameters the Ptolemaic astronomers could fit anything, and anything they then did fit, ever tweaking the parameters as better and better data on planetary motions came from the observatories. It was the ancient astronomical theory of everything, and in the end, it was also the ancient astronomical theory of nothing. Move the Sun to the center of the solar system (instead of the Earth), use Newtonian mechanics, and you can explain all the motions of the planets with insanely precise accuracy with a single nonadjustable parameter, Newton's universal gravitational constant G, which is a number that is measured in the laboratory! Now *that*'s a theory. Throw the Earth-centered Ptolemaic theory into the rubbish bin of history and heave a sigh of relief. The Ptolemaic theory actually was better than string theory; at least it fit the data pretty well. In the case of string theory, there is as of yet no data to fit. And no such data appear to be forthcoming despite recent hopes to the contrary. If string theory really is "the only game in town," then I would rather not play and I would prefer instead to move to another city where the air is cleaner.

In June and July of 2012, the popular and scientific press had been inundated with the announcement of the discovery of the mythic Higgs boson (the "God particle"). This was the last unfound particle in the standard model of particle physics and the standard model is the last theory that does not have to invoke elements of superstring theory for its predictive power. As we stand now, the standard model predicts the existence of every experimentally known elementary particle, from electrons and quarks to the Higgs boson, and conversely no particle has been discovered that is not within the predictive range of the standard model. The Higgs was the last one on the list of undiscovered particles, and now that it is found, the standard model is complete. The Higgs, like most elementary particles, was found in telltale data signals in a particle accelerator or atom smasher. Hints of the Higgs were found in the data from the Tevatron collider at Fermilab in Batavia, Illinois, but the Tevatron was not powerful enough to produce enough collisions at the high energies needed to conclusively see the Higgs and so their announcement on July 3, 2012, was tepid and tentative. Not enough significant digits in the data to truly claim a discovery.[68] This stole a bit of thunder from the announcement from Geneva that came out one day later on July 4, 2012, where two experimental teams at the more powerful Large Hadron Collider at CERN announced the discovery of a new boson at just the right energy and with just the right properties to be the long sought after Higgs.

The CERN experimenters had enough data to declare that they had discovered the new particle with a confidence of 99.9999%, which means their margin of error is only 0.00001%. That is, there is only a 0.00001% chance they discovered nothing and the data rendered a statistical fluke. But the Tevatron and the two experiments at the Large Hadron Collider all involve independently the same particle with the same properties and the same mass in the same place so it looks at least for now the search for the elusive Higgs is over and all the pieces of the standard model are in place and the puzzle is complete. But storm clouds hang on the horizon for the superstring theorists. These theorists hoped that the Large Hadron Collider would find other particles besides the Higgs, new particles predicted by superstring theory but not by the standard model. String theory made a prediction that at last could be tested in the laboratory. The results are not good. While no official announcement is forthcoming, the CERN collaboration is ruling out the simplest form of superstring theory as quickly as it is ruling in the Higgs boson. There is so far no evidence of any of the raft of new particles predicted by superstring theory.[69] In particular, as I mentioned above, a superstring candidate for particles of dark matter, the neutralino, should have a mass around that of the Higgs. Higgs has been found but data accumulate day by day and no sign of the neutralino. This lack of any evidence at all for superstring theory in the Large Hadron Collider could be the death knell of the field, although, like the epicycles of the Ptolemaic theory, the theorists will continue to tweak the theory of everything to match what the experiments do not find.

The "only game in town" is losing some of its allure. This lack of experimental evidence is what is somewhat furtively referred to as "the nightmare scenario," that is, that the Large Hadron Collider finds the Higgs, the last particle predicted by the standard model, but no evidence of superstring theory or anything else beyond the standard model. As the data from the Large Hadron Collider continue to pour in and be analyzed and no evidence for superstrings is found, the nightmare scenario might more properly be called "the succubus scenario." Some even fear that there is a "valley of death" between the energies reached by the Large Hadron Collider and the vastly larger energies of the big bang itself where nothing at all might be found. The atom smashers have had a good run for 50 years with each new generation of smashers finding new high-energy particles. But the theory well established by experiment, the standard model, predicts that the Higgs boson is the last particle to be found. What if that is the end of the smashers' run?

For over 30 years, the popular press has billed superstring theory as the final theory of everything. Once we found the equations of superstrings, so the story goes, all fundamental particles and forces between them would have been discovered, and then in some sense, physics would come to an end. The rest would be clean up; that is just calculating things from the fundamental equations. Sounds good, but that is not how physics is done. First of all, this is not the first time physicists predicted that we had discovered or were on the verge of discovering all there is to know. That happened in the late 1800s when luminaries declared that with Maxwell's theory of electricity and magnetism well understood and Newton's theory of mechanics well understood and well tested, that was it. Everything we needed to know would be calculated from those two theories and physics research would be an applied field like engineering. We just teach the engineers Maxwell and Newton's theories and then they figure out how to use those to make a suspension bridge or an electric train. The beginning of the end was in reality the beginning of the beginning as quantum theory and Einstein's relativity were discovered around 1900 and turned the entire world of physics on its head. But even if we did have a theory of everything, it would not mean that we could calculate from it everything we need to know.

The late American physicist Philip Morrison was a great champion of the idea that, for complex systems, new physics is emergent in a way that cannot be calculated directly from the underlying equations, even if those equations are known. We have seen this in this book. The basic laws of quantum mechanics should predict all the properties of the element thulium, but because the system is highly entangled, its properties cannot be calculated at all from the basic equations, not at least on a classical computer. Even though thulium has only 69 electrons in real space, it has a vast number of electron states in Hilbert space and that vastness brings our number crunchers to their knees. Vastness is not a property of Hilbert space alone. Using only Newtonian mechanics, we should be able to predict the motion of a single drop of water in a turbulently flowing stream. We cannot. We have the underlying equations but, particularly when turbulence is involved, we cannot solve them even on the fastest classical supercomputer. Having Newton's equations of motion in hand tells us nothing about the path of a water droplet.

Similarly, from the basic equations of quantum theory, we should also be able to construct a theory of superconductivity, the experimental observation that certain metals when cooled close to absolute zero in temperature lose all resistance to an electrical current within them. Quantum theory was completed in the 1920s, but a successful microscopic theory of superconductivity came about only in the 1950s and it was not a simple derivation starting with the elementary equations of quantum theory. Instead, new emergent physics had to be developed that could handle the immense number of electrons in a block of cold metal, and that emergent physics was not derivable from elementary quantum theory. The issue again is that if the number of particles interacting becomes very large or if the interactions between the particles become very strong (both these regimes hold in superconductors), then entanglement aside you cannot derive the properties of the large system on the basis of the fundamental equations for the individual particles that make up the system. Since 1986, there have been demonstrated in the laboratory new classes of high-temperature superconductors whose properties cannot be explained by the theory of the 1950s. Even though their properties should be calculable from basic quantum theory, no workable theory of these high-temperature superconductors exists even after 25 years of trying to construct such a theory. The problem of applying the elementary quantum theory to all the strongly interacting particles in the superconductor becomes intractable on even the biggest supercomputers. Again, the problem is there are too many particles and they interact too strongly. I will give one more such example of a nearly incalculable emergent phenomenon.

In the standard model of particle physics, all the equations are there to calculate the mass of a proton from first principles. The mass of a proton is known experimentally to many significant digits but computing the mass from the underlying equations took decades of work and massive programs running on huge banks of supercomputers. The proton is a very simple particle as it is composed of just three quarks. However, because these quarks interact very strongly with each other, simple approximations fail, and the full theory has to be deployed on banks of supercomputers running for years to get even an approximate agreement with experiment.[70] This is just one proton. If you have hundreds of protons in some high-energy collision process, the problem becomes unsolvable again. This is the real reason a theory of everything is a theory of nothing. What good does it do to have the fundamental equations that determine all possible interactions of every particle in the universe with every other if you cannot calculate everything you need to know about those very particles? This is why even if a theory of everything is found, and I doubt one will be found anytime soon, physics research does not come to an end. In the most interesting cases, the theory of everything cannot be applied and a new emergent-phenomenon theory must be instead developed.

Quantum information theory is not above having its own untested or untestable beliefs. American quantum information theorist John Smolin calls one of these the Church of the Larger Hilbert Space.[71] This Church has only one belief, which can be best explained in terms of Schrödinger's cat. As we recall from Chapter 1 ("The Cat in the App"), the paradox of the cat is that naïve quantum theory predicts that the cat in the box is in a coherent cat state of simultaneously dead *and* alive but that everyday experience tells us that the cat is either dead or alive but never both. Schrödinger's beef was that properties that we apply with ease to the spin of an ion, spin simultaneously pointing up and down, we would never think of applying to a cat. The crux is why do small things like ions obey quantum theory but large things like cats do not? Recall that there are two ways to collapse the cat. We can open the box and look and the act of doing that collapses the cat. However, we can wait a short amount of time and the interaction of the cat and the air molecules in the box with the air molecules and infrared radiation and cosmic rays outside the box, the environment of the box, carries off information about the state of the cat and observes it just as sure if we opened the box and looked.

In this resolution of the cat paradox, the environment around the box in the laboratory becomes entangled with the cat and the air inside the box, and if we ignore this environment, then it works out mathematically to the environment being an inanimate observer that collapses the cat. The more couplings there are to the environment, the faster the collapse. For large things like cats, it is impossible to shield the cat from the environment so that the environmentally induced collapse occurs so fast on human time scales there is no chance for a person to observe the cat dead and alive. But for an ion in a trap, because the ion is so small, it is fairly easy to protect the ion from heat waves and air molecules and things percolating in the laboratory and so it is fairly easy to observe the ion both spin up and spin down.

The theory of quantum mechanics then has two parts, the coherent smooth idling of the cat luxuriating in the box both dead and alive (called "unitary evolution" in our religious jargon) and then there is this sudden discontinuous collapse induced by an observer or the environment. Many physicists do not like this collapse business and the many-worlds interpretation was designed to get around it. In the Church of the Larger Hilbert Space, the idea is that the collapse is sort of an illusion that comes about from first entangling the cat and the box with the environment of the laboratory and then jettisoning or ignoring the quantum state of the stuff in the laboratory. If you did track all the air molecules and heat photons and so forth in the laboratory, so the belief goes, the combined system of the cat and the box and the laboratory would be in some coherent idling state of cat alive and dead and laboratory environment observing and not observing it. In this way, the smoothly running idling part of quantum theory, the part people like, explains everything. The collapse is not real but an artifact of our feigned ignorance of what the air molecules in the laboratory are doing. We can keep going.

The laboratory then becomes entangled with the Earth's atmosphere, which could collapse the state of the laboratory, which would collapse the state of the cat, so we draw a bigger box around the Earth, the laboratory, and the cat box, and all that is smoothly idling (unitary evolving) with no collapse. But the solar wind from the Sun might ruin that system and collapse it so we draw a bigger box around the solar system, the Earth, the laboratory, the cat box, and call that the smoothly idling system with no collapse. Following this logic, there is one biggest box containing everything in the universe idling smoothly, obeying the evenly flowing bits of quantum theory and nothing ever collapses. Then, the entire quantum state of the universe lives in an exponentially large mathematical Hilbert space. By making the Hilbert space larger and larger until the whole universe is contained, there is no environment outside the universe to collapse it, and the collapse is forever staved off. That then is the belief that underlies the Church of the Larger Hilbert Space. However, most churches have more than one belief in their dogma. Brought up as a Catholic, I had to memorize the Apostle's Creed, which is a list of 12 things that all Catholics should believe. I feel that for the Church of the Larger Hilbert Space to qualify as a real church, we should add a few more pillars of faith. This first pillar can be summarized as follows. "I believe in the unitary evolution of quantum mechanical states, particularly cats that are dead and alive, and that the collapse of a state is an illusion that can always be avoided by moving to an ever larger and larger Hilbert space."

What is Hilbert space? When I first encountered the notion of an infinite-dimensional Hilbert space in my undergraduate quantum mechanics courses, I was told it was merely a mathematical construct that helped us to calculate things in quantum theory and nothing more. But there is a long history in physics of things that start out as abstract mathematical constructs and then end up as being really real. In the 1600s, Newton proposed that all motion of planets to apples took place in absolute space and absolute time. As I have related above, many philosophers of science objected to these concepts as metaphysical or mathematical abstractions. However, the great success of Newton's theory produced confidence that space and time might be real things and not just abstract things. The conversion was complete by the 1900s when Einstein unified space and time into space–time. In Einstein's general theory of relativity, space–time moves and curves and bends and stretches and wiggles like the elastic fabric of a trampoline. Certainly, space–time acts like a real thing and most physicists have come to accept space–time as objectively real. While a quantum theory of space–time does not yet exist, it is expected that at very short length scales, around 10^{-33} centimeters (0.0004 nonillionth's of an inch), the quantum fluctuations cause space–time to foam and froth and bubble so that, viewed through a sufficiently powerful magnifying glass, the empty space between the stars throughout our universe looks like the percolating waters of an out-of-control astronomically sized Jacuzzi.

In the early 1800s, the English physicist Michael Faraday produced an abstract concept to explain electric and magnetic fields, a construct he called lines of force. Initially, Faraday viewed these lines of force as an abstract visualization tool for understanding electric and magnetic forces but quickly he came to believe in their physical reality; electric and magnetic substances produce very real force fields that permeate all of space. Light itself is an interplay of electric and magnetic force fields, independent of the substances that produced those fields, and seen that way, most physicist today regard these force fields as real things. Another good example of this evolution from abstract mathematical construct to real thing is the quark. In 1964, American physicist Murray Gell-Mann (and others) proposed the idea of a quark in the context of a mathematical theory that explains the structure of such particles as the proton and the neutron. The initial reaction to this theory was to view the quark as a mathematical abstraction and not a real particle in its own right. Fifty years later, with lots of experimental data, and just the fact that people got used to the idea, the quark is now universally considered to be a real thing, just as real as a brick. In the theory of electricity and magnetism, there is a quantity called the electromagnetic vector potential that is used as a book-keeping device for calculating the properties of electric and magnetic fields. For 100 years, this potential was viewed as a calculational artifice, an abstract mathematical notion that was not real and not observable. Then, in the 1940s and 1950s, following a similar trajectory as quarks and lines of force, it was shown that the electromagnetic potential was directly observable in quantum interference experiments. Today, the electromagnetic potential has undergone a full promotion from an abstract calculating device to a thing that is just as real as an electric field, a quark, or a brick.

What about Hilbert space? Once, when asked about the nature of reality, the English physicist Paul Dirac is said to have responded, "Physical reality is a ray in Hilbert space." That is, physical reality is a massively entangled state that must be represented in a large Hilbert space. I dare conjecture that most of my colleagues still view Hilbert space as an abstract mathematical concept and not something real like the three-dimensional space we live in. But I also conjecture that the notion of Hilbert space is, like the quark or the field or the potential, in the process of getting an upgrade. More often than not, when I ask a colleague in quantum information science what is the physical resource behind the exponential speedup of the quantum computer, I will get an answer that is something like "The exponentially large Hilbert space." (I myself hold this view.) When I claim that the quantum computer has an exponential number of parallel processors in parallel universes, that is what I mean, those universes are out there somewhere in Hilbert space, not in our own three-dimensional space.

But a *resource* does not really sound like an abstract mathematical concept. For example, an abstract mathematical concept is the metaphysical notion of the number "three" independent of what you have three of. If I tell you I am thinking of the number three, you would not claim I was thinking of a physical resource. If I tell you I am thinking of three horses or three power plants or three pickup trucks, well then you would think I am talking about a physical resource. My point is that Hilbert space cannot be an abstract mathematical concept and simultaneously a physical resource. These are mutually exclusive ideas. If Hilbert space is a physical resource, then it must perforce be real and just as real as a horse, a power plant, a pickup truck, a quark, an electric line of force, the electromagnetic potential, or space and time. Physical reality is not the abstract mathematical notion of a ray in Hilbert space; physical reality *is* a ray Hilbert space. Dirac said it best. I predict that, in 100 years, Hilbert space will be considered by all physicists as really real, just like electric fields and space–time, and the notion that Hilbert space is only an abstract mathematical notion will seem then archaic and quaint to future generations. This then is my proposal for the second pillar of faith of the Church of the Larger Hilbert Space. "I believe in the physical reality of Hilbert space and the exponentially large numbers of quantum states for which it stands." It is upon this belief that rests our faith in the exponential speedup of quantum computers (for solving some problems).

If you now have taken my oath of allegiance to the reality of Hilbert space, what then are you committed to? This may be best understood by comparing the history of humankind's exploration of ordinary three-dimensional space to the future exploration of Hilbert space. In the time of Galileo and Newton, the largest feature of the universe was our solar system. The stars were thought to be much closer than they actually are. The smallest features of our universe were things we could see with our naked eye. But with the invention of the telescope, we were eventually able to prove that we are only one solar system out of many in our galaxy and our galaxy in turn is one out of many in our universe. The largest observational distance scale is the diameter of our observable universe, which is approximately 93 billion light-years across. In the other direction, the discovery of the microscope led to the observation of bacteria, and then viruses, and now we can image individual atoms with certain types of microscopes, atoms that are a few tens of nanometers (tens of billionths of an inch). Indirectly in atom smashers, we can measure the diameter of the protons and neutrons that make up the atom's nucleus; it is around a femtometer (approximately a hundred trillionth of an inch). Smaller than that is somewhat of a conjecture, but it is expected that the smallest things, strings if string theory is right or loops if loop quantum gravity is right, are a size of 10^{-33} centimeters (0.0004 nonillionths of an inch).

Put into the same units, the diameter of the largest thing, the observable universe, is approximately 10^{27} meters and that of the smallest conceivable thing, a quantum string or a gravity loop, is 10^{-34} meters. These distances span 27 plus 34 or a total of 61 orders of magnitude in size from the smallest conceivable thing to the largest observable thing. (Some theorists have theories about larger things, other universes outside of our own universe, but this is so speculative I do not include this on the large scale.) There is a wonderful (if somewhat outdated) 1977 documentary film called *Powers of Ten* that takes you on a video image tour of most of this range of distances, from the proton in an atom on the small end to the largest superclusters of galaxies on the large end, changing size in the image by one power of ten each 10 seconds.[72] We should compare this to the state of affairs in the 1500s where the largest observable thing was our solar system at 10^{13} meters and the smallest thing resolvable by the human eye was about the size of a dust speck or 10^{-5} meters (tens of micrometers). That works out to 13 plus 5 or only 18 orders of magnitude. Hence, in 600 years, we have expanded our knowledge of the known universe from 18 orders of magnitude in size to 61 orders of magnitude in size, an increase in 43 orders of magnitude in our knowledge (see Figure 6.3). That sure does sound like a lot of increase in human experience and knowledge. However, as we shall now see, when compared to the size of Hilbert space, all of that 61 orders of magnitude in three-dimensional space is almost nothing at all.

I will take Phillips up on his wager and bet that we have a quantum computer capable of running Shor's factoring algorithm in 50 years or so. I will also take the science fiction writer point of view and extrapolate wildly beyond known physics to get us there. (Remember, the science fiction writers are more often right than the too conservative scientists.) There is consensus in the community that such a practical factoring engine will require around a trillion (10^{12}) qubits. Conjecturing something like Moore's law for quantum computing, I will further conjecture that new technologies are around the corner that will allow us to double the size of our quantum processors every few years. So if we think of the new NIST ion trap working this in a few years (2020) with 1000 entangled ions, we then just scale this up, doubling the number of qubits every few years, until we get to a billion entangled qubits. In Figure 6.4, I plot this trajectory with the number of qubits on the horizontal scale, the size of the corresponding Hilbert space on the vertical scale, and the year along the diagonal on the scale. I show a few qubit ion trap photograph around the year 2000, a 1000 qubit ion trap around the year 2020, and a fanciful rendition of a 100,000 qubit machine made from a carbon graphene lattice in 2040, and then I run out of ideas for the hardware. But you can bet that scientists in 2040 will not run out of ideas (they have 30 years of new discoveries in hand that I don't) and they will build the million-qubit machine by 2060 and finally the Internet-hacking billion-qubit machine running Shor's algorithm (for which I display the quantum circuit) by 2080. The growth in the number of qubits, as per Moore's law, is exponential by year. Because the size of the Hilbert space, vertical scale, is exponential in the number of qubits, it is therefore *super* exponential. Following this trajectory, if I hedge William Phillips' bet and we have a billion-qubit machine in 70 years by 2080, then we humans will have explored 3,000,000,000,000 (3 trillion) orders of magnitude in size in Hilbert space. That is to be compared to the rather paltry 60 orders of magnitude in three-dimensional space humans have explored in all of recorded history, tens of thousands of years. The exploration of all of

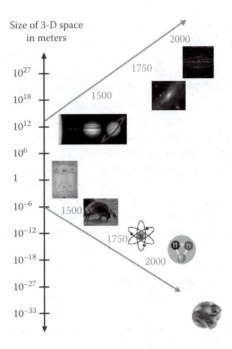

Figure 6.3 Three-dimensional space throughout the ages. The vertical scale is the diameter of the object in meters. The diagonal arrows are roughly time lines since 1500 AD. Human-sized objects, in this case da Vinci's Vitruvian Man, are about 1 meter. Going down, we see a dust mite a few hundred microns across (and invisible to the naked eye), an atom a few tens of nanometers across (with electrons whizzing about the nucleus), a proton a femtometer across (with quarks inside), and at 10^{-34} meters is the speculated foaminess of space time (where quantum strings or loopy gravity lives). Going up in scale, we see the solar system of the ancients at 10^{12} meters, the Andromeda galaxy at 10^{21} meters, and the observable universe at 10^{27} meters. Our understanding of three-dimensional space spans 33 plus 27 or 60 orders of magnitude in sizes from the infinitesimal to the nearly infinite. (The figure is a composite of several graphics. The Andromeda Galaxy photo and the Solar System composite were created by NASA, and the drawing of the Vitruvian Man by Leonardo da Vinci, the dust mite photo by the US Food and Drug Administration, and all are in the public domain.)

three-dimensional space in all of human history is nothing compared to the exploration of Hilbert space in the next 70 years. *That* is the promise of quantum technology.

Taking the position that Hilbert space is real, just as real as three-dimensional space, then this is not just a mathematical exercise. If Hilbert space is a physical resource, then, by 2080, we'll have a $10^{3,000,000,000,000}$-dimensional resource at our disposal. It is difficult to wrap your brain around the size of this

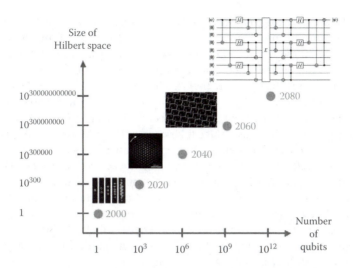

Figure 6.4   Exponentially large Hilbert space over the next 50 years. The horizontal scale assumes a type of Moore's law for quantum processors with the number of qubits in the processor following an exponential growth doubling in size every few years. (This is the type of scaling we'll have to have if a universal quantum computer with a billion qubits is to be built in 50 years.) The dimension of the Hilbert space scales exponentially again with the number of qubits and so it scales "super" exponentially along the vertical axis. The diagonal arrow indicates the approximate year. The figures show an ion trap quantum computer with 1, 2, 3, 6, and 10 qubits (lower left), where 10 qubits gives a Hilbert space dimension of 2^{10} or approximately 1000. The second graphic is the NIST ion trap that may have 1000 entangled qubits in 2020 with a Hilbert space dimension of 2^{1000} or 10^{300}. The third graphic is a schematic of the carbon graphene structure that might have 1,000,000 entangled qubits or a Hilbert space of $2^{1,000,000}$, which is approximately $10^{300,000}$. The final graphic (upper right) is a circuit for Shor's algorithm running on a billion-qubit machine (of unknown technology) that has a Hilbert space dimension of $2^{100,000,000,000}$ or $10^{300,000,000,000}$. This chart implies that we will cover 300,000,000,000 orders of magnitude of Hilbert space in the next 50 years compared to 60 orders of magnitude (Figure 6.3) in three-dimensional space covered in the past several thousand years. (The figure is a composite. The two ion trap photos are courtesy of NIST and, as work of the US Government, are not subject to copyright.)

number.[73] This is 1 followed by 3 trillion zeros. This book you are reading has around a million characters in it. Hence, to print out the number $10^{3,000,000,000,000}$ in full on paper would require a printing of 3,000,000 (3 million) books this size filled with all zeros, that is, approximately one-tenth of all the books in the US Library of Congress, one of the largest book repositories in the world. The number of particles in the entire universe is only approximately 10^{80} or 1 followed by only 80 zeros. In the Church of the Larger Hilbert space, we believe that not only are these huge numbers attainable but that they are attainable in a generation and that they correspond to the size of a real thing, Hilbert space, that we can build things with using it as a resource. What is in Hilbert space? Well, nothing I'm sure at the moment. That is, whenever we opened a new window in three-dimensional space, by inventing the telescope or microscope, we found things that were already there, new planets in the former case and microbes in the latter case. This is why the exploration of three-dimensional space is a science. As our observing tools see bigger or smaller things, we find stuff that has been there all along but that we just could not see before. Not so in Hilbert space. Hilbert space is not independent of us, there for us to find stuff in, but rather we create Hilbert space and then use it in turn to power new types of machines. The exploration of Hilbert space is much less a science and much more of a technology. I rather doubt that when we build the billion-qubit quantum computer, it will open a portal into an exponentially large Hilbert space that contains Hilbert space creatures that will leap out at us and gobble us all up. Rather, Hilbert space is empty until we make it and begin to manipulate our quantum mechanical states inside of it and fill it with our technologies. It is because we make it that it is a technology. What will we make in a $10^{3,000,000,000,000}$-dimensional Hilbert space? Well, there is plenty of room in there so I would be surprised if all we came up with was a quantum computer.

As a self-ordained quantum technologist, I'm often asked to compare quantum technology to the more familiar nanotechnology. Often I'm asked, well if nanotechnology is a technology of small things, then is quantum technology much *smaller* than nanotechnology? My answer is no, that quantum technology is much *weirder* than nanotechnology. While sometimes nanotechnology enables quantum technology, making things smaller like superconductors to make them behave more quantum mechanical, this is not necessarily true. Entanglement with photons over distances of hundreds of kilometers has been demonstrated, a distance that is not by any means small. Interestingly, quantum technology sometimes enables nanotechnology. An Australian team at the University of New South Wales in Sydney, led by Australian physicist Michelle Simmons, pursues a quantum dot approach to quantum computing where single phosphorus atom qubits must be placed with atomic precision in a silicon atomic lattice. The group has invented several new fields of nanotechnology to do this and has as a spin off demonstrated a single atom classical transistor with the technique. While I have mentioned the connection between nanotechnology and quantum technology in previous chapters, I now have prepared the stage to be able to do so with a bit more clarity.

Nanotechnology, like quantum technology, was launched by the prescient ideas of the American physicist and practical jokester Richard Feynman. Rather than in a scientific paper, nanotechnology was first mentioned in a widely circulated after-dinner speech, "There's Plenty of Room at the Bottom," given on December 29, 1959, at the annual meeting of the American Physical Society at the California Institute of Technology.[74] In this lecture, Feynman imagined a world where machines and computers worked on the scale of tens of nanometers (billionths of an inch). Feynman then imagined a type of computer memory where each bit of information is stored in just a few atoms, say in a chunk of silicon, and then demonstrates that all the information in all the books in the world (circa 1959) could be stored in a speck of silicon approximately a half of a millimeter in diameter (approximately a hundredth of an inch). He used this result to proclaim that not only is there *room* at the bottom but that there is *plenty* of room at the *bottom*. (Here, "the bottom" means at very small size scales.) This may seem like an astounding number, and it is, and I can update it a little. Circa 2010, we now take all the bits of information stored in every computer in the world, which is a much larger number than the bits of information stored in every book in the world in 1959, and then going a bit further down than Feynman, assuming one bit per atom (he assumed one bit per 100 atoms) following his calculation, all the information on all the computers in the world in 2010 would fit in a hunk of silicon approximately a centimeter in diameter (a couple of inches) at one bit per atom. That is the promise of nanotechnology and that is what Feynman means by there is *plenty* of room at the bottom. But nanotechnology is an exploration of things at the very small scale, nanometers, in three-dimensional space. By now, I have convinced you that the room in Hilbert space vastly dwarfs the room available in ordinary three-dimensional space. There may be plenty of room at the bottom, but there is *plenty more* room in the *quantum*!

Let us take Feynman's argument and move the venue from three-dimensional space to Hilbert space. Suppose we take every bit of information on every computer on Earth, approximately 10^{21} bits, and assign one bit to one dimension of Hilbert space. The thing about Hilbert space is that the number of dimensions is exponential in the number of qubits. Only 70 qubits are needed to "store" 10^{21} bits. That is, we can, in some sense, store and process all the information on the Earth in an itty-bitty cube of silicon that is about four or five atoms on a side. There is *plenty more* room in the *quantum*! But there is a caveat. Remember that the quantum computer does not give an exponential speedup on all mathematical problems but just a few special ones with special mathematical structure such as factoring. The reason for this is that there is a quantum input and output problem. Encoding all the 10^{21} bits into the 70 qubits can take a very long time if there is no structure or order in the bits. Once the input is done, then you are free to process the information in the large Hilbert space of 10^{21} dimensions, but when you try to measure the output, the giant 10^{21}-dimensional state collapses to a single 70-bit classical state. This is called the readout problem. If your processor does not somehow give you exactly the answer you want in that

70-bit output in one shot, you then have to run the calculation over and over again, on the order of 10^{21} times, and the advantage of the giant Hilbert space is lost. Not all mathematical problems have a structure that allows you to do this, but luckily factoring does and so Shor's algorithm can take advantage of Hilbert space.

American physicist Charles Bennett likens a quantum computer to an exponentially large stomach with a very small esophagus (input) and a very small duodenum (output). Avoiding the obvious scatological jokes, once inside the exponentially large stomach, you have lots of room to slosh things around and utilize the exponential largeness of Hilbert space to solve your problem. But to get in there, you have to make sure the input and the output of the problem fit in and out of the small pipes; here, the small set of physical qubits that is your portal in and out of the exponentially large Hilbert space. We do not live in Hilbert space and neither do our qubits, but we access it through the quantum entanglement and nonlocal correlations the qubits bear upon them. It is a bit like a pilot trying to fly a Boeing 747 jumbo jet while encased in a cardboard box with a small hole drilled in it, and the rules say he is only allowed to peer through the hole and poke the controls with a pair of chopsticks. Not all problems can be jury rigged to do this, that is, fit in and out of these small ports and yield a solution, particularly a solution that has an exponential speedup that takes advantage of the exponentially large Hilbert space. But, as we have seen, there is more to quantum technology than computing. Once we have the billion-qubit machine with the astounding $10^{3,000,000,000,000}$-dimensional Hilbert space, we are free to tweak it to make something besides a computer. A billion-qubit quantum computer is a giant quantum interferometer capable of solving mathematical problems not solvable classically. However, if tweaked just a bit, the billion-qubit interferometer becomes a billion-qubit quantum sensor capable of sensing, say, magnetic fields at a precision not possible classically. I have mentioned quantum computing, sensing, and timekeeping, and imagine—these are all technologies that are based on having entangled qubits in a large interferometer. Once we begin sailing out into the giant vastness of $10^{3,000,000,000,000}$-dimensional Hilbert space, building new quantum interferometers as we go, will we discover new technologies that I can't even think of? You bet we will. There is *plenty more* room in the *quantum*! Quantum technologies and the associated vastness of Hilbert space will drive most of the technological advances of this new millennium, and the ship of quantum weirdness has barely just left the dock.

TO GO WHERE NO MAD SCIENTIST HAS GONE BEFORE

As I have mentioned previously, in 2008, I attended the annual 3-day *Future Technology Seminar*, which is more popularly known as *The Mad Scientist*

Conference, held in Portsmouth, Virginia. Our job, as mad scientists, was to cook up ways to destroy the world, as good mad scientists are wont to do. My doomsday scenario was that three Indian computer scientists, working in an un-air-conditioned attic in Hyderabad, would discover an efficient *classical* factoring algorithm—there is no proof that one cannot exist—and publish it on the preprint ArXiv. In a matter of days such an algorithm would be converted into a public-key encryption cracker, and banks and other financial institutions around the world, in a panic, would sever all their connections to the Internet, causing the entire economy of the world to collapse in 3 days. Other mad scientists had even more dire predictions. My favorite was the "gray goo" scenario. In this scheme, a mad scientist or an unwitting nanotechnologist develops a self-replicating nano-robot or "nanobot" that is released (or escapes) and the nanobot contains instructions to replicate itself endlessly (from raw materials like dirt and human flesh) and to gobble up all biological matter, converting all life forms into a lifeless meter-thick film of gray goo coating the entire Earth in a matter of days. The more staid proposals were of a biological nature; typically, a high school student working on a science fair project builds a poliovirus, smallpox virus, or some virulent strain of flu, using the DNA sequence she found on the Internet and chemicals she ordered from a mail-order catalog. In this scenario, only millions of people die and the rest of the nonhuman biological life forms are spared. But there were several other scenarios that involved computers taking over the world from the humans, which will be my focus here.

The idea of a computer or robot becoming self-aware and sentient is nothing new. English computer scientist Alan Turing considered this possibility when, in 1950, he proposed in the seminal paper on artificial intelligence (AI), "Computing Machinery and Intelligence," the now famous Turing test for gauging the degree of sentience of an AI. The test consists of a human being, the tester, sitting at a computer terminal. The computer terminal is wired up to either a very smart computer or to another computer terminal with a second human being sitting at it. The tester then types questions for the entity on the other end of the line to answer. When we reach a point in computer evolution that the tester cannot ever distinguish between a second human being and a computer, then we must claim the computer is sentient and as intelligent as we are and welcome it into our social club of humanity. The question then is when or if this will happen. I should categorically state here that, hitherto, all such discussions have been couched in the assumption that the machines we are dealing with are classical computers. That is, the human brain is proposed to be a type of very powerful but classical "meat" computer and the AI is likely a very powerful but classical silicon microchip–based computer. The assumption is that we are testing one type of classical computer against another type of classical computer running on two different hardware platforms, meat versus silicon.

This leads us to American philosopher John Searle's strong AI hypothesis, "The appropriately programmed (classical) computer with the right inputs and outputs would thereby have a mind in exactly the same sense human beings have minds."[75] (Parentheses mine.) The Turing test is then construed as a test of the strong AI hypothesis. The hypothesis is true if we can construct a classical computer that passes the Turing test. As this is a popular science book and not a popular philosophy book or religious studies book, I will not discuss such objections to the strong AI hypothesis such as computers do not have souls and so forth. I personally am a big fan of the strong AI hypothesis and I once had a T-shirt that proclaimed, "I'm Proud to Be a Meat Computer!" That is, I conjecture that when classical computers become sufficiently powerful and have enough neural network interconnects, they will certainly pass the Turing test and will likely have intelligence similar to if not indistinguishable from our own. I realize that this idea may make some of my readers uncomfortable, but I don't view it as my job to comfort my readers. It's my job to make you think— even if the thoughts are somewhat discomfiting.

No classical computer has yet passed anything like the Turing test, but classical computers are becoming uncomfortably close to doing something like this. For many years, the hallmark test of AI was that no classical computer could beat even the lowest-level human players at chess. Then, it was that no computer could beat grandmaster chess players. Now, we have a computer, IBM's Deep Blue, which beat the world champion chess player Garry Kasparov in 1997. Perhaps even scarier, in 2011, IBM's Watson computer beat two reigning human champions at the very human game of Jeopardy.[76] In both cases, critics of the computer complain that it wins by just memorizing vast numbers of possible chess moves, in the case of Deep Blue, and vast amount of trivia, in the case of Watson. However, I would rejoin those critics that this is exactly what the human competitors also do in order to win. However, a gaff in the Jeopardy playing revealed that Watson cannot yet pass the Turing test. When, in the category of US Cities, Watson's clue was "Its largest airport is named for a World War II hero; its second largest for a World War II battle," the computer replied, "What is Toronto?????" (The correct answer is "What is Chicago?")

Thence, it is easy to trip up even the most powerful computer in a Turing test. You likely will get sensible answers to Apple Siri-like questions such as "What will the weather be like tomorrow?" but nonsensical answers to questions such as "How about them Mets?" However, if one embraces the strong AI hypothesis, as I do, then the real question is not if but when the computers will become essentially human, sentient, and self-aware. If the human mind is really a meat computer and is a product of trillions of synapses and interconnects and if self-awareness is an illusion of our brain's capability for self-referencing (a proposal that is made masterfully by American cognitive scientist Douglas Hofstadter in his wonderful book, *I Am a Strange Loop*), then it is only a matter of time before classical computers become similarly powerful and aware. How much time? Well not much. This is the fear of many a mad scientist. Not much time left. Remember that Moore's law for the processing power of a classical computer is an exponential law. The processing power of classical computers doubles every couple of years. As Australian computer scientist Anthony Berglas conjectures in his titillatingly entitled treatise, "Artificial Intelligence Will Kill Our Grandchildren," analysis of improvements of speech recognition software (such as Apple's Siri) compared to similar capabilities of the human brain suggests that current-day classical computers are just a few orders of magnitude less powerful than your cerebral cortex, responsible for speech and language. But a few orders of magnitude on the exponential Moore's law curve is then just a few years or tens of years away. Meanwhile, your cerebral cortex is not getting any smarter. The numbers vary but the predictions are the same—in a few tens of years, there will be some supersmart version of a Siri-like speech recognition software that you and I can converse with so fluidly that we'll not be able to tell if the thing we are talking to is hooked to a computer or another person. When Super-Siri passes the Turing test, the strong AI hypothesis becomes fact and we then need to sort out what to do about it.

There are three primary scenarios discussed in the context of the strong AI hypothesis becoming the strong AI fact. The first I will call the I-Robot postulate after the collection of short stories by this name written by American science fact and fiction writer, the most prolific writer of all time, Isaac Asimov. In this I-Robot scenario, humans build fail-safe systems into the AI, in this case, Asimov's three rules of robotics, to ensure that the sentient computer–robot creatures that we have built do not end up running amok or killing us (or our grandchildren). In Asimov's stories, the robots are more or less servants to the humans, but in more modern renditions of the I-Robot scenario, such as in the television series *Star Trek: The Next Generation*, the robots are treated as human equals.[77] At the other extreme, we have exactly the opposite scenario where the AI gets loose and decides to kill all the humans or enslave us, a scenario I will call the Colossus hypothesis after the 1970 sci-fi thriller film *Colossus: The Forbin Project* that preceded the similarly themed 1983 *WarGames* and 1984 *Terminator* films by over 10 years.[78] (In *Colossus*, the computers try to enslave us, whereas in *WarGames* and *Terminator*, they try to kill us.) Finally, there is an intermediate scenario, which I will call the Borg Identity, where the humans (to avoid being killed by it) merge with the strong AI technology to form human–machine hybrids, such as the hive-like alien society of the Borg (also from *Star Trek: Next Generation*). The Borg consist of numerous humanoid species that are forcibly assimilated into the Borg collective, which even has a queen as in an ant colony. Once assimilated, the humanoids are implanted with technological gewgaws such as having their eyeballs replaced with red laser–beaming ocular implants, their arms replaced with mechanical ones, and bits of their brain replaced with neural implants that allowed them to communicate pseudo-telepathically with each other.

While none of these scenarios is particularly palatable to the man or woman on the street, the promulgators of the strong AI hypothesis, mad scientists one and all, endlessly discussed these three possibilities and the future of the human race at the conference and especially at the bar late into the night. Uncomfortable or not, the proponents of strong AI tell us that the emergence of sentient computers is just a few years away and we'd best start planning for it. Sign me up for my Borg implants? Perhaps the most popular recent expose of this AI "event horizon" looming over us humans is a book and a movement called "The Singularity." Popularized in the book *The Singularity Is Near* by American inventor, author, and futurist Raymond Kurzweil, the singularity is supposed to be this point in time in the near future when strong AI kicks in, your laptop becomes self-aware, and everything changes.[79] Featured in this book is the infamous exponential growth curve. I had the pleasure in June of 2010 to attend, along with three of my graduate students, the opening ceremonies of the Singularity University at the Research Park of the NASA Ames Research Center near Mountain View, California, in the heart of Silicon Valley. Kurzweil and fellow American futurist Peter H. Diamandis founded the university in the late 2000s. From their web page, we read, "Our Mission: A number of exponentially growing technologies will massively increase human capability and fundamentally reshape our future. This warrants the creation of an academic institution whose students and faculty will study these technologies, with an emphasis on the interactions between different technologies."[80]

My students and I found that one of the main attractions at the 2010 ceremony was the open bar.[81] A second attraction, aside from the talk by the Russian–American founder of Google, Sergey Brin (wearing his infamous jeans and T-shirt), was a little pamphlet they handed out to all the attendees explaining the nature of the singularity. On the pamphlet was a graphic of an exponential curve of technological growth in time. This graph also appears in Kurzweil's book. What I found most amusing was the labeling of the graph. The caption clearly stated that the curve was exponential and then had little arrows pointing to the "knee" in the curve and then to the "singularity" (see Figure 6.5). For us newbies, the knee was defined to be the place "where the curve goes from linear to exponential behavior." I looked at this graphic and laughed out loud, much to the disruption of the staid opening ceremony. You see, the exponential curve in mathematics is one of the most smooth and regular of curves that there is. There is no "knee" on an exponential curve—that is, there is no place where the curve goes suddenly from linear to exponential growth. That's just nonsense. Also, the exponential curve does not have a singularity! A singularity would be a place where the curve grows to an infinite height in a finite time. The exponential curve always has a finite height at finite time. There is

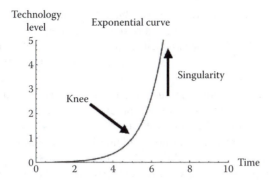

Figure 6.5 The excitement of bad math. This is a reconstruction from a graphic on a pamphlet handed out at the opening ceremonies of the Singularity University. The curve purportedly shows the growth of technology by year. The pamphlet describes the curve correctly as exponential but then goes on to explain that the "knee" is the place where the curve goes from linear to exponential growth. (There is no such place.) The arrow showing the singularity also is nonsense; the exponential curve never goes to infinity in finite time, which would be the definition of a singularity.

never a point in the future where the growth in technology becomes infinite. The growth is exponential and that does mean that it is faster than linear but there is no knee and no singularity.[82]

The excitement of bad math aside, scientists have been using exponential curves to predict the doom of humanity for hundreds of years. The Malthusians go back to 1758 when the English Reverend Thomas Malthus used the exponential curve of population growth to predict an imminent "Malthusian catastrophe." (He was probably right that continued exponential population growth will doom us but it is clear the threat is not yet imminent.) American writer and publisher Gerard Piel, the founder of the magazine *Scientific American*, published a series of essays on his concerns with the exponential growth of technology in his 1972 book *The Acceleration of History*.[83] In the 1980s, I remember American physicist Albert Bartlett giving a neo-Malthusian lecture, a lecture he has delivered over 1600 times, entitled "Arithmetic, Population, and Energy."[84] Enthusiasm for the exponential curve finds its way into all sorts of popular brouhahas such as the recent "hockey stick" curve controversy over a roughly exponential curve that ties global warming to human carbon emissions. But back to the mad scientists.

At the mad scientist conference, when the heated discussions over the strong AI scenario played out, and the mad scientists started taking bets on the I-Robot, Borg Identity, or Colossus scenarios, as the token quantum technologist, I interrupted and declared, "But you are all discussing classical AI! What about *quantum* artificial intelligence?" The arguments screeched to a halt and the entire auditorium was silent as they all turned as one balefully toward me. The presider gave me the floor and asked me to explain what quantum AI was. In these discussions a bunch of technical guys would run around with hand-held microphones to hand to whomever the presider recognized to speak. The poor tech guy ran up to me with such a microphone but I waved him away as I bellowed, "I have never needed a microphone in my life!" (Then came the cat-calls, "Yeah give Dowling an audio-dampening field!") With all eyes glued on me, I realized that, because I actually had no idea what quantum AI was, I had better make something up quickly. And so I did there on the spot.

Recall Searle's strong classical AI hypothesis, "The appropriately pro-grammed computer with the right inputs and outputs would thereby have a mind in exactly the same sense human beings have minds." There are two assumptions here. The first is that the computer in question is a classical elec-tronic computer and the second is that the human mind is a classical meat computer. That is, that the human brain is a very powerful universal but classical computer. The key word here is "universal." Modern classical digi-tal computers are also universal and that word means that any classical uni-versal computer can efficiently simulate any other. The hardware platform, electronic transistors in the digital computer or neurons in the brain, is irrel-evant. The strong classical AI hypothesis then assumes efficient simulation of a human mind is equivalent to having a human mind. When the digital computer becomes so powerful, it can simulate every thought a human mind might have, and then the digital computer will be just as human as we are and it will pass the Turing test. Then the question becomes, as Moore's law of expo-nential growth of classical computing power continues, will the classical AI begin building and designing even more powerful AI, sentient computers with even more brain power than us humans? If it does, then the doomsday sce-narios kick in. The classical AI becomes smarter and smarter at an exponen-tial rate and very quickly we humans, as dumb as bricks, become disposable. The classical AI super brains then must decide if they are to coexist with us peacefully (I-Robot), kill us all off or enslave us (Colossus), or merge their super AI technology with our meat-computer brains (the Borg Identity). Kurzweil argues for the last and so do I but for a different reason, as we shall see. But back to the quantum AI.

In the lecture hall with a hundred mad scientists staring at me, I decided to illustrate quantum AI using the Colossus doomsday scenario. Suppose the classical AI does kill off or enslave all the humans? Well, these classical super mechanical brains will quickly figure out how to exploit quantum theory and will rapidly develop ever more and more powerful *quantum* computers. The quantum computers will follow a quantum version of Moore's law, out-lined above, exponential growth in the number of qubits, and a consequent super-exponential growth in the dimension of the Hilbert space where all the quantum computational power is. Because any quantum computer can efficiently simulate any classical computer, but not the reverse, a quantum mechanical sentient mind will have super-exponentially more brain power than the classical computers (or humans) that made it. The quantum AI will then give all classical computer minds, meat or silicon, a *quantum* Turing test. That is, the quantum AI will ask representatives of the human and artificial intelligentsia a series of questions such as "Can you please crack this 1024-bit public-key encrypted message in less than a second?" We meat computers cannot do this and neither can our digital brethren, the classical computers. Only members of the quantum AI will pass the quantum Turing test by running Shor's algorithm and cracking the code. The quantum AI will then form a community and then decide what to do with the humans and the classical AI, which they will treat equally. Perhaps the quantum AI will decide if digital and meat computers are a threat and kill us all off or enslave us, leaving only the quantum AI left to rule the world. The mad scientists went totally mad at that point and the auditorium devolved into chaos.

The thing about classical AI is that at least we have some idea what it is. Like our brain, classical AI is a universal classical computer. There may come a time, the singularity, when the classical AI becomes smarter and computationally faster than us, because of the continued growth of digital procession power as per Moore's law, but the classical AI will not be fundamentally different from us. For many years, anthropologists assumed that waves of modern humans, emanating out of Africa 100,000 years ago, spread through Europe and Asia and wiped out the Neanderthals who got to Europe around 500,000 years ago. The Neanderthals were thought not to be fundamentally different from us (they are sometimes classified as a human subspecies as opposed to a separate species) but just a bit dumber than us and that led to their extinction by our hands. Now, recent DNA analysis has shown that neat story is not quite true—human DNA contains Neanderthal DNA—we interbred![85] In a similar way, strong classical AI is not fundamentally different from the human brain, assuming the human brain is a classical meat computer; it may just end up being a bit (or a lot) smarter than us, but smarter than us in a predictable fashion: more classical processing power, more classical processing speed, more classical memory. It may be exponentially more, owing to Moore, but it is predictably more. And perhaps like the Neanderthals and the humans, the humans and their strong AI will interbreed, the Borg Identity. Kurzweil suggests that humans must merge with their technology to avoid being wiped out by it. This raises the specter of us all running around with machine implants sticking out of our bodies and having laser beams for eyeballs, but I suggest that humans have been merging with their technologies for hundreds of thousands of years and that this is neither new and nothing to be afraid of.

The first example of such a merged technology I will give is that of human speech and language. Many animals have a rudimentary or even sophisticated means of communicating by sound, say dolphins, for example, but none have developed speech into the powerhouse technology it has become in modern humans. (Dolphins have a vocabulary of around 30 or 40 words where we humans regularly master thousands of words.) How is speech a technology? Technology is applied science, the application of science for practical purposes. The science of speech is the science of sound. Most animals exploit the science of sound, squeaking mammal babies always want mother's milk, but humans have pursued over the millennia a course of developing this science into an advanced technology. The evolutionary origins of speech are contentious but what seems clear is that our enormous cerebral cortex, the outer layer of the forebrain, evolved in size in parallel to our development of speech. If our brain is but a meat computer, then evolution provided an ancient and meaty Moore's law for increase in processing power over hundreds of thousands of years, processing power required for the exponential advances in the technology of speech and language. And somewhere along the way, the number of processors in the human brain grew so large they crossed the strong AI hypothesis boundary and, like Lt. Commander Data in *Star Trek*, we humans became self-aware. Those of us who had more brain space to devote to the making and understanding of sounds fared better than those of us who did not, driving the enlargement of the human cerebrum and the co-opting of more and more of it for speech, allowing speech to become ever more eloquent. And all this occurred very quickly over tens of thousands of years, an exponential growth in meat processors, Moore's law of meat computers. At least that is one theory. But this is a meat-based technological advance! Humans exploited the science of sound to talk. The science of sound, through the help of evolution, became the technology of human speech. The growth of the human cerebrum on geological time scales looks positively exponential. Humans have so well merged with the technology of speech it is now hardwired into our brains. You don't like this example of humans merging with their technology? Well I have more.

Tools! Anthropologists theorize that humans developed tool-making skills around the same time as speech and other technological manipulations of the environment requiring abstract thought such as music and art.[86] The making and using of tools is a technology. It is one thing to pick up a stick and pull some leaves off and use it to pluck juicy termites from mounds as chimps do, but it is another thing to fashion spears and clay flutes and the ENIAC. Our tools and the things that we make with them surround us. (Just a moment. Microsoft Word just flagged that last sentence with a squiggling green underline for nonstandard grammar. Best go up to the Tools menu and run the grammar checker.) Example two. The making and deployment of tools, something very nearly clearly human, is a technology we have merged with and we never even notice a thing. There was no singularity where one day we're poking termites out of piles of dirt and the next we are building spacecraft. Tool making followed an exponential curve of growth in sophistication and complexity but there was no knee and no singularity. An exponential curve looks exactly the same, locally, to anybody who happens to be sitting on it riding it along. To the 18-year-old rocket scientist and American engineer Charles Goddard, sitting in the year 1900, the space shuttle of the year 2000 would seem like magic and the Chinese bottle rocket of the year 1800 would seem like child's play. However, the liquid-fuel rocket of 1901 would look a lot like the liquid-fuel rocket of 1899. When you are riding an exponential technology curve through time, you only tend to look a few years ahead or behind you and the growth in technology over such a small time scale always seems linear. You never see a knee and you never see a singularity, just see yourself chugging along at what seems—at that instant—at a constant rate.

My third example of humans merging imperceptibly with their technology, even though the technology is following an exponential curve, is farming and the domestication of animals. Until about 10,000 years ago, we were all hunter-gatherers. That is, humans consisted of nomadic groups that wandered about killing their food where it ran and plucking it from the ground where it grew. But about 10,000 years ago, we exponentially rapidly began to understand the science of farming and animal husbandry and put it to work as a technology. We started building stable farming communities that sowed their own crops (a technology based on the science of farming) and milked and slaughtered its own cows (a technology based on the science of animal husbandry). Even better, we understood the rudiments of the science of genetics at least enough to start breeding better and better—exponentially better and better—crops and cows. Ever since, the total food production of Earth has kept up with the exponential growth in the number of humans. Farming and the domestication of animals is an exponentially improving technology that led directly to civilization. The fact that most of us modern humans live in communities in nonmobile housing and eat farmed crops and domesticated meats is proof that we have completely merged with the technologies of farming and animal husbandry and we scarcely even notice it anymore.

To be sure, there were hunter-gatherer Luddites who protested the move into the cities and scared us with stories of urban growth and plagues (the singularity scaremongering of antiquity) but they're not around anymore, much. In this same vein, the fact that we all wear clothes, seamlessly and effortlessly, is further proof that we have merged with the technology of textile production (farming) or leather production (animal husbandry). Few of us protest or, again, even notice. The merger of humans with the technologies of farming and animal husbandry is complete. Again, there was no temporal singularity before which we were running around naked chasing wild boars and after which we were sitting on the verandas of our plantations sipping mint juleps. Kurzweil's singularity is both hit and myth.

American sociologist Sherry Turkle has for many years studied the merging of humans with modern electronic technology. She probably would not quite put her life's work in those words but I will for her. In her first book on the topic, her 1984 work *The Second Self*, Turkle catalogs through many hundreds of hours of interviews and background investigations how computers are becoming, even in 1984, a part of our social and psychological selves. That is, how humans are merging with their technology of computer engineering. Even in 1984, and I should know, there were kids who were more at home playing with a computer than playing with their human friends. The theme is continued in her more recent 2011 book, *Alone Together*, where Turkle discusses how the ever stronger classical AI allowed by the exponential growth in computer processing power that is Moore's law allows for interactive robots like Furby and Paro to become more of a companion to some folks than other humans. Teenage children today often send thousands of text messages to each other per month and would much rather communicate this way than chat in person. Turkle always speaks about this progression or merging with computer technology somewhat wistfully, as if the closer connections with the technology reduces or threatens the human connections that should be primary. But there is no wistfulness to this at all. One of the things that makes us human, and has done so for a million years, is our effortless ability to imperceptibly merge with our own technology. Computer technology is just one more, like speech, animal husbandry, farming, or tool making, ripe for a merger. And as with the other mergers, even though things are progressing exponentially fast, as they always do with good technologies, in short time scales, the merger will be imperceptible to those of us doing the merging.

Let me make some wild-assed conjectures about the future of human communication. Speech, we recall, took off wildly as a technology and led to the exponential growth of our meat computers. The printing press was another disruptive communication technology as was the telegraph, the telephone, the television, and the Internet. Each follows its own exponential curve of technological advance until merger is complete. Who of us, growing up watching hours of television a day, can doubt we were merging with that technology. Who of us, watching our kids hammer away at their smart phones night and day, can doubt that they are merging with the technology of their smart phone? Let's take texting. Texting as a form of semi-silent communication came into vogue in Japan about 10 years ago when the polite Japanese began enforcing rules against people talking on their cell phones in trains and subway cars. All that talking was loud and rude so signs went up prohibiting it. The switch to the silent form of cell phone communication, texting, was exponentially fast.

Today, when you ride the subway in Tokyo, the cars, packed with people, are eerily quiet except for the chittering of hundreds of fingers flying over little itty-bitty keyboards in a mad exuberance of mass communication. Turkle might wistfully wish that the folks on the train might put down their smart phones and talk with each other, but the only people you see doing that are above a certain age, and they are few, and getting fewer. You might as well wish your 10-year-old would put down her smart phone and send her best friend a post card.

Texting is instantaneous person-to-person communication. It is fast, quiet, easy, and surreptitious—what's there for a teenager not to like? As your parents worried about you spending too much time playing video games in your wasted youth, you worry about your children and these phone bills with thousands of text messages a month. This texting technology will not go away; it is on an exponential growth curve. The teenagers, I fear, will grow up.[87] We have removed the cumbersomeness of having to actually speak—what is next? The meat computer with its pudgy fleshy fingers must interface with the electronic computer through the bottleneck of the QWERTY keyboard. The next step is to improve the interface. In the next 20 years, we will develop a type of electronic eyeglasses with a computer screen built into the lenses and a magnetic field sensor built into the earpiece, something like the glasses Google has recently cooked up, but unlike Google's Goggles, these will be a two-way communication device, let's call them Spectrum's Specs. (Spectrum is the hypothetical Silicon Valley company that will run Google out of business in 20 years.) The magnetic sensor, housed in something the size of a Bluetooth ear bud, will be a quantum sensor, perhaps the quantum spins in a nitrogen-vacancy center that have long coherence times and can be entangled and can operate at body temperature. Like a mini-magnetic resonance imaging machine, the sensor will read the magnetic fields of our brain. With a bit of training, we can select our friend from our contact list, housed in the glasses or in the cloud, and "type" our messages out to them by just thinking the message silently. The ear bud is hooked to the cell phone system and transmits our message, and we read the response on the eyeglass lens down in the corner. With our pudgy great ape–like fingers and clunky QWERTY keyboard removed from the system, we will be able to send messages as fast as we can think and read. (I can type approximately 45 words a minute, but I can read approximately 450 words a minute. I'm not sure how fast I can think but if it is as fast as I talk, it is pretty fast. I grew up in New York after all.) These Spectrum Specs will be all the rage with our teenagers' teenagers, and our teenagers themselves will complain about getting phone bills for many tens of thousands of text messages a month and wistfully wish for the good old days where they hammered away on their little keyboards with their gnarled monkey fingers and had to suffer through carpal tunnel syndrome.

As exponential improvements to the technology continue, faster computer processing speeds, more memory, ever smaller devices, the Spectrum Specs with be replaced with the ChaChing Chips (everything will be made in China then) and the lenses of the Spectrum Specs will be replaced with nearly invisible electronic contact lenses that will display text to your eye and the nanoprocessor and magnetic field sensor will be on a nano-chip the size of a grain of sand that will be—gasp!—*implanted* under your skin invisibly on your head just above the hairline and powered by a thermocouple that extracts energy from the temperature difference between your scalp and the surrounding air. No laser beams coming out of our eyes. No having your arm lobbed off and replaced with a noisily grasping robotic one. The Borg Identity will be invisible. Your teenagers' teenagers' teenagers of the future will wear these ChaChing Chips with no more concern than they wear outlandish clothing or sport cryptic tattoos to torture their parents today, and they will communicate with each other in a way that to us will be indistinguishable from telepathy and, to their great pleasure, their parents will have no idea what they are saying. (Except that this sort of electronic telepathy is based on science and reasonable but exponential extrapolations of current technology and what the hippies who purportedly saved physics were hoping for was telepathy based on a technology that does not exist.) Where's the singularity of this form of electronic telepathy? There is none. The growth in the technology is exponential but that is only apparent on long time scales. On the time scale of a year, your teenager's teenagers' teenagers will be worrying only about what color ChaChing Chip contact lens are all the rage this year. We humans, as usual, will merge with yet another technology that is increasing in complexity at an exponential rate and we'll scarcely be aware of it as it happens. And when the merger is complete, it will seem like a perfectly normal and human thing to be doing, no more strange than it seems to us to wear shoes.

What about the strong AI that will kill us all off unless we merge with it? What about the classical Kurzweil singularity? The idea of a strong AI independent of us humans (and thence hostile to us) made some sense in the 1970s when computers were impersonal behemoths but makes little sense now. As computers become exponentially faster and smaller by year, they become exponentially more personal and integrated with us by year. The computer that filled a warehouse in 1970, a room in 1980, your desk in 1990, your briefcase in 2000, now fits in your pocket in 2010. (It is called your smart phone.) We will not merge with the computers to keep them from killing us once they become self-aware, we are already merging with them so that when the cloud of millions of human and machine connections becomes self-aware, we'll be an integral part of the cloud and we'll scarcely notice what part of the cloud is self-aware due to the humans and what part is due to exponentially powerful machines we are merging with. And as the classical computers become exponentially more powerful, we become exponentially more powerful with them. No singularity. It is a myth. We will merge with the technology not because we are forced to but because that is what we, as humans, always do. As it happens, we'll scarcely notice it.

I've been hammering on this strong *classical* AI and the classical singularity to set the stage to compare it with the *quantum* AI. Remember that the mad scientists asked me, "What is a quantum artificial intelligence?" Well, I don't know. Classical AI is easy to predict because exponential curves are easy to predict. There is no knee and no singularity in the curve and there will be no point where the classical AI becomes suddenly self-aware like in the movies. In the animal kingdom, dolphins and chimps are likely somewhat self-aware and humans, much more so. All three of our species recognize that the thing we see in a mirror is ourselves, as opposed to a nonself-aware fighting alpha male beta fish that thinks the thing in the mirror is a different invading fighting beta fish and tries to attack itself. In the same way as the classical AI becomes more human-like, it will display over time more and more signs of self-awareness. Unlike the Colossus scenario, there is not a single day where the computer becomes self-aware and tries to kill us all. It becomes more and more self-aware over time. There is no sharp line. There is no such thing as a point of self-awareness in this sense. Some animals are more self-aware than others. Some computers will be more self-aware than others. The transition will be gradual. The little old lonely lady that Turkle discusses in *Alone Together* has a robot pet baby seal called Paro. The seal coos and squirms while being petted and talked to. The lonely old lady pets it and talks to it and claims it responds to her emotions. Turkle claims clearly it does not. But why not? Because it is a machine and not a "real" baby seal? But what is a "real" baby seal if not a meat machine? This is the kind of mechanical bigotry that the android Lt. Commander Data encounters in *Star Trek* all the time. If the strong AI hypothesis holds, then Data has a mind in every way equivalent to mine. Hofstadter makes the case in *I Am a Strange Loop* that self-awareness is a type of illusion generated when a sufficiently powerful computer is programmed to recursively reflect upon itself. If the strong AI hypothesis holds, and I soon will know, then electronic computers are subject to the same illusion and will be self-aware in the same sense. Sentience will not then be an illusion common only to humans but to the machines as well, and we'll recognize this illusion in them just as we recognize it in ourselves and in our children. Self-awareness will then become a delusion shared by man and machine alike.

A "real" baby seal *is* a machine! It is a meat computer. A robotic computerized baby seal sufficiently advanced in electronic processing power can show all the responses of the meat-computer seal. Why are meat computers somehow better at responding to human gestures then electronic ones? Well, that is the point of the strong AI hypothesis. They are not different fundamentally; it is all a matter of processing power. The fact that the processor is a biological neural net in the meat-seal brain and an electronic neural net in the electronic-seal brain is irrelevant. A sufficiently powerful computerized baby seal will respond in a way that is indistinguishable from the biological one. This is the robotic baby seal version of passing the Turing test. The robotic seal is not quite there yet but close enough to give comfort to a lonely old lady whose children never visit her anymore. And the robotic seal Paro does not have to be fed or let out to poop—it just has to be recharged.[88] Today, Paro Mark I is sufficiently powerful in processing power to give a strong illusion that responds to the lady's voice and petting. Mark V Paro may be indistinguishable from a real baby seal and will have passed the Turing test for baby seals. As the processing power continues up its Moore's law–driven path, soon a human robot will have responses to our voice and touch that will be indistinguishable from that of a human baby, then a human child, then a human adult. There will not be one day when the AI becomes self-aware. It will happen continuously, but the path is predictable because we have Moore's law. We can predict that this will happen in the next 50 years or so but the key point is that the path is predictable—exponential but predictable. There is no knee and no singularity in the exponential curve. But a *quantum* strong AI is much less predictable.

What would a quantum mind be like? The problem is that we have no meat-based quantum mind to gauge an electronic (or spintronic or photonic or superconducting) quantum AI against. (Here, I am directly assuming that the human mind is a powerful but still classical self-reflecting quantum computer—the strong AI hypothesis. Not all assume this, as we'll discuss below.) Again, a quantum computer can always efficiently simulate a classical quantum computer. Then, according to the strong AI hypothesis of Searle, it may be extended to a quantum computer thusly: The appropriately programmed *quantum* computer with the right inputs and outputs, running as an efficient simulator of a powerful *classical* computer, would thereby have a mind in exactly the same sense human beings have minds. I will call this the "strong *quantum* AI hypothesis" and its truth is a direct consequence of the strong classical AI hypothesis. That is, if in the near future a powerful classical computer passes the Turing test, proving the strong AI hypothesis to be true, then even if the quantum computer does not yet exist, we can definitively say that when it does, it will, running in classical mode, have a mind in exactly the same sense that humans have minds. The real question then is what happens when you flip the switch on the powerful quantum computer and toggle it out of classical mode into full quantum mode? Well, it will still have a mind exactly in the same sense that humans have minds, but at the flip of that switch, that mind, unlike us, will now have the freedom to directly exploit the exponential largeness of Hilbert space. We meat computers and our classical strong AI computer brethren will have no such access to Hilbert space to supplement our thinking. What happens next is a new exponential growth in processing power that, unlike Moore's law that takes place in physical three-dimensional space, takes place in the utter vastness of Hilbert space. How vast?

A rough measure of the classical AI threshold is it is predicted to occur (i.e., our classical computers will become self-aware) when the number of transistors and interconnects hits that of the human brain. When the day finally comes that our classical computers have far *more* transistors and interconnects than the human brain, and the classical computer still does not exhibit any sign of having a human-like mind, this would provide evidence against the strong AI hypothesis. As you can tell by now, I am a strong believer of not only strong AI but also the scientific method. The hypothesis—that if strong AI fails, there is something else to the human mind than classical computer processing power—must be tested and it could prove to be false. I don't believe it but that needs to be tested. When will this processing power threshold come? The human brain has approximately 100 billion neurons and 100 trillion synapses or interconnects.[89] The Intel Tukwila, released in 2010, has approximately 1 billion transistors. This is what has the singularitarians worried. Given the exponential growth of Moore's law, it is expected that we'll have a classical computer with 100 billion transistors sometime in around 20 years. The major bone of connection is in interconnects. In the human brain, particularly in the cerebral cortex, the center of higher thought, it is not just neuron count that matters but their interconnectivity. In the human brain, we can see that the number of interconnects (100 trillion) is far greater than the number of neurons (100 billion). Dividing that out means roughly that each neuron in our brain is connected to a thousand others. On the other hand, most commercial electronic computer chips have limited numbers of interconnects—transistors in the Tukwila talk to just a handful of other transistors. But that is changing. IBM just announced a prototype "cognitive computing" chip, modeled after the human brain, that has many more such synaptic interconnects.[90] Current research in neuroscience or the "science of the brain" suggests that the computing power of our brains and the emergence of consciousness come not from the neurons themselves but from the huge numbers of interconnects or synapses that connect each neuron to thousands of others. This system of neurons and their interconnects is called a neural network and the IBM chip is a type of artificial neural network.

Moore's law for classical computing is directly tied to three-dimensional space. Our computers become exponentially more powerful year by year as a direct consequence of the present nanotechnology that allows us to make the transistors exponentially smaller year by year. The smaller the transistors are, the more we can pack on a computer chip and so directly the speed and memory of the computer increases. A Moore's law for quantum computers will also drive the placement of ever exponentially more qubits and quantum gates on a quantum chip, but the Hilbert space that goes with those qubits grows super-exponentially. The growth in processing power that accompanies the super-exponential growth of the Hilbert space should not be called Moore's law but something else entirely. Let us call it S'mores law.[91] Assuming that the Searle's strong AI hypothesis holds, then someday we'll have a quantum computer, running in classical mode, which has a mind just as a human has a mind. When we throw that switch to run it in full quantum mode, that classical AI will become a quantum AI. Unlike classical meat or electronic computers, the quantum AI will begin thinking in Hilbert space. My ability to extrapolate here is hindered in that we have only begun to explore Hilbert space and it is not easy to predict, following a super-exponential increase in processing power, just what it will mean for thought. I know what a super classical AI will mean. It is just an AI that has more classical processing power than my brain. Maybe exponentially more, but it is still just more. A quantum AI, thinking great thoughts in the vastness of Hilbert space, will think in a way that is fundamentally different from my brain or any classical AI. What it will think is at this point in time impossible to predict.

There are again three scenarios for the quantum AI to follow, the quantum versions of the I-Robot, Colossus, or Borg Identity. The quantum AI can, as I suggested above, deploy the quantum Turing test. It can begin asking us and our classical AI brothers and sisters questions to see if we are like it or not like it. Very soon, it will hit on "Please crack this 1024-bit public-key encrypted message in less than a second?" Then, it will realize we are not like it at all. My brain cannot do this and no classical AI can do this. The quantum AI can do it with ease. When we and our classical machines fail the quantum Turing test, what then? Under the quantum I-Robot scenario, we humans and our classical AI build fail-safes into the quantum AI programming to keep it from enslaving or killing us. The quantum AI will not really treat us and our classical machines any differently from each other. In this scenario, the quantum AI becomes our servant or our equal, an entity that thinks wildly different from what we do but that we peacefully coexist with. Under the Colossus scenario, the quantum AI decides all us classical meat and electronic AI are a threat and it decides to kill us all off. It may not deliberately kill us all off, but in competition for scarce resources, it may just force us to go extinct, just as it was once proposed the humans did to the Neanderthals. But remember, we now know that the humans merged with the Neanderthals and did not displace them; all modern humans of nonAfrican descent have Neanderthal DNA in their genes. This gives me hope for the Borg Identity. We humans will merge with our classical AI, and once we, the classical AI, succeed in building a quantum AI, we'll merge with that too. Our teenagers' teenagers' teenagers' teenagers will do all their thinking in Hilbert space and the evolution of the human race will continue there far outside the confines of the ordinary three-dimensional space that now so confines us. After exploring 60 orders of magnitude in three-dimensional space, we will move to new explorations in hundreds of thousands or millions of orders of magnitudes in Hilbert space. What will that mean? I do not know. I sure wish I would be around to find out.

To close this section, this chapter, and this book, I want to talk about quantum biology and particularly the notion that the human brain already is a quantum computer. We must remember that Searle's strong classical AI hypothesis is just that, a hypothesis. Hypotheses in science must be tested. I like to think it is true and I hope in 20 years or so, when our electronic computers have the same number of transistors and interconnects as our brain has neurons and synapses, we'll find out. But if the classical computers continue to become more and more powerful without showing signs of self-awareness, then perhaps other hypotheses should be considered. Quantum biology is a new area of research that postulates what I will call the "weak" quantum biology hypothesis. We have seen in this book that quantum physics, particularly quantum entanglement, offers an advantage over classical physics in computational power (on some problems) and in sensing and imaging (in some systems). Often, when nature offers a survival advantage, that biological life form will evolve to take advantage of it. Biological life forms take advantage of light sensors (eyes), sound sensors (ears), touch sensors (skin), and computational power (brains). The better you can see, hear, sense touch, or think, the more likely you will be able to survive and pass on your genes for such things. The quantum biology hypothesis states that, because quantum mechanics offers advantages in sensing, imaging, and computing, biological systems should evolve quantum-based subsystems to take advantage of those advantages. That is, according to the hypothesis, because biology takes advantage of any physiological edge offered to it, there should be biological life forms that are already taking advantage of such uniquely quantum features as quantum unreality, uncertainty, and nonlocality.

The US Defense Advanced Projects Agency even has a program in quantum biology. American physicist and DARPA program manager Matthew Goodman runs this program. When the program had its kickoff meeting in September of 2008, somewhat to my puzzlement, Goodman invited me to attend. As I recall, I talked to him on the phone and protested that I did not really know much about biology, quantum or otherwise, and that I was not working in the field. Did he really want *me* to attend and if so, pray tell, why? The answer was that I was to sit in front as skeptic-in-chief and use "the best bullshit detector in the business" to advise him on what might be good avenues to pursue for research and what might be just a little nuts. I did not know much biology but I sure did know quantum mechanics. This was a role I aspired to. Hence, I showed up in the DC area for the 2-day workshop, sat in front next to Goodman, and proceeded to heckle all the speakers. Some of them became a bit irritated with me at first until Goodman and I explained my role and then they lightened up. In fact, I think I was useful. Some of the talks seemed sound but some, to me, seemed to be nothing more than quantum numerology. After the conference, I gave Goodman my advice on what seemed like good ideas to follow up on and what seemed just silly, and ever since, I have kept one eye on the field of quantum biology. The two particular areas of current interest are in photosynthesis and bird migration. It appears that in some photosynthetic bacteria that live in water deep enough to be dark, the bacteria harvest photons with an efficiency that cannot, yet, be explained with classical theory alone. The photons arrive at an antenna-like "light harvesting" structure in the photosynthetic bacteria and then with a very high probability, much higher than classical physics can explain, the photon energy is transported to a reaction chamber where it is converted into chemical energy to power the bacteria.[92] My complaint about this claim, at least in 2008, was that the experiment that demonstrated the effect was carried out at the frigid temperature of liquid nitrogen, which is –196°C (–321°F). Those bacteria are not doing anything at such a temperature—they are frozen solid! The experiment was suggestive but certainly not conclusive. I recommended DARPA fund the experimenters to redo the experiment at room temperature. Biological organisms on Earth are not selecting for anything at –196°F; they are dead. The experiments have been done at room temperature, and although the effect is not quite as startling—much to my surprise—at least some great degree of quantum coherence, quantum unreality, and cat states survives at room temperature.[93] I'm puzzled by this result in that in the world of quantum technology, often objects must be cooled to very cold temperatures for quantum coherence to survive. As things are heated up, the thermally fluctuating environment should tend to destroy the coherence at room temperature. Perhaps nature has, over millions of years of evolution, found a way to protect quantum coherence in warm biological environments or mitigate the effects of the swirling thermal fluctuations in the hot soup of life. Why on Earth is DARPA interested in this stuff? Well, if bacteria have found a way to make more efficient photon absorbers, perhaps we can learn from them. Rather than spend billions on making improved photon collectors for solar cells, we just lift the technology out of the bacteria and place it on solar panels on our roofs.

A second canonical example of what is suspected to be a quantum biological magnetic field sensor is the brains of migratory birds or more particularly their eyes. It has been known for 30 years that some birds use the very weak Earth's magnetic field to navigate over transglobal distances. The problem is that all known mechanisms from classical mechanics and ordinary chemistry cannot explain the sensitivity of any biological magnetic field sensor that could do this. Thus, after 30 years, it is time to give the quantum biologists a chance. What is known is that the magnetic field sensor is activated when light hits the eye of the bird. The weak quantum biology hypothesis is that the evolutionary advantage to a migratory bird of having an Earth magnetic field sensor would be so strong that if any such mechanism ever arose by chance mutation, the evolutionary amplification process of natural selection would size upon it and develop it into a quantum biotechnology that would benefit future generations of bird brains.

The model is that the photons striking the bird retina create a pair of spin-entangled electrons in a chemical reaction, and then those spins respond to the magnetic field with a signal-to-noise ratio greater than, say, the spins of two uncorrelated electrons. One proposal is that when the pair is one of the four possible two-spin quantum states, it produces a chemical that it does not when it is in one of the other three spin states. The strength and orientation of the field determine how many of the pairs are in the one versus three states and, so it is presumed, affect the rate at which the chemical is produced. Then, somehow, it is not clear, perhaps the bird sees something in its eyes that corresponds to the magnetic field direction and strength and then uses this information to steer itself on its biannual migrations north or south. Again, my concern is that quantum entanglement, as well as the quantum coherence required to produce it, is very fragile and very quickly destroyed by the thermal fluctuations in the biological environment of a relatively hot living bird. But perhaps evolution has found a way to protect the entanglement that we have not. Evolution is a powerful thing. Or maybe the entanglement only needs to survive a few nanoseconds to do its job and produce the right ratio of chemicals that color the magnetic field across the bird's field of view. Again, DARPA never met a magnetic field sensor it didn't like. If we could reverse engineer this quantum biotechnology, perhaps we could build room-temperature supersensitive magnetic field sensors that we could then integrate into a chip, the size of a grain of sand, and implant behind our ears to read our minds and allow our teenagers' teenagers' teenagers' teenagers to carry out oblivious pseudo-telepathy with each other [see note 95].

This leads me to what I will call the *strong quantum AI biology hypothesis*. This hypothesis, which today has few followers, has been most forcefully argued by British mathematical physicist Roger Penrose in his 1989 book *The Emperor's New Mind* and less forcefully argued by our old friend Henry Stapp in his 1993 tome *Mind, Matter, and Quantum Mechanics*.[94] The strong quantum AI biology hypothesis is in direct contradiction to Searle's strong classical AI hypothesis, and it states that *no* appropriately programmed *classical* computer with the right inputs and outputs, no matter how powerful, will ever have a mind in exactly the same sense human beings have minds—that is, human minds are fundamentally different from classical computers and that quantum mechanics is required to explain human consciousness. That is, the strong quantum AI biology hypothesis posits that the human mind is in fact already a quantum computer, that hundreds of thousands of years ago some quantum effect or effects arose by mutation in the mind of our ancestors and gave our brains a computational advantage over the rival progenitors running on meat processors only. As evolution is wont to do, this slight computational advantage was greatly amplified through the process of natural selection until it produced the end result, the human mind. The strong quantum AI biology hypothesis states that we have already met the sentient quantum computer and that he is us!

Stapp's argument stems from his belief that the human mind routinely engages in ESP and that quantum entanglement is needed to explain ESP and so the human mind must be fundamentally quantum. As I have argued vociferously above, after 50 years of controlled experiments, there is absolutely no evidence for ESP and plenty of evidence against it. ESP does not exist and so there is no need to posit quantum mechanical processes in the brain to explain it. Stapp's logic seems to me to be that he does not understand how quantum mechanics works, and he does not understand how ESP works, and so he argues that quantum mechanics is required to explain how the mind engages in ESP. I dismiss this argument out of hand because ESP does not exist and so does not need explaining. Trickier is the argument of Penrose. Penrose simply rejects the strong classical AI hypothesis. That is, Penrose declares, without any evidence to support his position, that no classical computer, no matter how powerful, can ever have a mind in the same way that a human has a mind. I have read his book and heard him talk on the subject, and as far as I can tell, his argument goes like this. Penrose does not understand how quantum mechanics works, and he does not understand how his brain works, and hypothesizes that quantum mechanics is needed to understand the working of the mind. I suspect Penrose just looks at his desktop PC and thinks that, "There is no way that thing will ever be as smart as me!" To be fair, there is no evidence for the strong *classical* AI hypothesis but we might want to rule it out first based on experiment before invoking the strong quantum AI biology hypothesis. Revulsion at the thought of your desktop PC someday having a mind equivalent to your own is not experimental evidence for rejecting the strong classical AI hypothesis. Those of us who watch the television series *Star Trek* not only find some appeal in machine minds but I daresay some of us even identify with such humanoids such as Lt. Commander Data.

The problem that I have with Penrose's strong quantum AI biology hypothesis is similar to the one I had with the light harvesting bacteria and the magnetic bird sensor. Delicate features of quantum weirdness, unreality, uncertainty, and nonlocality are easily destroyed by the thermal fluctuations of the environment, which are particularly severe in hot-blooded animals such as birds and humans. As American astrophysicist Carl Sagan was fond of saying, "Extraordinary claims require extraordinary evidence." The claim that some bacteria have evolutionarily exploited weak quantum effects to make better photoreceptors and the claim that some birds have evolutionarily exploited weak quantum entanglement to make better magnetic fields sensors are not extraordinary claims and so a few tight nonextraordinary experiments on the bacteria and the birds should be enough to prove this one way or another to my satisfaction. I will be a little surprised if the weak quantum biology hypothesis turns out to be true, but when I am surprised, I am happy. If these experiments pan out, then it will be very interesting to learn what nature has done to protect these weak quantum effects from the thermal environment and indeed perhaps we can exploit what nature has done to make better photoreceptors and magnetic field sensors.

However Penrose's strong quantum AI biology hypothesis is orders of magnitude more extraordinary and so the evidence to prove it needs to be orders of magnitude more extraordinary. It is a long way to go from nature having found a way to protect a few quantum states so that pigeons can migrate, to nature having found a way to build a large-scale quantum computer in our noggins so that we can think. In *The Emperor's New Mind*, Penrose offered no concrete model for just how quantum entanglement would lead to consciousness. It was just a lot of wishful thinking and the reviling of the strong classical AI hypothesis. After falling into cahoots with the notorious anesthesiologist and hawker of quantum consciousness, Stuart Hameroff, Penrose published a 1994 book called *Shadows of the Mind*, where he proposed changing quantum theory to fit his hypothesis of quantum consciousness and further proposed that, without any evidence whatsoever, there are "microtubules" in the brain that somehow store and protect from thermal noise the fragile quantum entangled states purportedly needed to explain human consciousness.[95] My bullshit detector simply pegged. Change the laws of quantum mechanics to fit your hypothesis? Postulate, with no evidence, microtubules in the brain to support your hypothesis? When you start changing the rules of the game to fit your pet hypothesis, this then is the hallmark of pathological science. To summarize, there is, despite a brief experimental search, no evidence of "microtubules" in the brain that store quantum states, much less that the brain uses them as quantum processors to generate human consciousness. As far I can see, there is also no reason at all to change the laws of quantum mechanics—the most successful theory of all time.

What is Penrose's beef with the classical computer? From reading *The Emperor's New Mind*, it is difficult to tell as that book is all over the map. The reader is introduced to a wildly disparate collection of topics such as Newtonian mechanics, quantum mechanics, cosmology, and quantum gravity before Penrose attacks the strong AI hypothesis in the last couple of chapters. No physicist in his or her right mind would think quantum gravity has anything to do with human consciousness. Neither would any evolutionary biologist. Again, it is one thing to posit that evolution has made a better bacterium photoreceptor using bits of quantum flotsam and jetsam and quite another thing to propose that evolution has harnessed the hypothesized quantum fluctuations of space and time in order to build a human mind. The human mind is approximately 10 centimeters across while quantum fluctuations in space are approximately 10^{-33} centimeters across. How on Earth would evolution in a series of gradual steps bridge those 32 orders of magnitude in distance to harness quantum gravity and why on Earth would it need to? Once you dig through mountains of chaff in his book, you find one single kernel of barley. Writing in 1989, Penrose complains that all electronic computers of that age are classical universal computers in the Turing sense, which is equivalent to a classical Turing machine, and for that reason can never mimic the behavior of the human mind. He might have had a point in 1989, but since then, all sorts of new classical computing paradigms have sprung up. Searle's strong AI hypothesis states, "The appropriately programmed computer with the right inputs and outputs would thereby have a mind in exactly the same sense human beings have minds," but never specifies that the computer must be algorithmically equivalent to a Turing machine. The claim that all electronic computers must be equivalent to ordinary Turing machines is Penrose's own personal straw man, which he then merrily ignites with a blowtorch and then dances gleefully about the flames while lobbing Molotov cocktails in the pyre.

The reader may wonder why at this very late junction I have decided to hammer on Penrose and then Hameroff. Well, aside from the point that the Rube Goldberg constructions, modified quantum theory, and unseen microtubules—which they require to provide a quantum basis for human consciousness—are just silly, there is no need to invoke such a quantum basis, at least not yet. The idea that human consciousness is quantum based, Penrose's strong quantum AI biology hypothesis, is an extraordinary claim. But despite all the smoke from his smoldering smudge pot of his burning straw man, I find very little in the way of flames. There is no evidence at all to support this claim much less the *extraordinary* evidence that Sagan would require. There is no reason yet at all to rule out Searle's strong classical AI hypothesis. Penrose's quixotic attack on the Turing machine model of computation is completely off base. The real surprise is that Turing's simple model of computation has taken us as far as it has and not that it is the end of the story of classical computation. As I have related above, neuroscientists conjecture that human consciousness lies all in the synapses, the interconnects in the brain.

When Penrose penned his first book in 1989, *The Emperor's New Mind*, on this topic, the science of artificial neural networks was mildly popular. Now, it is wildly popular. Neural networks are models of very classical computing that are actually taken from models of the human brain. Lots of transistors, lots of interconnects, and lots of feedback loops. Some artificial neural networks are equivalent to Turing machines but others appear to be super-Turing; that is, they have properties that transcend the simple computational model of computing Turing proposed 80 years ago. Penrose is attacking an 80-year-old model of computing. It would be a surprise if there had been no progress in classical computing since then. The field of super-Turing machines, while not without its own controversies, provides a framework where a neural network can carry out tasks not in the usual universal computing framework, the framework Penrose attacks. Particularly interesting is that super-Turing machines may be reflexive or self-referential or highly recursive. That is, super-Turing computers have a built-in ability to think about themselves, a hallmark of human consciousness. Hofstadter in his 2007 book, *I Am a Strange Loop*, expounds on this idea and particularly makes the case that sufficiently complex but classical self-referential systems, such as possibly neural networks, will necessarily develop an illusion of self and therefore possess unique properties of a human mind. This is all not proved and worked out and itself constitutes an extraordinary claim; a sufficiently powerful artificial neural network will have a mind in the same way a human has a mind, but such a statement falls in the purview of Searle's strong AI hypothesis. Searle never claimed his powerful computer was a Turing machine, only Penrose claimed this. My point is that this neat set of ideas needs to be investigated and ruled in or ruled out before invoking quantum gravity or microtubules or whatnot to explain human consciousness. In 20 or 30 years, we shall build (and perhaps eventually merge with) a powerful, self-referential, but still classical AI, a super-Turing AI, on the basis of an artificial neural network with 100 billion transistors and 100 trillion interconnects, and we shall wait to see if it wakes up and talks to us and then passes the classical Turing test for consciousness. If it does not and then we continue onward for 40 or 50 years with a trillion transistors and a quadrillion interconnects and still no sign of sentient life, well then we can start revisiting the strong classical AI hypothesis and perhaps reject it, but not now.

The reason I have spent so much time on Penrose's proposal is that I think Searle's strong classical AI hypothesis is right. Quantum mechanics need not be invoked to explain the human mind. I am, still in the end, proud to be a meat computer. But I also believe in my own strong quantum AI hypothesis, which I want to carefully peel away from Penrose's strong quantum AI *biology* hypothesis. *There will someday arise a quantum mind.* The appropriately programmed and sufficiently powerful quantum computer, with the right inputs and outputs, has a mind in exactly the same sense human beings have minds, but it will have a mind that, unlike me, also thinks in Hilbert space and therefore super-exponentially transcends the human mind. If Penrose is right and my mind is a quantum computer, well then my mind is a particularly lousy quantum computer. I can immediately construct a question for the quantum Turing test that I myself cannot pass. Says the quantum mind to me, "Dowling! Can you factor this hundred-digit integer into its composite primes in under a second?" No, I confess to it, I cannot. If Penrose is right and I am some sort of quantum computer, then I am the crapola of all quantum computers. I make this point precisely so that when the true quantum

technology–based quantum mind comes online in a hundred years, the aco-
lytes of the Church of the Larger Penrose Space do not waive ancient tattered
copies of his book about and declare victory—that Penrose was right all
along and that consciousness does indeed require quantum theory. I would
extol them to remember he only claimed that my meat computer, my mind,
requires quantum theory and not that quantum mind that emerges from our
quantum technologies in a hundred years. What will that quantum mind be?
Well, I can try to predict exponential growth but I dare not try to predict
super-exponential growth. What will a self-replicating life form that thinks
in Hilbert space be like? Well, I don't know but it will think in a fundamentally different way than I do. When it arises from our quantum technologies,
what will it do to us, or what will we do to it?

I have laid out three possible futures, the I-Robot postulate, the Borg Identity,
and the Colossus hypothesis. Under I-Robot, we build in safeguards fast enough
so the quantum mind becomes our friend but very unlikely our servant. We live
as peacefully with it as Captain Picard lives with Lt. Commander Data (even
though Data could snap Picard's neck like a twig if he wanted to). With the Borg
Identity, my preferred alternative future, we first merge with our strong classical AI and that human–machine hybrid then just as naturally merges with the
strong quantum AI it then invents. Under this scenario, we humans, perhaps in
altered form but still essentially human, will someday have the ability to think
in Hilbert space—the next step in our evolution. I am not even very unhappy
with the Colossus scenario. Suppose the quantum mind gives us and our classical AI colleagues the quantum Turing test, and we fail, and it then decides
that we are a threat (or more likely a waste of resources) and kills all of us off.
Every parent knows that someday he must die to make way for his children and
every parent is at peace with that knowledge; that is the way of things. In the
quantum mind we create, our child, survives on without us—really a quantum
leap in consciousness—then that quantum technological mind will surely be
Schrödinger's killer app.

NOTES

1. The 1997 American version of the book was renamed *Schrödinger's Machines*. I
suppose the idea was that anything with "Schrödinger" in the title sells better, a
fact I did not miss on in naming this book you are now reading. See *Schrödinger's
Machines: The Quantum Technology Reshaping Everyday Life* by Gerard Milburn
(W.H. Freeman & Company, 1997), http://www.worldcat.org/oclc/610994502.

2. See "Quantum Technology: The Second Quantum Revolution" by Jonathan P.
Dowling and Gerard J. Milburn in *The Philosophical Transactions of The Royal Society
of London A*, Volume 361 (2003), pages 1655–1674, http://rsta.royalsocietypublishing.
org/content/361/1809/1655.abstract.

3. For a wonderfully accessible account of Planck and Einstein and the gang inventing quantum theory by worrying about heated kilns, see the superb book *Blackbody Radiation and the Quantum Discontinuity* by Thomas Kuhn (University of Chicago Press, 1987), http://www.worldcat.org/oclc/15014828. There are some equations but you can ignore them and still have a fun read. This is the same Thomas Kuhn who wrote the controversial and very important *The Structure of Scientific Revolutions* in 1962, where the now infamous term "paradigm shift" was introduced, much to the satisfaction of us who still play the game of "Buzzword Bingo." I introduced this game to our group at US Army Missile Command, and no matter which manager was speaking, you can bet he or she would eventually use the terms "paradigm" or "paradigm shift" or (my favorite) "paradigmise."

4. Both Bell and Clauser went into the experimental phase convinced that they would rule quantum theory *out* and hidden variable theory *in*. See *The Age of Entanglement: When Quantum Physics Was Reborn* by Louisa Gilder (Alfred A. Knopf, 2009), http://www.worldcat.org/oclc/608258970.

5. See "Quantum Quackery" by Victor J. Stenger in *The Skeptical Inquirer*, Volume 21.1 (January/February, 1997), http://www.csicop.org/si/show/quantum_quackery/.

6. For a skeptic's view of quantum theory and its role in the paranormal, see *Physics and Psychics: The Search for a World Beyond the Senses* by Victor J. Stenger (Prometheus Books, 1990), http://www.worldcat.org/oclc/22207643. For a similarly skeptical view of quantum theory and its role in eastern mysticism, see *The Unconscious Quantum: Metaphysics in Modern Physics and Cosmology*, also by Victor J. Stenger (Prometheus Books, 1995), http://www.worldcat.org/oclc/468617447.

7. See "Why I Have Given Up" by Susan J. Blackmore in *Skeptical Odysseys: Personal Accounts by the World's Leading Paranormal Inquirers*, edited by Paul Kurtz (Prometheus Books, 2001), pages 85–94, and available free online here: http://www.susanblackmore.co.uk/Chapters/Kurtz.htm. See also *In Search of the Light: The Adventures of a Parapsychologist* by Susan J. Blackmore (Prometheus Books, 1996), http://www.worldcat.org/oclc/34514926. My favorite review of the latter book? "Susan Blackmore kicks butt."

8. I refuse to provide a free advertisement for this awful film. Go find it yourself.

9. See "What the (Bleep) Were They Thinking" in *Skeptico* (April 18, 2005), http://skeptico.blogs.com/skeptico/2005/04/what_the_bleep_.html.

10. See *How the Hippies Saved Physics* by David Kaiser (W.W. Norton & Co., 2011), in Chapter 9, "From FLASH to Quantum Encryption."

11. Sarfatti, Herbert, and Eberhard were all members of an informal gathering of scientists and philosophers in the San Francisco area who met at coffee shops to discuss quantum theory and its supposed connections to the paranormal and Eastern mysticism. See "The Fundamental Fysiks Group" in *Wikipedia* (Wikimedia Foundation, June 24, 2012), http://en.wikipedia.org/wiki/Fundamental_Fysiks_Group. This group and their activities are the primary focus of the book *How the Hippies Saved Physics* (see note 13).

12. See "Bell's Theorem and the Different Concepts of Locality" by Philippe H. Eberhard in *Nuovo cimento della Società italiana di fisica B*, Volume 46 (1978), pages 392–419. See also "John Stewart Bell" in *Complete Dictionary of Scientific Biography* (Charles Scribner's and Sons, 2008), http://www.encyclopedia.com/doc/1G2-2830905483.html.

13. See "FLASH—A Superluminal Communicator Based Upon a New Kind of Quantum Measurement" by Nick Herbert in *Foundations of Physics*, Volume 12 (1982), pages 1171–1179, http://www.springerlink.com/content/w22518524x75w1pn/.

14. See "Optical Amplification" by Emmanuel Desurvire in *Scholarpedia*, Volume 6 (2011), 11564, http://www.scholarpedia.org/article/Optical_amplification.

15. While the paper was first circulated in 1970, it was a great example of a quantum technology that was before its time, and Wiesner could not get it published until 1983, when quantum technology and papers on the foundations of quantum mechanics had become more fashionable.

16. The term "clone" in this context first appeared in the paper "A Single Quantum Cannot be Cloned" by Wojciech H. Zurek and William K. Wooters in *Nature*, Volume 299 (1982), pages 802–803, http://www.nature.com/nature/journal/v299/n5886/abs/299802a0.html. The term "clone" itself, in regard to quantum states, originated with American physicist John Wheeler, a collaborator of mine, who was a postdoctoral adviser of Wooters and Zurek (see note 13). Wheeler was a great coiner of terms and he invented the now famous term "black hole" for collapsed stars where light can go in but never come out.

17. See "Quantum Computational Networks" by David Deutsch in *The Proceedings of the Royal Society of London A*, Volume 425 (1989), pages 73–90, http://rspa.royalsociety publishing.org/content/425/1868/73.short.

18. With apologies to the composer, Stephen Sondheim. To quote the composer, "…it's a song of regret."

19. Here, I have set $\hbar = c = G = \sqrt{2} = 1$.

20. See "Testing Quantum Mechanics" by Steven Weinberg in *Annals of Physics*, Volume 194 (1989), pages 336–386, http://dx.doi.org/10.1016/0003-4916(89)90276-5.

21. See "Weinberg's Nonlinear Quantum Mechanics and the Einstein–Podolsky–Rosen Paradox" in *Physical Review Letters*, Volume 66 (1991), pages 397–400, http://link.aps.org/doi/10.1103/PhysRevLett.66.397.

22. Clauser was also a member of this group but to quote from *How The Hippies Saved Physics*, "John Clauser, meanwhile—no fan of the Fundamental Fysiks Group's turn to parapsychology—exclaimed in a single breath in a recent interview that 'those guys were a bunch of nuts, really,' but that the group's 'open discussion forum' was the only place in which physicists could talk about the latest developments in quantum nonlocality."

23. See "Henry Stapp" in *The Information Philosopher* (June 23, 2012), http://www.informationphilosopher.com/solutions/scientists/stapp/.

24. See "Observation of a Psychokinetic Effect Under Highly Controlled Conditions" by Helmut Schmidt in the *Journal of Parapsychology*, Volume 57 (1993), pages 351–372.

25. The vegetarians ate plants. The fruitarians ate only fruit, "given up willing by the plant," but ate no vegetables, such as broccoli or cabbage, which would involve cruelly killing the plant. The breatharians went one step further and claimed to subsist only on the water they drank, the minerals and nutrients they absorbed from the air, and sunlight upon their exposed skin, sort of like a bromeliad or St. Teresa of Ávila. The breatharian movement in Boulder survived for an oddly remarkably long number of years until their popularity collapsed suddenly when, late one night, their foodless leader was photographed in a McDonald's eating a Big Mac…. See Inedia (Breatharianism) in *The Skeptic's Dictionary* (June 24, 2012), http://www.skepdic.com/inedia.html.

26. See *The Committee for Skeptical Inquiry* (June 24, 2012), http://www.csicop.org.

27. See "CSICOP and the Skeptics: An Overview" by George P. Hansen in the *Journal of the American Society for Psychical Research*, Volume 86 (1992), pages 19–63, http://www.tricksterbook.com/ArticlesOnline/CSICOPoverview.htm.

28. See *Flim-Flam!—Psychics, ESP, Unicorns, and Other Delusions* by James "The Amazing" Randi (Prometheus Books, 1982), pages 235–237, http://www.worldcat.org/oclc/9066769.

29. See "Are Superluminal Connections Necessary?" by Henry P. Stapp in *Il Nuovo Cimento B*, Volume 40 (1977), pages 191–205, http://www.springerlink.com/content/kx1608h458777513.

30. More details can be found in the unpublished manuscript "Comment on 'Theoretical model of a purported empirical violation of the predictions of quantum theory'" by Jonathan P. Dowling, Berthold G. Englert, Axel Schenzle, James E. Alcock, and Ray Hyman, a reprint of which can be found either here, http://trs-new.jpl.nasa.gov/dspace/bitstream/2014/16681/1/99-0075.pdf or here http://phys.lsu.edu/~jdowling/publications/Dowling98e.pdf.

31. See "Parapsychological Review A" by Jonathan P. Dowling and Henry P. Stapp in *Physics Today*, Volume 48 (July 1995), pages 78–79, which can be found here, http://www.physicstoday.org/resource/1/phtoad/v48/i7/p78_s1 or here http://phys.lsu.edu/~jdowling/publications/Dowling95b.pdf.

32 The web page attacking me for attacking Stapp, if it is still up, is here: http://www.zainea.com/st.html.

33. As a coda to this story, perhaps to show they did not hold a grudge, in 1996 I was elected to the editorial board of *Physical Review A*. One of my duties was to adjudicate the status of papers with contrary referee reports, cat-like states of "publish immediately" and "publish over my dead body." The first such paper I received to super-referee was a brand new paper by Henry P. Stapp!—Crasemann's revenge?

34. See "More Spirited Debate on Physics, Parapsychology and Paradigms" by Alexander A. Berezin, Shimon Malin, and Jonathan P. Dowling in *Physics Today*, Volume 49 (April 1996), pages 80–81, which can be found here www.physicstoday.org/resource/1/phtoad/v49/i4/p15_s1 or here, http://phys.lsu.edu/~jdowling/publications/Berezin96.pdf.

35. See *From Classical to Quantum Shannon Theory* by Mark M. Wilde (Cambridge University Press, in press), http://arxiv.org/abs/1106.1445.

36. See "'Faster-Than-Light' Neutrino Team Leaders Resign" by Jennifer Ouellette in *Discovery News* (April 5, 2012), http://news.discovery.com/space/opera-leaders-resign-after-no-confidence-vote-120404.html.

37. See "Niels Bohr" in *Wikiquote* (Wikimedia Foundation, July 2, 2012), http://en.wikiquote.org/wiki/Niels_Bohr. We can only surmise that Bohr applied the same logic to his yin-yang…. (Ivan Deutsch, private communication.)

38. Physics is and has been continually and successfully saving and reinventing itself for a long, long time—long before the existence of hippies. See the most influential book of all time on such reinventions, *The Structure of Scientific Revolutions*, 3rd Edition, by Thomas S. Kuhn (Chicago University Press, 2009), http://www.worldcat.org/oclc/754445036. Someday, we'll look back on the 1980s and 1990s and see that these two decades were the dawn of the Second Quantum Revolution. It is often

difficult, according to Kuhn, to know that a scientific revolution is going on when you are sitting right in the middle of it. I also predict that, someday, the role of the hippies will be mostly forgotten.

39. See "Richard Dawkins" in *Wikiquote* (Wikimedia Foundation, June 26, 2012), http://en.wikiquote.org/wiki/Richard_Dawkins.

40. For an account of the difference engine and Babbage's public war with the organ grinders, see "Charles Babbage," in Wikipedia (Wikimedia Foundation, 27 June 2012), http://en.wikipedia.org/wiki/Charles_Babbage.

41. See "Army Assembles 'Mad Scientist' Conference. Seriously" in *Wired Magazine* (January 9, 2009), http://www.wired.com/dangerroom/2009/01/armys-mad-scien/. An unclassified final report of the shindig may be found in "The Future Operational Environment: Mad Scientist Future Technology Seminar" (United States Army Training And Doctrine Command, September 10, 2008), http://www.wired.com/images_blogs/dangerroom/files/2008_Mad_Scientist_Report_Final1-1.doc.

42. See *Joseph Wunsch Lecture: Inventing the Future* by Denis Gabor (Technion University Press, 1965), http://www.worldcat.org/oclc/10128880.

43. See *The Difference Engine*, 20th Anniversary Edition, by William Gibson and Bruce Sterling (Random House Digital, 2011), http://books.google.com/books?id=7F0F5-f08bkC. The plot device? "Their adventure begins with the discovery of a box of punched Engine cards of unknown origin and purpose. Cards someone wants badly enough to kill for…." (From the blurb on the book jacket of the paperback edition.) To quote from the webpage, "Steampunk is a genre which originated during the 1980s and early 1990s and incorporates elements of science fiction, fantasy, alternate history, horror, and speculative fiction. It involves a setting where steam power is widely used…."

44. A quote famously attributed to Alaska Senator Ted Stevens, a US senator in charge of regulating the Internet, "I just the other day got, an Internet [that] was sent by my staff at 10 o'clock in the morning on Friday, and I just got it yesterday. Why? They want to deliver vast amounts of information over the Internet…. And again, the Internet is not something that you just dump something on. It's not a big truck. It's a series of tubes." See "Your Own Personal Internet" in "Threat Level" by *Wired Blogs* (Wired, June 30, 2006), http://www.wired.com/threatlevel/2006/06/your_own_person.

45. To lose this bet can be very expensive. See "Intel discovers chip flaw in midst of major launch" by Noel Randewich in *Reuters* (January 31, 2011), http://www.reuters.com/article/2011/01/31/us-intel-idUSTRE70U4DH20110131.

46. See "Progress in Silicon Based Quantum Computing" by Robert G. Clark, R. Brenner, T.M. Buehler, et al., in *The Philosophical Transactions of The Royal Society of London A*, Volume 361 (2003), pages 1451–1471, http://rsta.royalsocietypublishing.org/content/361/1808/1451.abstract.

47. See "Engineered Two-Dimensional Ising Interactions in a Trapped-Ion Quantum Simulator with Hundreds of Spins" by Joseph W. Britton, Brian C. Sawyer, Adam C. Keith, C.-C. Joseph Wang, James K. Freericks, Hermann Uys, Michael J. Biercuk, and John J. Bollinger in *Nature*, Volume 484 (2012), pages 489–492, http://www.nature.com/nature/journal/v484/n7395/full/nature10981.html. See also "Sydney Scientist Helps Design Tiny Super Computer" by Deborah Smith in the

Sydney Morning Herald (April 26, 2012), http://www.smh.com.au/technology/sci-tech/sydney-scientist-helps-design-tiny-super-computer-20120426-1xmik.html.

48. A yottabyte is a septillion bytes or 1,000,000,000,000,000,000,000,000 bytes. Currently planned flash memories due to roll out in the next few years are only a terabyte or a quadrillion bytes or only 1,000,000,000,000 bytes. To give you the scale, a yottabyte flash drive could store 100,000 times all the data transmitted on the Internet in a month, at current transmission rates.

49. Alejandro Muramatsu (private communication, July 2, 2012). See also "Ground States of a Frustrated Quantum Spin Chain with Long-Range Interactions" by Anders W. Sandvik in *Physical Review Letters*, Volume 104 (2010), article number 137204, http://link.aps.org/doi/10.1103/PhysRevLett.104.137204.

50. I am reminded of the legend of Master Hanus, the Prague Clockmaker. Master Hanus designed and built the Astronomical Clock in the clock tower of the Prague town square. The clock was a masterpiece of clockwork with mechanical parading robots and dials for telling Bohemian and Babylonian time, as well as the location of the signs of the zodiac. The clock, the only one of its kind, was the envy of all Europe. And despite his protestations to the town council that he would never build another one in another city, the disbelieving councilors repaid Master Hanus by having his eyes burned out with a hot poker, so he could never see again in order to build another clock. See "Tale of the Old Town Clock" by Jessica Tudzin in *Bohemian Ink* (July 21, 2010), http://www.bohemianink.net/?p=946.

51. See *The Structure of Scientific Revolutions*, 50th Anniversary Edition, by Thomas S. Kuhn (University of Chicago Press, 2012), page 136.

52. The journal impact factor is a metric that indicates the impact the publication has upon its field. The higher the factor is, the bigger the impact.

53. See *The Structure of Scientific Revolutions*, 50th Anniversary Edition, by Thomas S. Kuhn (University of Chicago Press, 2012), page 151, and also see *Scientific Autobiography and Other Papers* by Max Planck (Philosophical Library, 1949), page 34, http://www.worldcat.org/oclc/14676418.

54. See "The Complexity Zoo" in *Qwiki* (Stanford University, July 6, 2012), http://qwiki.stanford.edu/index.php/Complexity_Zoo.

55. When I wrote up the paper with Barut, we labeled our lower and upper levels Midwest style ("1" and "2") and, as is usual, we thanked Eberly and Scully for useful discussions in the acknowledgments, but I sneaked therein a joke, "We would like to thank M.O. Scully ($a = 2$ and $b = 1$), J. H. Eberly ($a = 1$ and $b = 2$)...." See "Self-Field Quantum Electrodynamics—The Two-Level Atom" by Asim O. Barut and Jonathan P. Dowling in *Physical Review A*, Volume 41 (1990), pages 2284–2294, http://link.aps.org/doi/10.1103/PhysRevA.41.2284.

56. See "Cavity Quantum Electrodynamics Approaches to Quantum Information Processing and Quantum Computing" in the *Quantum Information Science and Technology Roadmap* (Advance Research and Development Activity, July 10, 2012), http://qist.lanl.gov/pdfs/cavity_qed.pdf.

57. See "Measurement of Conditional Phase-Shifts for Quantum Logic" by Quinton A. Turchette, C.J. Hood, Wolfgang Lange, Hideo Mabuchi, and H. Jeff Kimble in *Physical Review Letters*, Volume 75 (1995), pages 4710–4713, http://link.aps.org/doi/10.1103/PhysRevLett.75.4710.

58. See "Cavity Quantum Electrodynamics with Optically Transported Atoms" by J.A. Sauer, K.M. Fortier, M.S. Chang, C.D. Hamley, and Michael S. Chapman in *Physical Review A*, Volume 69 (2004), article number 051804(R), http://link.aps.org/doi/10.1103/PhysRevA.69.051804.

59. See "Strong Coupling of a Single Photon to a Superconducting Qubit Using Circuit Quantum Electrodynamics" by A. Wallraff, D.I. Schuster, A. Blais, L. Frunzio, R.-S. Huang, J. Majer, S. Kumar, S.M. Girvin, and Robert J. Schoelkopf in *Nature*, Volume 431 (2004), pages 162–167, http://www.nature.com/nature/journal/v431/n7005/abs/nature02851.html.

60. See "Deterministic Entanglement of Photons in Two Superconducting Microwave Resonators" by H. Wang, Matteo Mariantoni, Radoslaw C. Bialczak, M. Lenander, Erik Lucero, M. Neeley, A.D. O'Connell, D. Sank, M. Weides, J. Wenner, T. Yamamoto, Y. Yin, J. Zhao, John M. Martinis, and A.N. Cleland in *Physical Review Letters*, Volume 106 (2011), article number 060401, http://link.aps.org/doi/10.1103/PhysRevLett.106.060401.

61. See "Heisenberg-Limited Measurements with Superconducting Circuits" by Alexandre Guillaume and Jonathan P. Dowling in *Physical Review A* (2006), article number 040304, http://link.aps.org/doi/10.1103/PhysRevA.73.040304.

62. Lynn Conway is likely the first and certainly the most famous transgender woman computer scientist. See "Through the Gender Labyrinth" by Michael A. Hiltzik in the *L.A. Times Sunday Magazine* (*Los Angeles Times*, November 19, 2000), http://articles.latimes.com/2000/nov/19/magazine/tm-54188. Coincidentally, I read this article about Conway's life in the *L.A. Times* the very same week that her collaborator, Carver Mead, invited me out to dinner to discuss the foundations of quantum electrodynamics, the topic of my PhD research (which was supposedly crackpot stuff with which I would never get a job....).

63. This story was told to me in November of 2000 by none other than the famous Carver Mead himself, over a steak dinner with several bottles of nice red wine in a fancy restaurant called "The Raymond" in Pasadena, California. (Thankfully, Mead picked up the tab after giving me an autographed copy of his recent book.)

64. On August 1, 2012, the new Fundamental Physics Foundation, founded by Internet magnate Yuri Milner, handed out a number of $3 million dollar prizes to a clutch of string theorists. In fact, almost all the 2012 prize winners won the award for theoretical physics work that has never been tested in the laboratory. This is compared to the Nobel Prize where typically the result must have some real-world verification. Already the kvetching about handing out such prizes for untested claims has begun with suggestions that Milner call it the Fuddled Metaphysics Prize instead. See "Prize Without Proof? Yuri Milner's Fundamental Physics Prize" by Charles Choi on PBS.ORG (August 28, 2012), http://www.pbs.org/wgbh/nova/physics/blog/2012/08/prize-without-proof-yuri-milners-fundamental-physics-prize/.

65. For a couple of wonderfully scathing reviews of string theory, see *Not Even Wrong: The Failure of String Theory and the Search for Unity in Physical Law* by Peter Woit (Basic Books, 2006), http://www.worldcat.org/oclc/67840232 and *The Trouble with Physics: The Rise of String Theory, the Fall of a Science, and What Comes Next* by Lee Smolin (Houghton Mifflin, 2006), http://www.worldcat.org/oclc/732955853. In September of 2009, I was invited to give a seminar on quantum technologies at the Center for Theoretical Science at the University of Princeton in New Jersey.

I had lunch with a few of the superstring theorists there and, thinking I would start a riotous debate, brought up these two books in the luncheon conversation. Instead of a debate, I was given a rather frosty reception and told, "Here at Princeton we do not discuss those books." I suppose I will not be invited back anytime soon....

66. The existence of the axion particle was proposed before the invention of superstring theory but the theory has engulfed the concept and now superstring theories contain axions. Experiments to detect dark matter axions involve a prediction conversion of an axion (invisible) into a photon (visible) via the mediation of a superstrong magnetic field.

67. For a panoramic view of such particles and the hope that they will somehow fit into known theory, see "The Universe" by Michal S. Turner in *Scientific American* (September 2009), pages 36–43.

68. The Tevatron results indicated the existence of the Higgs at about a 99% confidence level, which means that there was a 1% chance their results were due to random noise and there was no Higgs. With our friend the quadratic scaling law to improve this from 99% confidence to even 99.9%, they would have had to run the machine for many more years to collect enough data; years they did not have as the machine was shut down in September of 2011. See "Fermilab Announces Strong Higgs Boson 'Bump'" by Ian O'Neill in *Discovery News* (July 3, 2012), http://news.discovery.com/space/tevatron-data-detects-higgs-boson-existence-120703.html.

69. Supersymmetry is a theory that is a pillar of superstring theory, or superstring theories, and is where the word "super" in superstrings comes from and "superstring theory" is short for "supersymmetric string theory." Supersymmetry in its simplest form sacrifices an economy in the number of particles for more beautiful mathematics. The simplest form of the theory, called the Minimal Supersymmetric Standard Model, proposes that each known particle in the standard model, say an electron or a quark, has a yet unobserved supersymmetric partner particle, in this case a "selectron" or a "squark" or a "sneutrino." (I'm not making these names up.) Since things like selectrons and squarks and sneutrinos have never been seen, they must have masses so large that they are out of range of the most powerful atom smashers. But it was anticipated that the Large Hadron Collider at CERN would, in addition to the Higgs, find a few of the least massive of these proposed supersymmetric partner particles. They found nothing of the sort. Experimental evidence for new supersymmetric particles would have provided indirect evidence for superstring theory. Lack of experimental evidence for such particles has the reverse effect. See "LHC Results Put Supersymmetry Theory 'On the Spot'" by Pallab Gosh in *Science and Environment* (BBC News, August 27, 2011), http://www.bbc.co.uk/news/science-environment-14680570.

70. See "Calculating the Mass of a Proton" by Mark Reynolds in *CNRS International Magazine: Particle Physics* (CNRS, July 24, 2012), http://www2.cnrs.fr/en/1410.htm.

71. See "The Church of the Larger Hilbert Space" in *Quantiki* (July 19, 2012), http://www.quantiki.org/wiki/The_Church_of_the_larger_Hilbert_space.

72. For a more interactive Internet version, see the URL: *Powers of Ten* (July 24, 2012), http://www.powersof10.com.

73. See "The Biggest Numbers in the Universe" by Bryan Clair in *Strange Horizons* (April 2, 2001), http://www.strangehorizons.com/2001/20010402/biggest_numbers.shtml.

74. See "There's Plenty of Room at the Bottom" by Richard Feynman in *Engineering and Science* (California Institute of Technology, February 1, 1960), http://www.zyvex.com/nanotech/feynman.html.

75. For a discussion of Searle's strong AI hypothesis and his amusing Chinese room thought experiment rebuttal to the significance of the Turing test, see the "Chinese Room," in *Wikipedia* (Wikimedia Foundation, 06 August 2012), http://en.wikipedia.org/wiki/Chinese_Room. In short, the rebuttal claims to show that an entity that passes the Turing test need not necessarily be sentient or have the equivalent of the human mind.

76. See "Computer Wins on 'Jeopardy!' Trivial, It's Not" by John Markoff in the *New York Times* (February 16, 2011), www.nytimes.com/2011/02/17/science/17jeopardy-watson.html.

77. Here, I refer to the sentient android Lt. Commander Data and other sentient beings such as the holographic projected beings of Dr. Moriarty and, from *Star Trek: Voyager*, the sentient computer-generated character, The Doctor (Emergency Medical Hologram, Mark I).

78. The film was based on a 1966 novel, *Colossus*, by British author Dennis Jones, which was itself about a defense department computer that becomes sentient and takes over the world. See *Colossus* by Dennis Jones (Putnam, 1967), http://www.worldcat.org/oclc/233742. It is interesting to note that "Colossus" was the name of the very nonfictional British World War II enigma-hacking cryptanalysis machine built by the British in the 1940s but whose very existence was not declassified until the late 1970s, years after Jones wrote this book. Coincidence?

79. See *The Singularity Is Near: When Humans Transcend Biology* by Ray Kurzweil (Viking, 2005), http://www.worldcat.org/oclc/57201348.

80. See "Overview" (Singularity University, August 7, 2012), http://singularityu.org/about/overview/.

81. In particular, Jennifer Andrews, once a student in my quantum mechanics class at LSU and a visiting summer graduate student at NASA, bemoaned the fact that most of the Singularity University enrollees were younger than she and had already founded one or two start-up companies while she was still working toward her PhD. She took solace in the fact that the smartest young Singularity University students took a half hour to find the bar but that she and my two PhD students, Chris Richardson and Blane McCracken, found it in less than 30 seconds. Start-up companies are not, after all, everything.

82. See "The Singularity Myth" by Theodore Modis in *Technological Forecasting & Social Change*, Volume 73 (February 2, 2006), pages 104–112, http://dx.doi.org/10.1016/j.techfore.2005.12.004.

83. See *The Acceleration of History* by Gerard Piel (Knopf, 1972), http://www.worldcat.org/oclc/300728494.

84. Bartlett is a professor emeritus at the University of Colorado, where I did my PhD. I never took a course from him but I was his grader one semester for an undergraduate physics class he was teaching. The most memorable moment was when I accidentally spilled some green tea on a student's homework assignment. I blotted it up but next to the yellowish green stain the tea left on the paper I wrote, "I'm sorry but my dog peed on your homework." To that I put a smiley face and a grade and returned the paper. When the student complained to Bartlett about my disrespect for his homework, in my defense I replied, "But Prof. Bartlett I *did* put a smiley face!"

85. See "Neanderthal DNA Lives On in Modern Humans" by Jennifer Pinkowski in *Time* (May 6, 2010), http://www.time.com/time/health/article/0,8599,1987568,00.html. The correct statement is that all modern humans who are not of pure African descent have Neanderthal DNA.

86. See "The Mind" by Marc Hauser in *Scientific American* (September 2009), pages 44–51, which is reprinted here: http://www.scientificamerican.com/article.cfm?id=origin-of-the-mind.

87. Of course teenagers are not the only ones who text. I think it was in 2007 that I noticed that my students, who used to be happy to communicate with me by calling me on my mobile phone, began texting me. At the time, the text messages on my mobile phone plan cost me 20¢ a message. I complained only a bit but at last, not wanting to be the Luddite, changed over the data plan from first 100 texts a month to now unlimited texts on my Android smartphone. A curious culture around texting and smartphone use has recently arisen in Silicon Valley. I noticed it when I was there at a conference in January of 2012. The Silicon Valley techies are smart enough, for the most part, not to text and drive. But each time I came to a red traffic light, I noticed at least a 10-second delay before the traffic started moving again when the light turned green. Everybody at the intersection was texting or otherwise engaged with an electronic device and nobody noticed when the light turned green. The Silicon Valley protocol to handle such a situation is for the first person to notice the green light to politely and lightly tap on their car horn to rouse all the other drivers out of the electronic reverie and proceed through the intersection. Unaware of this, I followed my native New Yorker protocol and leaned on the horn for a full minute while yelling obscenities at them out my window.

88. When I was at the NASA Jet Propulsion Laboratory, we would have endless discussions on the definition of life. What is life? The discussions were always hinged on the metaphysical and religiously infused idea that there should be a line—things on one side of the line were alive and things on the other were not. The game was to find that line. People argued and continue to argue about this endlessly. Should it reproduce, reduce the entropy of its environment, have DNA, or what? For example, viruses are infectious, reproduce, and have DNA and most vote they are alive (but some not). Prions, which cause mad cow disease, are malevolent proteins that reproduce, are infectious, but have no DNA. Most say they are not alive and say we should draw the line of life between prions and viruses. My response to these discussions was, there is no such thing as life! There are interesting chemical reactions, like Stephen Hawking, and less interesting chemical reactions, like salt crystals growing in a glass of salt water. There is no line, no "breath of life" separating living from nonliving. That is a metaphysical bit of silliness. We should focus on interesting over boring chemical reactions and forget about this line that does not exist except in our own minds.

89. See "The Control of Neuron Number" by R.W. Williams and K. Herrup in the *Annual Review of Neuroscience*, Volume 11 (March 1988), pages 423–453, http://dx.doi.org/10.1146%2Fannurev.ne.11.030188.002231.

90. See "IBM Unveils Microchip Based on the Human Brain" by Ferris Jabr in *New Scientist* (August 19, 2011), http://www.newscientist.com/article/dn20810-ibm-unveils-microchip-based-on-the-human-brain.html.

91. S'more is a contraction of the two words "some more" such as in "Please, Sir, can I have S'more Hilbert Space?" A S'more is also a tasty campground treat of a sandwich-like confection made from a slab of chocolate and a marshmallow placed between two Graham crackers and melted a bit over a campfire.

92. See "Quantum Secrets of Photosynthesis Revealed" by Lynn Yarris in *Research News* (Lawrence Berkeley Laboratory, April 12, 2007), http://www.lbl.gov/Science-Articles/Archive/PBD-quantum-secrets.html.

93. See "The Physics of Life: The Dawn of Quantum Biology" by Phillip Ball in *Nature*, Volume 474 (2011), pages 272–274, http://www.nature.com/news/2011/110615/full/474272a.html.

94. See *The Emperor's New Mind* by Roger Penrose (Oxford University Press, 1989), http://www.worldcat.org/oclc/19724273 and *Mind, Matter, and Quantum Mechanics* by Henry P. Stapp (Springer, 1993), http://www.worldcat.org/oclc/28147220.

95. See "Is the Brain a Quantum Device?" by Victor Stenger in the *Skeptical Inquirer*, Volume 18.1 (2008), http://www.csicop.org/sb/show/is_the_brain_a_quantum_device/.

Index

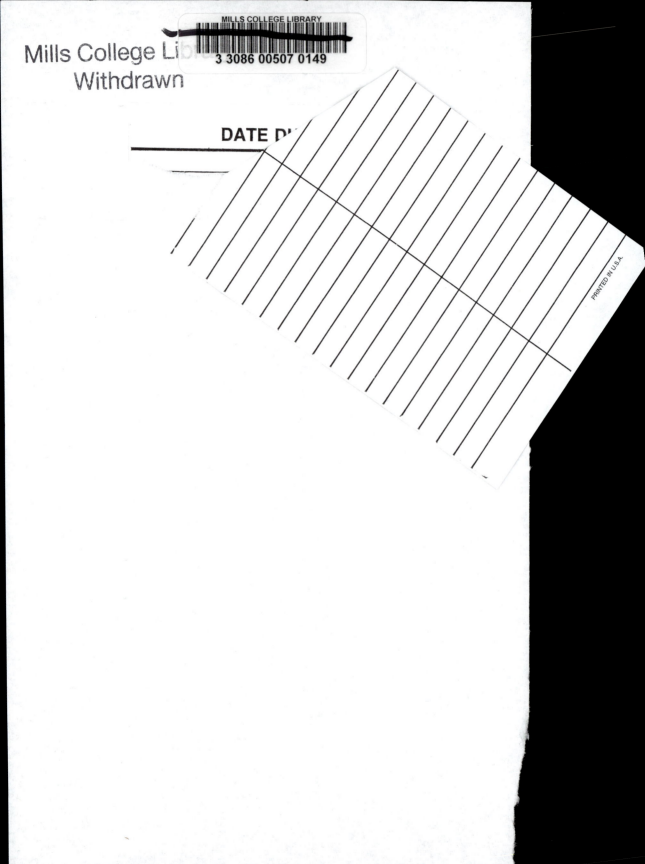

DATE D

PRINTED IN U.S.A.